INTRODUCTION TO
AC MACHINE DESIGN

IEEE Press
445 Hoes Lane
Piscataway, NJ 08854

INTRODUCTION TO AC MACHINE DESIGN

THOMAS A. LIPO
Emeritus Professor
University of Wisconsin
Madison, WI

Research Professor
Florida State University
Tallahassee, FL

IEEE PRESS SERIES ON POWER ENGINEERING

IEEE PRESS

WILEY

Published by John Wiley & Sons, Inc., Hoboken, New Jersey.
Published simultaneously in Canada.

Library of Congress Cataloging-in-Publication Data is available.

ISBN: 978-1-119-35216-7

MIX
Paper from responsible sources
FSC® C013604

This book is dedicated to my many students who have taken the course ECE 713 *Electromagnetic Design of AC Machines* over the past 35+ years. You have helped prove that such a course remains a vital, important issue in a modern power systems/power electronics graduate program.

We are as dwarfs seated on the shoulders of giants that we might see more and further than they. Yet not by virtue of the keenness of our eyesight nor the breadth of our vision but alone because we are raised aloft on that giant mass.

Bernard of Chartres d. ca. 1130 AD

CONTENTS

CHAPTER 2 *THE MMF AND FIELD DISTRIBUTION OF AN AC WINDING* 51

CHAPTER 3 *MAIN FLUX PATH CALCULATIONS USING MAGNETIC CIRCUITS* 79

CHAPTER 4 *USE OF MAGNETIC CIRCUITS IN LEAKAGE REACTANCE CALCULATIONS* 125

PREFACE AND ACKNOWLEDGMENTS

THE MATERIAL IN THIS BOOK is derived from notes for a graduate level course on the design of AC electrical machines that I taught semi-annually at the University of Wisconsin since 1980. These notes were produced prior to this text in limited distribution form for teaching purposes over these many years. Upon retirement, the evolution of this material has now come to an end.

While given short shrift in most engineering curricula, electrical machine design remains one of the most challenging disciplines within electrical engineering. Since literally all of the significant texts dealing with this subject have long been out of print, it was decided to offer this material as a means to enable those interested in this subject to acquire the necessary basic knowledge.

It is the belief of the writer that, while the finite element method is a powerful analytic technique, the basic principles of machine theory cannot be taught well if finite elements are introduced too early. In this spirit, the subject of finite elements has been included only at the end of the book. Hopefully, with a deeper understanding of the basic principles garnered from this book, the student can go forth with modern mathematical tools such as finite elements and put them to work in a more enlightened manner.

It will be evident to the reader from the outset that the author has been deeply influenced by the writings of Alger, Liwschitz-Garik, and many others. Hopefully, sufficient new material exists to warrant the writer as being an "author" and not simply the "editor" of profound insights from these titans of the past.

My special thanks to four of my students, Michael Klabunde, Xiaogang Luo, Wenbo Liu, and Wen Ouyang, who made important contributions to Chapters 5, 6, 9, and 10, respectively. Also, I wish to thank Professor Katsuji Shinohara of Kagoshima University, Japan and his colleagues for their careful reading of an earlier version of this manuscript. Finally, special thanks to Rich Schiferl for his masterful editing of the final manuscript.

THOMAS A. LIPO

Madison, WI

LIST OF PRINCIPAL SYMBOLS

Note: Additional subscripts s and r generally denote stator and rotor values of the quantity, respectively. The subscript 1 denotes the fundamental component. Boldface denotes a three-dimensional vector.

Symbol	Meaning	Page Number First Used
$\mathbf{A}$	magnetic vector potential vector (Wb/m)	5
A	area (m^2)	21
$\mathbf{B}$	magnetic field or magnetic flux density vector (Wb/m^2)	2
$b_{1/3}$	breadth of slot 1/3 of the way from narrow portion (m)	83
B_c	core flux density (Wb/m^2)	97
B_g	flux density in the air gap (Wb/m^2)	53
$B_{g,ave}$	average value of air gap flux density over one tooth and slot (Wb/m^2)	87
B_{gm}	flux density in the gap produced by the magnet (Wb/m^2)	374
B_{gm1}	fundamental component of air gap flux density produced by the magnet (Wb/m^2)	376
B_{g1}	peak fundamental air gap flux density (Wb/m^2)	87
b_o	slot opening (m)	81
B_{top}, B_{mid}, B_{root}	flux density at top, midpoint, and root of a tooth (Wb/m^2)	87
C	number of parallel circuits	73
C_f	number of parallel field winding circuits	412
C_s	number of parallel stator circuits	403
C_h	thermal capacitance (J/°K)	353
C_{ir}	loss coefficient (W/m^3)	233
c_p	specific heat (J/kg-°K)	307
$\cos \phi_{gap}$	power factor as measured at the air gap	260
d_{cs}, d_{cr}	depth of the stator, rotor core (m)	101
d_e	equivalent depth of solid copper (m)	
d_p	penetration constant (m)	233
d_m	magnet depth (magnet thickness) (m)	373
D_{is}, D_{ir}	inner diameter of stator, rotor punching (m)	101

D_{os}, D_{or}	outer diameter of stator, rotor punching (m)	101
d_{ss}, d_{sr}	depth of stator, rotor slot (m)	101
d_t	depth of tooth (m)	88
$\boldsymbol{E}$	electric field intensity vector (V/m)	9
e_b	induced voltage in a rotor bar (V)	172
$\boldsymbol{F}$	force vector (N)	1
$\mathcal{F}$	magnetomotive force (MMF) (A-turn)	21
$\mathcal{F}_{cr}$	rotor core MMF drop (A-turn)	80
$\mathcal{F}_{cs}$	stator core MMF drop (A-turn)	80
f_e	stator applied frequency (Hz)	236
$\mathcal{F}_g$	air gap MMF drop (A-turn)	80
$\mathcal{F}_{s1}$	peak fundamental component of stator MMF	79
$\mathcal{F}_{t(ave)}$	MMF drop over the entire length of a tooth (A-turn)	89
$\mathcal{F}_{ts}, \mathcal{F}_{tr}$	stator, rotor tooth MMF drop (A-turn)	80
g	mechanical gap (m)	30
g_c	equivalent gap including fringing and saturation (m)	119
h	harmonic index	55
h_k	slot harmonics	62
$\boldsymbol{H}$	magnetic field intensity vector (A/m)	14
H_{top}, H_{mid}, H_{root}	magnetic field intensity at the top, midpoint, and bottom section of a tooth (A/m)	89
$H_{t(ave)}$	average field intensity over entire length of a tooth (A/m)	89
I	steady (DC) current (A)	1
i_a, i_b, i_c	instantaneous current in phases a, b, and c (A)	74
i_b	current in a rotor bar (A)	172
i_e	current in an end-ring segment (A)	172
I_d, I_q	direct and quadrature current components	403
I_{mr}	peak rotor bar current (A)	175
$I_{r,har}$	equivalent current accounting for rotor space harmonics	178
I_s	peak AC stator current (A)	75
$\boldsymbol{J}$	current density vector (A/m^2)	3
J_m	volumetric polarization current (A/m^2)	14
k_c	Carter factor	82
k_{ch}	winding factor for harmonic *h* for concentrated windings	63
k_{cu}	copper space factor	274
k_{dh}	distribution factor for *h*th harmonic	60

k_h	winding factor for *h*th harmonic	72
k_{hys}	Steinmetz coefficient	220
k_i	lamination space factor	87
K_m	surface polarization current	14
K_p	factor to include the effect of slot leakage flux on saturation of the core	98
K_{pf}	pole face factor	234
k_{ph}	pitch factor for *h*th harmonic	55
k_{sr}	ratio of rotor to stator surface current density	265
k_s, k_m, k_{sl}	slot factors	143
k_d, k_q, k_f	pole face factors	404, 406, 412
k_{sh}	skew factor for *h*th harmonic	72
$\boldsymbol{K}$	surface current density vector (A/m)	16
K_s, K_r	stator, rotor surface current density (A/m)	265
$K_{s(rms)}$	stator surface current density assuming rms amps (Arms/m)	262
K_{s1}	peak fundamental component of stator surface current density (A/m)	259
$k_{\chi h}$	slot opening factor for the *h*th harmonic	66
L	inductance (H)	27
L_b	inductance of one rotor bar (H)	172
$L_{b(har)}$	harmonic leakage inductance of the rotor per phase (H)	180
L_{be}	equivalent bar inductance including the end-ring inductance (H)	175
l_{cs}, l_{cr}	length of the path of one pole pitch as measured at the midpoint of the stator, rotor core (m)	101
l_e	stator effective length including fringing due to ducts (m)	81
L_{ew}	end-winding leakage inductance (H)	153
l_i	physical axial length of the stator iron (not including air ducts and fringing) (m)	85
L_{lk}	leakage inductance arising from belt leakage (H)	158
L_{ls}	stator leakage inductance per phase (H)	184
L_{lT}, L_{lB}	slot leakage inductance of coil side in top, bottom of slot (H)	138
L_{lTB}	mutual inductance between slot leakage components of top and bottom coil (H)	138
L_{lr}	rotor leakage inductance (H)	184
L'_{lr}	rotor leakage inductance referred to the stator turns and phases (H)	185

L_{lsk}	skew leakage inductance (H)	188
L_{lzz}	zigzag leakage inductance per phase (H)	163
L_{ms}	stator magnetizing inductance (H)	119
L_{md}	direct axis magnetizing inductance	405
L_{mq}	quadrature axis magnetizing inductance	406
L_{mr}	magnetizing inductance of one rotor mesh (H)	183
l_{os}, l_{or}	axial length of one stator, rotor air duct (m)	85
$L_{r,har}$	equivalent rotor leakage inductance accounting for rotor space harmonics (H)	180
l_s	stator physical length including ducts (m)	53
L_{slot}	slot leakage inductance of a single slot (H)	139
L_{sls}, L_{slm}	self, mutual component of slot leakage (H)	142
L_{sr}	mutual inductance between stator phase and one rotor mesh (H)	182
L_{lzz}	zigzag leakage inductance per phase (H)	164
$\boldsymbol{m}$	magnetic dipole moment (A/m^2)	12
$\boldsymbol{M}$	magnetic polarization vector (A/m)	13
m	number of phases	62
N	number of turns	20
N_s	number of series connected turns for one phase	119
N_{se}	effective number of series connected turns per phase	120
$N(\theta)$	winding function	76
n_s	turns per slot	128
N_t	total number of turns of one phase	52
O	contour bounding surface S (m)	1
P	total number of poles	73
p	winding pitch	141
$\mathcal{P}$	permeance (H)	21
p	specific permeance (H/m)	128
p_e	eddy current power loss per unit volume (W/m^3)	219
p_{ew}	specific permeance corresponding to end-winding leakage inductance (H/m)	153
$\mathcal{P}_g$	gap permeance (H)	81
p_{hys}	hysteresis loss per unit volume (W/m^3)	220
p_h	heat gradient (W/m^2)	308
p_i	eddy current plus hysteresis loss per unit volume (W/m^3)	220
$\mathcal{P}_p$	permeance of one pole	119
P_{mech}	mechanical output power (W)	262
$P_{r(surf)}$	surface power loss (W/m^2)	234
p_{rad}	heat gradient resulting from radiation (W/m^2)	328
$\mathcal{P}_s$	slot permeance (H)	83

Q_b	number of coils per pole per phase	
q	number of coil sides in a phase belt (slots/pole/phase)	57
q_c	charge (coulomb or C)	18
q_h	heat generation rate (W/m^3)	307
Q_h	thermal power flow due to conduction (W)	308
Q_c	thermal power flow due to convection (W)	326
Q_r	thermal power flow due to radiation (W)	328
R	distance between two points (m)	1
$\mathcal{R}$	reluctance (H^{-1})	22
R_{ac}	rotor bar AC resistance (Ω)	199
R_b	resistance of one rotor bar (Ω)	172
R_{be}	equivalent value of bar resistance including the effect of end-ring resistance (Ω)	175
R_e	resistance of one rotor end-ring segment (Ω)	172
R_h	thermal resistance to conduction heat flow (K/W)	309
r_{is}	radius of stator inner surface (m)	53
r_p	resistance of the *p*th layer of a rotor bar (Ω)	212
R_{rad}	thermal resistance to radiation (K/W)	328
$\mathcal{R}_t$	reluctance of one tooth (H^{-1})	89
S_1, S_2	total number of stator, rotor slots	62
S	per unit slip	174
$\mathbf{S}$	surface area (m^2)	8
T_e	electromagnetic torque (N-m)	261
t_o	tooth thickness (m)	81
t_t, t_m, t_b	tooth width at the tooth top, midpoint and bottom of a tooth (m)	88
$\mathbf{u}$	unit vector	1
V	volume (m^3)	3
v	velocity (m/s)	18
VA_{gap}	volt-amperes at the air gap (VA)	260
V_L, V_R	reactive, resistive voltage drop across on rotor bar (V)	198
V_m	air gap voltage across the magnetizing inductance (V-pk)	123
w	equivalent span of a distributed winding (m)	55
W_m	magnetic field energy (J)	29
$W_{m,ave}$	average energy stored in stator rotor tooth region (J)	162
W_{mp}	magnetic field energy per pole (J)	118
WTHD	weighted total harmonic distortion	63
X_{ac}	rotor bar AC reactance (Ω)	199
X_{ls}, X_{lr}	total stator, rotor leakage reactance (Ω)	97, 186
X_m	magnetizing reactance per phase (Ω)	97

Z	arc length occupied by the q coils in a phase belt (m)	59
Z_b	base impedance (Ω)	190

Greek Symbols

α	per unit pole arc	402
α_o	inverse of the skin depth (m^{-1})	199
α_c	convection heat transfer coefficient (W/m^2-K)	326
α_m	width of one magnet (radians)	374
α_r	radiation heat transfer coefficient (W/m^2-K)	328
α_s	skew angle (radians)	71
β	MMF angle (radians)	371
β_o	slot opening parameter (radians)	225
χ	slot opening (radians)	66
∇	gradient operator (m^{-1})	5
∇^2	laplacian (m^{-2})	11
$\nabla\times$	curl operator (m^{-1})	5
$\nabla\bullet$	divergence operator (m^{-1})	8
Δ	area of a triangle	466
ε	permittivity ($coul^2/Nm^2$)	9
Φ	magnetic flux (Wb)	8
Φ_c	magnetic flux in the core (Wb)	96
Φ_g	flux crossing the air gap (Wb)	98
Φ_p	flux per pole (Wb)	259
Φ_t	flux in one tooth (Wb)	89
γ	pitch angle (radians)	55
γ_o	skin effect constant (m^{-1})	196
η_{gap}	efficiency as measured at the gap (not including stator copper and iron losses)	260
κ_a	acceleration factor	39
λ	flux linkage (Wb-turns)	27
Λ	thermal conductivity (W/m-K)	307
λ_{ms}	total air gap flux linkages of one stator phase (Wb-turns)	118
λ_{mr}	air gap flux linkages of one rotor mesh (Wb-turns)	183
λ_p	flux linkages per pole (Wb-turns)	118
μ_0	permeability (H/m)	1
υ	reluctivity—inverse of permeability (m/H)	457
θ	angular position along the air gap (radians)	52
Θ	temperature (K)	307
ω_e	stator applied angular frequency (radians/s)	75
ω_{rm}	mechanical angular speed (radians/s)	174
Ω_s	synchronous mechanical speed (revolutions/s)	260
ρ_c	charge density ($coulombs/m^3$)	9
ρ	resistivity (Ω-m)	194
σ	conductivity (Ω^{-1}/m)	18

σ_m	magnetic shear stress (N/m^2)	262
σ_o	slot opening parameter	239
τ_p	pole pitch (m)	54
τ_s, τ_r	stator, rotor slot pitch (m)	81
υ_e	ratio of end-winding length to core length	283
ξ_o	$D_{os}{}^3 l_e$ output coefficient	273
ξ_r	$D^{2.5} l_e$ output coefficient	278
ζ	Lagrange multiplier	275

ABOUT THE AUTHOR

THOMAS A. LIPO is a native of Milwaukee, WI. Dr. Lipo has spent his entire career in the technical field of solid-state AC motor drives. He has BEE and MSEE degrees from Marquette University and a Ph.D. from the University of Wisconsin. From 1969 to 1979, he was an Electrical Engineer in the Power Electronics Laboratory, Corporate Research and Development, General Electric Company, Schenectady, NY, where he participated in some of the earliest work in this field. In 1979, he left GE to take a position as Full Professor at Purdue University. In 1981, he joined the University of Wisconsin, Madison, where he co-founded the industry consortium WEMPEC and served for 28 years as its Co-Director and as the W. W. Grainger Professor for Power Electronics and Electrical Machines. He is presently both an Emeritus Professor at the University of Wisconsin and a Research Professor at Florida State University.

Dr. Lipo's contributions in the field of electrical machinery and power electronics are extensive, having published over 700 technical papers as well as 50 patents, 5 books and 8 book chapters. According to Google Scholar, he has had well over 40,000 citations referring to his work. His citation h-index is 106. He has made pioneering contributions to emerging electrical machine topologies including flux-switched machines, axial flux machines of many types, self-excited synchronous machines, open-winding machines, vernier machines, and novel permanent magnet machines. His efforts in solid-state power converters include work on resonant converters, matrix converters, low switch count and reduced cost rectifiers and inverters, to name a few.

Dr. Lipo is a Life Fellow of IEEE and has received the Outstanding Achievement Award from the IEEE Industry Applications Society in 1986 for his work in motor drives, the William E. Newell Award of the IEEE Power Electronics Society in 1990 for contributions to power electronics, and the Nicola Tesla IEEE Field Award from the IEEE Power Engineering Society in 1995 for his work on electrical machinery. Dr. Lipo is a Fellow of the Royal Academy of Engineering (UK), a Member of the National Academy of Engineering (USA) and a Charter Member of the National Academy of Inventors. In 2014, he received the IEEE Medal in Power Engineering, the highest award presented by IEEE for research in the field of power engineering.

CHAPTER 1

MAGNETIC CIRCUITS

OUR KNOWLEDGE OF MAGNETIC phenomena is literally as old as science itself. According to the Greek philosopher Aristotle, the attractive power of magnets was known even at that time. However, it was not until the sixteenth century that experimental work on magnetism began in earnest. Notable, among scientists active at that time, were the works of Gilbert, who discovered the earth's magnetism, Volta, who developed the voltaic cell, and Oersted, who related the magnetic field to the flow of current. However, it is on the works of Biot and Savart, Ampere, and finally Faraday that the modern theory of magnetism is based. In their experiments, the force on a current-carrying wire due to the flow of current in another wire was carefully measured and forms the experimental basis for the entire theory of magnetism.

1.1 BIOT–SAVART LAW

Using modern notation, the experiments of these pioneers can be expressed compactly in a single vector equation. With reference to Figure 1.1, let O_1 and O_2 be two very thin closed conducting current loops in which steady (DC) currents flow. The coordinates along the loop O_1 can be designated by x_1, y_1, z_1 and the coordinates along the second loop O_2 as x_2, y_2, z_2. The arc lengths along the loops O_1 and O_2 are denoted as vector quantities $\boldsymbol{dl}_1$ and $\boldsymbol{dl}_2$, respectively. From the experiments of Biot, Savart, and Ampere, the differential force in Newtons expressed as a vector exerted on a small piece of loop 2 carrying current I_2 due to the current I_1 in a small piece of loop 1 can be expressed, in modern notation and units as

$$\boldsymbol{dF}_{21} = \left(\frac{\mu_0}{4\pi}\right) \frac{I_2 \boldsymbol{dl}_2 \times [I_1 \boldsymbol{dl}_1 \times \boldsymbol{u}_{r12}]}{R^2} \quad \text{newtons(N)} \tag{1.1}$$

where

$$R = \sqrt{(x_2 - x_1)^2 + (y_2 - y_1)^2 + (z_2 - z_1)^2}$$

and $\boldsymbol{u}_{r12}$ is a unit vector pointing from $\boldsymbol{dl}_1$ to $\boldsymbol{dl}_2$. Essentially, this force acts to align the two differential elements (i.e., make $\boldsymbol{dl}_1$ and $\boldsymbol{dl}_2$ collinear). This expression can

Introduction to AC Machine Design, First Edition. Thomas A. Lipo.

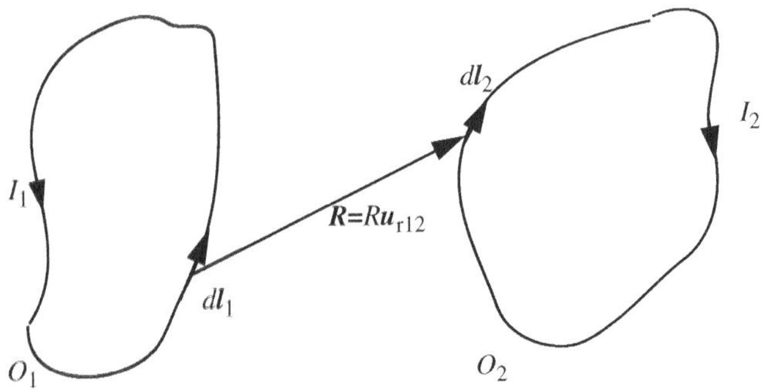

Figure 1.1 Illustration of Biot-Savart's law.

be integrated around coil 1 to find the total force exerted on the differential element of coil 2 as

$$dF_{21} = \frac{\mu_0}{4\pi} \oint_{O_1} \frac{I_2 dl_2 \times (I_1 dl_1 \times u_{r12})}{R^2} \tag{1.2}$$

To find the total force on wire 2, one can simply integrate a second time to form what is called Biot–Savart law or, alternatively, Ampere's law of force,

$$F_{21} = \frac{\mu_0}{4\pi} \oint_{O_2} \oint_{O_1} \frac{I_2 dl_2 \times (I_1 dl_1 \times u_{r12})}{R^2} \quad N \tag{1.3}$$

When the force F_{21} is measured in newtons and the currents are in amperes with the tests made in a vacuum the proportionality constant μ_0 is equal to $4\pi \times 10^{-7}$ newtons per ampere squared (eventually defined as henries per meter). Thus, the proportionality constant μ_0, is called the *permeability of free space*. It can be shown that reciprocity holds, that is, $F_{12} = -F_{21}$

1.2 THE MAGNETIC FIELD *B*

One of the great philosophical contributions of mathematics to science was the use of so-called "fields" to explain the action at a distance, a concept justly troubling to these early researchers. Upon examination of equation (1.3), one can define an incremental magnetic field vector dB_{21} at point 2 due to a current element at point 1 as

$$dB_{21} = \frac{\mu_0}{4\pi} \frac{I_1 dl_1 \times u_{r12}}{R^2} \tag{1.4}$$

The magnetic field resulting from the entire circuit 1 is then

$$B_{21} = \frac{\mu_0}{4\pi} \oint_{O_1} \frac{I_1 dl_1 \times u_{r12}}{R^2} \tag{1.5}$$

whereupon, the force equation, equation 1.3, becomes the much simpler "*BIl*" form,

$$\boldsymbol{F}_{21} = \oint_{O_2} I_2 d\boldsymbol{l}_2 \times \boldsymbol{B}_{21} \tag{1.6}$$

In contrast to equation (1.3), this formulation evaluates the force on a current loop in terms of the interaction of this current with a *magnetic field* $\boldsymbol{B}$. It is important to remember that the basic unit of magnetic field is *newtons per ampere-meter* which will lead to the interpretation of magnetic flux lines as "lines of force." Note that there need be no restriction on the value of $\boldsymbol{u}_{r12}$ and R in equation (1.5). That is, these quantities need not be concerned with the actual distance between two current elements on two circuits. In this case, $\boldsymbol{B}$ is well defined everywhere in space and thereby constitutes what is called a *vector field*.

One of the advantages of the field formulation is that when $\boldsymbol{B}$ is known, the relation permits one to evaluate what would be the force exerted on a current-carrying conductor placed anywhere in the $\boldsymbol{B}$ field without consideration as to what are the system of currents actually giving rise to this field.

An alternative expression for the vector $\boldsymbol{B}$ can be obtained if the current loops cannot be considered to have negligibly small cross-sectional areas, that is

$$\boldsymbol{B}_{21} = \frac{\mu_0}{4\pi} \int_V \frac{\boldsymbol{J} \times \boldsymbol{u}_{r12}}{R^2} dV \tag{1.7}$$

where V is the volume and the vector $\boldsymbol{J}$ is the volumetric current density in *amperes per meter*^{2}.

1.3 EXAMPLE—COMPUTATION OF FLUX DENSITY *B*

The computation of flux density within an electrical machine forms the basic principle behind the machine design process. Consider here the simple example in which a short segment of wire of length L carries a current I as shown in Figure 1.2. Since the

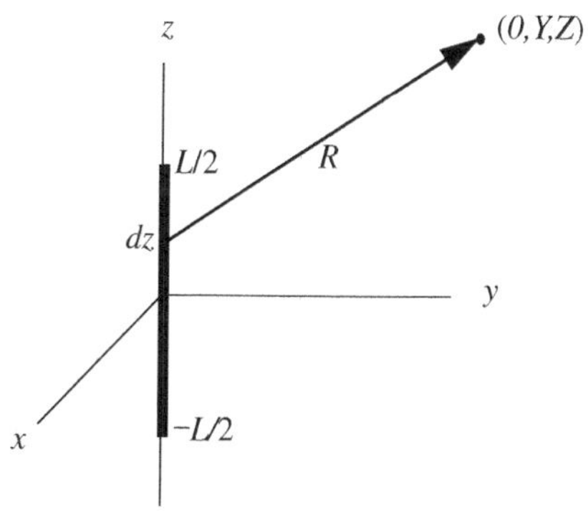

Figure 1.2 Magnetic field of a short wire.

current flows in the z-axis, equation (1.5) becomes

$$d\boldsymbol{B}_{21} = \frac{\mu_0}{4\pi}\frac{Idz\boldsymbol{u}_z \times \boldsymbol{u}_{r12}}{R^2} \tag{1.8}$$

Since the cross product cannot result in a z component and is also normal to the unit vector $\boldsymbol{u}_{r12}$, the flux density must also be normal to the plane containing the $\boldsymbol{u}_z$ as well as the vector $\boldsymbol{u}_{r12}$.

The magnitude of the total field at any point on the $x = 0$ plane is

$$B_{21} = \frac{\mu_0 I}{4\pi}\int_{-\frac{L}{2}}^{\frac{L}{2}} \frac{|\boldsymbol{u}_z \times \boldsymbol{u}_{r12}|dz}{R^2} \tag{1.9}$$

On the $x = 0$ plane, the unit vector $\boldsymbol{u}_{r12}$ is given by

$$\boldsymbol{u}_{r12} = \frac{Y}{\sqrt{Y^2 + (Z-z)^2}}\boldsymbol{u}_y + \frac{Z}{\sqrt{Y^2 + (Z-z)^2}}\boldsymbol{u}_z \tag{1.10}$$

where Y and Z designate a specific point on the $x = 0$ plane. After taking the cross products, equation (1.9) becomes

$$\boldsymbol{B}(0, Y, Z) = \frac{\mu_0 I}{4\pi}\int_{-\frac{L}{2}}^{\frac{L}{2}} \frac{YdZ}{\sqrt[3]{Y^2 + (Z-z)^2}}\boldsymbol{u}_x \tag{1.11}$$

Upon integrating,

$$\boldsymbol{B}(0, Y, Z) = \frac{\mu_0 I}{4\pi Y}\left[\frac{Z + \frac{L}{2}}{\sqrt{Y^2 + \left(Z + \frac{L}{2}\right)^2}} - \frac{Z - \frac{L}{2}}{\sqrt{Y^2 + \left(Z - \frac{L}{2}\right)^2}}\right]\boldsymbol{u}_x \tag{1.12}$$

Of general interest is the magnetic field on the plane perpendicular to the wire and at the center line of the conductor, where $Z = 0$. Here,

$$\boldsymbol{B}(0, Y, 0) = \frac{\mu_0 I}{4\pi Y}\left[\frac{L}{\sqrt{Y^2 + \left(\frac{L}{2}\right)^2}}\right]\boldsymbol{u}_x \tag{1.13}$$

For a current of infinite length, $L \rightarrow \infty$ and equation (1.13) becomes

$$\boldsymbol{B}(0, Y, 0) = \frac{\mu_0 I}{2\pi Y}\boldsymbol{u}_x \tag{1.14}$$

and becomes independent of Z. When the wire is infinitely long, the general result is clearly, from symmetry,

$$\boldsymbol{B} = \frac{\mu_0 I}{2\pi R} \quad \text{Wb/m}^2 \tag{1.15}$$

where R is the radial distance from the wire and the direction of $\boldsymbol{B}$ is normal to the plane containing both the wire and the length R.

1.4 THE MAGNETIC VECTOR POTENTIAL A

The expression for magnetic field can be further simplified by introducing the concept of the magnetic vector potential. It can be easily shown that the following expression is an identity:

$$\frac{\boldsymbol{u}_{r12}}{R^2} = -\nabla\left(\frac{1}{R}\right) \tag{1.16}$$

where ∇ is the gradient operator defined by

$$\nabla = \frac{\partial}{\partial x}\boldsymbol{u}_x + \frac{\partial}{\partial y}\boldsymbol{u}_y + \frac{\partial}{\partial z}\boldsymbol{u}_z \tag{1.17}$$

Using equation (1.16), equation (1.7) can be written as

$$\boldsymbol{B}_{21} = \frac{\mu_0}{4\pi}\int_V \boldsymbol{J} \times \nabla\left(-\frac{1}{R}\right) dV \tag{1.18}$$

The vector differential operator affects only the variables at the point at which $\boldsymbol{B}_{21}$ is evaluated while the integral is taken over the region for which the current density $\boldsymbol{J}$ is defined. However, another identity states that if f is any scalar function of x, y, and z, and $\boldsymbol{v}$ is any vector

$$\nabla \times (f\boldsymbol{v}) = f\nabla \times \boldsymbol{v} + \nabla f \times \boldsymbol{v} \tag{1.19}$$

where $\nabla\times$ denotes the curl operator.

Then, letting f be $(1/R)$ and setting $\boldsymbol{v}$ to $\boldsymbol{J}$

$$\nabla \times \left(\frac{\boldsymbol{J}}{R}\right) = \left(\frac{1}{R}\right)\nabla \times \boldsymbol{J} + \nabla\left(\frac{1}{R}\right) \times \boldsymbol{J} \tag{1.20}$$

In Cartesian coordinates, the curl of any vector $\boldsymbol{F}$ is expressed as the determinant of the matrix

$$\nabla \times \boldsymbol{F} = \det\begin{bmatrix} \boldsymbol{u}_x & \boldsymbol{u}_y & \boldsymbol{u}_z \\ \dfrac{\partial}{\partial x} & \dfrac{\partial}{\partial y} & \dfrac{\partial}{\partial z} \\ F_x & F_y & F_z \end{bmatrix} \tag{1.21}$$

Upon defining the concept of the electric field (later in Section 1.7) it will become evident that the first term on the right-hand side of equation (1.20) is zero by virtue of equations (1.41) and (1.42). (This result assumes that the current distribution is not time-dependent or that the frequency is sufficiently low as is typically the case with electrical machinery.) Thus, equation (1.18) becomes

$$\boldsymbol{B}_{21} = \nabla \times \frac{\mu_0}{4\pi}\int_V \frac{\boldsymbol{J}}{R} dV \tag{1.22}$$

The curl operation can be brought outside of the integral since the two operations are independent; that is, the integral is taken over the volume containing $\boldsymbol{J}$ while the differential operator operates at the point defining $\boldsymbol{B}_{21}$. The field $\boldsymbol{B}_{21}$ has now been defined by the curl of a function that can be designated as $\boldsymbol{A}$,

$$\boldsymbol{B} = \nabla \times \boldsymbol{A} \tag{1.23}$$

which can be formally defined as

$$\boldsymbol{A} = \frac{\mu_0}{4\pi}\int_V \frac{\boldsymbol{J}}{R}dV \quad \text{Wb/m} \tag{1.24}$$

The use of the subscript "21" on $\boldsymbol{B}$ has now been dropped for simplicity. The quantity $\boldsymbol{A}$ is called the *magnetic vector potential* and must be formally evaluated by decomposing the integrand into components along the three coordinates. That is, for the x component of $\boldsymbol{A}$,

$$A_x = \frac{\mu_0}{4\pi}\int_V \frac{J_x}{R}dV \tag{1.25}$$

and so forth for A_y and A_z. Note that the vector potential in the x direction is caused only by the current distribution in the x direction. Hence, the problem of computing the magnetic field $\boldsymbol{B}$ has been reduced to solving three decoupled scalar integrals.

The circuit equivalent to equation (1.24) is

$$\boldsymbol{A} = \frac{\mu_0}{4\pi}\int_L \frac{\boldsymbol{I}}{R}dl \tag{1.26}$$

and is generally more useful when currents flow through wires having negligible cross section.

1.5 EXAMPLE—CALCULATION OF MAGNETIC FIELD FROM THE MAGNETIC VECTOR POTENTIAL

Consider that it is necessary to determine the vector potential and the resulting flux density at a distance Y from the center of a current element of length L and on a line perpendicular to its midpoint as shown in Figure 1.3.

Since the current is directed solely in the z direction, the magnetic vector potential will have only a z component. By symmetry, equation (1.26) reduces to

$$A_z = \frac{\mu_0}{4\pi}\int_0^{\frac{L}{2}} \frac{I}{R}dz \tag{1.27}$$

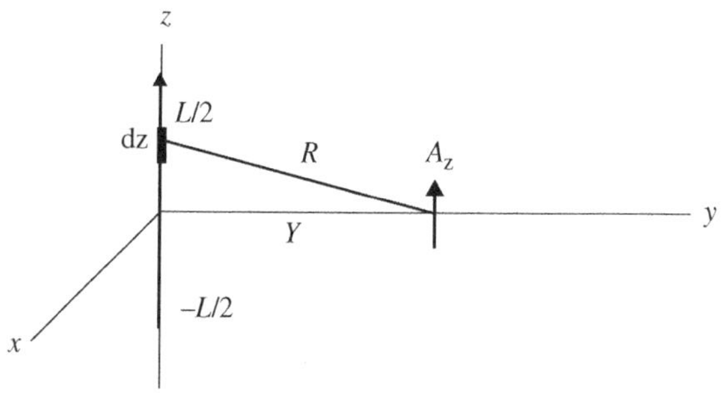

Figure 1.3 Magnetic vector potential of a current element.

so that

$$A_z = \frac{\mu_0 I}{4\pi} \int_0^{\frac{L}{2}} \frac{1}{\sqrt{Y^2 + z^2}} dz \tag{1.28}$$

which becomes ultimately

$$A_z = \frac{\mu_0 I}{4\pi} \ln\left(\frac{L}{2Y} + \sqrt{1 + \left(\frac{L}{2Y}\right)^2} \right) \tag{1.29}$$

where ln denotes the natural logarithm. The solution for magnetic vector potential can now be used to obtain the magnetic flux density at the same point. From equation (1.23),

$$\boldsymbol{B} = \nabla \times \boldsymbol{A} \tag{1.30}$$

and from equation (1.21), if $\boldsymbol{A}$ has only a z-directed component,

$$\boldsymbol{B} = \frac{\partial A_z}{\partial y} \boldsymbol{u}_x - \frac{\partial A_z}{\partial x} \boldsymbol{u}_y \tag{1.31}$$

Since A_z does not vary with x, the second term is zero and

$$B_x = \frac{\partial A_z}{\partial Y} = \frac{\mu_0 I}{2\pi Y} \left(\frac{\frac{L}{2}}{\sqrt{\left(\frac{L}{2}\right)^2 + Y^2}} \right) \tag{1.32}$$

which is the same as equation (1.13).

1.6 CONCEPT OF MAGNETIC FLUX

It has been determined that the magnetic field $\boldsymbol{B}$ can be expressed in terms of the curl of an auxiliary vector potential function $\boldsymbol{A}$. However, again from vector calculus, the divergence of the curl of any function is always zero, that is

$$\nabla \bullet (\nabla \times \boldsymbol{A}) = 0$$

where, in the Cartesian system, the divergence operator is defined as

$$\nabla \bullet \boldsymbol{F} = \frac{\partial F_x}{\partial x} + \frac{\partial F_y}{\partial y} + \frac{\partial F_z}{\partial z} \tag{1.33}$$

From the definition of $\boldsymbol{A}$, it follows that the divergence of $\boldsymbol{B}$ must be identically zero.

$$\nabla \bullet \boldsymbol{B} = 0 \tag{1.34}$$

If equation (1.34) is integrated over a volume

$$\int_V \nabla \bullet \boldsymbol{B} dV = 0 \tag{1.35}$$

whereupon, from Gauss' law, one obtains the result that

$$\int_V \nabla \bullet \boldsymbol{B} dV = \oint_S \boldsymbol{B} \bullet d\boldsymbol{S} = 0 \tag{1.36}$$

in which the surface S encloses the volume V.

In many cases, it is advantageous to think of a vector field as the "flow" of a quantity and in the case of the magnetic field, as suggested from equation (1.36), it is useful to now think of $\boldsymbol{B}$ as a density of flow of "something" per unit area. In the SI system of units, it has been agreed to term this "something" as *magnetic flux* with unit *webers*. Consequently, $\boldsymbol{B}$ can now be considered to have units *webers per square meter* and when the webers per unit area are integrated over a closed surface, the total amount of magnetic flux enclosed is identically zero. The modern SI unit for $\boldsymbol{B}$ is the *tesla* which is identically equal to a weber per square meter. However, both these terms will be used interchangeably throughout this book.

For an arbitrary surface S, bounded by a closed contour O as shown in Figure 1.1, the total magnetic flux Φ passing through the surface S is expressed by

$$\Phi = \int_S \boldsymbol{B} \bullet d\boldsymbol{S} \quad \text{webers} \tag{1.37}$$

The flux which passes through the surface S is said to link the contour O and is generally referred to as the *flux linkage* of the contour. The flux which links a contour O may also be expressed in terms of the vector potential $\boldsymbol{A}$. Since $\boldsymbol{B}$ is the curl of $\boldsymbol{A}$, one can write

$$\Phi = \int_S \boldsymbol{B} \bullet d\boldsymbol{S} = \int_S \nabla \times \boldsymbol{A} \bullet d\boldsymbol{S}$$

This expression can be transformed to a contour integral by using Stokes' theorem, in which case,

$$\Phi = \int_S \nabla \times \boldsymbol{A} \bullet d\boldsymbol{S} = \oint \boldsymbol{A} \bullet d\boldsymbol{l} \tag{1.38}$$

This expression is sometimes more convenient to evaluate than equation (1.37) and will be particularly useful when finite element analysis is investigated later in this text.

1.7 THE ELECTRIC FIELD *E*

In a manner similar to the magnetic field discussion above, the force impressed on one electric charge by another located some distance away can be described by an *electric field* acting directly on the charge. The force is usually expressed as the force on a unit "test" charge as

$$\frac{\boldsymbol{F}_{21}}{q_2} = \frac{1}{4\pi\varepsilon}\int_V \frac{\boldsymbol{u}_{r12}\rho_c(V)}{R^2}dV \tag{1.39}$$

where ρ_c is the charge density in coulombs per meter^3, R is the distance from the differential charge $\rho_c dV$ to the point at which $\boldsymbol{E}$ is evaluated, $\boldsymbol{u}_{r12}$ is the corresponding unit vector, ε is the permittivity of the material. The unit vector $\boldsymbol{u}_{r12}$ again points from the location of the charge, point 1, to the point at which $\boldsymbol{E}$ is to be evaluated, point 2.

Whereas the force exists only on the test charge q_2, a field can again be said to exist everywhere in space given by the vector

$$\boldsymbol{E}_{21}(x,y,z) = \frac{1}{4\pi\varepsilon}\int_V \frac{\boldsymbol{u}_{r12}\rho_c(V)}{R^2}dV \tag{1.40}$$

The electric field is usually given in fundamental units of newtons per coulomb. The free space value of permittivity is $\varepsilon_0 = (1/36\pi) \times 10^{-9}$ coulombs^2/Nm^2.

Finally, when the electric field exists inside a conducting material, the presence of the field establishes a current according to Ohm's law, that is,

$$\boldsymbol{J} = \sigma\boldsymbol{E} \tag{1.41}$$

From the form of the definition of $\boldsymbol{E}$, equation (1.40), and by using equation (1.16) and the vector identity $\nabla \times \nabla(1/R) = 0$, it can be readily shown that

$$\nabla \times \boldsymbol{E} = 0 \tag{1.42}$$

if ρ_c is not time-dependent. Equation (1.42) remains valid for DC current flow in a conductor since the charge at each point in the wire is always the same.

From Stokes' theorem, equation (1.42) has the property that

$$\int_S (\nabla \times \boldsymbol{E}) \bullet d\boldsymbol{S} = \oint_O \boldsymbol{E} \bullet d\boldsymbol{l} = 0 \tag{1.43}$$

where O bounds the surface area S. Equation (1.43) essentially implies that the line integral of $\boldsymbol{E}$ between any two points is independent of the path resulting in the electrical field being said to be conservative. That is, no energy is lost or gained in moving a charged particle around a closed path in an electromagnetic field produced by static (non-moving charges) or by steady currents. In a practical sense, this statement implies that if a conductor is placed in a steady field, no current will flow in the conductor in the steady state.

Upon examining equation (1.43), it is apparent that the electric field has the properties of a "gradient," that it is expressed in terms of "some quantity" per meter. This quantity is formally defined as a *volt* in which case, the electric field is defined to have unit volts per meter for which a volt has fundamental units of newton-meter per coulomb. The unit of permittivity ε_0 in terms of the voltage as a unit is coulomb per volt meter.

1.8 AMPERE'S LAW

Ampere's law forms the fundamental basis upon which all machine design begins. While often presented as a separate law to that of Biot and Savart, its basis is, in actuality, embedded in the definition of the magnetic field $\boldsymbol{B}$. Upon taking the curl of $\boldsymbol{B}$ as defined by equation (1.7), and replacing the curl-curl operator by the equivalent expression, the gradient of the divergence minus the Laplacian, it is possible to obtain

$$\nabla \times \boldsymbol{B} = \nabla \times \nabla \times \frac{\mu_0}{4\pi} \int_V \frac{\boldsymbol{J}}{R} dV \tag{1.44}$$

$$= \frac{\mu_0}{4\pi} \int_V \left[\nabla\nabla \bullet \frac{\boldsymbol{J}}{R} - \boldsymbol{J}\nabla^2 \left(\frac{1}{R} \right) \right] dV \tag{1.45}$$

The differential operators have been taken behind the integral since these operators are taken with respect to the point at which $\boldsymbol{B}$ is desired whereas the integral is taken over the region where the current density $\boldsymbol{J}$ exits.

The first term in equation (1.45) can be written alternatively as

$$\int_V \nabla\nabla \bullet \frac{\boldsymbol{J}}{R} dV = \nabla \int_V \nabla \bullet \frac{\boldsymbol{J}}{R} dV \tag{1.46}$$

where the gradient operator has been brought out from under the integral sign since the gradient and integral operations can again be interchanged. However from Gauss' theorem, this integral can be replaced by the expression

$$\nabla \int_V \nabla \bullet \frac{\boldsymbol{J}}{R} dV = -\nabla \oint_S \frac{\boldsymbol{J}}{R} \bullet d\boldsymbol{S} \tag{1.47}$$

The minus sign appears in this expression since the divergence operator is taken with respect to the point defining $\boldsymbol{B}$ whereas the integral is taken over the volume defining the current density $\boldsymbol{J}$. However, the surface S describes the outer surface of the conductor over which a net current clearly is not escaping. Thus the dot product $\boldsymbol{J} \bullet d\boldsymbol{S}$ is zero on this surface and the first term in equation (1.45) is, therefore, zero.

The expression for the curl of $\boldsymbol{B}$ reduces to

$$\nabla \times \boldsymbol{B} = -\frac{\mu_0}{4\pi} \int_V \boldsymbol{J}\nabla^2 \left(\frac{1}{R} \right) dV \tag{1.48}$$

where $\nabla^2 = (\nabla \bullet \nabla)$ is the Laplacian operator. That is,

$$\nabla^2 = \frac{\partial^2}{\partial x^2} + \frac{\partial^2}{\partial y^2} + \frac{\partial^2}{\partial z^2} \tag{1.49}$$

where the set (x, y, z) denotes the point at which $\boldsymbol{B}$ is defined. Also, R is the distance from the point where $\boldsymbol{B}$ is evaluated at the differential element dV locating $\boldsymbol{J}$. In a similar manner, one can define the set (x', y', z') as denoting the point at which the differential volume dV is defined and the corresponding Laplacian operator

$$\nabla'^2 = \frac{\partial^2}{\partial (x')^2} + \frac{\partial^2}{\partial (y')^2} + \frac{\partial^2}{\partial (z')^2} \tag{1.50}$$

By formal differentiation, it can be shown that the expression $\nabla^2(1/R)$ is identically zero everywhere except at the point of singularity, namely where $R \to 0$. At the point of singularity, the points (x,y,z) and (x',y',z') coincide so that $\nabla^2(1/R) = \nabla'^2(1/R)$ where the prime indicates differentiation with respect to the prime variables. Since ∇^2 is equivalently written as $\nabla \bullet \nabla$, equation (1.48) can also be expressed as

$$\nabla \times \boldsymbol{B} = -\lim_{R\to 0} \frac{\mu_0}{4\pi} \int_V \boldsymbol{J}\nabla' \bullet \nabla' \left(\frac{1}{R}\right) dV \tag{1.51}$$

which, by Gauss' theorem becomes

$$\nabla \times \boldsymbol{B} = -\lim_{R\to 0} \frac{\mu_0}{4\pi} \oint_S \boldsymbol{J}\nabla' \left(\frac{1}{R}\right) d\boldsymbol{S} \tag{1.52}$$

The differential surface area in spherical coordinates is $\boldsymbol{u}_n(R^2 \sin\theta d\phi d\theta)$, where $\boldsymbol{u}_n$ is the unit normal to the surface $d\boldsymbol{S}$. However, the gradient of $1/R$ is equal to $-\boldsymbol{u}_n/R^2$ so that the integral becomes

$$\nabla \times \boldsymbol{B} = \lim_{R\to 0} \frac{\mu_0}{4\pi} \oint_S \boldsymbol{J} \sin\theta d\phi d\theta \tag{1.53}$$

Since the radius of the small sphere approaches zero, the current density vector $\boldsymbol{J}$ can be removed from the integrand since it becomes a constant. The remaining integral can now be evaluated as simply 4π. Equation (1.44) finally reduces to *Ampere's law*

$$\nabla \times \boldsymbol{B} = \mu_0 \boldsymbol{J} \tag{1.54}$$

The integral form of Ampere's law can be obtained by integrating equation (1.54) over an arbitrary finite open surface which includes the region where the current density $\boldsymbol{J}$ is flowing, whereupon,

$$\int_S \nabla \times \boldsymbol{B} \cdot d\boldsymbol{S} = \mu_0 \int_S \boldsymbol{J} \bullet d\boldsymbol{S} \tag{1.55}$$

The right-hand side of equation (1.55) is clearly proportional to the current I flowing through the surface S. With the use of Stokes' theorem, the left hand side can be altered to the form

$$\int_S (\nabla \times \boldsymbol{B}) \bullet d\boldsymbol{S} = \oint_O \boldsymbol{B} \bullet d\boldsymbol{l} \tag{1.56}$$

where the path O corresponds to the outer edge of the surface S. Thus, the integral form of Ampere's law is

$$\int_O \boldsymbol{B} \bullet d\boldsymbol{l} = \mu_0 I \tag{1.57}$$

1.9 MAGNETIC FIELD INTENSITY *H*

Thus far, the behavior of the magnetic field in a vacuum has been considered. When dealing with material bodies, orbiting electrons of each individual atom can be considered as a current loop. With no external magnetic field, the orbiting atoms are randomly positioned so that they do not produce, themselves, a magnetic field (except for permanent magnets). The presence of the magnetic field influences the orbits of the individual atoms creating what is called a *magnetic dipole moment* $\boldsymbol{m}$. The dipole moment is defined as equal to the product of the area of the circular loop defined by the orbiting electron times the magnitude of the circulating current and with a direction perpendicular to the plane of the loop in the direction of a right-hand screw. That is, if the current loop is located in the x,y plane and orbiting in a counterclockwise direction as shown in Figure 1.4, the magnetic dipole moment is defined by

$$\boldsymbol{m} = \pi r^2 I \boldsymbol{u}_z \quad \text{A-m}^2$$

The vector potential for this small electric circuit is

$$\boldsymbol{A} = \frac{\mu_0 I}{4\pi} \oint_O \frac{d\boldsymbol{l}}{R_1} \tag{1.58}$$

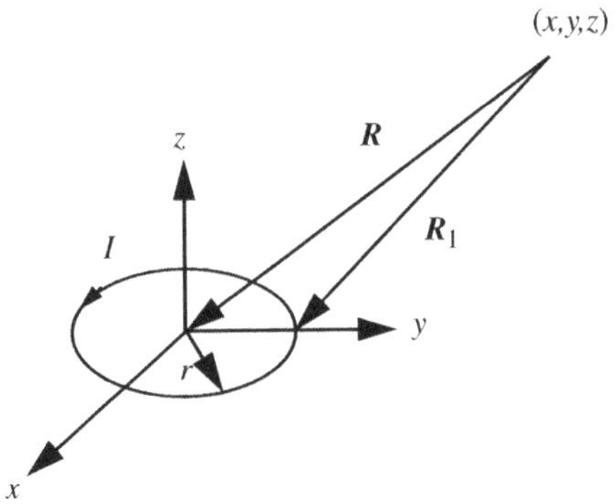

Figure 1.4 The magnetic dipole.

If R^2 is much greater than r^2, then

$$\frac{1}{R_1} \approx \frac{1}{R}\left(1 + \frac{rx}{R^2}\cos\phi' + \frac{ry}{R^2}\sin\phi'\right) \tag{1.59}$$

Equation (1.58) then integrates to yield

$$\boldsymbol{A} = \frac{\mu_0 I r^2}{4R^3}(-y\boldsymbol{u}_x + x\boldsymbol{u}_y) \tag{1.60}$$

However,

$$\boldsymbol{u}_z \times \boldsymbol{R} = \boldsymbol{u}_z \times (x\boldsymbol{u}_x + y\boldsymbol{u}_y + z\boldsymbol{u}_z) = -y\boldsymbol{u}_x + x\boldsymbol{u}_y$$

so that the vector potential can be expressed in vector form as

$$\boldsymbol{A} = \frac{\mu_0 m}{4\pi}\boldsymbol{u}_z \times \frac{\boldsymbol{u}_r}{R^2} \tag{1.61}$$

In general, if the magnetization axis direction is arbitrary,

$$\boldsymbol{A} = \frac{\mu_0 m}{4\pi}\boldsymbol{u}_m \times \frac{\boldsymbol{u}_r}{R^2} \tag{1.62}$$

or, alternatively, from equation (1.16), as

$$\boldsymbol{A} = -\frac{\mu_0 m}{4\pi}\boldsymbol{u}_m \times \nabla\left(\frac{1}{R}\right) \tag{1.63}$$

A mathematical representation of the overall magnetic dipole moment of a finite body can be obtained by multiplying $m\boldsymbol{u}_m$ by the number of atoms per unit volume N_a to obtain the *magnetic polarization vector* $\boldsymbol{M}$

$$\boldsymbol{M} = N_a\boldsymbol{m} = N_a m\boldsymbol{u}_m \quad \text{A/m} \tag{1.64}$$

so that it is possible to write, in place of (1.63),

$$\boldsymbol{A}(x, y, z) = -\frac{\mu_0}{4\pi}\int_V \boldsymbol{M}(x', y', z') \times \nabla\left(\frac{1}{R}\right)dV' \tag{1.65}$$

where $\boldsymbol{M}$ is considered here as varying within the body (i.e., a function of (x', y', z') and R represents the distance between the external point (x,y,z) and the internal point (x', y', z'). Now,

$$\boldsymbol{M} \times \nabla\left(\frac{1}{R}\right) = -\boldsymbol{M} \times \nabla'\left(\frac{1}{R}\right) \tag{1.66}$$

and

$$\boldsymbol{M}(x', y', z') \times \nabla'\left(\frac{1}{R}\right) = \left(\frac{1}{R}\right)\nabla' \times \boldsymbol{M}(x', y', z') - \nabla' \times \frac{\boldsymbol{M}(x', y', z')}{R} \tag{1.67}$$

Equation (1.65) can now be written as

$$\boldsymbol{A} = -\frac{\mu_0}{4\pi}\int_V \left(\frac{1}{R}\right)\nabla' \times \boldsymbol{M}(x', y', z')dV' + \frac{\mu_0}{4\pi}\int_V \nabla' \times \frac{\boldsymbol{M}(x', y', z')}{R}dV' \tag{1.68}$$

It can be shown as a homework problem that a corollary to Stokes' theorem is the fact that

$$\int_V \nabla' \times \frac{\boldsymbol{M}(x',y',z')}{R} dV' = -\oint_S \frac{\boldsymbol{M} \times \boldsymbol{u}_{\mathrm{n}}}{R} dS \qquad (1.69)$$

for any vector field $\boldsymbol{M}$ wherein $\boldsymbol{u}_{\mathrm{n}}$ is the unit normal to the surface dS. Thus, finally,

$$\boldsymbol{A} = \frac{\mu_0}{4\pi} \oint_S \frac{\boldsymbol{M} \times \boldsymbol{u}_{\mathrm{n}}}{R} dS + \frac{\mu_0}{4\pi} \int_V \left(\frac{1}{R}\right) \nabla' \times \boldsymbol{M}(x',y',z') dV' \qquad (1.70)$$

If one compares this expression of the vector potential with that for true currents, it is apparent that one can interpret the term $\boldsymbol{M} \times \boldsymbol{u}_{\mathrm{n}}$ as an equivalent surface polarization current $\boldsymbol{K}_m$. Similarly, the curl of the magnetization vector $\nabla \times \boldsymbol{M}$ is an equivalent volumetric polarization current $\boldsymbol{J}_m$. The expression for vector potential becomes

$$\boldsymbol{A} = \frac{\mu_0}{4\pi} \left(\oint_S \frac{\boldsymbol{K}_m}{R} dS + \int_V \frac{\boldsymbol{J}_m}{R} dV \right) \qquad (1.71)$$

Note that this equation again generates three "decoupled" equations involving only x, y, and z components of $\boldsymbol{A}$, $\boldsymbol{K}_m$, and $\boldsymbol{J}_m$, respectively.

Consider now the magnetic flux density at any point within a material body having both true current $\boldsymbol{J}$ and polarization current $\boldsymbol{J}_m$. From equation (1.22),

$$\boldsymbol{B} = \nabla \times \left(\frac{\mu_0}{4\pi} \int_V \frac{\boldsymbol{J}_m + \boldsymbol{J}}{R} dV \right) \qquad (1.72)$$

In Section 1.8, it was shown that $\nabla \times \boldsymbol{B} = \mu_0 \boldsymbol{J}$. In an analogous manner it is evident that with polarization currents,

$$\nabla \times \boldsymbol{B} = \mu_0 (\boldsymbol{J} + \boldsymbol{J}_m) \qquad (1.73)$$

Since the volumetric polarization current $\boldsymbol{J}_m$ is equal to the curl of the magnetization $\boldsymbol{M}$, this expression can be written as

$$\nabla \times \left(\frac{\boldsymbol{B}}{\mu_0} - \boldsymbol{M} \right) = \boldsymbol{J} \qquad (1.74)$$

in which the vector on the left-hand side of the equation $\boldsymbol{B}/\mu_0 - \boldsymbol{M}$ has as its source, only the true currents $\boldsymbol{J}$. It is, therefore, useful to define a new quantity, the *magnetic field intensity* $\boldsymbol{H}$ as

$$\boldsymbol{H} = \frac{\boldsymbol{B}}{\mu_0} - \boldsymbol{M} \quad \mathrm{A/m} \qquad (1.75)$$

from which it can be established that

$$\mu = \frac{\mu_0}{1 - \mu_0 \left(\frac{\boldsymbol{M}}{\boldsymbol{B}} \right)} \qquad (1.76)$$

in which case

$$\boldsymbol{B} = \mu \boldsymbol{H} \tag{1.77}$$

or, alternatively,

$$\boldsymbol{B} = \mu_r \mu_0 \boldsymbol{H} \tag{1.78}$$

where $\mu_r = \mu/\mu_0$ is defined as the *relative permeability*. The differential form for Ampere's law is finally obtained, namely

$$\nabla \times \boldsymbol{H} = \boldsymbol{J} \tag{1.79}$$

Since $\nabla \bullet \boldsymbol{H} = -\nabla \bullet \boldsymbol{M}$, the divergence of the magnetic field intensity is not zero as is the case for the divergence of $\boldsymbol{B}$. The magnetic field intensity is also sometimes called the *magnetic potential gradient*.

Starting with equation (1.79), and applying Stokes' theorem, results in the usual integral form of Ampere's law,

$$\int_S (\nabla \times \boldsymbol{H}) \bullet d\boldsymbol{S} = \oint_O \boldsymbol{H} \bullet d\boldsymbol{l} = \int_S \boldsymbol{J} \bullet d\boldsymbol{S} = I \tag{1.80}$$

where I is the total current enclosed by the path defined by O. If current I is confined to a conductor and flows N times through the loop O, I is replaced by NI in equation (1.80).

1.10 BOUNDARY CONDITIONS FOR *B* AND *H*

In the derivation of the differential form for Ampere's law, points within the material were specified and not points on the boundary, where an additional polarization current component, $\boldsymbol{K}_m$ exists. Hence, for points on the boundary, the results obtained must be modified to take account of this current which results from a discontinuity in the magnetization vector $\boldsymbol{M}$. Consider now an interface between two material bodies with permeability μ different from μ_0 as illustrated in Figure 1.5.

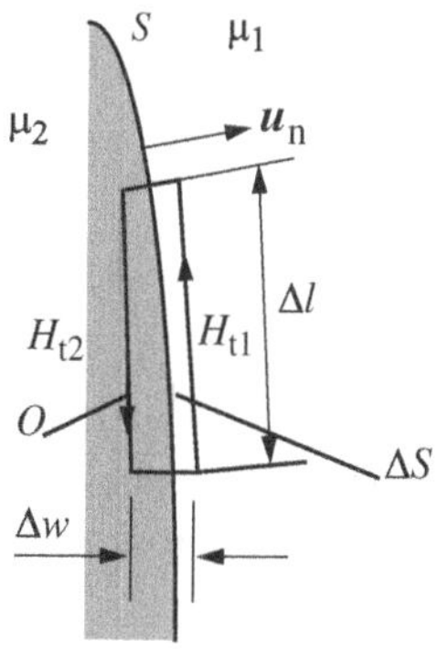

Figure 1.5 Determination of boundary condition for the tangential component of $\boldsymbol{H}$.

Let H_{t1} and H_{t2} be the components of $\boldsymbol{H}$ tangent to the interface surface in the material body and air, respectively. From equation (1.79), $\nabla \times \boldsymbol{H} = \boldsymbol{J}$ and if the path O is chosen as shown in Figure 1.5, the integral form for Ampere's law gives

$$\int_{\nabla S} (\nabla \times \boldsymbol{H}) \bullet d\boldsymbol{S} = \oint_{\Delta O} \boldsymbol{H} \bullet d\boldsymbol{l} = \int_{\Delta S} \boldsymbol{J} \bullet d\boldsymbol{S} \tag{1.81}$$

where ΔO is the outer contour of the surface ΔS. Since no physical current is enclosed, these expressions are equal to zero. If one shrinks Δw to a negligibly small value, then $H_{t1}\Delta l - H_{t2}\Delta l = 0$ so that

$$H_{t1} = H_{t2} \tag{1.82}$$

which states, in effect, that the tangential components of $\boldsymbol{H}$ must be continuous across a boundary not carrying a true surface current $\boldsymbol{K}$. It is clear that if the boundary supports a physical current surface current $\boldsymbol{K}$, then equation (1.82) must be replaced with

$$H_{t1} = H_{t2} = K \tag{1.83}$$

where positive current is taken in the direction made by a right-hand screw when rotated in the direction defined by the path O. In vector form, the equivalent expression is written as

$$\boldsymbol{u}_{\text{n}} \times (\boldsymbol{H}_1 - \boldsymbol{H}_2) = \boldsymbol{K} \tag{1.84}$$

Although true surface currents are essentially impossible, equation (1.83) is often used to approximate physical situations in an electrical machine design.

As a corollary to equation (1.82), it is apparent that the tangential components of B are discontinuous across a boundary separating materials with different permeabilities, that is

$$\frac{B_{t1}}{\mu_1} = \frac{B_{t2}}{\mu_2} \tag{1.85}$$

when no surface currents flow on the boundary.

The behavior of the normal components of B and H can also be determined as shown in Figure 1.6.

In this case, from equation (1.36),

$$\oint_S \boldsymbol{B} \bullet d\boldsymbol{S} = 0$$

If this expression is applied to the pill box shape of Figure 1.6, then

$$B_{n1}\Delta S - B_{n2}\Delta S = (B_{n1\text{s}} + B_{n2\text{s}})(\pi \Delta r \Delta w)$$

where B_{n1}, B_{n2} pertains to the top and bottom of the pill box and $B_{n1\text{s}}$ and $B_{n2\text{s}}$ pertains to the sides of the pillbox in materials 1 and 2, respectively. If the sides of the pillbox are made arbitrarily small, then $B_{n1}\Delta S = B_{n2}\Delta S$, or finally across any boundary,

$$B_{n1} = B_{n2} \tag{1.86}$$

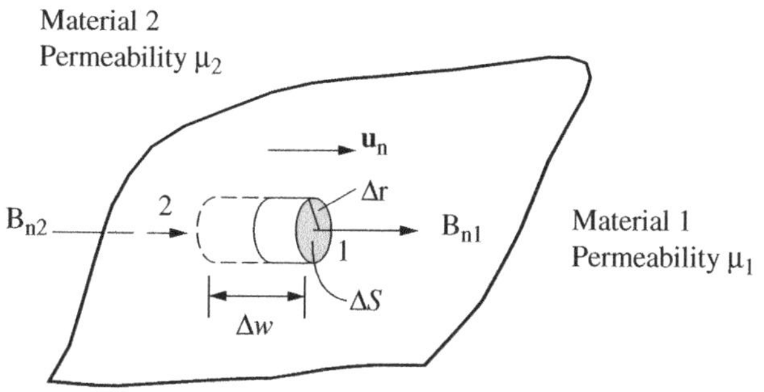

Figure 1.6 Determination of the boundary condition for the normal component of ***B***.

In vector form, this expression is equivalent to

$$\boldsymbol{u}_n \bullet (\boldsymbol{B}_1 - \boldsymbol{B}_2) = 0 \tag{1.87}$$

The corresponding boundary condition for ***H*** is clearly

$$\mu_1 H_{n1} = \mu_2 H_{n2} \tag{1.88}$$

It is observed that the normal component of ***B*** is continuous but not so for ***H***.

1.11 FARADAY'S LAW

It was Michael Faraday and Joseph Henry who jointly discovered that the electric field becomes non-conservative when the line integral of ***E*** is evaluated in a magnetic field which is non-steady, that is, when the magnetic field linking the line integral path varies with time. In this case, equation (1.43) must be modified to form

$$\oint_O E \bullet dl = -\frac{d}{dt}\int_S \boldsymbol{B} \bullet dS = -\frac{d\Phi}{dt} \tag{1.89}$$

where O bounds surface S. In a practical sense, this expression shows that an additional electric field is produced by a time-changing magnetic field and consequently, a voltage is produced in a closed short-circuited coil when placed in this field. The strength of this voltage is proportional to the time rate of change of flux linking the coil and, in turn, induces a current in the conducting loop. The negative sign indicates that the voltage is directed in such a manner so as to produce a current which produces a consequent magnetic field which reduces the net flux linking the loop.

The differential form of equation (1.89) may be obtained by using Stokes' theorem to replace the line integral by a surface integral so that

$$\oint_O E \bullet dl = \int_S \nabla \times \boldsymbol{E} \bullet dS = -\frac{d}{dt}\int_S \boldsymbol{B} \bullet dS \tag{1.90}$$

or

$$\int_S \left(\nabla \times \boldsymbol{E} + \frac{d\boldsymbol{B}}{dt} \right) \bullet d\boldsymbol{S} = 0 \tag{1.91}$$

from which,

$$\nabla \times \boldsymbol{E} = -\frac{d\boldsymbol{B}}{dt} \tag{1.92}$$

The current which flows in the conducting loop creates also an electric field within a conducting material which is proportional to the current, the basis for Ohm's law. Expressed at a point rather than averaged over a finite body, Ohm's law at a point is

$$\boldsymbol{J} = \sigma \boldsymbol{E} \tag{1.93}$$

where, σ is the conductivity of the material in amperes per volt-meter $(ohm\text{-}meters)^{-1}$.

1.12 INDUCED ELECTRIC FIELD DUE TO MOTION

Since a changing magnetic field linking a conducting coil can also be produced by simply moving it physically through a stationary, non-uniform field, equation (1.89) is equally valid for this condition as well. Movement of the coil through the field, however, produces an accompanying phenomenon, the Lorentz force which states that moving charges in a magnetic field experience a force proportional to the velocity of the charge and the strength of the magnetic field according to the vector equation

$$\boldsymbol{F} = q_c \boldsymbol{v} \times \boldsymbol{B} \tag{1.94}$$

The force is seen acting in a direction perpendicular to both $\boldsymbol{v}$ and $\boldsymbol{B}$. Note that this is just an equivalent form of the Biot–Savart law since an ampere flowing through a meter of wire is essentially the same as 10^{-6} coulombs of electrons traveling at 10^6 m/s. This force is proportional to the electric field since

$$\frac{\boldsymbol{F}}{q_c} = \boldsymbol{E} = \boldsymbol{v} \times \boldsymbol{B} \tag{1.95}$$

The corresponding field, in turn, induces a voltage in the coil resulting in current flow according to Ohm's law, equation (1.93). This induced voltage is typically called the *electromotive force* (a misnomer since the actual units of the quantity are fundamentally newtons/coulomb). Note from Figure 1.7 that this voltage is in such a direction so as to produce a current which resists any change in the flux linking the coil. The degree to which this is accomplished, depends on the resistance of the coil. If superconducting, flux linking the coil will not change at all. The interrelationship between the force on a moving coil and the resulting current (or vice versa) is the key component in the principle of *electromechanical energy conversion*.

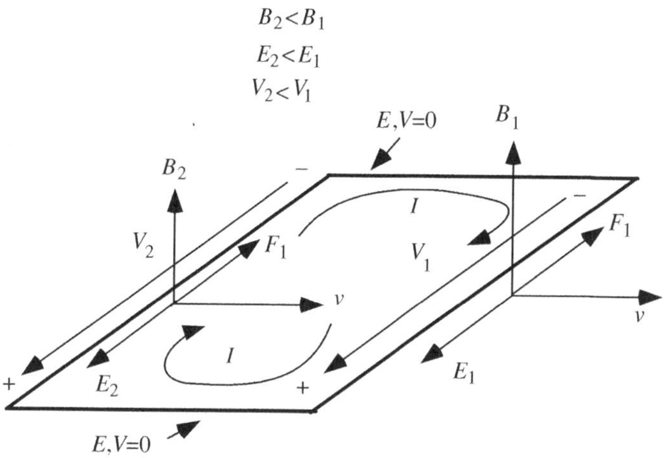

Figure 1.7 Induced voltage in a coil moving in a direction so as to increase the flux linking the coil. Force assumed to be impressed on a negative charge (electron).

1.13 PERMEANCE, RELUCTANCE, AND THE MAGNETIC CIRCUIT

The solution to the general magnetostatic boundary-value problem involving conduction currents in the presence of magnetic material is very difficult to obtain analytically. Fortunately, applications involving electric machine design allow for good approximate solutions to be obtained. The analysis procedure parallels that of DC circuits which are composed of series and parallel resistors. Consider, for example, the field in the region of a toroid with a rectangular cross section wound with N turns with the coil current I as illustrated in Figure 1.8.

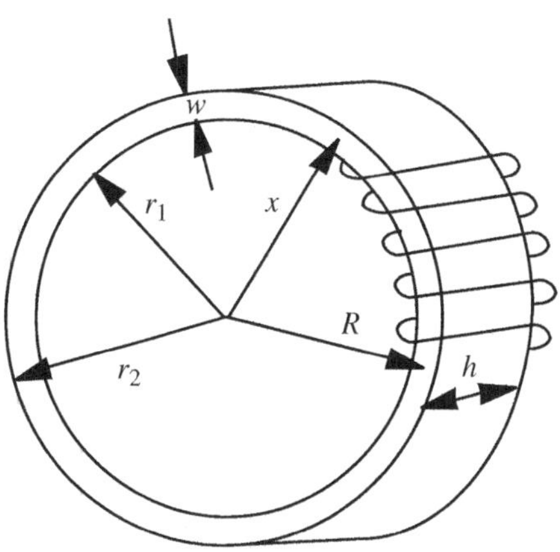

Figure 1.8 Flux distribution of a toroid with a rectangular cross section.

Due to symmetry, the magnetic field intensity has only a circumferential component. At any point in the core x meters from its center, the magnetic potential gradient is

$$H = \frac{NI}{2\pi x} \tag{1.96}$$

where N is the number of turns enclosed by the path for H. The flux density at the distance x is therefore

$$B = \mu H = \frac{\mu NI}{2\pi x} \tag{1.97}$$

However, the flux density at any point is equal to $d\Phi/dA$. The total flux over a cross-sectional area ($h\,dx$) is

$$d\Phi = BdA = Bh\,dx = \mu \frac{NIh}{2\pi}\frac{dx}{x} \tag{1.98}$$

The total flux over the area A is given by

$$\begin{aligned}\Phi &= \int_{r_1}^{r_2} \mu \frac{NIh}{2\pi}\frac{dx}{x} \\ &= \mu \frac{NIh}{2\pi} \ln(r_2/r_1) \\ &= \mu \frac{NIh}{2\pi} \ln\left(\frac{R + w/2}{R - w/2}\right) \\ &= \mu \frac{NIh}{2\pi} \ln\left(\frac{1 + \frac{w}{2R}}{1 - \frac{w}{2R}}\right) \\ &= \mu \frac{NIh}{2\pi} \ln\left(1 + \frac{w}{R} + \frac{w^2}{2R^2} + \frac{w^3}{4R^3} + \cdots\right)\end{aligned} \tag{1.99}$$

so that finally,

$$\Phi = \mu \frac{NI}{2\pi} h \frac{w}{R} \quad \text{if } w/R \ll 1 \tag{1.100}$$

When $w/R = 0.2$, then the natural log of the expansion $1 + w/R + w^2/2R^2 + \cdots$ is equal to w/R to within 0.3%. Therefore, for cases where the core width is small in comparison with the mean radius R, one can assume the flux density to be uniform. Thus

$$\Phi = \frac{\mu NIhw}{2\pi R} \tag{1.101}$$

Considering the core width w to be small compared with R enables one to assume that H is constant at all points of the toroid and equal to its value at the center. In this case, from equation (1.97)

$$\frac{\mu NI}{2\pi R} = B \tag{1.102}$$

and

$$hw = A \tag{1.103}$$

Therefore,

$$\Phi = \frac{\mu NI}{2\pi R}A \quad (= BA) \tag{1.104}$$

Note that a rectangular toroid has purposely been chosen for this example. The case of a circular toroid reduces in the same manner but the exact solution involves Bessel functions. It is useful noting that in this book, in deference to convention, the symbol A will be used for both vector potential and cross-sectional area. Similarly, S will be used for surface area and per unit slip of an induction motor, hopefully without much confusion.

It can be observed that equation (1.104) may be written as

$$\Phi = \left(\frac{\mu A}{2\pi R}\right) NI \tag{1.105}$$

where $A = hw$. Note that the coefficient of NI is a constant depending upon the geometry of the magnetic circuit and its permeability. Defining this constant as the permeance

$$\mathcal{P} = \frac{\mu A}{2\pi R} \tag{1.106}$$

Since $2\pi R$ is the length of the magnetic path, it may be replaced for purposes of generalization by l. Therefore, in general, when the flux is uniformly distributed over a constant cross-sectional area,

$$\mathcal{P} = \frac{\mu A}{l} \tag{1.107}$$

The permeance $\mathcal{P}$ is given in units of *webers per ampere-turn* or *henries*.

It is also useful to define

$$\mathcal{F}_{12} = \int_1^2 H \bullet dl \tag{1.108}$$

The quantity $\mathcal{F}_{12}$ is said to express the magnetomotive force (MMF) acting between points 1 and 2 which has the SI unit *amperes*. The units used in this text will be *ampere-turns* as a reminder of the important influence of the number of winding turns on this quantity. When the closed path O encloses a circuit of N turns carrying I amperes, it is clear from Ampere's law that

$$\mathcal{F} = NI \tag{1.109}$$

Therefore, in general, the flux in a magnetic circuit can be expressed as

$$\Phi = \mathcal{P}\mathcal{F} \tag{1.110}$$

In almost all practical cases, the permeance is not so easy to find. Therein lies the art and science of electrical machine design. When the flux density varies over the cross-sectional area, the differential form of equation (1.110) is often useful. In this case

$$d\Phi = \mathcal{F}d\mathcal{P} \tag{1.111}$$

where

$$d\mathcal{P} = \frac{\mu dA}{l}$$

The total permeance is found by taking

$$\mathcal{P} = \int_0^A \mu \frac{dA}{l} \tag{1.112}$$

where l is frequently a function of A.

In the case of the example rectangular toroid, $dA = hdx$ and $l = 2\pi x$. Equation (1.112) becomes

$$\mathcal{P} = \int_{r_1}^{r_2} \mu \frac{hdx}{(2\pi x)}$$

whereupon,

$$\mathcal{P} = \frac{\mu h}{2\pi} \ln\left(\frac{r_2}{r_1}\right)$$

directly.

The reciprocal of permeance also has great utility in magnetic circuit analysis. Formally, by definition, reluctance is

$$\mathcal{R} = 1/\mathcal{P} \tag{1.113}$$

and if the cross-sectional area A and the permeability are constants, independent of the length of the circuit,

$$\mathcal{R} = \frac{l}{\mu A} \tag{1.114}$$

The quantity reluctance carries units of ampere-turns per weber. In the MKS unit system, this corresponds to inverse henries (H^{-1}). In SI units, reluctance has not been formally given a unique name such as *siemens* for Ω^{-1} so that it is generally described in terms of its basic units. In this text, inverse henries is adopted as the preferred unit. The reluctance is often found to be a more useful quantity in the analysis of electrical machines than permeance.

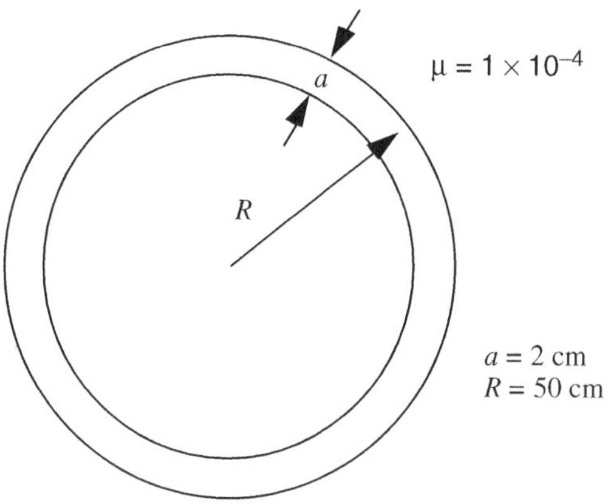

Figure 1.9 Square toroid for Example.

1.14 EXAMPLE—SQUARE TOROID

1. Find the reluctance of a toroid of square cross section and radius R to the center of the core. The core is designed such that $\mu = 1 \cdot 10^{-4}$, $R = 50$ cm, and $a =$ 2 cm. Refer to Figure 1.9. The reluctance $\mathcal{R}$ is given by

$$\begin{aligned}\mathcal{R} &= \frac{l}{\mu A} \\ &= \frac{2\pi(0.5)}{1 \cdot 10^{-4}(0.02)^2} \\ &= 7.85 \cdot 10^7 \quad H^{-1}\end{aligned}$$

2. The toroid has 1570 ampere-turns uniformly wound around its core. What is the value of the flux inside the toroid?

$$\begin{aligned}\Phi &= \mathcal{F}/\mathcal{R} \\ &= \frac{1570}{7.85 \cdot 10^7} = 2 \cdot 10^{-5} \quad \text{Wb}\end{aligned}$$

3. What is the flux density in the core of the toroid?

$$B = \frac{\Phi}{A} = \frac{2 \cdot 10^{-5}}{(0.02)^2} = 0.05 \quad \text{Wb/m}^2$$

1.15 MULTIPLE CIRCUIT PATHS

When the same flux is set up in a magnetic circuit made of the same material but with parts of different cross sections or when the parts of the circuit are of materials of different permeabilities but of the same or different cross sections, the flux can be found by finding the reluctance of the different parts, adding them to give the total reluctance of the circuit, and finally dividing the magnetomotive force by the total

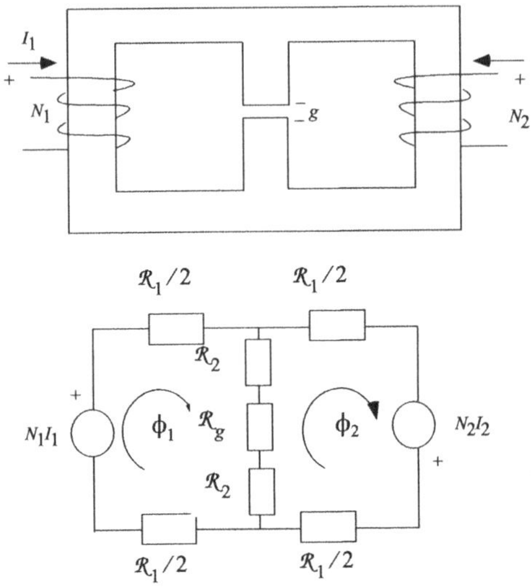

Figure 1.10 A type of iron core transformer and its equivalent magnetic circuit.

reluctance. As in the case of the resistances of an electric circuit, the reluctances in a magnetic circuit when in series are added to give the total reluctance of the circuit.

When the magnetic circuits are in parallel, their total permeance is equal to the sum of the permeances of the parallel circuits. The total reluctance is simply the reciprocal of the total permeance. When the circuit is more complicated, the usual application of Kirchhoff's laws will generally yield an answer. For example, consider the iron-core transformer with an air gap in the center leg as shown in Figure 1.10. (a) The problem is to compute the flux linkage for the coil with N_2 turns when a current I_2 flows in the other coil. The equivalent magnet circuit is illustrated in Figure 1.10b. The magnetomotive force $N_1 I_1$ is applied to the circuit and it acts in series with the reluctance $\mathcal{R}_1$ of the left leg. The reluctance can be split into two portions of magnitude $\mathcal{R}_1/2$. Following the usual DC circuit analysis approach, the following two equations may be obtained:

$$\begin{aligned} N_1 I_1 &= \Phi_1(\mathcal{R}_1 + 2\mathcal{R}_2 + \mathcal{R}_g) - \Phi_2(2\mathcal{R}_2 + \mathcal{R}_g) \\ N_2 I_2 &= \Phi_2(\mathcal{R}_1 + 2\mathcal{R}_2 + \mathcal{R}_g) - \Phi_1(2\mathcal{R}_2 + \mathcal{R}_g) \end{aligned} \tag{1.115}$$

Solving for Φ_2 yields

$$\Phi_2 = [N_1 I_1(2\mathcal{R}_2 + \mathcal{R}_g) + N_2 I_2(\mathcal{R}_1 + 2\mathcal{R}_2 + \mathcal{R}_g)]/[\mathcal{R}_1(\mathcal{R}_1 + 4\mathcal{R}_2 + 2\mathcal{R}_g)] \tag{1.116}$$

1.16 GENERAL EXPRESSION FOR RELUCTANCE

Assume now a body of homogeneous permeability but of an arbitrary shape. If flux lines can be approximated, flux tubes containing a specified number of flux lines can be identified which take the general shape of the sketch in Figure 1.11.

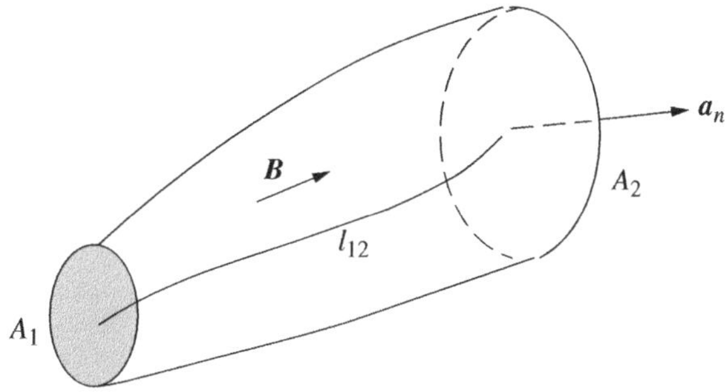

Figure 1.11 An arbitrary flux tube.

The difference of magnetic potential between the two faces A_1 and A_2 can be evaluated by again taking

$$\int_{l_{12}} H \bullet dl = \mathcal{F}_1 - \mathcal{F}_2 \tag{1.117}$$

where l_{12} is the path from A_1 to A_2 along the side or within the flux tube. The flux which is confined within the flux tube is

$$\int_{A} B \bullet dS = \Phi \tag{1.118}$$

where A is the area A_1, A_2, or any cross-sectional area across the flux tube. By definition, the reluctance between the two surfaces A_1 and A_2 is

$$\mathcal{R} = \frac{\int_{l_{12}} H \bullet dl}{\mu \int_{A} H \bullet dS} \tag{1.119}$$

Although this expression is rather simple in form, the values of the integrals cannot be established easily since the location of the flux lines must be known before the integrals can be carried out. Evaluation of the reluctance can be made more accurate if the cross section of the flux tubes are considered as curvilinear squares or rectangles. That is, it is assumed that the corners of the cross-sectional area of a flux tube are always 90 degrees but the sides of the rectangles are allowed to be curved lines. This behavior of the cross-sectional area is a natural consequence of the fact that the lines of constant magnetic potential must be at right angles to the lines of magnetic flux. Plots of the magnetic field using "curvilinear squares" is a traditional method which can yield remarkably accurate results when care is taken to always maintain a curvilinear (right angle) relationship between the potential and flux lines when sketching the field plot.

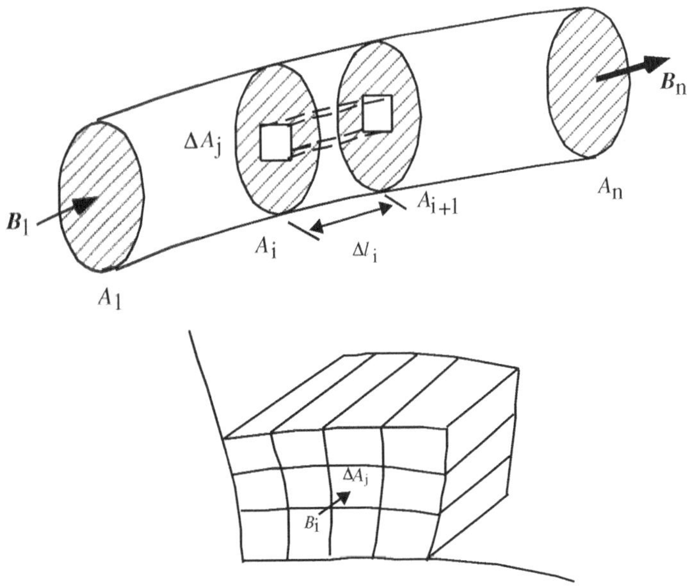

Figure 1.12 Orthogonal curvilinear squares used to portray a magnetic flux tube.

Consider the more accurate flux plot of Figure 1.12, where two equipotential surfaces A_i and A_{i+1} are identified. Let the potential difference between these surfaces be $\Delta\mathcal{F}_s$.

The region between A_i and A_{i+1} can be decomposed into a number of more elementary flow tubes of length Δl_i and cross section ΔA_{ij}. The reluctance of an arbitrary elementary flow tube is

$$\mathcal{R}_i = \frac{\Delta\mathcal{F}_i}{\Delta\Phi_{ij}} = \frac{\Delta l_i}{\mu\Delta A_{ij}} \tag{1.120}$$

Since permeances add directly in parallel, the total permeance between surfaces A_i and A_{i+1} is

$$\Delta\mathcal{P}_i = \sum_j \frac{\mu\Delta A_{ij}}{\Delta l_i} \tag{1.121}$$

The corresponding reluctance is

$$\Delta\mathcal{R}_i = \frac{1}{\sum_j \frac{\mu\Delta A_{ij}}{\Delta l_i}} \tag{1.122}$$

The total reluctance is obtained by adding up all such reluctances over the total length of the flux tube. The result is

$$\mathcal{R} = \sum_i \frac{1}{\sum_j \frac{\mu \Delta A_{ij}}{\Delta l_i}} \tag{1.123}$$

It is important to note that the equation remains a function of the geometry of the magnetic structure only. For more details on the method of curvilinear squares for flux plotting, the reader is referred to any good basic text on electromagnetic fields.

1.17 INDUCTANCE

In most practical cases the magnetic flux links a number of circuit loops N or "turns" in which case one defines the flux linkage λ as

$$\lambda = N\Phi \tag{1.124}$$

The inductance of a coil is defined as "the number of flux linkages in weber turns per ampere of current flowing in the coil." Flux linkages per ampere is formally defined as a *henry*. Interpreted in mathematical form

$$L = \frac{\lambda}{I} = \frac{N\Phi}{I} \quad \text{weber-turns/ampere or henries} \tag{1.125}$$

where

L is the inductance in henries

N is the number of turns of the coil

Φ is the flux in webers linking the turns

I is the current in the turns in amperes

If the current I and flux Φ correspond to the same circuit, then the resulting inductance is termed self-inductance. When the current I and flux Φ correspond to different circuits, a mutual inductance can be defined.

The self-inductance can be written in several other useful forms:

$$L = \frac{N^2\Phi}{\mathcal{F}} \tag{1.126}$$

$$= N^2\mathcal{P} \tag{1.127}$$

and if the cross-sectional area and μ are constant,

$$L = \mu\frac{N^2A}{l} \tag{1.128}$$

Note that the inductance is proportional to the square of the number of turns. Since the inductance has been formally defined in henries, the permeability μ formally takes on the alternate unit of *henries per meter*.

In the above expressions for inductance, it was assumed that the magnetic path or magnetic circuit is defined. When the path of the magnetic flux is not defined as in a solenoid, formulas for inductances are derived from field theory, flux plotting, experimentation, or numerical solution of Laplace's or Poisson's equation.

1.18 EXAMPLE—INTERNAL INDUCTANCE OF A WIRE SEGMENT

It was shown in Example 1.1 that the flux density of an infinitely long wire was given by

$$B = \frac{\mu_0 I}{2\pi R} \tag{1.129}$$

Consider a more detailed view of the wire as shown in Fig. 1.13 showing now a circular inner portion of the wire with radius r. If it is assumed that the current density J in A/m^2 is directed along the wire length and is uniform over the wire cross section, then by Ampere's law, the integral along a circular path C with radius r within the wire will yield

$$\oint_C \boldsymbol{H} \bullet d\boldsymbol{l} = H_\phi 2\pi r = J\pi r^2 \quad 0 < r < R_w, \tag{1.130}$$

where R_w is the outer radius of the wire. Since the wire is assumed to be non-magnetic,

$$B_\phi = \mu_0 H_\phi = \frac{\mu_0 J r}{2} \tag{1.131}$$

Since J is constant, the current in the wire is $I = J\left(\pi R_w^2\right)$ and thus

$$B_\phi = \mu_0 I \frac{r}{2\pi R_w^2} \tag{1.132}$$

where R_w is the radius of the wire.

Consider now a circular inner portion of the wire with radius r. The flux in an annular portion of length l and thickness dr will be

$$d\Phi(r) = B_\phi l dr \tag{1.133}$$

$$= \frac{\mu_0 I l}{2\pi} \frac{r}{R_w^2} dr \tag{1.134}$$

This flux links the only the current $I\frac{r^2}{R_w^2}$ so that the corresponding flux linkages

$$d\lambda = \frac{\mu_0 I l}{2\pi} \frac{r^3}{R_w^4} dr \tag{1.135}$$

and the total flux linkages become

$$\lambda = \frac{\mu_0}{2\pi} I l \int_0^{R_w} \frac{r^3}{R_w^4} dr \tag{1.136}$$

$$= \frac{\mu_0}{8\pi} I l \tag{1.137}$$

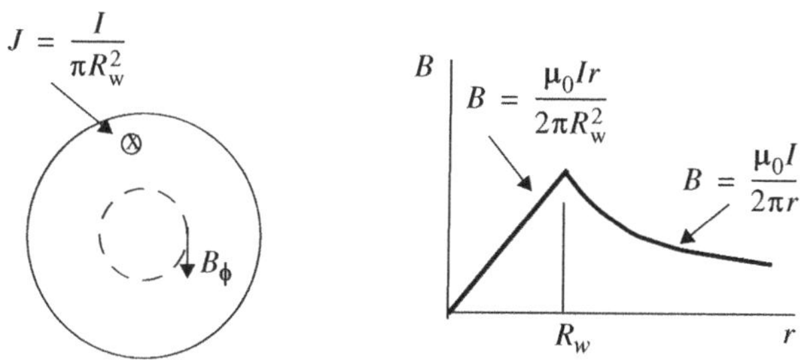

Figure 1.13 Calculation of flux density and inductance within a circular conductor.

Thus, the internal inductance of a wire of length L is

$$L_{\text{internal}} = \frac{\lambda}{I} = \frac{\mu_0}{8\pi}L \tag{1.138}$$

1.19 MAGNETIC FIELD ENERGY

Utilization of the magnetic field energy is often a convenient method for determining the inductance. The energy stored in a magnetic field can be expressed as

$$W_{\text{m}} = \frac{1}{2}\int_V (\boldsymbol{B} \bullet \boldsymbol{H})\, dV \tag{1.139}$$

since, also,

$$W_{\text{m}} = \frac{1}{2}LI^2 \tag{1.140}$$

then, when $\boldsymbol{B}$ and $\boldsymbol{H}$ arise from the same current,

$$L_{\text{self}} = \frac{1}{I^2}\int_V (\boldsymbol{B} \bullet \boldsymbol{H})\, dV \tag{1.141}$$

Alternative forms of equation (1.141) are useful. In machine analysis, it is frequently possible to assume that the field intensity and flux density are only radially directed in the air gap and that they vary only circumferentially. Furthermore, since $H = B/\mu$, if $\mu \to \infty$, H can be assumed as zero in the iron. Alternatively, the relatively small MMF drop in the iron can be corrected by appropriately increasing the MMF in an equivalent gap g_e. If θ denotes the angular measure in the circumferential direction, r the radial direction, and l the axial direction, after performing integration in the radial and axial directions of a typical cylindrically shaped geometry, equation (1.141) can be written as

$$L_{\text{self}} = \frac{grl}{I^2}\int_0^{2\pi} B_g(I,\theta)\, H_g(I,\theta)\, d\theta \tag{1.142}$$

Since H can be assumed constant in the gap then

$$H_g g = \mathcal{F}_g \tag{1.143}$$

where $\mathcal{F}_g$ represents the MMF acting across the air gap. One can now write this equation as either

$$L_{\text{self}} = \frac{rl}{I}\int_0^{2\pi} B_g(I,\theta)\frac{\mathcal{F}_g(I,\theta)}{I}d\theta \tag{1.144}$$

or

$$L_{\text{self}} = \mu_0\frac{rl}{g}\int_0^{2\pi}\left[\frac{\mathcal{F}_g(I,\theta)}{I}\right]^2 d\theta \tag{1.145}$$

Since the gap g is comprised of air, the MMF must vary linearly with the current so that it can be expressed as the product of current times a second function which only depends on θ. Thus, the self-inductance can be obtained from either

$$L_{\text{self}} = \frac{rl}{I}\int_0^{2\pi} B_g(I,\theta)N(\theta)d\theta \tag{1.146}$$

or the expression,

$$L_{\text{self}} = \mu_0\frac{rl}{g}\int_0^{2\pi} N(\theta)^2 d\theta \tag{1.147}$$

The new quantity $N(\theta) = \mathcal{F}_g(I,\theta)/I$ is called the *winding function* and is frequently employed in the circuit analysis of AC machines.

The field representation of stored energy can also be used to calculate mutual inductance. When $\boldsymbol{B}$ and $\boldsymbol{H}$ arise from currents in two different circuits,

$$W_m = \frac{1}{2}\int_V (\boldsymbol{B}_1 + \boldsymbol{B}_2)\bullet(\boldsymbol{H}_1 + \boldsymbol{H}_2)dV \tag{1.148}$$

However, it is also true from circuit theory that

$$W_m = \frac{1}{2}L_1 I_1^2 + L_{12}I_1 I_2 + \frac{1}{2}L_2 I_2^2 \tag{1.149}$$

Comparing equations (1.148) and (1.149), the terms involving the mutual inductance can be equated whereupon,

$$L_{12}I_1 I_2 = \frac{1}{2}\int_V (\boldsymbol{B}_1\bullet\boldsymbol{H}_2 + \boldsymbol{B}_2\bullet\boldsymbol{H}_1)dV \tag{1.150}$$

When the $\boldsymbol{B}$s and $\boldsymbol{H}$s are collinear, the product exists only in the gap and is only a function of θ, equation (1.150) clearly reduces to

$$L_{12} = \mu_0 \frac{rl}{g} \int_0^{2\pi} N_1(\theta) N_2(\theta)\, d\theta \tag{1.151}$$

where $(B_1/\mu_0)g = H_1 g = \mathcal{F}_1$, $(B_2/\mu_0)g = H_2 g = \mathcal{F}_2$, and

$$N_1(\theta) = \frac{\mathcal{F}_1(\theta)}{I_1}; \quad N_2(\theta) = \frac{\mathcal{F}_2(\theta)}{I_2} \tag{1.152}$$

When the flux density produced by one of the two windings is known, the following expression for the mutual inductance L_{12} is also convenient.

$$L_{12} = \frac{rl}{I_1} \int_0^{2\pi} B_1((\theta) N_2(\theta)) d\theta \tag{1.153}$$

1.20 THE PROBLEM OF UNITS

One of the facts of life concerning the electromagnetic design of an electric machine is inconsistency regarding physical units. This inconsistency is a consequence of the long history associated with this discipline. In practice, three unit systems are used based on the SI or MKS system (Europe originally and now worldwide), the CGS unit system (small transformers, permanent magnet or PM machines, and small sub-fractional horsepower (HP) machines) and the English unit system (fractional HP machines and above, during early work in the United States). The English unit system is a throwback to the use of inches and pounds whereas the CGS system came into use via the physicists. Clearly, the SI system is the unit system of choice today. However, in view of the tremendous work done in the past incorporating the other units, it is important to be equally familiar with all three sets of units.

In this text, a particular relationship will be derived using SI units and then the result converted, if desired, to the different units. As an example, consider the *constituent equation* for magnetic materials.

$$B(\text{Wb/m}^2) = \mu_r \mu_0\, (\text{H/m})\, H\, (\text{A/m}) \tag{1.154}$$

Multiplying this equation by 10^4 and substituting explicitly for μ_0,

$$10^4 B(\text{Wb/m}^2) = \mu_r [4\pi \cdot 10^{-3} H\, (\text{A/m})] \tag{1.155}$$

If one defines

$$B\,(\text{G}) = 10^4 B(\text{Wb/m}^2) \tag{1.156}$$

and

$$H\,(\text{Oe}) = 4\pi \cdot 10^{-3} H\, (\text{A/m}) \tag{1.157}$$

then in CGS units

$$B\,(\mathrm{G}) = \mu_r H\,(\mathrm{Oe}) \tag{1.158}$$

where the new units of B and H are the *gauss* and the *oersted*, respectively.

Consider now the possibility of converting this equation to English units. Multiplying equation (1.154) by 10^8, the result is

$$10^8 B(\mathrm{Wb/m^2}) = 10^8 \mu_r \mu_0, H\,(\mathrm{A/m}) \tag{1.159}$$

One can now define a new unit of flux called the *maxwell* or *line* such that

$$1 \text{ weber} = 10^8 \text{ lines or maxwells} \tag{1.160}$$

Making this substitution in equation (1.159), and explicitly substituting for B and μ_o,

$$B\left(\frac{\text{lines}}{\mathrm{m^2}}\right) = \mu_r\,(40\pi)\,H\,(\mathrm{A/m}) \tag{1.161}$$

Now,

$$B\left(\frac{\text{lines}}{\mathrm{m^2}}\right) = B\left(\frac{\text{lines}}{\text{in.}^2}\right)\left(\frac{1 \text{ in.}}{2.54 \text{ cm}}\right)^2\left(\frac{100 \text{ cm}}{\text{m}}\right)^2 \tag{1.162}$$

and

$$H\,(\mathrm{A/m}) = H\,(\mathrm{A/in.})\left(\frac{1 \text{ in.}}{2.54 \text{ cm}}\right)\left(\frac{100 \text{ cm}}{\text{m}}\right) \tag{1.163}$$

so that in the English system,

$$\begin{aligned} B\left(\frac{\text{lines}}{\text{in.}^2}\right) &= \mu_r 40\pi\left(\frac{2.54}{100}\right)^2\left(\frac{100}{2.54}\right) H\left(\frac{\mathrm{A}}{\text{in.}}\right) \\ B\left(\frac{\text{lines}}{\text{in.}^2}\right) &= \mu_r\left(\frac{4\pi}{10}\right) 2.54\, H\left(\frac{\mathrm{A}}{\text{in.}}\right) \end{aligned} \tag{1.164}$$

The quantity $(4\pi/10)2.54 = 3.192$ is sometimes called the "free space" permeability in the English system.

A similar development can be carried out for the magnetic circuit equation, equation (1.110), that is

$$\Phi(\mathrm{Wb}) = \mathcal{P}(\mathrm{H})\mathcal{F}(\text{A-t}) \tag{1.165}$$

Multiplying both sides by 10^8 and making use of equations (1.160) and (1.107)

$$\begin{aligned} \Phi\,(\text{lines}) &= (10^8)(4\pi \cdot 10^{-7})\mu_r \frac{A(\mathrm{m^2})}{l\,(\mathrm{m})}\mathcal{F}(\text{A-t}) \\ &= 40\pi\mu_r \mathcal{F}(\text{A-t})\,\frac{A(\mathrm{m^2})}{l\,(\mathrm{m})} \end{aligned} \tag{1.166}$$

Now

$$A(\mathrm{m^2}) = A(\mathrm{cm^2})\left(\frac{1 \text{ m}}{100 \text{ cm}}\right)^2 \tag{1.167}$$

$$l(\mathrm{m}) = l(\mathrm{cm})\left(\frac{1 \text{ m}}{100 \text{ cm}}\right) \tag{1.168}$$

TABLE 1.1 Comparison of magnetic circuit equations with various systems of units

Constit. Eq.	MKS (SI) units $B = \mu_0\mu_r H$	CGS units $B = \mu_r H$	English units $B = \mu_0\mu_r H$
Magnetic Ohm's law	$\Phi = \mu_0 \frac{\mu_r A}{l}\mathcal{F}$	$\Phi = \frac{\mu_r A}{l}\mathcal{F}$	$\Phi = \mu_0 \frac{\mu_r A}{l}\mathcal{F}$
Faraday's law	$v = N\frac{d\Phi}{dt}$	$v = N\frac{d\Phi}{dt} \cdot 10^{-8}$	$v = N\frac{d\Phi}{dt} \cdot 10^{-8}$
Free space permeability	$\mu_0 = 4\pi \cdot 10^{-7}$	$\mu_0 = 1$	$\mu_0 = \frac{4\pi}{10} \cdot 2.54$ $= 3.192$
	B in Wb/m^2 H in A-t/m Φ in Wb $\mathcal{F}$ in A-t/m	B in gauss H in Oe Φ in lines (Maxwells) $\mathcal{F}$ in Gb	B in lines/in.^{2} H in A-t/in. Φ in lines $\mathcal{F}$ in A-t/m

so that equation (1.166) becomes

$$\Phi\,(\text{lines}) = 0.4\pi\mu_r \frac{A(\text{cm}^2)}{l\,(\text{cm})}\mathcal{F}(\text{A-t}) \tag{1.169}$$

In the CGS system the *gilbert* is defined as

$$\mathcal{F}(\text{gilberts}) = 0.4\pi\mathcal{F}(\text{A-t})$$

so that in the CGS system, the magnetic circuit equation, equation (1.169), becomes

$$\Phi\,(\text{maxwells}) = \mu_r \frac{A(\text{cm}^2)}{l(\text{cm})}\mathcal{F}(\text{gilberts}) \tag{1.170}$$

In the English system, it is necessary to convert lengths to inches:

$$\Phi\,(\text{maxwells}) = \frac{(0.4\pi)\,\mu_r A(\text{in.}^2)}{(2.54)\,l\,(\text{in.})}\mathcal{F}(\text{A-t}) \tag{1.171}$$

The key equations in magnetic circuit analysis in the three systems are summarized in Table 1.1. The reader should get to know these equations well. In order to help sort out the various conversion factors between the three systems, "flow charts" for the important variables are provided in Figure 1.14.

1.21 MAGNETIC PATHS WHOLLY IN IRON

The analogies between the electric circuit and the magnetic circuit seem to indicate, upon first consideration, that the Ohm's law type of relationship among MMF, flux, and reluctance or permeance ought to provide a straightforward method for solving magnetic circuit problems. However, the problem is much more difficult than the simple examples thus far considered. The direct application of the method is made difficult in practice by the relatively large flux leakage encountered in magnetic circuit

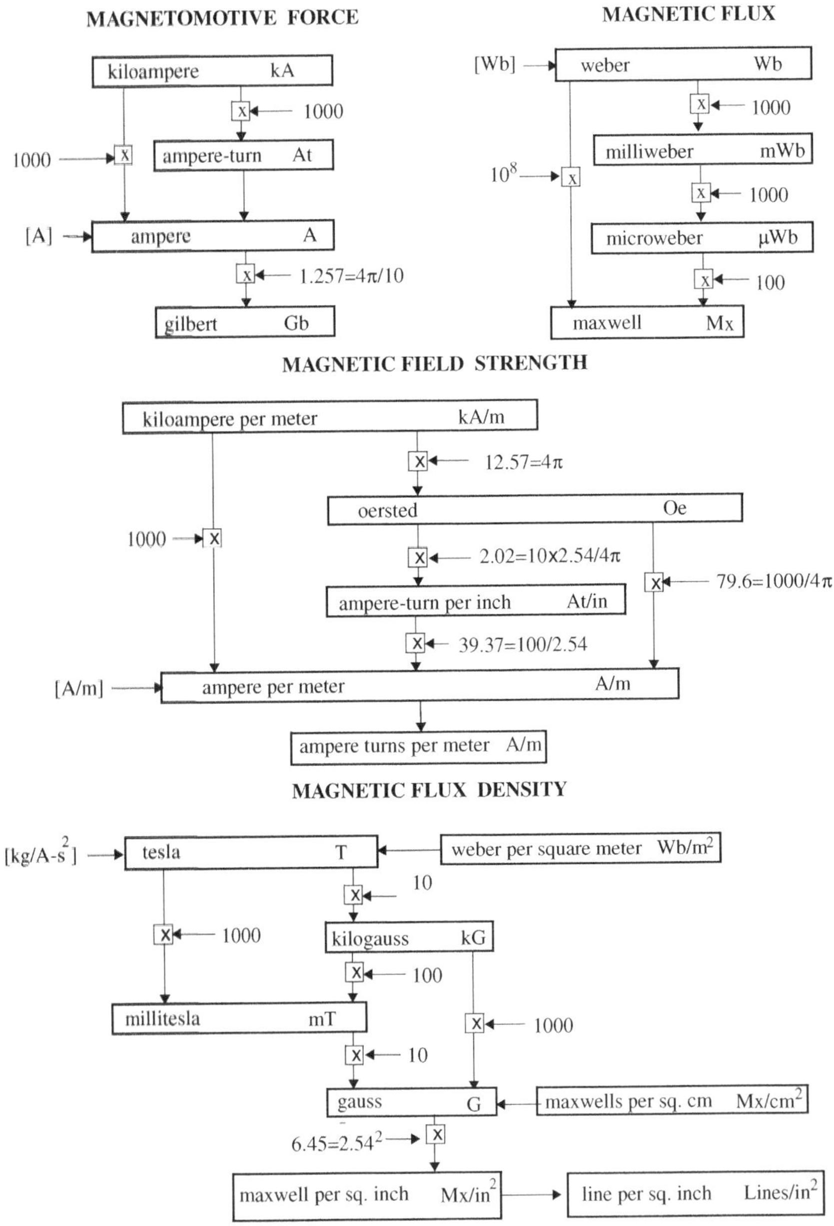

Figure 1.14 Conversion factors for magnetic field quantities.

problems and by the dependence of the reluctance of a ferromagnetic material upon the flux density; that is, the problem is nonlinear.

In general, solutions of magnetic circuit problems are solved to resolve two key questions: (a) the determination of the MMF required to produce a desired flux or flux density in a specified part of a structure, (b) the determination of the flux or flux density produced at specified places in a magnetic structure brought about by MMFs impressed at various places throughout the structure. Strictly speaking, the magnetic circuit method of analysis does not yield flux densities except as averages of total fluxes over the cross-sectional areas of the circuit so that the exact determination of the flux distribution becomes a field problem.

When the problem is to determine the MMF required to produce a desired total flux or flux density, that is, the calculation as in (a), the procedure is direct, provided that the leakage flux is neglected or estimated. In each portion of a series magnetic path having a cross-sectional area A, the average value of flux density B is equal to the ratio of the total flux Φ to the area A. The value of magnetizing force H required to establish this value of B is determined as a curve of B plotted as a function of H for the particular material. This value of H is then multiplied by the length of that portion of the path for which B is assumed constant, to give the magnetic potential difference $\mathcal{F}_{ab}$ between the ends of that portion of the path, that is

$$\mathcal{F}_{ab} = Hl_{ab} \tag{1.172}$$

where the distance a to b is the length of the path of uniform material and cross-sectional area. If the path includes portions of different kinds of ferromagnetic material, the value of H for each material is multiplied only by the length of the path in that material to give the magnetic potential difference for that portion of the path. The sum of the magnetic potential differences for all such portions of paths a–b, b–c, c–d, etc. taken around the series circuit gives the total MMF required, that is

$$\mathcal{F} = \mathcal{F}_{ab} + \mathcal{F}_{bc} + \mathcal{F}_{cd} + \cdots + \mathcal{F}_{na} \tag{1.173}$$

If the construction of the circuit is such that the average flux density differs markedly from the extremes of flux density on the cross section, more elaborate magnetic circuits must be employed.

When the problem is to determine the total flux or flux density produced by MMFs impressed at various places, that is, the calculation as in (b) above, the procedure is not so straightforward even if leakage fluxes are neglected. In certain simple combinations of paths, graphical methods are applicable. These are illustrated in the examples to follow. In complicated combinations of paths, a successive-approximation method leads rapidly to a solution. For such problems, the MMF required to produce an assumed value of flux Φ_1, is first calculated. If the calculated MMF does not approach the assumed impressed value within limits, a second trial value Φ_2 is chosen, greater or less by the amount required to equal the magnetic potential drop produced by the assumed flux Φ_1. After a few iterations, a solution is easily obtained. A Newton–Raphson iteration procedure using a digital computer is convenient for this purpose.

1.22 MAGNETIC MATERIALS

Historically, electric motors have been constructed from magnetic steels usually in the form of thin laminations, electrical conductors (either copper or aluminum), insulation for the conductors and slots, high tensile strength steel for shafts, and steel or copper alloys for bearings. The laminations used in most general-purpose motors have been "common iron" or low carbon steel. Although low in cost, this material typically produces machines of only modest efficiency. More recently, high efficiency motors often feature higher quality silicon steels at a correspondingly higher cost. The percent of silicon in the steel has a beneficial effect in reducing losses in the steel but at the same time tends to reduce the saturation flux density. The percent of silicon in motor steels typically range from 1% to 4%. The corresponding 60 Hz AC losses range from 0.6 watt per pound of core for 3.25% silicon steel to 1.0 watt per pound for 1% silicon steel at a peak flux density of 15,000 gauss (1.5 tesla). Nickel alloys, such as permalloy, have low losses but are very expensive and have low saturation flux density. The cobalt alloys such as Supermendur (49% iron, 49% cobalt, and 2% vanadium) have peak flux densities over 2 teslas, but are also very expensive ($7–$8 per pound) and have higher losses.

When the magnetic structure is assembled by means of stacking laminations punched from thin sheet material, the volume occupied by the stacked laminations does not truly represent the volume of iron that supports the magnetic flux. A region whose permeability is that of air exists between the laminations because of the presence of irregularities in the laminations or due to a thin coat of insulating varnish applied to avoid circulating current flow between laminations (eddy currents). In order to allow for this effect, the effective cross-sectional area of iron is equal to the cross-sectional area of the stack, times a factor called the *stacking factor*. The stacking factor, defined as the ratio of the cross-sectional area of the iron to the cross-sectional area of the stack, ranges between about 0.95 and 0.90 for lamination thickness between 0.025 in. and 0.014 in. (25 and 14 mils), respectively. For thinner laminations, for example, 1 mil to 5 mil thick, the stacking factor can be in the range of 0.4–0.75. Thinner laminations than 14 mils are generally not used unless iron loss is a severe problem. This choice typically occurs when the machine operates at high frequencies, for example an aircraft generator.

A new group of alloys has been developed, grouped under the generic title of amorphous metal alloys. These materials represent a new state of matter for electromagnetic materials, the so-called amorphous or non-crystalline state. Ordinary window glass is a typical example of an amorphous material. Some of these new amorphous alloys have magnetic properties which surpass the properties of conventional alloys. Thus, they appear to be a potentially useful new class of soft magnetic material. These alloys contain about 80% ferritic elements such as iron, nickel, and cobalt, and 20% glasseous elements such as silicon, phosphorous, boron, and carbon. A good example of an amorphous alloy having 80% iron and 20% boron by atomic weight is Fe80B20 (*Metglas* from Metglas Inc.). Major advantages of amorphous metal include low cost (roughly $0.30 per pound vs. $0.50 for silicon steel), very low core loss (one fifth that of the best silicon steels), low annealing temperature, and high tensile strength. Unfortunately, this new material has not yet been successfully used in a large scale because the high tensile strength also makes the material difficult to punch. Also, amorphous materials are presently only available in thicknesses of 1–2 mils (0.001″ to 0.002″) which results in a poor stacking factor and creates problems during assembly.

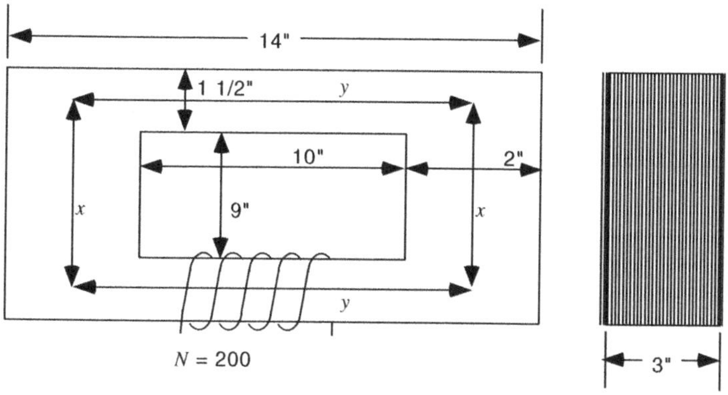

Figure 1.15 Core type transformer structure with two different cross-sectional areas.

1.23 EXAMPLE—TRANSFORMER STRUCTURE

The magnetic structure shown in in Figure 1.15 is similar to that of a core type transformer. The core is made of 29 gauge (14 mils) fully processed steel. The *B–H* curve for this material is shown in Figure 1.16. The sheets are stacked into a 3-inch stack. The stacking factor is 0.91. The exciting winding has 200 turns. Compute the current required in the exciting winding to produce a maximum core flux density of 1.2 tesla. Leakage flux is to be neglected.

Solution.

The cross-sectional area of the iron portion of legs x is $2 \times 3 \times 0.91 = 5.46$ in.^{2} and of the legs y is $1.5 \times 3 \times 0.91 = 4.1$ in.^{2}

- The maximum flux density will clearly occur in the y member having the smallest cross section. If $B_y = 1.2$ tesla then in the x leg $B_x = 1.2 \times 4.1/5.46 = 0.9$ tesla.
- From Figure 1.16, the magnetizing force is 2.9 Oe or 5.8 A-t/in. for legs y and 1.4 Oe or 2.85 A-t/in for legs x.
- The mean length of the flux paths in Figure 1.15 is estimated as 21 in. for the two x legs and 24 in. for the two y legs.
- The sum of the MMFs acting on the two y legs is $5.8 \times 24 = 140$ A-t and for the x legs $2.85 \times 21 = 60$ A-t. The total ampere-turns for the entire magnetic circuit is therefore $140 + 60 = 200$.
- The excitation current required to produce a flux density of 1.2 T in the transformer core is thus $200/200 = 1.0$ A.
- The flux in the core is clearly $\Phi = B_y \times A_y = 1.2 \times 3 \times 1.5 \times 0.91 \times (0.0254)^2$ $= 3.17$ mWb.
- The saturated inductance is then $L = N\Phi/i = 200 \times 3.17 \times 10^{-3}/1.0 = 0.63$ H.

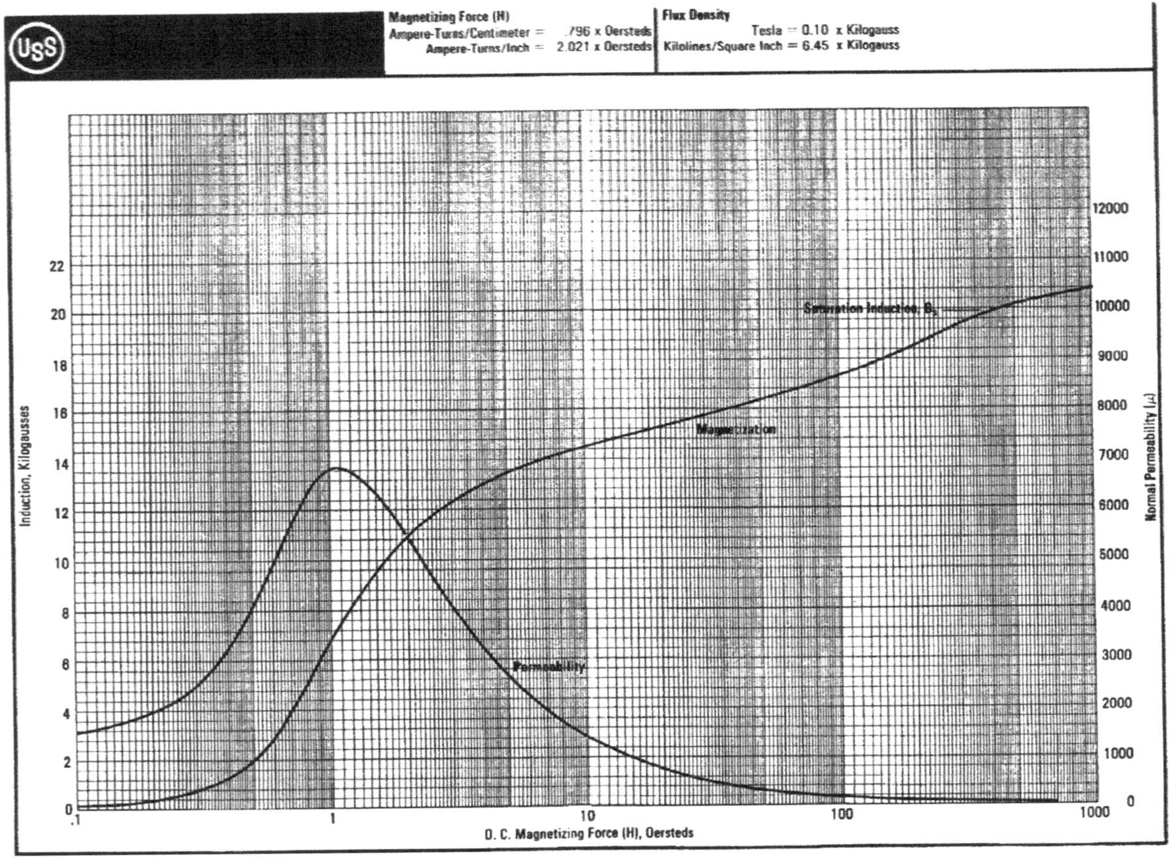

Figure 1.16 *B-H* curves for 29 Gage M27 fully processed steel.

When a specified MMF acts on a core, the inverse problem of calculating the fluxes is not simple. Assume, for example, that the results of the previous calculations are not known and that the core is excited with 200 A-t. In this case, it is necessary to estimate the probable magnetic potential difference between the ends of each core portion.

- Since the cross-sectional area of the y legs is much smaller than that of the x legs, the flux density is much larger in the y legs and would consume the major part of the MMF drop. As a first approximation, all the MMF is assumed to drop along the y legs. The resulting potential gradient is therefore 200/24 = 8.3 A-t/in. or 4.1 Oe.
- From Figure 1.16, the flux density in the y legs is then about 1.3 T. By proportionality, the flux density in the x legs is (4.1/5.46)1.3 = 0.97 T. Again from Figure 1.16, the x legs require an MMF of $1.6 \times 2.021 \times 21 = 68$ A-t. The MMF required by the entire circuit is 200 + 68 = 268, which is, of course, too much MMF to satisfy Ampere's law.
- As a second approximation, the MMF drops in the x and y legs can be estimated by taking ratios. For the y legs assume $\mathcal{F}_y = (200/268)200 = 149$ A-t so that $H_y =$ 6.2 A/in. or 3.1 Oe. The second iteration yields $B_y = 1.22$ T resulting in a flux density in the x leg of $B_x = (4.1/5.46)1.22 = 0.91$.
- The corresponding field intensity in the x leg becomes 1.45 Oe or 2.9 A-t/in.
- The MMF drop in the x legs of $2.9 \times 21 = 61$ A-t.
- The total MMF drop around the circuit is now estimated to be 149 + 61 = 210 A-t which is now just slightly greater than the correct value of 200 A-t. As a third iteration, it is now possible to assume that $\mathcal{F}_y = (149/210)149 = 106$ A-t.

Note that the method oscillates about the correct solution, but nonetheless converges rapidly if implemented on a digital computer since the B–H curve is a simple monotonically increasing function. The iteration method for the y leg MMF can be made less oscillatory by changing the new estimate by only a fraction of the error from the last iteration by using the algorithm,

$$\mathcal{F}_i = \mathcal{F}_{i-1} + \kappa_a(\mathcal{F}_{i(\text{est})} - \mathcal{F}_{i-1}) \tag{1.174}$$

The quantity κ_a is an acceleration factor which can be taken as roughly 0.5 and $\mathcal{F}_{i(\text{est})}$ is the estimated MMF for the ith iteration using ratios as described above.

For simpler problems, one can also resort to a graphical method. The procedure is to first determine the relationship between the total flux and the total MMF for each of the two nonlinear portions of the circuit. The curves of Φ_x as a function of $\mathcal{F}_x$ and Φ_y as a function of $\mathcal{F}_y$ is plotted in Figure 1.17 in such a manner that the abscissa for the x leg runs from left to right and for the y leg from right to left. The plot for the y legs, turned end for end, is called a negative magnetization curve and its origin is put at the point where $\mathcal{F}$ equals 200 on the plot for the x legs. The point 200 is chosen because it is equal to the applied MMF. Since the same total flux is present in both legs x and y, the solution for the impressed value of 200 A-t. is the intersection of the two curves.

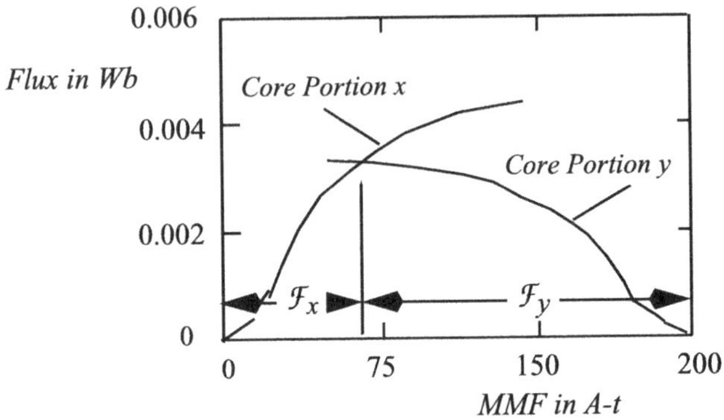

Figure 1.17 Graphical solution of Example.

1.24 MAGNETIC CIRCUITS WITH AIR GAPS

Because electrical machines involve magnetic circuits in relative rotation, an air gap must exist between the stator and rotor. In addition, other air gaps frequently occur because of limitations inherent in the construction. Air gaps are often introduced into iron-core inductors in order to make the inductance of the element essentially independent of the current in the coil throughout its working range, but at the same time to make the inductance larger than if the inductor had the same coil and only an air core.

When an air gap is inserted in a magnetic circuit, the flux spreads out, or fringes, around the gap as shown by the sketch of Figure 1.18 and the flux density in the gap assumes a non-uniform distribution. The flux that terminates near the edges of the gap is called the *fringing flux*. Because of the spreading of the flux, the apparent reluctance of the gap is not that of an air space of the same dimensions as the gap. Since the permeability of iron is several thousand times that of air the reluctance of

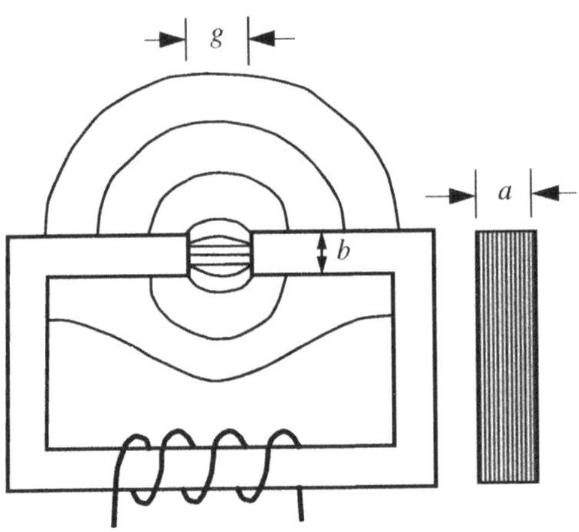

Figure 1.18 Magnetic circuit showing fringing flux.

even a short air gap is usually large compared to that of the iron portion making the magnetic potential between the stator and roor teeth relatively large. Relatively large magnetic potentials may also exist between iron parts not immediately near the gap. For example, in a synchronous machine, the main flux that traverses the gap fringes at the pole tips and because of the large reluctance of the air gap, considerable flux goes directly from rotor pole to rotor pole, constituting rotor leakage flux. This flux is often as much as 25% of the flux in the core of the field pole and contributes considerably to the saturation of the pole body.

When the air gap is short compared with its cross-sectional dimensions and has parallel faces, the fringing effect can be incorporated into the analysis by the use of simple correction factors. If the cross-sectional dimensions of the core are the same on both sides of the gap, the equivalent gap is assumed to have a length g equal to the actual air gap, but to have an equivalent cross-sectional area

$$A = (a + g)(b + g) \tag{1.175}$$

where a and b are the cross-sectional dimensions of the core faces. If one of the faces of the gap has a cross-sectional dimension much larger than the corresponding dimensions of the other, a correction of $2g$ should be used. Experience has shown that these rules give satisfactory results if the correction applied does not exceed about 1/5 of the physical cross-section.

If the total MMF applied is known a successive approximation solution can again be used. The first approximation can be obtained by considering that all the ampere-turns are required to overcome the reluctance of the air gap. A direct graphical method can also be used. The required solution is again obtained by superposing a plot of Φ_s as a function of $\mathcal{F}_s$ with a plot of Φ_a as a function of $\mathcal{F}_a$, where s denotes the steel portion and a the air portion of the flux path. The construction is shown in Figure 1.19, where $\mathcal{F}_t$ denotes the total impressed MMF. Note that the ordinate intersection of the air gap line is readily determined since

$$\Phi_a = \frac{\mathcal{F}_a}{\mathcal{R}_a} = \frac{\mu_0 A}{g}\mathcal{F}_a \tag{1.176}$$

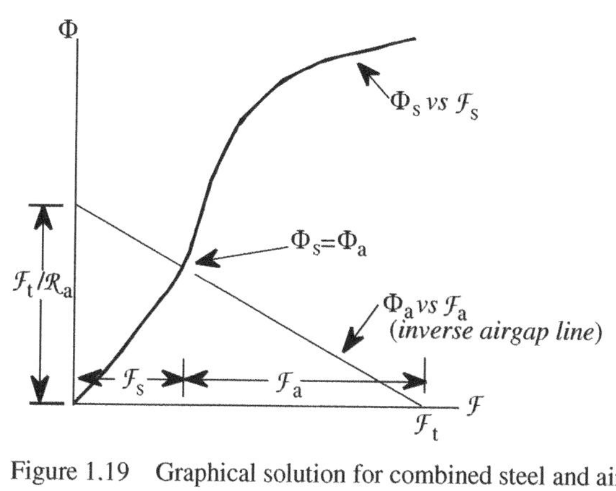

Figure 1.19 Graphical solution for combined steel and air magnetic circuit.

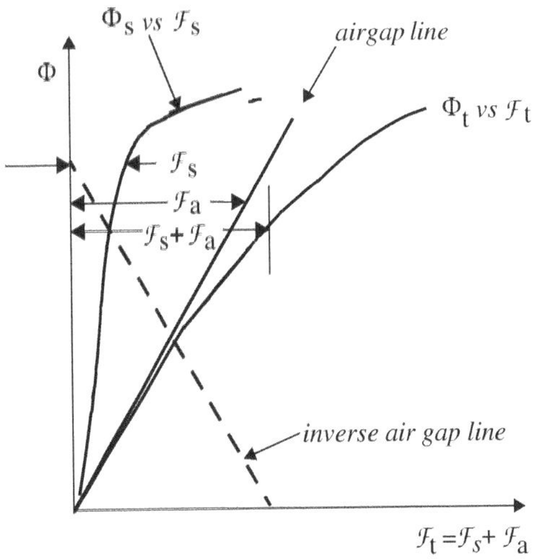

Figure 1.20 Illustrating the concept of the air gap line.

The intersection of the negative air gap line with the saturation characteristic of the steel for all values of $\mathcal{F}_t$ will generate the flux versus MMF characteristics for the overall device, that is, the "sat curve." The net saturation curve can be readily visualized as the sum of the iron and air saturation curves at each value of flux. Figure 1.20 shows such a construction.

1.25 EXAMPLE—MAGNETIC STRUCTURE WITH SATURATION

A magnetic structure similar to Figure 1.18 is made of 29 gage sheet steel laminations 0.014 in. thick stacked 2 in. thick. Dimension b is 2.5 in. The air gap length g is 0.10 in.

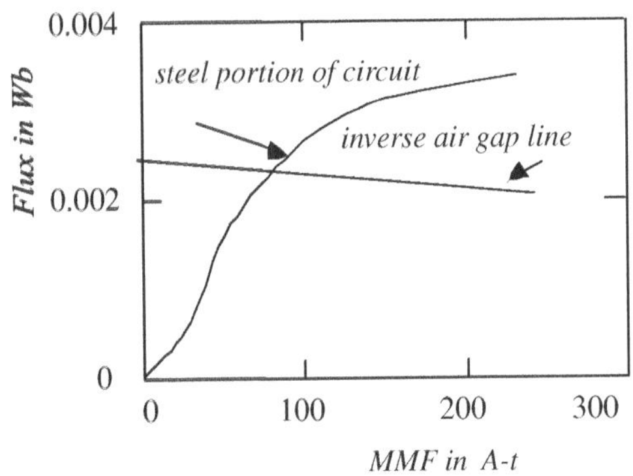

Figure 1.21 Graphical solution of Example.

The mean length of the steel part of the circuit is 30 in. Find the resultant flux if the applied MMF is 1400 A-t.

The equivalent air gap area, using a $2g$ correction is (2.0 × 0.91 + 0.2) (2.5 + 0.2) = 5.45 in.^{2} The negative air gap line intersects the abscissa at $\mathcal{F}$ = 1400 A-t. The intersection on the ordinate is found by solving

$$\Phi_a = (\mu_0)\frac{A}{l}\mathcal{F}_a \tag{1.177}$$

or

$$\Phi_a = 4\pi 10^{-7}\left(\frac{5.45}{0.1}\right)\frac{1}{39.37}(1400) = 2.44\,\text{mWb} \tag{1.178}$$

The saturation curve for the iron is found by neglecting the MMF drop of the gap. A plot of the steel saturation curve and the negative air gap line is shown in Figure 1.21. The point of intersection of the two curves is read as 2.2 mWb.

1.26 EXAMPLE—CALCULATION FOR SERIES–PARALLEL IRON PATHS

A type of core construction used frequently for certain transformers in which a relatively small magnetic coupling between primary and secondary coils is desired is shown in Figure 1.22. This type of geometry will also be used in the subsequent analysis of a machine for calculating the flux entering the core through both the tooth and the slot. Because of the nonlinearity of the core material the fraction of the total flux bypassed through the leg y varies with the amount of magnetic saturation.

For illustration, assume a flux of 3.8 mWb is set up in the leg x by a coil wound around this leg. The ampere-turns required are to be calculated. The magnetic material is again fully processed steel, 14 mil thick. The stacking factor is 0.91. The paths axb and azb are assumed to have a mean length of 21 in. The mean length of core in the center leg is 7.9 in.

- In order to solve this problem, the cross-sectional area of the core and air gap is first calculated. For the core, the area is 2 × 2 × 0.91 = 3.64 in.^{2} For the air gap, including fringing, 2.1 × 2.1 = 4.41 in.^{2}

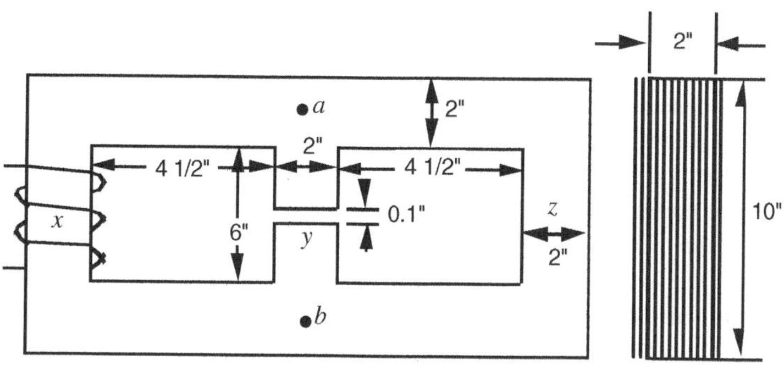

Figure 1.22 Magnetic circuit with series-parallel paths.

- The flux density in the x leg is calculated to be $0.00380/(3.64 \times 0.0254^2) = 1.62$ tesla. The MMF required to set up this flux density in the path *axb* is calculated as $830 \times 21 = 1743$ A-t. Figure 1.16 as $83 \times 21 = 1743$ A-t.

The total flux of 3.8 mWb in the path *axb* divides between the z and y legs in such a manner that the MMF from a to b is the same using either path *azb* or *ayb*. The division of flux must be calculated by assuming a tentative flux distribution and then correcting. The flux in the air gap or y leg is assumed so that the MMF from a to b necessary to produce this flux can be calculated. Since this MMF also acts on leg z, the flux in the z leg is now calculated and added to the flux assumed to exist in the y leg. The result is compared to the assumed total value of flux and if the result differs, the amount of flux assumed in the air gap is corrected accordingly. The procedure continues iteratively until convergence occurs.

- For example, assume that 0.80 mWb of flux exists in the y leg of the transformer. The corresponding value of flux density in the y leg iron is $0.00080/(3.64 \times 0.0254^2) = 0.34$ tesla. Since this is a relatively small value of flux density, the MMF drop in the iron can be neglected relative to the drop in the air.
- The MMF consumed in the air gap is, from equation (1.110), the result $(0.0008)\ 0.1/[0.0254 \times (2 + 0.1)^2 \mu_0] = 568$ A-t. This MMF drop acts on the z leg producing a magnetizing force of $568/21 = 27$ A-t/in. which from the data of Figure 1.16 establishes a flux density of 1.47 tesla.
- The corresponding flux in the z leg is then $(1.47)(3.64)(0.0254^2)$ or 3.5 mWb. The total flux in the two legs is 4.3 mWb which is somewhat more than the specified value of 3.8 mWb.
- If necessary, a second trial is now made using the value $0.8 \times 3.8/4.3 = 0.71$ as the value of flux assumed to flow in the y leg. The procedure proceeds iteratively to about 0.47 mWb in the y leg and 3.33 mWb in the z leg.
- The MMF drop in the y leg is now computed to be 333 A-t resulting in a total MMF drop around the path *xaybx* corresponding to the ampere-turns required to produce 3.8 mWb in the x leg, namely $\mathcal{F} = 1743 + 333 = 2076$ A-t.

1.27 MULTIPLE WINDING MAGNETIC CIRCUITS

In many cases, the magnetic circuit includes the effect of two or more sources of MMF. This was, for example, the case when the transformer of Section 1.15 was examined. This is also typically the case in many electrical machines which have not only an excitation component but also a separate load component typically on different members of the machine. When the iron is allowed to saturate the problem is now not so simple.

The basic issue can be demonstrated by the simple electromagnet shown in Figure 1.23. This device can be represented by the equivalent magnetic circuit shown in Figure 1.24.

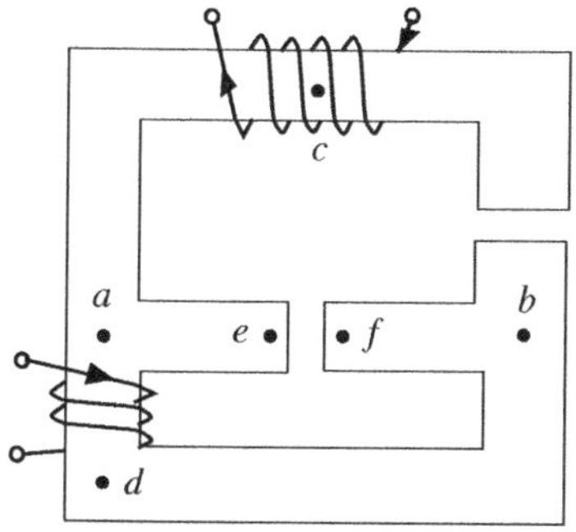

Figure 1.23 Electromagnetic with two sources of excitation.

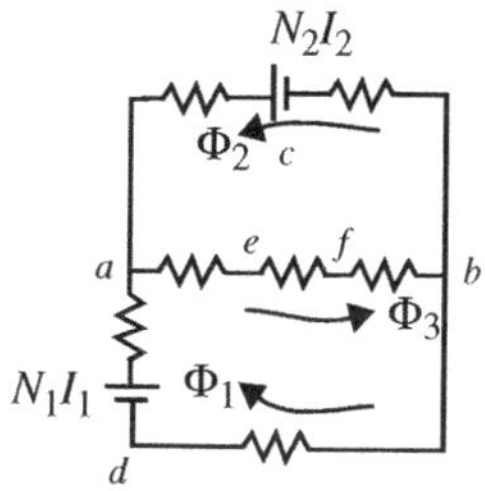

Figure 1.24 Equivalent magnetic circuit of Figure 1.23.

In this case, the MMF drop from point a to point b can be written in terms of three equations, namely

$$\mathcal{F}_{ab}(\Phi_1) = N_1I_1 - \mathcal{R}_{adb}(\Phi_1)\Phi_1 \tag{1.179}$$

$$\mathcal{F}_{ab}(\Phi_2) = N_2I_2 - \mathcal{R}_{acb}(\Phi_2)\Phi_2 \tag{1.180}$$

$$\mathcal{F}_{ab}(\Phi_3) = \mathcal{R}_{aefb}(\Phi_3)\Phi_3 \tag{1.181}$$

where

$$\Phi_1 + \Phi_2 = \Phi_3 \tag{1.182}$$

Three MMF versus flux curves can be constructed as shown in Figure 1.25.

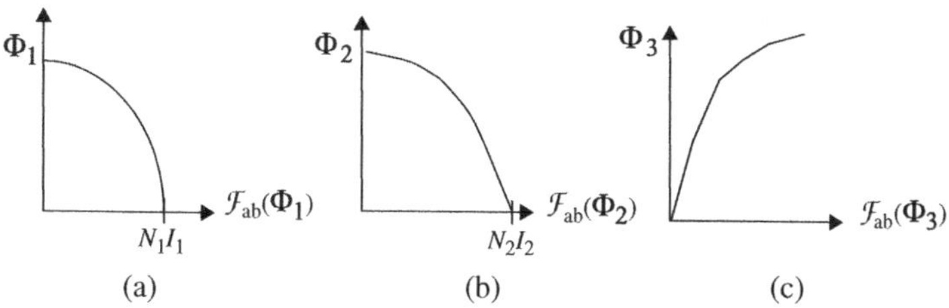

Figure 1.25 MMF vs. flux curves for the three magnetic paths: (a) lower member, (b) upper member, (c) middle member.

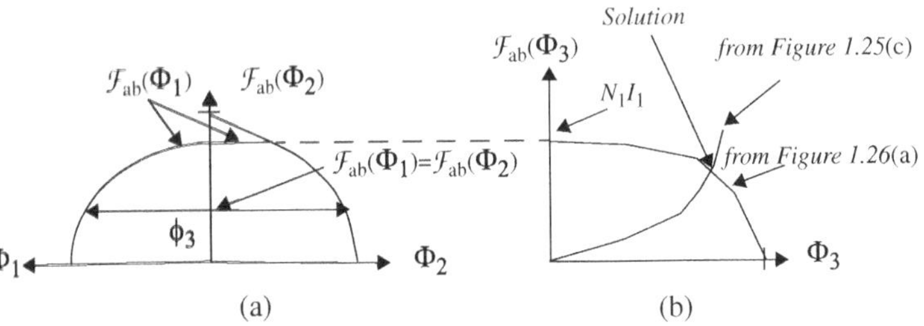

Figure 1.26 Graphical solution of the dual excitation problem.

Clearly, $\mathcal{F}_{ab}(\Phi_1) = \mathcal{F}_{ab}(\Phi_2)$. If both Φ_1 and Φ_2 are plotted versus $\mathcal{F}_{ab}$, then the flux Φ_3 can be determined for every value of $\mathcal{F}_{ab}$. The resulting construction is shown in Figure 1.26a. The actual solution is the point where the MMF as determined by this curve matches the MMF as determined by the function $\mathcal{F}_{ab}(\Phi_3)$ as computed in Figure 1.25c.

1.28 MAGNETIC CIRCUITS APPLIED TO ELECTRICAL MACHINES

Although the methods and simplifying assumptions outlined in the preceding sections yield results of reasonable accuracy for simple geometries, electrical machines present a considerably more complicated problem. Nonetheless, the basic principles which have been discussed form the basis for analysis and design of any machine structure. For illustration, Figure 1.27 shows the magnetic circuit of a typical two-pole DC machine. The center-slotted member is the rotor which carries the rotor winding in the slots. It is usually assembled from laminated steel punchings having 1–3% silicon. The outer member is the field structure, the cylindrical portion being the yoke or frame which frequently is of cast iron or cast steel. The protruding

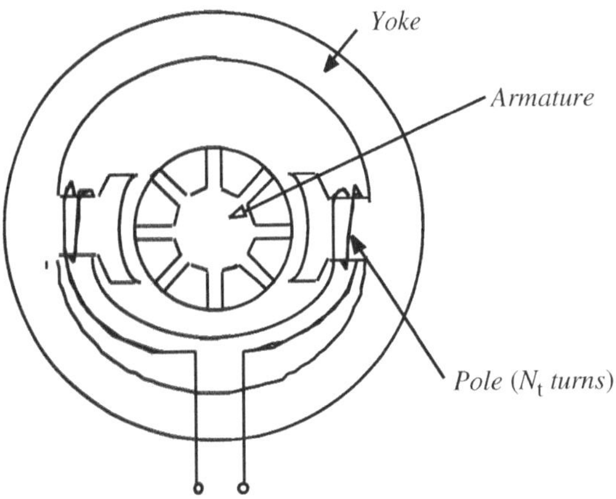

Figure 1.27 Magnetic circuit of a DC machine.

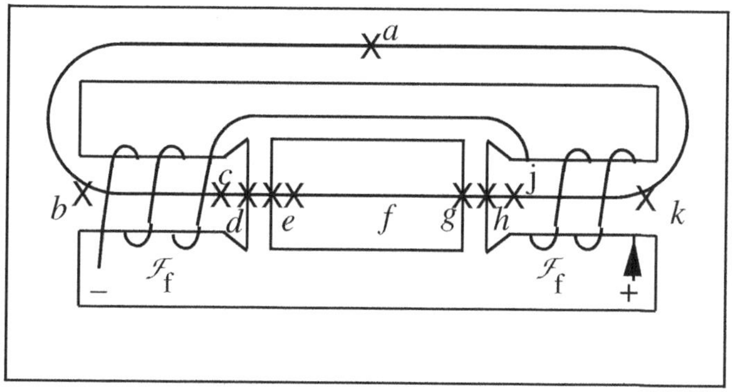

Figure 1.28 Simplified magnetic equivalent circuit corresponding to Figure 1.27.

portions of the field structure of *salient poles* are the poles which are usually made of laminated steel.

With a DC voltage applied to the field windings, the steady-state value of field current is determined entirely by the resistance of the circuit. For the polarity shown, the field winding produces an MMF in the direction to establish a flux from left to right through the field poles, air gaps, and armature. The flux path is then completed through the frame. The magnetic circuit is redrawn schematically in Figure 1.28. In order to determine the portions of the magnetic circuit whose properties predominate, a graph of the relative MMF of various points around the circuit is sketched in Figure 1.29. The MMFs are given with respect to an arbitrary point at the center of the yoke, point a. The magnetic potential drop from a to b is shown as a negatively sloped line in Figure 1.29. From b to c, similar conditions hold but since the material is different a slightly different slope is shown. The air gap d–e offers a very large

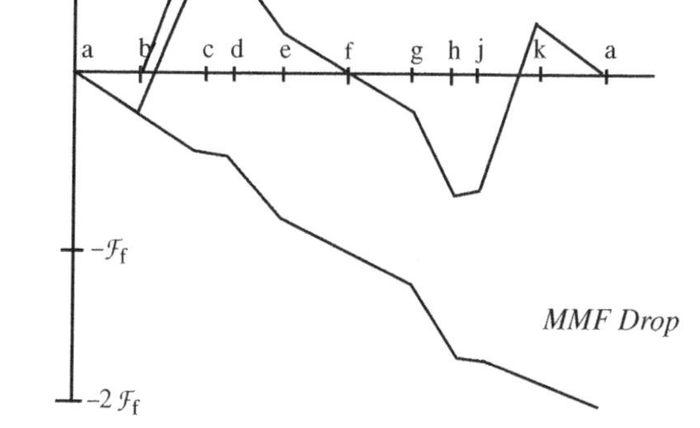

Figure 1.29 Magnetic potential drops and rises in a two-pole DC machine.

MMF drop relative to the iron. If the plot is completed for the remainder of the path, returning to the point *a*, the total reluctance drop is found to equal twice the MMF drop from *a* to *f*.

For the drop from *a* to *f* to exist, an equal and opposite rise in MMF must exist somewhere in the circuit. The required potential rise is established by the MMF of the field windings, the magnitude of the contribution being the same as the MMF drop from *a* to *f* for each of the two field windings. The upper curve of Figure 1.29 shows the rise in potential given by the windings. The next MMF curve shows the sum of the rises and drops at each point in the circuit and thus gives the actual MMF at each point with respect to point *a*.

Several important points can be deduced from Figure 1.29. First, note that the points *a* and *f* are at the same potential and the MMFs from *f* to *a* and *a* to *f* are mirror images of each other. Therefore, it can be deduced that all calculations can be made on a "per pole" basis. Second, the resultant curve shows that all points on the yoke are nearly at the same potential. Therefore, the leakage flux from *b* to *k* through the air will be relatively small. Third, the same is not true for the tips of the field poles. The potential difference through the pole body increases until at the pole tips the MMF is a maximum. In particular, the difference in potential from the top of one pole to the tip of the other is $2N_tI$. Since the distance between the tops of adjacent pole shoes is relatively small, the leakage flux from pole to pole is likely to be appreciable. Indeed, even for a good design, this leakage flux is generally 10 to 20% of the useful flux.

1.29 EFFECT OF EXCITATION COIL PLACEMENT

Up to this point, little attention has been paid to the exact placement of the core coil making up an inductor. In practice, the location of the exciting coil has a considerable effect on the overall losses as well as the exact value of the inductance obtained. Consider again the simple air-gapped core of Figure 1.18. Figure 1.30 shows three cases in which the coil is placed (a) on the limb farthest from the gap, (b) on the limb with the gap and (c) on the upper and lower limbs. In each case, the MMF is plotted from point *a* to point *f* as identified in the figure. The potential is plotted with respect to point *a*. While the remainder of the flux path is not plotted (from *f* back to *a* in the lower portion of the core), it is identical (mirror image) to the upper half. In case (a) where the coil is on the side away from the gap, the difference in MMF potential between the upper half of the core and the lower half is large over the distance *c* to *d* resulting in a large flux passing from the top limb to the bottom limb which closes through the left-hand limb. Since the useful flux is presumably the flux in the air gap region, this additional flux could be considered as leakage flux.

In case (b) of Figure 1.30, MMF is large only in the region near the air gap (*d* to *e*) resulting in "leakage" flux concentrated in this region. Clearly, if the purpose of the inductor design is to create a specified amount of flux in the air gap region, then this design would wind up being the smallest and lightest since most of the iron path does not have to support these additional leakage flux lines. Unfortunately, this option is often not a good choice since heating of the copper conductors around the

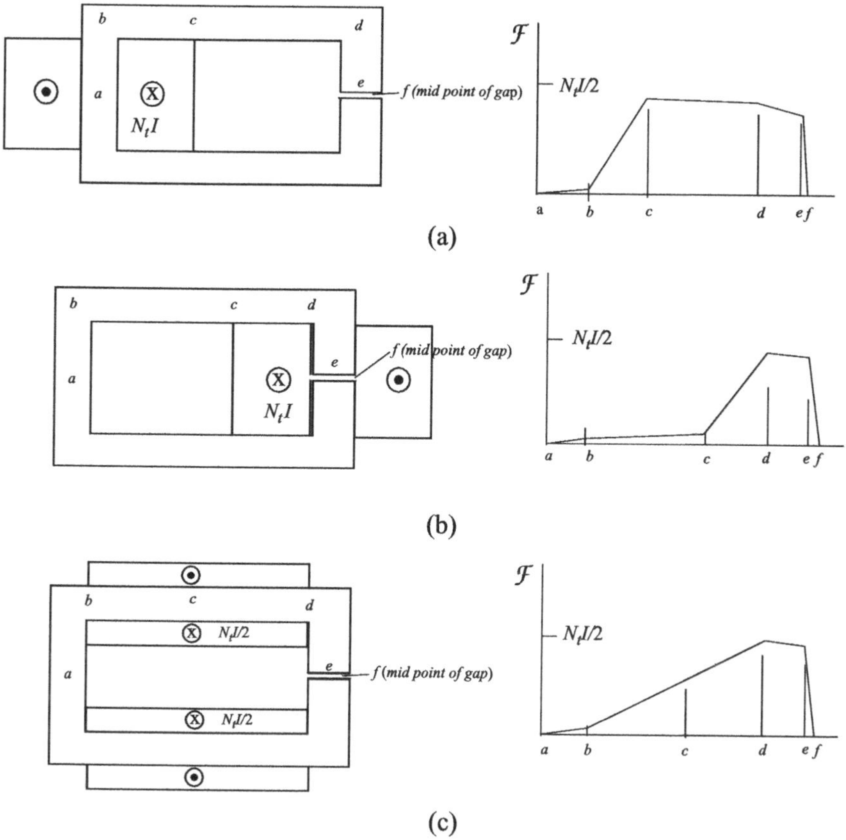

Figure 1.30 Three different methods of winding a simple gapped core. MMF potential plotted with respect to point *a*.

air gap will occur due to eddy current effects. The issue of losses will be taken up in more detail in Chapter 5.

With case (c), a compromise can be reached concerning these additional flux lines. Since the MMF in this case increases linearly over the entire length of the upper and lower limbs, the overall average difference in potential decreases resulting in leakage flux lines somewhere between cases (a) and (b).

Although a poor choice for inductor design, case (b) clearly does produce the desirable effect of creating the maximum number of flux lines in the air gap. This result is also a valid and important observation in the design of electrical machines where the process of creating a maximum amount of flux lines in the air gap for a given amount of ampere-turns is of critical importance. The design of case (b) teaches that it is important to design machines in which the copper exciting the main magnetic circuit be as near to the air gap as possible. Hence, many shallow slots are preferable to fewer deep slots containing the same amount of ampere-turns. Also, magnets far away from the air gap (buried magnets) are poorer in creating torque than a machine with magnets simply fixed on the rotor surface, that is, in the air gap. Much more will be presented concerning these interesting aspects of machine design in future chapters.

1.30 CONCLUSION

This chapter has served as a brief, but intensive, review of electromagnetic fields as applied to electric machine design. Although the math appears formidable, fortunately, with reasonable approximations, most calculations required in the design process can be carried out without an advanced mathematical treatment. Generally, the simple concepts presented in Section 1.13 provide a good starting point for the design process.

REFERENCE

[1] R. Plonsey and R. E. Collin, *Principles and Applications of Electromagnetic Fields*, McGraw-Hill Book Company, Inc., 1961.

CHAPTER 2

THE MMF AND FIELD DISTRIBUTION OF AN AC WINDING

IN THIS CHAPTER, THE "forcing function" of the magnetic field of a typical AC induction machine, namely the stator MMF, will first be introduced. The stator MMF primarily acts to magnetize the machine and is accounted for, in the per phase equivalent circuit, by the magnetizing inductance. This MMF secondly attempts to synthesize a sinusoidally distributed rotating flux density in the air gap so as to induce a sinusoidal current in the rotor of an induction motor and thereby produce a smooth, non-fluctuating torque. A sinusoidal flux density can only be produced by a sinusoidal MMF. However, limitations imposed on winding distribution by slotting and the effect of discrete numbers of turns in the slots make this goal a challenging one as shall be seen in this chapter.

2.1 MMF AND FIELD DISTRIBUTION OF A FULL-PITCH WINDING FOR A TWO POLE MACHINE

To begin, a single-phase full pitch winding having one slot per pole is assumed. Figure 2.1 shows a schematic representation of such a winding. The directions of the current are shown for a single instant of time. It will be assumed that the length of the air gap along the surface of the armature is constant. Figure 2.1 also shows the direction of the lines of force (flux lines) produced by the current-carrying conductors. The only lines of force possible are those which link one coil. No lines of force are possible which link two coils at the same time because the MMF producing such lines would be zero. Since all flux lines enclose $N_t\, I$ ampere-turns, the MMF corresponding to each field line is also $N_t I$, where N_t is the total number of turns with which the lines of force are linked. Since every line of force consists of two symmetrical parts taking opposite directions in the air gap, teeth and core the MMF can be divided into two portions corresponding to single crossings of the gap, that is, corresponding to one-half the path of the magnetic circuit (one magnetic pole).

Introduction to AC Machine Design, First Edition. Thomas A. Lipo.

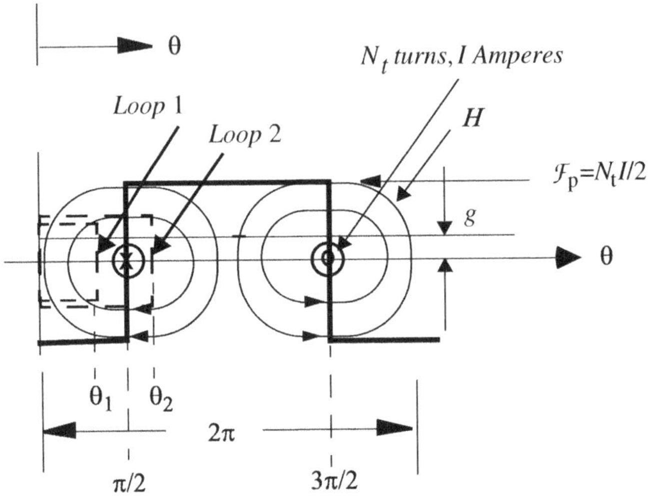

Figure 2.1 MMF and field intensity of a full-pitch coil as a function of position along the air gap.

The line integral of the magnetic field intensity H along a closed path is equal to the ampere-turns linked by this path. In general, the integral has to be divided into a sum corresponding to the stator core, rotor core, stator teeth, rotor teeth, and the air gap. However, assuming no saturation in the iron, the field intensity in the iron is very small and can first be neglected. Taking the path defined as loop 1 in Figure 2.1 in a clockwise direction.

$$\oint \boldsymbol{H} \bullet dl = 0 \tag{2.1}$$

Since the gap is small the field intensity can be assumed constant over the portion of the path leading from the stator to the rotor resulting in,

$$[H(0) - H(\theta_1)]g = 0 \tag{2.2}$$

or

$$H(0) = H(\theta_1) \quad 0 \le \theta_1 < \frac{\pi}{2} \tag{2.3}$$

where $H(0)$ and $H(\theta_1)$ represent the field in the gap encountered when traversing the two vertical sides of the test loop 1. While it is not possible to solve for the field intensity at this point, equation (2.3) indicates that it is uniformly constant over the region where $0 < \theta < \pi/2$ since the same result will be obtained over any path which closes before encircling any current.

Upon taking loop 2 clockwise,

$$\oint \boldsymbol{H} \bullet dl = N_t I \tag{2.4}$$

which yields

$$[H(0) - H(\theta_2)]g = N_t I \quad \frac{\pi}{2} \leq \theta_2 < \frac{3\pi}{2} \tag{2.5}$$

where $H(\theta_2)$ is constant over the air gap region since $H(0)$ and $N_t I$ are effectively constant.

Finally, if one integrates over a closed path that crosses the gap vertically upward where $\theta = 0$ and returns across the gap where $3\pi/2 < \theta < 2\pi$ then, again

$$[H(0) - H(\theta_3)]g = 0 \quad \frac{3\pi}{2} < \theta_3 < 2\pi \tag{2.6}$$

However, from Gauss' law the integral of the flux density over a closed surface is zero, or

$$\oint_s \boldsymbol{B} \bullet d\boldsymbol{S} = 0 \tag{2.7}$$

If this integral is taken over a cylinder passing through the air gap whose top and bottom are outside the region of the machine, the portion of the integral associated with the regions outside the air gap proper can be neglected whereupon,

$$\int_{\text{airgap surface}} \mu_0 H(\theta) l_s r_{is} d\theta = 0 \tag{2.8}$$

where l_s is the length of the machine (gross length of the iron stack plus air ducts) and r_{is} is the radius of the inner stator surface. Taking the integral along the surface bounded by angles θ_1 and θ_2 and, since both $H(\theta_1)$ and $H(\theta_2)$ are constants, this integral becomes,

$$[H(\theta_1) + H(\theta_2)]\pi l_s r_{is} = 0 \tag{2.9}$$

so that

$$H(\theta_2) = -H(\theta_1) \tag{2.10}$$

and, from equations (2.3) and (2.5),

$$H(\theta_1)g = \frac{N_t I}{2} \tag{2.11}$$

Since the quantity $H(\theta)g$ represents the magnetic potential between the stator and rotor surfaces at any point θ along the air gap, it is useful to define

$$\mathcal{F}_p(\theta) = H(\theta)g \tag{2.12}$$

so that in the gap

$$B_g(\theta) = \frac{\mu_0 \mathcal{F}_p(\theta)}{g} \tag{2.13}$$

The quantity $\mathcal{F}_p(\theta)$ corresponds to the MMF distribution over one-half of the magnetic circuit, that is, the MMF of one pole of a complete two-pole magnetic circuit. Equation (2.13) indicates that, for the idealized situation considered, the shape of the flux density curve is essentially proportional to the MMF at every point in the gap.

If the origin for the measure of angular distance along the air gap periphery is taken instead as the point at which the winding is located ($\theta = \pi/2$ in Figure 2.1) the square wave can be expanded in a Fourier series wherein all cosine terms and even sine terms are zero. The result is, for a two pole machine,

$$\mathcal{F}_{\mathrm{p}}(\theta) = \left(\frac{4}{\pi}\right)\left(\frac{N_{\mathrm{t}}I}{2}\right)\left[\sin\theta + \frac{1}{3}\sin 3\theta + \frac{1}{5}\sin 5\theta + \cdots\right] \tag{2.14}$$

2.2 FRACTIONAL PITCH WINDING FOR A TWO-POLE MACHINE

When the span of the coil sides of an individual coil is less than the pole pitch, the winding as a whole is said to be a *fractional pitch* winding. Such windings are extensively used for the reason that the MMF waveform is more nearly sinusoidal than with full-pitch windings, and because of the saving in copper and the greater stiffness of the coils due to the shorter end connections. The latter reason is especially important in the case of two-pole high speed turbo-alternators because of stresses produced in the end connections under short circuit conditions.

Fractional pitch windings are ordinarily of the two-layer lap wound type, although single-layer lap windings are also used. The use of fractional pitch makes it possible to use a number of slots which is not an exact multiple of the number of poles, thus tending to suppress pulsations of flux as the teeth move relative to the pole faces and so largely eliminate tooth ripple in the voltage waveform.

As a simple example, consider the MMF distribution of the fractional pitch winding of Figure 2.2 in which the coil of N_{t} turns of Figure 2.1 has been split into two coils and both have been moved back and forth (pitched) by a net electrical angle

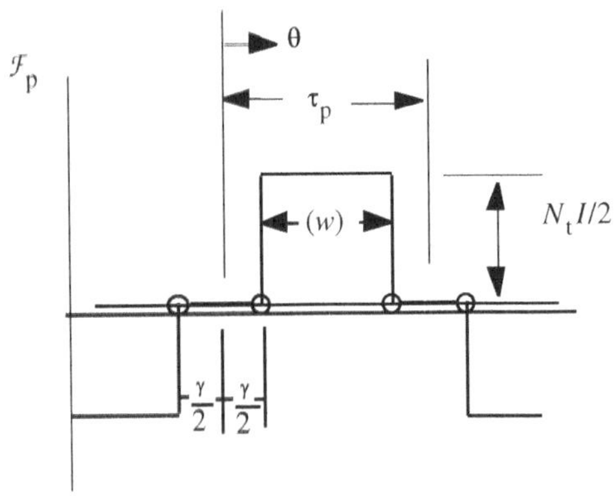

Figure 2.2 MMF distribution of a coil with fractional pitch.

γ. The amplitude of the corresponding MMF distribution for the hth harmonic is readily calculated in electrical radians (as opposed to mechanical radians) as

$$\mathcal{F}_{\text{ph}} = \frac{2}{\pi}\left[\int_0^{\gamma/2} 0 \sin h\theta d\theta + \int_{\gamma/2}^{(\pi-\gamma/2)} \frac{N_t I}{2} \sin h\theta d\theta + \int_{(\pi-\gamma/2)}^{\pi} 0 \sin h\theta d\theta\right]$$

$$= \left(\frac{4}{\pi}\right)\left(\frac{N_t I}{2}\right)\frac{\cos h\frac{\gamma}{2}}{h} \quad (h \text{ odd})$$

Hence the fractional pitch, rectangular distribution of Figure 2.2, can be represented by the series

$$\mathcal{F}_{\text{p}}(\theta) = \sum_{h=1,3,5,\ldots}^{\infty} \mathcal{F}_{\text{ph}} \sin h\theta$$
$$\mathcal{F}_{\text{p}}(\theta) = \frac{4}{\pi}\frac{N_t I}{2}\left[\cos\frac{\gamma}{2}\sin\theta + \frac{1}{3}\cos\frac{3\gamma}{2}\sin 3\theta + \frac{1}{5}\cos\frac{5\gamma}{2}\sin 5\theta + \cdots\right] \tag{2.15}$$

Upon comparing equation (2.15) with equation (2.14), the additional factor introduced by short pitching the winding is clearly, for the hth harmonic,

$$k_{\text{ph}} = \cos\frac{h\gamma}{2} \quad (h \text{ odd}) \tag{2.16}$$

It is conventional to write the series of equation (2.15) in a slightly different form. If τ_{p} denotes the length of one pole pitch measured along the air gap periphery and w is the actual width of the coil span, the corresponding angle γ is then

$$\gamma = \frac{\tau_{\text{p}} - w}{\tau_{\text{p}}}\pi = \pi\left(1 - \frac{w}{\tau_{\text{p}}}\right) \tag{2.17}$$

and

$$\cos\frac{h\gamma}{2} = \cos\left[h\left(\frac{\pi}{2} - \frac{w}{\tau_{\text{p}}}\frac{\pi}{2}\right)\right] = \sin\frac{h\pi}{2}\sin h\left(\frac{w}{\tau_{\text{p}}}\frac{\pi}{2}\right) \quad (h \text{ odd})$$

so that

$$\mathcal{F}_{\text{ph}} = \frac{4}{\pi}\left(\frac{N_t I}{2h}\right)\sin\frac{h\pi}{2}\sin\left[h\left(\frac{w}{\tau_{\text{p}}}\right)\frac{\pi}{2}\right] \tag{2.18}$$

The quantity

$$k_{\text{ph}} = \sin\left[\frac{hw}{\tau_{\text{p}}}\left(\frac{\pi}{2}\right)\right]\sin\left(\frac{h\pi}{2}\right) \quad (h \text{ odd}) \tag{2.19}$$

is called the *pitch factor* for the hth harmonic of MMF. The term ± 1 denoted by $\sin(h\pi)/2$ is usually not included in the definition and causes some confusion but will be included in the definition here. The relative magnitudes of the harmonic components of MMF as determined by the pitch factor k_{ph} is shown in Figure 2.3.

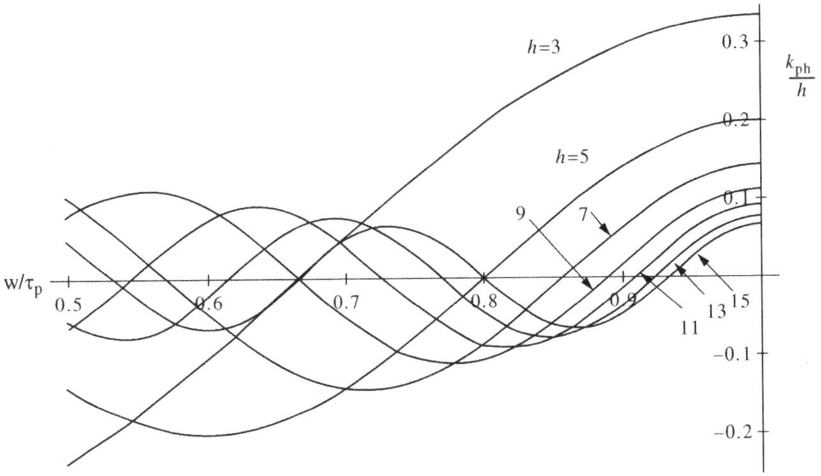

Figure 2.3 Pitch factor for harmonics $h = 3, 5, 7\ldots15$ as a function of per unit coil pitch w/τ_p.

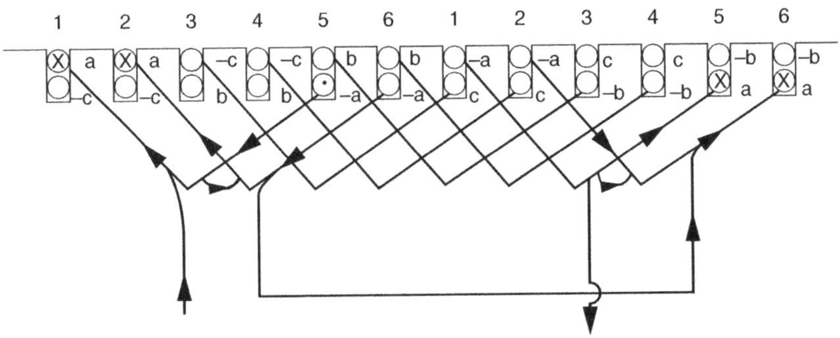

Figure 2.4 Two layer fractional pitch lap winding for a three-phase machine having windings with 2/3 pole pitch.

Figure 2.4 illustrates a practical winding layout for the case of six slots per pole. It can be noted that each winding is short-pitched by two slots resulting in a pitch angle of 60 electrical degrees or pitch of 2/3. It can be noted that all windings are identical making for economical construction. The overlapping nature of this winding layout has prompted the use of the term *lap winding* for this coil configuration. It can be noted that with a 2/3 pitch each slot contains coil sides belonging to different phases, which has an effect upon the leakage reactance of the winding and also upon the resultant MMF of the winding.

2.3 DISTRIBUTED WINDINGS

In addition to pitching the winding in order to remove undesirable harmonics, windings are generally distributed in several slots in order to better utilize the available

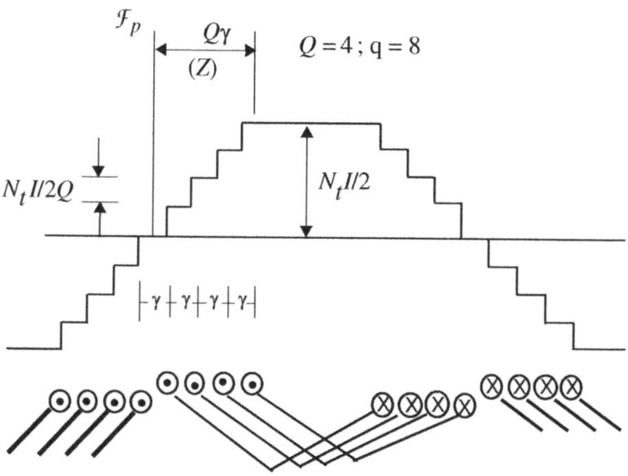

Figure 2.5 MMF of distributed winding having four coils in a phase belt.

space around the periphery of the machine. Figure 2.5 shows one phase of a two layer winding in which the concentrated coil of N_t turns has been distributed into four coils per pole each having a full pitch. It should be noted here that the winding layout is not necessarily a practical one. For the purpose of analysis, it is convenient to reconnect the coils as shown in Figure 2.6. Note that the MMF distribution in the gap remains unchanged.

It can be now noted that each pair of coils connected as shown contributes a rectangular distribution of the type discussed in Section 2.2 (see Figure 2.2). The maximum MMF of each individual component is a rectangular distribution and will be the same, namely, $N_t\, I/2Q$ where Q is the number of coils per pole (four in this case). The $q = 2Q$ coil sides are said to comprise a *phase belt* and is a key parameter for the representation of a winding layout. Note that the displacement of each coil side with respect to its neighbor is again taken to be γ. From Section 2.2, it follows

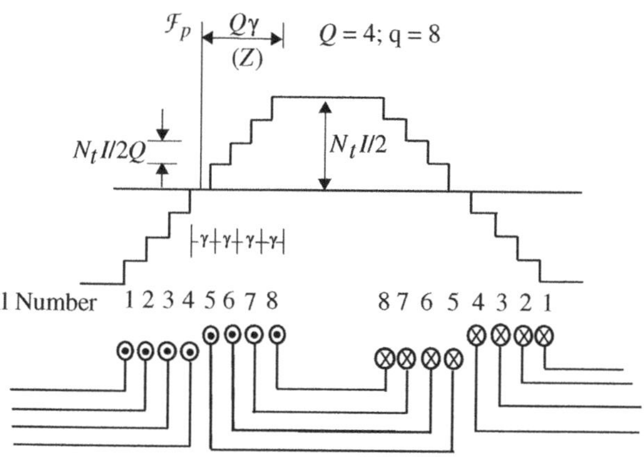

Figure 2.6 MMF of distributed winding viewed as four coils per pole of different pitch.

that the amplitudes of the MMF harmonics of the successive coils, beginning with the outer coil, are, respectively,

$$
\begin{array}{ll}
\text{Coil} & \text{Amplitude of } h\text{th harmonic} \\
1(\text{outer}) & \dfrac{4}{\pi}\left(\dfrac{N_t I}{2Q}\right)\dfrac{\cos\left(\frac{h\gamma}{2}\right)}{h} \\
2 & \dfrac{4}{\pi}\left(\dfrac{N_t I}{2Q}\right)\dfrac{\cos\left(\frac{3h\gamma}{2}\right)}{h} \\
3 & \dfrac{4}{\pi}\left(\dfrac{N_t I}{2Q}\right)\dfrac{\cos\left(\frac{5h\gamma}{2}\right)}{h} \\
\vdots & \vdots \\
Q(\text{inner}) & \dfrac{4}{\pi}\left(\dfrac{N_t I}{2Q}\right)\dfrac{\cos h[\gamma/2+(Q-1)\gamma]}{h}
\end{array}
\tag{2.20}
$$

The Q amplitudes can be written in the form

$$
\begin{array}{ll}
\text{Coil} & \text{Amplitude of } h\text{th harmonic} \\
1(\text{outer}) & \dfrac{4}{\pi}\left(\dfrac{N_t I}{2Q}\right)\dfrac{\cos h(\gamma/2)}{h} \\
2 & \dfrac{4}{\pi}\left(\dfrac{N_t I}{2Q}\right)\dfrac{\cos h(\gamma/2+\gamma)}{h} \\
3 & \dfrac{4}{\pi}\left(\dfrac{N_t I}{2Q}\right)\dfrac{\cos h(\gamma/2+2\gamma)}{h} \\
\vdots & \vdots \\
Q(\text{inner}) & \dfrac{4}{\pi}\left(\dfrac{N_t I}{2Q}\right)\dfrac{\cos h[\gamma/2+(Q-1)\gamma]}{h}
\end{array}
\tag{2.21}
$$

The net amplitude of the hth harmonic is found by summing the Q terms. After using the trig identity $\cos(A+B) = \cos A \cos B - \sin A \sin B$, this summation can be written as

$$
\begin{aligned}
\mathcal{F}_{\text{ph}} = \left(\frac{4}{\pi}\right)\left(\frac{N_t I}{2Qh}\right)\bigg\{&\cos\frac{h\gamma}{2}[1+\cos h\gamma+\cos 2h\gamma+\cdots+\cos(Q-1)h\gamma] \\
&-\sin\frac{h\gamma}{2}[\sin h\gamma+\sin 2h\gamma+\cdots+\sin(Q-1)h\gamma]\bigg\}
\end{aligned}
\tag{2.22}
$$

It can be shown that the sum of this cosine series is (Reference [1], Entry 420.2),

$$1 + \cos h\gamma + \cos 2h\gamma + \cdots + \cos(Q-1)h\gamma = \sin\left[\frac{(Qh\gamma)}{2}\right] \frac{\cos\left[\dfrac{(Q-1)h\gamma}{2}\right]}{\sin\left[\dfrac{(h\gamma)}{2}\right]} \tag{2.23}$$

and that the sum of the sine series is (Reference [1], Entry 420.1),

$$\sin h\gamma + \sin 2h\gamma + \cdots + \sin(Q-1)h\gamma = \frac{\sin\left(\dfrac{Qh\gamma}{2}\right)\sin\left[(Q-1)\dfrac{h\gamma}{2}\right]}{\sin\left(\dfrac{h\gamma}{2}\right)} \tag{2.24}$$

Substituting these values into equation (2.22), one has

$$\mathcal{F}_{\text{ph}} = \left(\frac{4}{\pi}\right)\left(\frac{N_t I}{2Qh}\right)\frac{\sin\left(\dfrac{hQ\gamma}{2}\right)}{\sin\left(\dfrac{h\gamma}{2}\right)}\left\{\left[\cos\left(\frac{h\gamma}{2}\right)\cos\left[(Q-1)\frac{h\gamma}{2}\right]\right.\right.$$
$$\left.\left. - \sin\left(\frac{h\gamma}{2}\right)\sin\left[\frac{(Q-1)h\gamma}{2}\right]\right]\right\} \tag{2.25}$$

which reduces to

$$\mathcal{F}_{\text{ph}} = \left(\frac{4}{\pi}\right)\left(\frac{N_t I}{2h}\right)\frac{\sin\left(\dfrac{hQ\gamma}{2}\right)}{Q\sin\left(\dfrac{h\gamma}{2}\right)}\cos\left(\frac{hQ\gamma}{2}\right) \quad (h \text{ odd}) \tag{2.26}$$

The quantity

$$k_{\text{dh}} = \frac{\sin\left(\dfrac{hQ\gamma}{2}\right)}{Q\sin\left(\dfrac{h\gamma}{2}\right)} \quad (h \text{ odd}) \tag{2.27}$$

is defined as the harmonic distribution factor of the winding. Note that the quantity $Q\gamma$ represents the angular measure in radians corresponding to that portion of the air gap occupied by the Q coil sides.

It is convenient to let Z denote the arc length occupied by the Q coils of the winding group. Then, if τ_p denotes one pole pitch,

$$Q\gamma = \frac{Z}{\tau_p}\pi \tag{2.28}$$

The distribution factor for the hth harmonic then can also be written as

$$k_{dh} = \left(\frac{1}{Q}\right) \frac{\sin\left(\frac{hZ}{2\tau_p}\pi\right)}{\sin\left(\frac{hZ}{\tau_p}\frac{\pi}{2Q}\right)} \quad (h \text{ odd}) \tag{2.29}$$

Consider now the cosine term in equation (2.26). Examining Figure 2.5 it is apparent that the angle $Q\gamma/2$ is the equivalent pitch of the coil which is now considered to be distributed over Q slots. Again it is convenient to let

$$Q\gamma = \left(1 - \frac{w}{\tau_p}\right)\frac{\pi}{2} \tag{2.30}$$

where W is the equivalent span of the distributed winding The cosine term becomes

$$\cos\left(\frac{hQ\gamma}{2}\right) = \sin\left(\frac{hw}{\tau_p}\frac{\pi}{2}\right)\sin\left(h\frac{\pi}{2}\right) \tag{2.31}$$

which is recognized as the pitch factor for the hth harmonic, which has already been defined by equation (2.19). The final expression for the hth harmonic of MMF corresponding to the coil of Figure 2.5 becomes

$$\mathcal{F}_{ph} = \frac{4}{\pi}\frac{N_t I}{2h} k_{ph} k_{dh} \quad (h \text{ odd}) \tag{2.32}$$

where k_{ph} and k_{dh} are given by equations (2.19) and (2.29) respectively. The overall product of the two winding factors can also be written as

$$k_{pdh} = k_{ph} k_{dh} = \frac{\sin(hQ\gamma)}{2Q\sin\left(\frac{h\gamma}{2}\right)} = \left(\frac{1}{2Q}\right) \frac{\sin\left(\frac{hZ}{\tau_p}\pi\right)}{\sin\left(\frac{hZ}{\tau_p}\frac{\pi}{2Q}\right)}$$

In most cases, winding distributions are nearly always symmetrical if the number of slots per pole per phase is two or greater. This is usually the case for three-phase machines having six poles or less ranging up to several hundred horsepower. Hence, the harmonic components of the MMF distribution of the winding can typically be written as the product of a distribution factor times a pitch factor times the harmonic component of MMF for a full-pitch coil in which all the turns are concentrated in one slot.

In principle, phase belts spanning any portion of a pole pitch are possible. Unless used for a special purpose such as for two-speed motors, phase belts greater than one pole pitch are considered as impractical since the end turn length is excessive. However, practical considerations for accommodating symmetrical phase groups result in 60° and 120° phase belts the most practical for three-phase system and 90° and 180° the best for two-phase systems. Table 2.1 shows the distribution factor for the fundamental component of MMF for these four phase belts. Note that the difference between two slots per phase belt and a continuous or an infinite number of slots is relatively small, particularly for the 60° phase belt. Considerably more

TABLE 2.1 Fundamental winding distribution factor k_{d1} for four typical phase belts

Slots or coil sides per phase belt q	Phase belt in electrical degrees 60°	90°	120°	180°
1	1.000	1.000	1.000	1.000
2	0.966	0.924	0.866	0.707
3	0.960	0.911	0.844	0.667
4	0.958	0.906	0.837	0.654
5	0.957	0.904	0.833	0.648
∞	0.955	0.900	0.827	0.636

voltage is generated by the 60° phase belt compared to the 120° belt and hence is the most widely used winding configuration.

Distribution factors for the first 26 harmonics of MMF of three-phase windings with 60° phase belts is given in Table 2.2. Note that the distribution factors of the harmonics decrease much more rapidly with increasing values of q than does the

TABLE 2.2 Harmonic distribution factors for a winding with 60° phase belt

h	k_{dh} – *Harmonic distribution factors* $q=2$	$q=3$	$q=4$	$q=5$	$q=6$	$q=\infty$
1	0.966	0.960	0.958	0.957	0.957	0.955
3	0.707	0.667	0.654	0.646	0.644	0.636
5	0.259	0.217	0.205	0.200	0.197	0.191
7	−0.259	−0.177	−0.158	−0.149	−0.145	−0.136
9	−0.707	−0.333	−0.270	−0.247	−0.236	−0.212
11	−0.966	−0.177	−0.126	−0.110	−0.102	−0.087
13	−0.966	0.217	0.126	0.102	0.092	0.073
15	−0.707	0.667	0.270	0.200	0.172	0.127
17	−0.259	0.960	0.158	0.102	0.084	0.056
19	0.259	0.960	−0.205	−0.110	0.084	−0.059
21	0.707	0.667	−0.654	−0.247	−0.172	−0.091
23	0.966	0.217	−0.958	−0.149	−0.092	−0.041
25	0.966	−0.177	−0.958	0.200	0.102	−.038
27	0.707	−0.333	0.654	0.646	0.236	0.071
29	0.259	−0.177	−0.205	0.957	0.145	0.033
31	−0.259	0.217	0.158	0.957	−0.197	−0.031
33	−0.707	0.667	0.270	0.646	−0.644	−0.058
35	−0.966	0.960	0.126	0.200	−0.957	−0.027
37	−0.966	0.960	−0.126	−0.149	−0.957	0.026
39	−0.707	0.667	−0.270	−0.247	−0.644	0.049
41	−0.259	0.217	−0.158	−0.110	−0.197	0.023
43	0.259	−0.177	0.205	0.102	0.145	−0.022
45	0.707	−0.333	0.654	0.200	0.236	−0.042
47	0.966	−0.177	0.958	0.102	0.102	−0.020
49	0.966	0.217	0.958	−0.110	−0.092	0.019
51	0.707	0.667	0.654	−0.247	−0.172	0.038

distribution factor of the fundamental. The distribution factor for the fundamental is always positive. The distribution factor of a harmonic can be negative which means that the harmonic is in phase opposition to the fundamental. Also in the table of distribution factors, some harmonics have the same amplitude as the corresponding fundamental component. These harmonics, h_k, are called *slot harmonics*. Their orders are

$$h_k = 6kQ \pm 1 \tag{2.33}$$

$$= 2kmQ \pm 1 \tag{2.34}$$

$$= \frac{2kS_1}{P} \pm 1 \tag{2.35}$$

where S_1 represents the total number of stator slots and m is the number of phases ($m = 3$ in this case). The analysis applies equally well to a wound rotor winding distribution in which case S_1 is replaced by S_2. The "first-order" slot harmonics corresponding to $k = 1$ are amongst the most troublesome resulting in noise as well as contributing to "stray losses," a subject treated in more detail in Chapter 5.

2.4 CONCENTRIC WINDINGS

Although lap windings are historically the configuration of choice, they are generally used today only for large AC machines over a few hundred horsepower having coils made from copper bars (*form wound coils*) where manual insertion of the coils into the slots still predominates. For smaller machines, the coils are made from copper wire, *random wound coils*, which is more amenable to factory automation. However, the overlapping nature of the lap-wound construction make machine insertion of the coils difficult. An alternative to lap winding is the concentric coil arrangement shown in Figure 2.7.

In this case, the coils are nested in such a manner that the entire phase group can be inserted in slots in one operation.

Calculation of the winding factor for concentric windings proceeds in the same manner as for lap windings. The amplitude of the hth harmonic is found by summing q terms since the concentrated winding can, in general, span one pole pitch ($3q$ slots per pole for a three phase winding). From a similar process used to obtain equation (2.22)

$$\mathcal{F}_{ph} = \left(\frac{4}{\pi}\right)\left(\frac{I}{2Qh}\right)\left\{\cos\frac{h\gamma}{2}[N_1 + N_2\cos h\gamma + N_3\cos 2h\gamma + \cdots + N_Q\cos(Q-1)h\gamma] - \sin\frac{h\gamma}{2}[N_2\sin h\gamma + N_3\sin 2h\gamma + \cdots + N_Q\sin(Q-1)h\gamma]\right\} \tag{2.36}$$

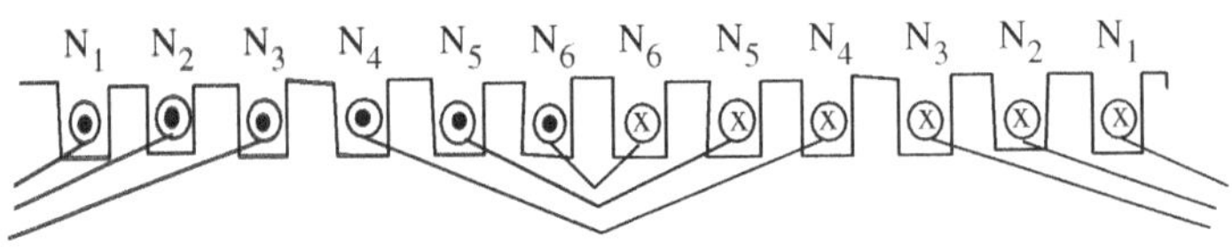

Figure 2.7 Concentric winding arrangement having a 180° phase belt (one phase only shown).

By normalizing to the total number of turns $N_t = (N_1 + N_2 + \cdots + N_Q)$, this result can be written as,

$$\mathcal{F}_{ph} = \left(\frac{4}{\pi}\right)\left(\frac{N_t I}{2Qh}\right)\left\{\cos\frac{h\gamma}{2}\left[\frac{N_1}{N_t} + \frac{N_2}{N_t}\cos h\gamma + \frac{N_3}{N_t}\cos 2h\gamma + \cdots + \frac{N_q}{N_t}\cos(Q-1)h\gamma\right] - \sin\frac{h\gamma}{2}\left[\frac{N_2}{N_t}\sin h\gamma + \frac{N_3}{N_t}\sin 2h\gamma + \cdots + \frac{N_Q}{N_t}\sin(Q-1)h\gamma\right]\right\} \tag{2.37}$$

The hth harmonic winding factor for concentrated windings is therefore,

$$k_{ch} = \cos\frac{h\gamma}{2}\left[\frac{N_1}{N_t} + \frac{N_2}{N_t}\cos h\gamma + \frac{N_3}{N_t}\cos 2h\gamma + \cdots + \frac{N_Q}{N_t}\cos(Q-1)h\gamma\right] - \sin\frac{h\gamma}{2}\left[\frac{N_2}{N_t}\sin h\gamma + \frac{N_3}{N_t}\sin 2h\gamma + \cdots + \frac{N_Q}{N_t}\sin(Q-1)h\gamma\right] \tag{2.38}$$

Figure 2.7 shows the case where the coil sides of the windings of one phase occupy all of the slots of the machine (180° phase belt). In general, the concentric windings occupy in any number of slots per pole. Assuming three phases, the phase belts can take on values corresponding from one slot to a value equal to the number of slots per pole. When a phase belt between 120° and 180° is chosen, a fraction of the slots contains coils from two phases while the remainder contain coils from three phases. Conversely, when a phase belt between 60 and 120° is used, some of the coils contain windings from one phase while the remainder contain windings from two phases. If phase belts of 60° or 120° are chosen, all of the slots contain windings from one or two phases. Hence, these configurations can be easily wound and are most frequently encountered.

The presence of the winding ratios N_1/N_t, N_2/N_t, etc. clearly opens the door for selecting these ratios in an optimal manner so as to minimize the harmonic content in the resulting MMF waveform. This work has been carried out in Reference [2] by minimizing the total harmonic distortion given by

$$\text{WTHD} = \frac{\sqrt{\sum_{h=2}^{\infty} \mathcal{F}_{ph}^2}}{\mathcal{F}_{p1}} \tag{2.39}$$

Tables 2.3–2.6 show optimal winding patterns for typical 60° and 120° phase belt windings with q ranging from 3 to 6 and N_t ranging from 4 to 25. The tables show the number of turns progressing from left to right across the phase belt. For example, with $N_t = 16$ and $q = 6$, the number of turns N_1 of coil 1 of Figure 2.6 is 1, the turns of coil 2 is 3, coil 3 is 4, and coils 4–6 are 4, 3, and 1 respectively. The distribution of turns in the other phases is appropriately phase-shifted by ±120 electrical degrees. Since the winding is assumed to be one of a three-phase set, third

TABLE 2.3 Optimal number of turns for concentric windings with 60° phase belts and with $q = 3$

N_t	N_1	N_2	N_3	k_{c1}	WTHD
9	3	3	3	0.9598	0.1134
10	3	4	3	0.9638	0.1183
11	4	3	4	0.9561	0.1151
12	4	4	4	0.9598	0.1134
13	4	5	4	0.9629	0.1166
14	5	4	5	0.9569	0.1142
15	5	5	5	0.9598	0.1134
16	5	6	5	0.9623	0.1156
17	6	5	6	0.9574	0.1138
18	6	6	6	0.9598	0.1134
19	6	7	6	0.9619	0.1151
20	7	6	7	0.9578	0.1136

harmonic components have been omitted in the calculation of the WTHD. Additional data for other concentric winding patterns is contained in [2] and [3].

2.5 EFFECT OF SLOT OPENINGS

Although it has been assumed to this point that the MMF changes abruptly when each coil is encountered when traversing the inner surface of the stator, the slot opening has a filtering effect on the MMF. While the exact variation of MMF across an open

TABLE 2.4 Optimal number of turns for concentric windings with 60° phase belts and with $q = 4$

N_t	N_1	N_2	N_3	N_4	k_{c1}	WTHD
12	3	3	3	3	0.9577	0.0898
13	3	4	3	3	0.9603	0.0947
14	4	3	3	4	0.9528	0.0909
15	4	4	3	4	0.9554	0.0911
16	4	4	4	4	0.9577	0.0898
17	4	4	5	4	0.9597	0.0930
18	5	4	4	5	0.9539	0.0900
19	5	4	5	5	0.9559	0.0904
20	5	5	5	5	0.9577	0.0898
21	6	4	5	6	0.9528	0.0918
22	6	5	6	6	0.9546	0.0896
23	6	5	6	6	0.9562	0.0901
24	6	6	6	6	0.9577	0.0898
25	7	5	6	7	0.9536	0.0908

TABLE 2.5 Optimal number of turns for concentric windings with 120° phase belts and with $q = 4$

N_t	N_1	N_2	N_3	N_4	k_{c1}	THD
4	1	1	1	1	0.8365	0.1633
5	1	2	1	1	0.8624	0.1743
6	1	2	2	1	0.8797	0.1480
7	1	2	3	1	0.8920	0.1582
8	1	3	3	1	0.9012	0.1468
9	1	3	4	1	0.9084	0.1540
10	1	4	4	1	0.9142	0.1480
11	1	4	5	1	0.9189	0.1531
12	2	4	4	2	0.8797	0.1480
13	2	4	5	2	0.8863	0.1506
14	2	5	5	2	0.8920	0.1468
15	2	5	6	2	0.8969	0.1493
16	2	6	6	2	0.9012	0.1468
17	2	6	7	2	0.9050	0.1490
18	2	7	7	2	0.9084	0.1473
19	3	6	7	3	0.8842	0.1490
20	3	7	7	3	0.8883	0.1470
21	3	8	7	3	0.8920	0.1481
22	3	8	8	3	0.8953	0.1467
23	3	8	9	3	0.8984	0.1478
24	3	9	9	3	0.9012	0.1468
25	3	9	10	3	0.9018	0.1479

slot is difficult to determine, the effect is well approximated by assuming that the MMF varies linearly across the slot as shown in Figure 2.8 for a single concentrated coil. Any MMF harmonic h can now be represented by the Fourier coefficient,

$$\mathcal{F}_{\text{ph}} = \left(\frac{4}{\pi}\right)\frac{N_t I}{2}\left[\int_0^{\chi/2} \frac{\theta}{\frac{\chi}{2}} \sin h\theta d\theta + \int_{\chi/2}^{\pi/2} \sin h\theta d\theta\right] \tag{2.40}$$

which when solved becomes, assuming only odd harmonics,

$$\mathcal{F}_{\text{ph}} = \left(\frac{4}{\pi}\right)\left(\frac{N_t I}{2h}\right)\frac{\sin\left(\frac{h\chi}{2}\right)}{\left(\frac{h\chi}{2}\right)} \tag{2.41}$$

The added factor, the *slot opening factor*, can be defined as

$$k_{\chi h} = \frac{\sin\left(\frac{h\chi}{2}\right)}{\left(\frac{h\chi}{2}\right)} \tag{2.42}$$

TABLE 2.6 Optimal number of turns for concentric windings with 120° phase belts and with $q = 6$

N_t	N_1	N_2	N_3	N_4	N_5	N_6	k_{c1}	THD
6	0	1	2	2	1	0	0.9452	0.1018
7	1	1	1	2	1	1	0.8532	0.1159
8	1	1	2	2	1	1	0.8696	0.1045
9	0	2	2	3	2	0	0.9320	0.1089
10	0	2	3	3	2	0	0.9373	0.0997
11	1	2	2	3	2	1	0.8794	0.1026
12	1	2	3	3	2	1	0.8882	0.0961
13	1	2	3	4	2	1	0.8956	0.1001
14	1	2	4	4	2	1	0.9020	0.0988
15	1	3	4	3	3	1	0.8917	0.1013
16	1	3	4	4	3	1	0.8975	0.0962
17	1	3	4	5	3	1	0.9027	0.0974
18	1	3	5	5	3	1	0.9072	0.0957
19	1	3	6	5	3	1	0.9113	0.0984
20	2	3	5	5	3	2	0.8808	0.0980
21	1	4	6	5	4	1	0.9070	0.0975
22	1	4	6	6	4	1	0.9105	0.0957
23	1	4	6	5	4	1	0.9138	0.0969
24	2	4	6	6	4	2	0.8882	0.0961
25	2	4	6	7	4	2	0.8921	0.0971

Since this term is common to all coils, it simply becomes a coefficient in the final expression for MMF harmonics. With slot opening accounted for, the MMF harmonics, previously given by equation (2.32), can be altered to form,

$$\mathcal{F}_{\text{ph}} = \left(\frac{4}{\pi}\right)\left(\frac{N_t I}{2h}\right) k_{\text{ph}} k_{\text{dh}} k_{\chi h} \quad (h \text{ odd}) \tag{2.43}$$

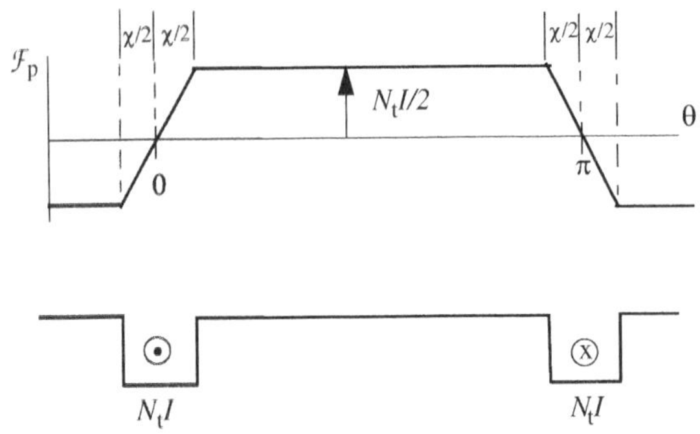

Figure 2.8 Concentrated full-pitch coil including the effects of slot opening.

2.6 FRACTIONAL SLOT WINDINGS

From the previous section, it is clear that it is desirable to increase q, the number of slots per pole per phase to as large a value as possible so as to reduce the slot harmonic frequency. An increase in q implies a corresponding increase in the number of slots for a machine with a specified number of poles. In some cases where the pole number is large, a value of even $q = 2$ cannot be reached without the number of slots reaching a prohibitively large number for the machine rating. Rather than select a value of $q = 1$ resulting in strong slot harmonics at the fifth and seventh harmonic, a fractional value of q can be chosen. In some cases, for example, in very large water wheel generators, values of q less than even unity is often chosen. Fractional slot windings are also becoming popular for permanent magnet machines in which the absence of rotor windings allows for a relatively large harmonic distortion of the air gap MMF.

Although the number of slots of a fractional slot winding arrangement is not a multiple of the number of poles, it must be a multiple of the number of phases in order to maintain phase symmetry. In the case of a three-phase machine, this constraint stipulates that the number of slots must be at least a multiple of three. Consider, for example, a three-phase winding for a 10-pole machine. In this case, the machine must have at least 30 slots ($q = 1$) to maintain phase symmetry. If $q = 2$ is desired, this would imply that the machine must have 60 slots, too large a number for most small machines. Instead, if one chooses 42 stator slots, then $q = 42/(3 \times 10) = 7/5$ slots per pole per phase. Thus, there are $(7/5) \times 3 = 21/5$ slots per pole, suggesting that the winding arrangement must fit into 21 slots distributed over 5 poles and that the winding pattern will begin to repeat after five poles are wound. The fact that $q = 7/5$ implies that the slots occupied by one of the phases over consecutive poles must alternate in some manner between 2 and 1 slots occupied per pole. For example, the pattern 2, 1, 1, 2, 1 can potentially satisfy this requirement as can the pattern 1, 1, 2, 2, 1.

In order to determine which one of several possible patterns is most desirable, it is first necessary to choose the pitch of the winding which is assumed to be the same for all coils. For this example, the phase shift in electrical degrees between slots is $10 \times (180/42) = 42^6\!/\!_7{}^\circ$. Hence, if the coil spans four slots, the pitch angle in electrical degrees is $180 - 4 \times (42^6\!/\!_7) = 8^4\!/\!_7{}^\circ$ which is probably the best choice. The pitch factor for the fundamental component of the winding becomes, from equation (2.16),

$$k_{p1} = \cos\left(\frac{8.57}{2}\right) = 0.997 \tag{2.44}$$

Since the winding pattern repeats after 21 slots, a sketch of the first 21 slots of the machine can be constructed as shown in Figure 2.9 with the first coil of one of the three phases inserted in slots 1 and 5 (coil short-pitched by $4^2\!/\!_7{}^\circ$ and spanning four slots). This coil is designated as coil 1.

Coil 1 can now be considered as the reference coil and a second coil inserted in slots 2 and 6. If it is assumed that these coils are excited by a sinusoidal 10-pole traveling wave in the air gap, the voltage induced in this second coil can be considered to be phase-shifted (phase-delayed) by $42^6\!/\!_7{}^\circ$ from coil 1. These sinusoidal induced

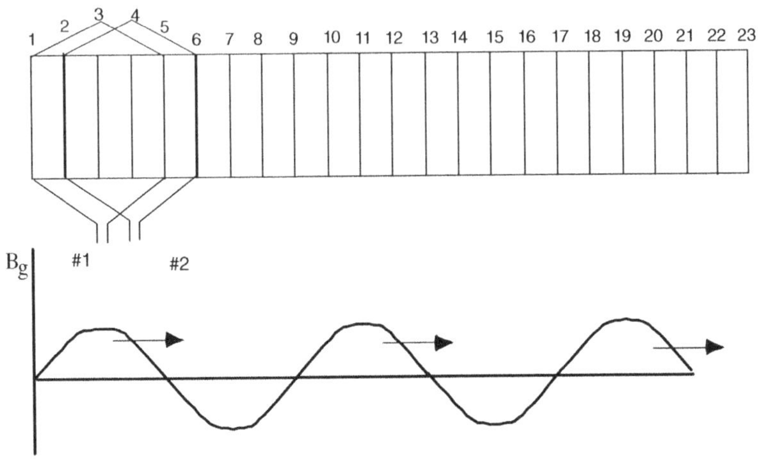

Figure 2.9 Showing first 21 slots of a machine with first and second coils inserted in slots 1 and 5 and also 2 and 6.

voltages can be expressed as phasors as shown in Figure 2.10. Proceeding now with a third coil placed in slots 3 and 7, the voltage induced in this coil will lag the voltage induced in coil 1 by $85\frac{5}{7}°$. It is possible to now proceed with all such coils placed in slots 4-8, 5-9,…, until 21-25. The 21 phasors form the phasor diagram of Figure 2.10.

It is now possible to collect all of the coil voltages which are intended to contribute to one of the three phases, say phase a. If coil 1 is considered to be associated with phase a, then all of the coils encountered clockwise over the next 60 degrees

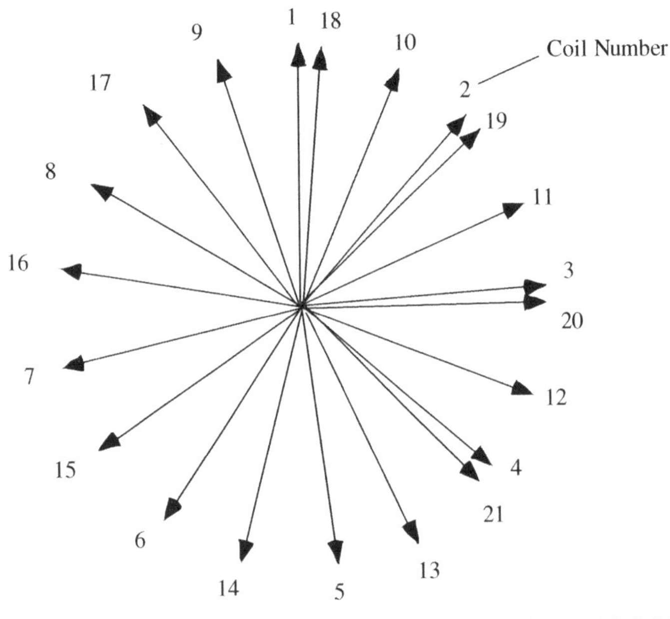

Figure 2.10 Showing EMF of the 21 Coils inserted in slots 1-5, 2-6, 3-7,…, 21-25.

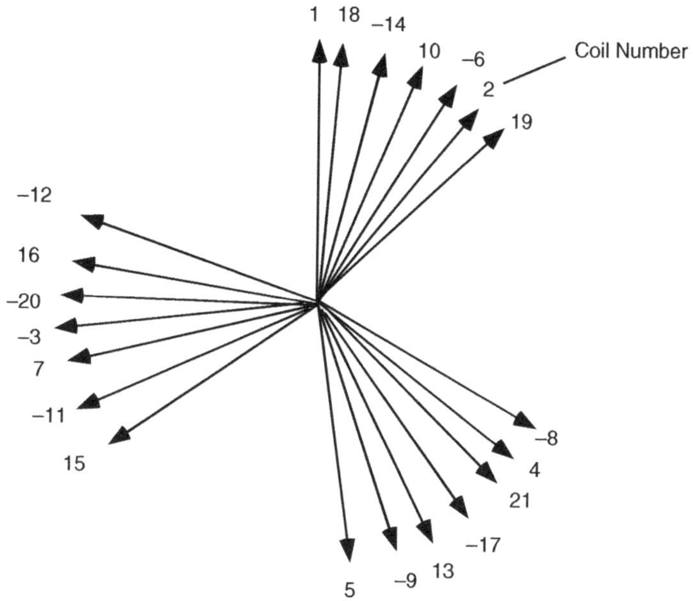

Figure 2.11 EMFs of coils making up the three phases after reversing the polarities of coils to obtain 60° *phase belts*.

can be claimed for phase *a*. This includes coils 1, 18, 10, 2, and 19. In addition, if the polarities of coils 6 and 14 are reversed, then one can also claim these for phase *a* making a total of 7 of the 21 coils (as required for one phase of the three-phase system. Phase *b* can now be made up of coils 4, 5, −8, −9, 13, −17, and 21. Phase *c* is made up of coils −3, 7, −11, −12, 15, 16, and −20. (The minus signs indicate the reversal of the polarity of the coil). Figure 2.12 shows the winding layout of the three phases over the five-pole, 21-slot portion of the machine. The winding pattern for the remaining five poles is, of course, identical. Note that each slot is occupied by two coil sides. Even though the coils are taken from different poles, it can be said that the winding forms an equivalent 60° phase belt distributed over seven slots ($\gamma = 8^4\!/_7°$). The distribution factor is therefore

$$k_{d1} = \frac{\sin\left(7 \cdot \dfrac{60}{7 \times 2}\right)}{7 \sin\left(\dfrac{60}{7 \times 2}\right)} = 0.956 \tag{2.45}$$

The product of the pitch and distribution factor for this winding layout is consequently,

$$k_{p1} k_{d1} = 0.997 \times 0.956 = 0.953 \tag{2.46}$$

Similar results can be shown to apply for the other odd winding harmonics. The harmonics clearly will result in a much greater leakage inductance than a machine with a true 60 degree phase belt with $q = 7$. It is also important to mention that when

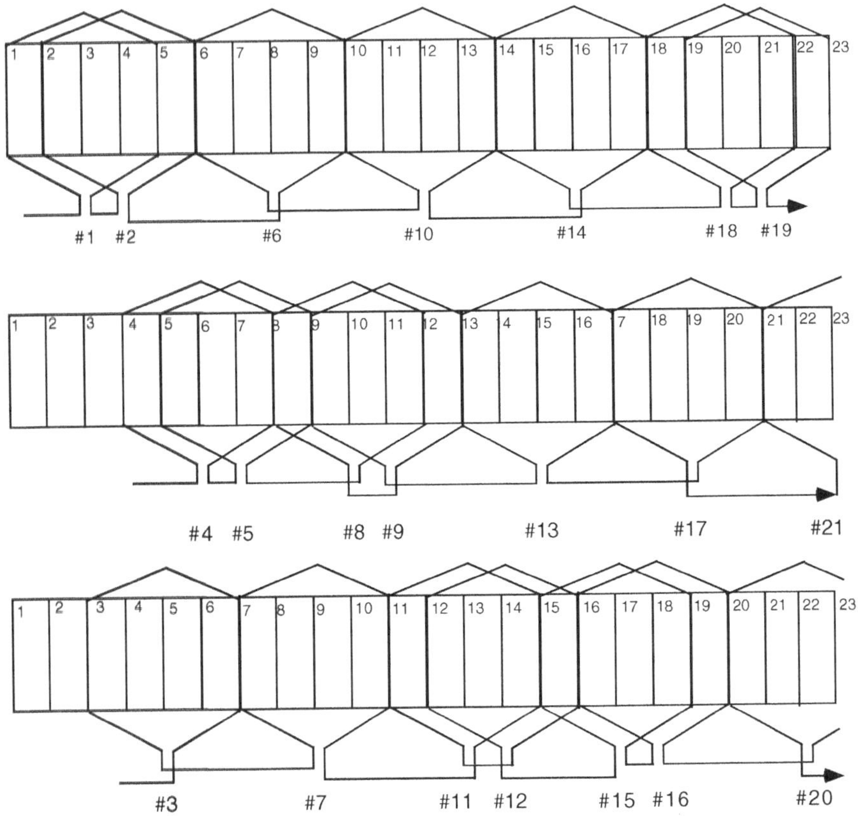

Figure 2.12 Winding layout of the three phases over the first five poles.

the approach is used for machines in which induced currents can flow on the other member, that is, squirrel cage induction and synchronous machines with amortisseur windings, additional losses occur during starting. These additional losses will be caused by the fractional slot approach as the short-circuited bars progressively move through the magnetic field set up by the unbalanced MMF pattern set up by the fractional slot winding.

2.7 WINDING SKEW

While pitching and distributing the winding has the effect of reducing the harmonics, substantial harmonics typically remain primarily associated with the slot harmonics. Such harmonics of higher order can be further reduced by "skewing" the winding. Skewing is invariably done to the rotor of a squirrel cage induction machine in order to reduce the so-called cogging or sub-synchronous torques which occur without skewing. In addition, the stator slots are sometimes skewed in smaller salient pole synchronous generators or the poles in larger slow speed salient pole generators. In

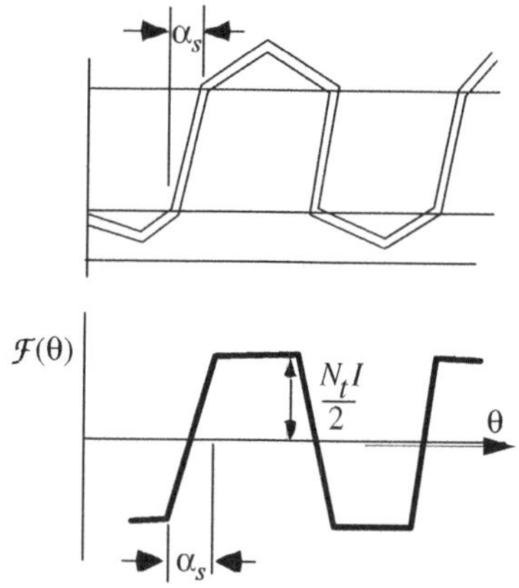

Figure 2.13 Concentrated full-pitch coil skewed by a skew angle α_s.

this case, the skewing also reduces the flux variation in the fringing of the flux at the pole tips due to the slots entering and leaving the poles which lead to acoustic noise problems.

In order to illustrate the effect of skew, consider a single full-pitch concentrated coil having one slot per pole. In this case, the slot is constructed so that the winding is twisted circumferentially in the slot "skewed" by an angle α_s. The corresponding MMF acting in the gap becomes dependent on the axial position. However, "on the average," the effective MMF is again a rectangular function but with sloping sides as shown. The harmonic components of the resulting MMF can again be calculated by means of Fourier series. However, with proper interpretation, the harmonic coefficients can be obtained directly from equation (2.27). In particular, note that as the number of slots (coil sides) q approaches infinity the resulting shape assumes the continuous function of Figure 2.13. Although q becomes infinite, it is clear that the product $q\gamma$ remains constant. In particular,

$$q\gamma = \alpha_s/2 \tag{2.47}$$

so that

$$\sin\left(\frac{hq\gamma}{2}\right)\cos\left(\frac{hq\gamma}{2}\right) = \sin\left(\frac{h\alpha_s}{4}\right)\cos\left(\frac{h\alpha_s}{4}\right) \tag{2.48}$$

Also, as γ approaches zero the denominator becomes

$$\begin{aligned}\lim_{\gamma\to 0}\left[q\sin\left(\frac{h\gamma}{2}\right)\right] &= \frac{qh\gamma}{2}\\ &= \frac{h\alpha_s}{4}\end{aligned} \tag{2.49}$$

Hence, as the number of slots tend to infinity, equation (2.26) becomes

$$\mathcal{F}_{\text{ph}} = \left(\frac{4}{\pi}\right)\left(\frac{N_{\text{t}}I}{2h}\right)\frac{\sin\left(\frac{h\alpha_{\text{s}}}{4}\right)\cos\left(\frac{h\alpha_{\text{s}}}{4}\right)}{\frac{h\alpha_{\text{s}}}{4}} \tag{2.50}$$

However,

$$\sin\left(\frac{h\alpha_{\text{s}}}{4}\right)\cos\left(\frac{h\alpha_{\text{s}}}{4}\right) = \frac{1}{2}\sin\left(\frac{h\alpha_{\text{s}}}{2}\right)$$

so that equation (2.50) reduces to

$$\mathcal{F}_{\text{ph}} = \frac{4}{\pi}\left(\frac{N_{\text{t}}I}{2h}\right)\frac{\sin\left(\frac{h\alpha_{\text{s}}}{2}\right)}{\left(\frac{h\alpha_{\text{s}}}{2}\right)} \tag{2.51}$$

The quantity

$$k_{\text{sh}} = \frac{\sin\left(\frac{h\alpha_{\text{s}}}{2}\right)}{\frac{h\alpha_{\text{s}}}{2}} \tag{2.52}$$

is called the *skew factor* for the hth harmonic. This result can also be written in another useful form. If one again lets Z be the arc which the coil sides occupy per pole as in Figure 2.5 and again letting τ_{p} be the pole pitch measured on the surface of the armature, then

$$\frac{\alpha_{\text{s}}}{2} = \frac{Z}{\tau_{\text{p}}}\pi \tag{2.53}$$

and k_{sh} can be expressed as

$$k_{\text{sh}} = \frac{\sin\left[h\left(\frac{Z}{\tau_{\text{p}}}\right)\pi\right]}{h\left(\frac{Z}{\tau_{\text{p}}}\right)\pi} \tag{2.54}$$

The MMF for a distributed, fractional pitch winding with skew can again be computed by considering these effects separately. The combined effect can then be obtained by taking the product of the four winding factors.

$$\mathcal{F}_{\text{ph}} = \frac{4}{\pi}\frac{N_{\text{t}}I}{2h}k_{\text{ph}}k_{\text{dh}}k_{\chi\text{h}}k_{\text{sh}} \tag{2.55}$$

The product

$$k_{\text{h}} = k_{\text{ph}}k_{\text{dh}}k_{\chi\text{h}}k_{\text{sh}} \tag{2.56}$$

is often termed simply the *winding factor*.

2.8 POLE PAIRS AND CIRCUITS GREATER THAN ONE

Thus far, MMF distributions for two-pole, single-circuit machines have been discussed. The actual number of electrical poles, however, depends on design considerations which will be discussed later. This number is nearly always greater than two since two-pole machines are particularly wasteful of stator conductors due to large return paths at the end of the machine (end windings). A bulky end winding also results in a machine with high leakage reactance which is often an undesirable design. Also, because of the large amount of flux per pole, the amount of steel that must be used in the stator core tends to increase. Finally, it is also difficult to find sufficient room in the rotor of the machine to contain the large flux per pole. In general, for one phase of a P pole machine, with N_t total turns and with all poles connected in series, the MMF for the hth harmonic, equation (2.55), becomes

$$\mathcal{F}_{\text{ph}} = \frac{4}{\pi}\left(\frac{N_t I}{P}\right)\frac{k_{\text{ph}}k_{\text{dh}}k_{\chi\text{h}}k_{\text{sh}}}{h} \tag{2.57}$$

As a final complication, phase windings are often not all connected in series to form one continuous path for each phase. Practical considerations such as limited wire diameters often require that two or more poles be connected in parallel rather than in series in order to achieve the required number of turns with the wire diameters available. The number of parallel circuits must always result in equal voltage across each circuit in order to avoid circulating currents. In order to avoid this problem, the ratio of poles to circuits must clearly be an integer. The final general expression for the hth harmonic of a phase of a P pole, C circuit winding is

$$\mathcal{F}_{\text{ph}} = \frac{4}{\pi}\left(\frac{N_t I}{CP}\right)\frac{k_{\text{ph}}k_{\text{dh}}k_{\chi\text{h}}k_{\text{sh}}}{h} \qquad h = 1, 3, 5, 7\ldots \tag{2.58}$$

where the current I is now taken to be the net current into the C circuits, that is, the current for each circuit is I/C.

2.9 MMF DISTRIBUTION FOR THREE-PHASE WINDINGS

When the machine is excited with three symmetrically wound coils the net MMF acting on the magnetic circuits of the machine is simply the superposition of the three MMFs of the individual coils. Assuming that the three coils are wound identically but physically placed in the machine at 120° intervals, the MMFs of the three windings a, b, c can be written

$$\mathcal{F}_{\text{a}} = \frac{4}{\pi}\left(\frac{N_t i_a}{CP}\right)\sum_{h=1,5,7,11\ldots} k_h \frac{\sin h\frac{P}{2}\theta}{h} \tag{2.59}$$

$$\mathcal{F}_b = \frac{4}{\pi}\left(\frac{N_t i_b}{CP}\right) \sum_{h=1,5,7,11\ldots} k_h \frac{\sin h\frac{P}{2}\left(\theta - \frac{4\pi}{3P}\right)}{h} \tag{2.60}$$

$$\mathcal{F}_c = \frac{4}{\pi}\left(\frac{N_t i_c}{CP}\right) \sum_{h=1,5,7,11\ldots} k_h \frac{\sin h\frac{P}{2}\left(\theta + \frac{4\pi}{3P}\right)}{h} \tag{2.61}$$

where

$$k_h = k_{ph} k_{dh} k \chi_h k_{sh}$$

With the aid of trig identities, equations (2.59) to (2.61) can also be written in the form

$$\mathcal{F}_a = \frac{4}{\pi}\left(\frac{N_t i_a}{CP}\right) \sum_{h\ \mathrm{odd}} k_h \frac{\sin[(hP\theta)/2]}{h} \tag{2.62}$$

$$\mathcal{F}_b = \frac{4}{\pi}\left(\frac{N_t i_b}{CP}\right) \sum_{h\ \mathrm{odd}} k_h \left[\frac{\sin[(hP\theta)/2]}{h}\cos\left(\frac{2h\pi}{3}\right) - \frac{\cos[(hP\theta)/2]}{h}\sin\left(\frac{2h\pi}{3}\right)\right] \tag{2.63}$$

$$\mathcal{F}_c = \frac{4}{\pi}\left(\frac{N_t i_c}{CP}\right) \sum_{h\ \mathrm{odd}} k_h \left[\frac{\sin[(hP\theta)/2]}{h}\cos\left(\frac{2h\pi}{3}\right) + \frac{\cos[(hP\theta)/2]}{h}\sin\left(\frac{2h\pi}{3}\right)\right] \tag{2.64}$$

The three MMFs can be summed to form

$$\mathcal{F}_a + \mathcal{F}_b + \mathcal{F}_c = \frac{4}{\pi}\left(\frac{N_t}{CP}\right) \sum_{h\ \mathrm{odd}} \frac{k_h}{h} \left[i_a + i_b\cos\left(\frac{2h\pi}{3}\right) + i_c\cos\left(\frac{2h\pi}{3}\right)\right]\sin[(hP\theta)/2] \tag{2.65}$$
$$+ \left(i_c \sin\frac{2h\pi}{3} - i_b \sin\frac{2h\pi}{3}\right)\cos[(hP\theta)/2]$$

Now

$$\cos\frac{2h\pi}{3} = -1/2 \quad \text{for } h = 1, 5, 7, 11 \ldots$$

$$\cos\frac{2h\pi}{3} = 1 \quad \text{for } h = 3, 9, 15, \ldots$$

$$\sin\frac{2h\pi}{3} = \frac{\sqrt{3}}{2} \quad \text{for } h = 1, 7, 13, \ldots$$

$$\sin\frac{2h\pi}{3} = -\frac{\sqrt{3}}{2} \quad \text{for } h = 5, 11, \ldots$$

$$\sin\frac{2h\pi}{3} = 0 \quad \text{for } h = 3, 9, 15, \ldots$$

If it is assumed that the machine has no neutral return, then

$$i_a + i_b + i_c = 0 \tag{2.66}$$

Equation (2.65) reduces to the form

$$\mathcal{F}_\mathrm{a} + \mathcal{F}_\mathrm{b} + \mathcal{F}_\mathrm{c} = \frac{4}{\pi}\left(\frac{N_\mathrm{t}}{CP}\right) \sum_{h=1,5,7,11,\ldots} \frac{k_h}{h}\left[\left(\frac{3i_\mathrm{a}}{2}\right)\sin[(hP\theta)/2] + \frac{\sqrt{3}}{2}(\pm 1)(i_\mathrm{c} - i_\mathrm{b})\cos[(hP\theta)/2]\right] \tag{2.67}$$

where

$$(\pm 1) = 1 \quad \text{for } h = 1,\ 7,\ 13, \ldots$$

and

$$(\pm 1) = -1 \quad \text{for } h = 5,\ 11,\ 17, \ldots$$

Note that the third harmonic components have been eliminated if the machine has no neutral return.

Consider now the special case of balanced sinusoidal currents, that is,

$$i_\mathrm{a} = I_\mathrm{s}\cos\omega_e t \tag{2.68}$$

$$i_\mathrm{b} = I_\mathrm{s}\cos\left(\omega_e t - \frac{2\pi}{3}\right) \tag{2.69}$$

$$i_\mathrm{c} = I_\mathrm{s}\cos\left(\omega_e t + \frac{2\pi}{3}\right) \tag{2.70}$$

It can be shown that

$$\frac{\sqrt{3}}{2}(i_\mathrm{c} - i_\mathrm{b}) = -\frac{3}{2}I_\mathrm{s}\sin\omega_e t \tag{2.71}$$

Equation (2.67) can now be written in the form

$$\mathcal{F}_\mathrm{a} + \mathcal{F}_\mathrm{b} + \mathcal{F}_\mathrm{c} = \left(\frac{3}{2}\right)\left(\frac{4}{\pi}\right)\left(\frac{N_\mathrm{t}I_\mathrm{s}}{CP}\right) \sum_{h=1,5,7,\ldots} \frac{k_h}{h}\{\sin[(hP\theta)/2]\cos\omega_e t - (\pm 1)\cos[(hP\theta)/2]\sin\omega_e t\} \tag{2.72}$$

which reduces to

$$\mathcal{F}_\mathrm{a} + \mathcal{F}_\mathrm{b} + \mathcal{F}_\mathrm{c} = \mathcal{F}_\mathrm{s} = \left(\frac{3}{2}\right)\left(\frac{4}{\pi}\right)\left(\frac{N_\mathrm{t}I_\mathrm{s}}{CP}\right)\left[\sum_{h=1,7,13,\ldots} \frac{k_h}{h}\sin\left(\frac{hP\theta}{2} - \omega_e t\right) + \sum_{h=5,11,17,\ldots} \frac{k_h}{h}\sin\left(\frac{hP\theta}{2} + \omega_e t\right)\right] \tag{2.73}$$

Hence a balanced, three-phase operation of a machine with a practical winding distribution containing odd harmonics results in an MMF which produces positively and negatively rotating MMFs in the machine. Each of these MMFs is of constant amplitude and rotate in the forward direction for $h = 1, 7, 13, 19, \ldots$ and in the negative direction for $h = 5, 11, 17, \ldots$ The harmonic components corresponding to $h = 3, 9, \ldots$ are not present if the machine has no neutral return, that is, no zero-sequence component. It is important to note that the amplitude of each MMF harmonic is 3/2 times the amplitude of the MMF for an individual phase. The MMFs corresponding to n greater than one are called the *space harmonics* of the machine winding. The velocity of rotation of these MMFs can be found by differentiating the argument of the sine function. Hence, the velocity of harmonic h of the MMF is

$$\frac{d\left(\frac{hP\theta}{2} - \omega_e t\right)}{dt} = \left(\frac{hP}{2}\right)\frac{d\theta}{dt} - \omega_e$$

or

$$\frac{d\theta}{dt} = \frac{2\omega_e}{hP} \tag{2.74}$$

so that the synchronous mechanical speed for each harmonic is an integer fraction of the speed of rotation of the fundamental component of MMF.

2.10 CONCEPT OF AN EQUIVALENT TWO-PHASE MACHINE

It is interesting to note that the form of equation (2.73) has implications concerning modeling of three-phase machines. In particular since the coefficient of the bracketed term is precisely 3/2 greater than the coefficient of an individual phase as in equations (2.59) to (2.61), it is possible to replace the physical three-phase windings by an equivalent two-phase winding which results in the same instantaneous MMF in the gap of the machine. The form of the equivalent winding can be discerned if the form of equation (2.67) is examined. This equation is a general expression for the instantaneous MMF in the gap for arbitrary values of currents i_a, i_b, and i_c. Note that, in effect, the MMF has at this point been decomposed into orthogonal components since the arguments of the angle θ involve sin $h\theta$ and cos $h\theta$. Although the concept can be extended to machines with a neutral, this issue is not of concern here. Hence, if no neutral return is assumed, equation (2.67) results. If a three-halves factor is taken out of the square bracket there results

$$\mathcal{F}_a + \mathcal{F}_b + \mathcal{F}_c = \left(\frac{3}{2}\right)\left(\frac{4}{\pi}\right)\left(\frac{N_t}{CP}\right)\sum_{h=1,5,7,11\ldots}\left[i_a\frac{k_h}{h}\sin\frac{hP\theta}{2} + \frac{1}{\sqrt{3}}(i_c - i_b)(\pm 1)\frac{k_h}{h}\cos\frac{hP\theta}{2}\right] \tag{2.75}$$

Equation (2.75) suggests that if one defines two fictitious currents i_x and i_y

$$i_x = i_a \tag{2.76}$$

$$i_y = \frac{1}{\sqrt{3}}(i_c - i_b) \tag{2.77}$$

which flow through winding distributions

$$N_x(\theta) = \frac{3}{2}\left(\frac{4}{\pi}\frac{N_t}{CP}\right)\sum_{h=1,5,7,11\ldots}\left(\frac{k_h}{h}\right)\sin\frac{hP\theta}{2} \tag{2.78}$$

and

$$N_y(\theta) = \frac{3}{2}\left(\frac{4}{\pi}\frac{N_t}{CP}\right)\sum_{h=1,5,7,11\ldots}(\pm 1)\left(\frac{k_h}{h}\right)\cos\frac{hP\theta}{2} \tag{2.79}$$

then the instantaneous MMF in the gap will remain unchanged. Alternatively, if the right-hand sides of equations (2.76) and (2.77) are multiplied by 3/2 and those of equations (2.78) and (2.79) divided by 3/2, one could choose currents

$$i_x = \frac{3}{2}i_a \tag{2.80}$$

$$i_y = \frac{\sqrt{3}}{2}(i_c - i_b) \tag{2.81}$$

which flow through winding distributions

$$N_x(\theta) = \left(\frac{4}{\pi}\frac{N_t}{CP}\right)\sum_{h=1,5,7,11\ldots}\left(\frac{k_h}{h}\right)\sin\frac{hP\theta}{2} \tag{2.82}$$

and

$$N_y(\theta) = \left(\frac{4}{\pi}\frac{N_t}{CP}\right)\sum_{h=1,5,7,11\ldots}(\pm 1)\left(\frac{k_h}{h}\right)\cos\frac{hP\theta}{2} \tag{2.83}$$

Again the MMF in the gap would be unchanged. This principle forms the basis of *d–q* axis theory which is normally studied in quite some detail in other texts. Use of the former concept will be required later in Chapter 9 involving the analysis of synchronous machines.

2.11 CONCLUSION

This chapter has presented the details of the forcing function for the magnetic field problem associated with the design of an induction motor. A solution for the resulting flux distribution and its associated circuit parameter, the magnetizing inductance, will be dealt with as the next topic.

REFERENCES

[1] H. B. Dwight, *Tables of Integrals and Other Mathematical Data*, 4th edition, Macmillan, 1961.

[2] W. Ouyang, "Optimization of Winding Turns for Concentric Windings with 60 and 120 Degree Phase Belts," Technical Report, University of Wisconsin, May 2004.

[3] W. Ouyang, T. A. Lipo and A. El-Antably, "Analysis of Optimal Stator Concentric Winding Pattern Design," International Conference on Electrical Machines and Systems (ICEMS) Vol. 1, 2005, pp. 94–98.

CHAPTER 3

MAIN FLUX PATH CALCULATIONS USING MAGNETIC CIRCUITS

HAVING DETERMINED THE MMF forcing function, it is now necessary to calculate the fluxes and MMF drops in the main magnetic circuit of an induction machine. It has already been shown in Chapter 2 that a balanced, sinusoidal set of currents results in an infinite series of constant amplitude, rotating MMFs. It can further be shown that torque varies with the square of the current, so that the corresponding higher harmonic fluxes will produce negligible average torque. However, these harmonics do interact with the main fundamental component of flux to produce primarily pulsating components of torque. Hence, these unwanted components of flux can be relegated to the category of "leakage" fluxes. In general, the higher space harmonics of flux within the machine are said to produce belt leakage and the corresponding inductance is termed the *belt leakage inductance*. More will be said regarding these leakage inductances in Chapter 4.

3.1 THE MAIN MAGNETIC CIRCUIT OF AN INDUCTION MACHINE

If only the fundamental components of MMF are considered, the fluxes produced by balanced sinusoidal stator currents form symmetrical flux paths within the machine. For purposes of illustration, Figure 3.1 shows the situation for a four pole machine. The peak fundamental stator MMF per pole of a three- phase machine is given from equation (2.73) as

$$\mathcal{F}_{s1} = \frac{3}{2}\left(\frac{4}{\pi}\right)\left(\frac{N_t}{CP}\right)k_{p1}k_{d1}k_{s1}k_{\chi 1}I_s \tag{3.1}$$

where k_{p1}, k_{d1}, k_{s1}, and $k_{\chi 1}$ are the pitch, distribution, skew, and slot opening factors for the fundamental component of MMF respectively, $N_t/C = N_s$ is the number of series connected turns of the C circuits and I_s is the peak current per phase which is equal to the peak line current in the case of a wye-connected machine.

Introduction to AC Machine Design, First Edition. Thomas A. Lipo.

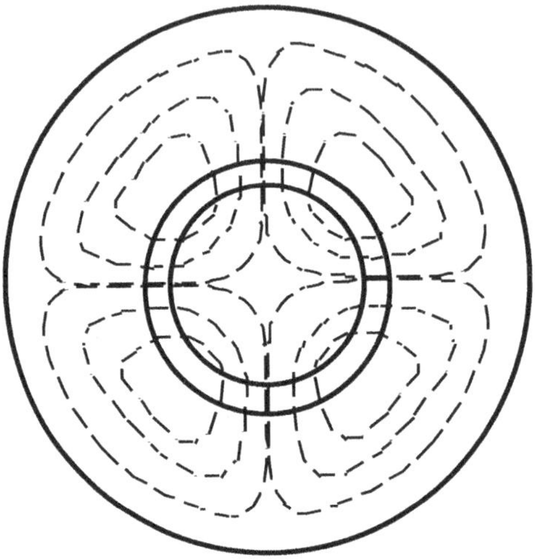

Figure 3.1 Flux distribution in a four-pole machine.

It can be recognized from Figure 3.1 that the flux for each circuit traverses two rotor teeth, two stator teeth, two air gaps, a portion of the stator core, and finally a portion of the rotor core. For the present, the possibility of flux bypassing a portion of the teeth and instead passing through the slot on its way to the core will be neglected. It is convenient to write Ampere's law for this case as

$$2\mathcal{F}_{s1} = 2\mathcal{F}_{ts} + 2\mathcal{F}_{tr} + 2\mathcal{F}_{g} + \mathcal{F}_{cs} + \mathcal{F}_{cr} \tag{3.2}$$

where the subscript "ts" denotes the stator tooth drop, "tr" the rotor tooth drop, "cs" the stator core drop, "cr" the rotor core drop, and "g" the MMF drop across the gap. In Chapter 1, it was assumed that the exciting ampere-turns were concentrated in one portion of the core. However, in the case of an electrical machine the exciting MMF is distributed along the magnetic path. Also, the presence of slots along both the stator and rotor surfaces make the correction for the effect of the air gap considerably more complicated than in the case of a rectangular core. Nonetheless, the same principles set forth in Chapter 1 can be applied with proper attention to these effects. For this purpose it is useful to consider in detail each of the three key areas of the magnetic circuit, the air gap, the teeth, and the core.

3.2 THE EFFECTIVE GAP AND CARTER'S COEFFICIENT

Although the permeability of the gap region is constant, it is bounded on either side by iron surfaces which, far from being smooth, are indented with slots in the circumferential direction and by cooling ducts in the axial direction. Rather than using more complicated expressions for reluctance, it is conventional to derive equivalent values of cross-sectional area and length and to continue to use the same expression derived in Chapter 1 for a magnetic path of uniform cross section.

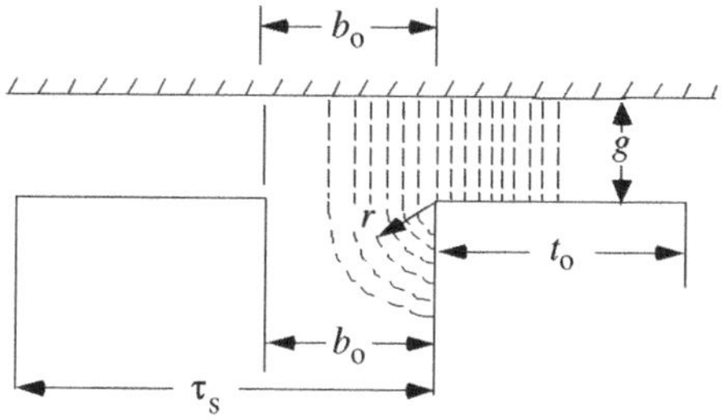

Figure 3.2 Approximate shape of flux lines entering a slotted armature.

In order to explain how the permeance of the air paths between a smooth iron surface and a slotted surface may be calculated, the flux lines will, for simplicity, be assumed to follow the paths indicated in Figure 3.2. The tooth is drawn, for convenience, with parallel sides and the magnetic lines entering the sides of the tooth are assumed to follow a path consisting of a straight portion of length g equal to the actual air gap, and a circular arc of radius r as indicated in the figure. Consider a portion of the air gap l_e effective inches long axially (i.e., in a direction normal to the plane of the section shown). The meaning of the term "effective" will be explained shortly. Note that the permeance over the slot pitch of width τ_s is made up of two parts: (a) the permeance $\mathcal{P}_1$ between the unslotted surface and the top of the tooth, and (b) the permeance $\mathcal{P}_2$ between the unslotted surface and the slot opening b_o on one side of the tooth.

In SI units, the permeance $\mathcal{P}_1$ is simply

$$\mathcal{P}_1 = \mu_0(\tau_s - b_o)\frac{l_e}{g} \quad \text{henries(H)} \tag{3.3}$$

The permeance of any small section of thickness dr and length l_e meters in the region over the slot is

$$d\mathcal{P}_2 = \frac{\mu_0 l_e dr}{g + \left(\frac{\pi r}{2}\right)} \tag{3.4}$$

Hence, the total permeance over one side of the tooth is

$$\mathcal{P}_2 = \frac{2\mu_0}{\pi}\int_0^{b_o/2} \frac{l_e dr}{\frac{2g}{\pi} + r} = \frac{2\mu_0 l_e}{\pi} \ln\left[\frac{g + \frac{\pi b_o}{4}}{g}\right] \tag{3.5}$$

The total effective permeance of both the slot and the tooth is therefore

$$\mathcal{P}_g = \mathcal{P}_1 + 2\mathcal{P}_2 = \mu_0 l_e \left\{\frac{\tau_s - b_o}{g} + \frac{4}{\pi}\ln\left[1 + \frac{\pi}{4}\frac{b_o}{g}\right]\right\} \tag{3.6}$$

The effective reluctance is the reciprocal of this quantity.

It is now possible to suppose that the actual slotted surface can be replaced by an equivalent unslotted surface having the same cross section but with a modified "equivalent" gap. Equating the permeance of the equivalent unslotted surface to the actual permeance

$$\frac{\mu_0 \tau_s l_e}{g_e} = \mu_0 l_e \left\{ \frac{\tau_o - b_o}{g} + \frac{4}{\pi} \ln\left[1 + \frac{\pi}{4}\frac{b_o}{g}\right] \right\} \tag{3.7}$$

Equation (3.7) can now be solved for the equivalent gap g_e as

$$g_e = \frac{\tau_s}{\dfrac{\tau_s - b_o}{g} + \dfrac{4}{\pi} \ln\left[1 + \dfrac{\pi}{4}\dfrac{b_o}{g}\right]} \tag{3.8}$$

Alternatively, a so-called *Carter factor* can be defined such that

$$g_e = k_c g \tag{3.9}$$

where

$$k_c = \frac{\tau_s}{\tau_s - b_o + \dfrac{4g}{\pi} \ln\left[1 + \dfrac{\pi}{4}\dfrac{b_o}{g}\right]} \tag{3.10}$$

The equation that has been derived assumes that the flux is evenly distributed along the side of the tooth to the depth $b_o/2$. In reality, the flux density is nonuniform with the highest flux density near the surface of the tooth and drops off as the flux penetrates into the tooth. A more exact formula, obtained by conformal mapping which takes account of the fact that the flux lines do not exactly follow the paths assumed in Figure 3.2, is given by

$$k_c = \frac{\tau_s}{\tau_s - \dfrac{2b_o}{\pi}\left\{ \operatorname{atan}\dfrac{b_o}{2g} - \dfrac{g}{b_o} \ln\left[1 + \left(\dfrac{b_o}{2g}\right)^2\right] \right\}} \tag{3.11}$$

It can be shown that for large b_o/g, this equation can be approximated by

$$k_c = \frac{\tau_s}{\tau_s - b_o + \dfrac{4g}{\pi} \ln\left[\dfrac{b_o}{2g}\right]} \tag{3.12}$$

Note that this result is remarkably similar to the result previously obtained. In many cases a simple approximation to equation (3.11) proposed by Metzler suffices

$$k_c \approx \frac{\tau_s}{\tau_s - \dfrac{b_o^2}{(5g + b_o)}} \tag{3.13}$$

The approximation is within 10% for b_o/g between one and infinity.

The effective gap g_e can be as much as 70–80%greater than g for machines with open slots. For most practical machines, however, g_e is generally found to be 15–25% larger than g. If there are slot openings on both sides of the air gap, the effective gap length is typically found by multiplying the actual gap by the product

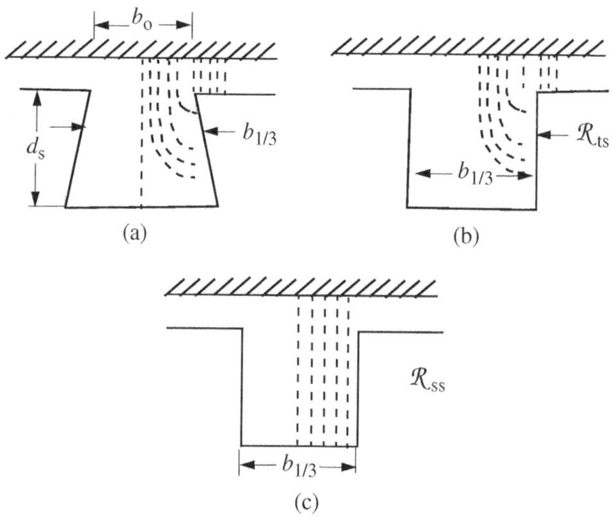

Figure 3.3 Modification of air gap permeance for finite iron permeability for deeply saturated tooth. (a) ≅ (b) + (c).

of the two Carter factors, each calculated separately from either equation (3.11) or (3.12). That is, the Carter factor for a doubly slotted machine is

$$k_c = k_{cs} k_{cr} \tag{3.14}$$

where k_{cs} and k_{cr} are the Carter factors for the stator and rotor slotting, respectively.

Thus far, the calculation of air gap permeance has been based on the assumption of infinite iron permeability. Consider now Figure 3.3a, which illustrates the case of a highly saturated tooth. Note that as the tooth saturates, there will be more and more flux passing directly from the gap to the bottom of the slot in addition to the flux lines directed to the side of the tooth as previously calculated from Figure 3.2. A closer approximation to the actual conditions which exist in a saturated tooth may be obtained by assuming a parallel field in the slot superimposed on the field of Figure 3.2 as illustrated in Figure 3.3b and Figure 3.3c. The resultant field in the slot and air gap will then be a closer approximation to that of Figure 3.3a. With small values of tooth flux density, the MMF between the tooth tops and the bottom of the slots will be small and only a few flux lines will pass from the pole face into the armature core without entering the teeth. With higher tooth densities, however, the MMF to overcome the increasing tooth reluctance becomes larger and larger and more flux will be diverted into the parallel path and pass directly to the bottom of the slot.

The permeance of the slot portion of a single tooth pitch is simply

$$\mathcal{P}_s = \frac{\mu_0 l_e b_{1/3}}{g/2 + d_s} \tag{3.15}$$

where $b_{1/3}$ is the breadth of the slot taken 1/3 of the way from the narrow portion of the slot and d_s is the slot depth. This value of $b_{1/3}$ is used to partially account for cases in which the slot is not exactly rectangular. Most often this condition occurs

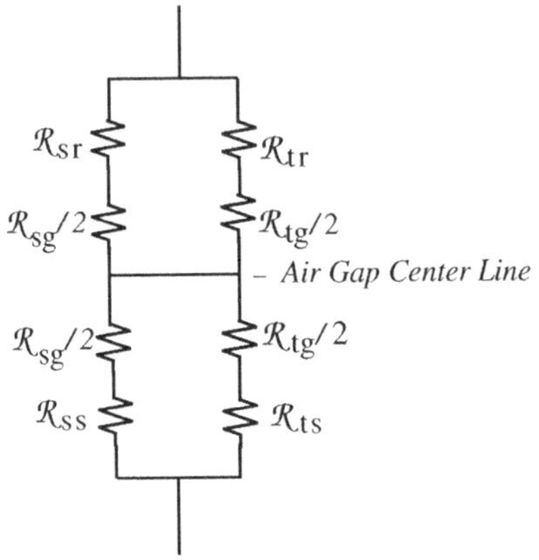

Figure 3.4 Magnetic circuit for region of one stator and one rotor tooth assuming the same tooth and slot width.

with rotor slots which often have a trapezoidal or "coffin" shape. In large machines, the shape of the stator slot is rectangular with the tooth being nonuniform due to the cylindrical geometry. However, in small random wound machines, the reverse is typically true and the same breadth factor can be used. It can be observed that only half the gap has been used in equation (3.15) with the other half of the gap being used to calculate the permeance of the corresponding rotor (or stator) tooth. The magnetic equivalent circuit for the region of one tooth is given in Figure 3.4. Note that the equivalent circuit is given in terms of reluctances, which are simply the inverse of the permeances in this section. The tooth reluctance $\mathcal{R}_t$ is as yet unknown at this point.

3.3 THE EFFECTIVE LENGTH

The presence of slots to accommodate windings is only one type of disturbance in the stator and rotor air gap surfaces. Another type of slotting occurs in machines with cooling ducts spaced along the core length. When forced ventilation is used, a fan or centrifugal blower is attached at one end of the rotor. The cooling action is assisted by either radial ventilating ducts spaced along the air gap or by axial ducts which are constructed by punching holes in the core area of the laminations. Since the thermal conductivity of a lamination is from 40 to 50 times greater in the direction parallel to the plane of the laminations than in a direction perpendicular to this plane, radial air ducts are often preferred. However, axial ventilation must also be used in machines with long core lengths due to the difficulty in supplying the central parts of the core with cool air. When radial ducts are employed, duct widths are typically 3/8–1/2 in. In order to maintain adequate cooling, the spacing between the core ducts do not generally exceed 6 in.

The approach to incorporating the effect of radial ducts is similar to that for slotting and can be readily derived. The following approximate equations, similar to the simplified expression for slotting, equation (3.14), can be used to calculate the effective length of a machine without ducting. If desired, a more accurate expression for ducting similar to equation (3.11) can be employed [1]. The effective length as calculated from these equations is used together with the effective gap, equation (3.8), to calculate the permeance of an equivalent machine with smooth iron surfaces in both the circumferential and longitudinal directions.

In general, it can be noted that the stator and rotor iron lengths need not be equal, as in case #2 below. However, for simplicity, for the remainder of the book, it will be assumed that $l_{is} = l_{ir} = (= l_i)$. If not so, the proper use of one quantity or the other for stator or rotor quantities is implied by the context.

1. Stator and Rotor Same Length, No Ducts

$$l_e = l_i + 2g; \quad l_i = l_{is} = l_{ir}$$

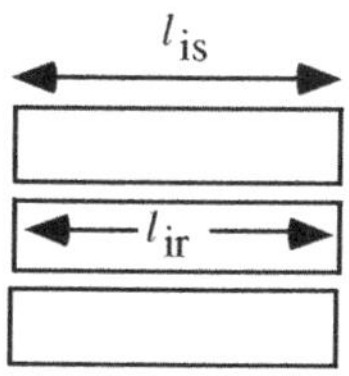

2. Stator and Rotor with Slightly Different Core Lengths, No Ducts

$$l_e = \frac{l_{is} + l_{ir}}{2} \quad (l_{is} \neq l_{ir})$$

$$\text{if } l_{is} < l_e < l_{is} + 8g$$

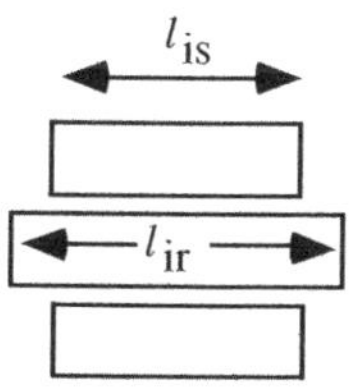

3. Stator and Rotor with Same Stack Length, Same Number of Ducts

$$l_e = l_i + 2g + n\left(\frac{l_{0s} + l_{0r}}{2}\right)\left[\frac{5}{5 + \frac{l_{0s}}{g}}\right]\left[\frac{5}{5 + \frac{l_{0r}}{g}}\right]$$

where $l_i = l_1 + l_2 + l_3 + \cdots + l_{n+1}$

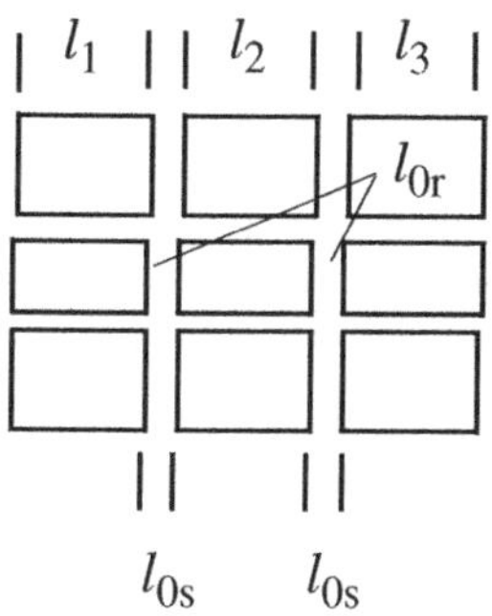

(n=2)

4. Stator Has *n* Ducts, Rotor is Smooth

$$l_e = l_e + 2g + nl_{0s}\left[\frac{5}{5+\frac{l_{0s}}{g}}\right]$$

where $l_i = l_1 + l_2 + l_3 + \cdots + l_{n+1}$

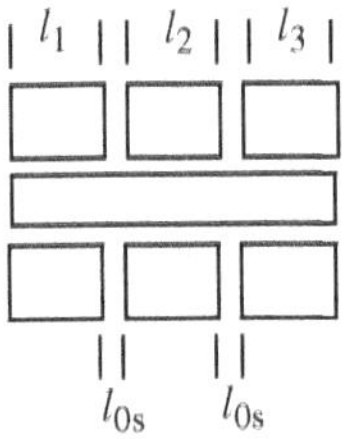

(n=2)

5. Stator and Rotor Have *n* Ducts but Are Not Aligned

$$l_e = l_i + 2g + \frac{n}{2}\left\{l_{0s}\left[\frac{5}{5+\frac{l_{0s}}{g}}\right] + l_{0r}\left[\frac{5}{5+\frac{l_{0r}}{g}}\right]\right\}$$

where $l_i = l_1 + l_2 + l_3 + \cdots + l_{n+1}$

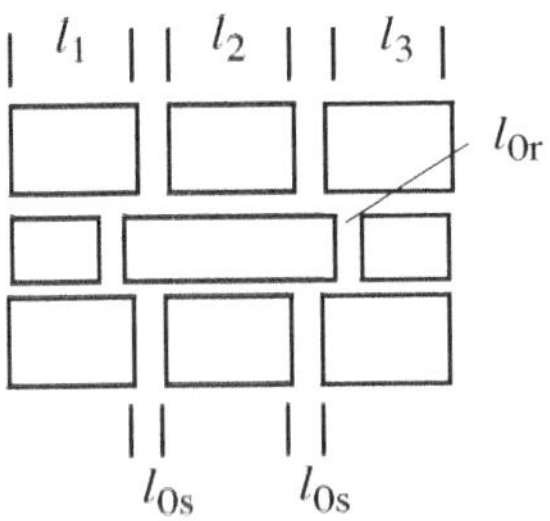

(n=2)

3.4 CALCULATION OF TOOTH RELUCTANCE

Thus far, only the air portion of the magnetic circuit has been considered. It is now time to calculate the permeance of a typical stator (or rotor) tooth. Since the iron permeance is assumed to be nonlinear, the calculation of this quantity depends upon a knowledge of the actual flux density in the tooth. For low flux densities in the iron (up to 1.4 teslas), the tooth density is readily computed by assuming that all of the flux passes through the tooth. That is to say, practically all the flux entering the armature over one tooth pitch will pass into the core through the root of the tooth.

This component of flux is accounted for by the Carter factor. For densities exceeding 1.4 teslas and even for lower values, when the depth of slot is small in relation to the air gap, the calculations should take account of a component of the total flux which goes straight from the gap to the bottom of the slot without entering the teeth as discussed in Section 3.3. Since the Carter factor assumes that the slot is infinitely deep, these flux lines are not included in the Carter factor.

It is now useful to define the following symbols: $B_{g,ave}$ = the average gap flux density along the center line of the air gap over one slot pitch; B_{g1} = the peak fundamental component of air gap flux density at the center line of the air gap; B_{mid} = the flux density of the tooth at the center point midway down the tooth; B_{top} the flux density just at the surface of the tooth, B_{root} the flux density at the root of the tooth (where the flux enters the stator or rotor core).

When the tooth is not greatly saturated, it can simply be assumed that all of the flux over one slot pitch enters the surface of the tooth. In this case, the flux entering the core from the slot portion is essentially neglected. Then,

$$B_{top}k_i t_t l_i = B_{g,ave}\tau_s l_e \tag{3.16}$$

or

$$B_{top} = B_{g,ave}\left(\frac{\tau_s}{t_t}\right)\left(\frac{l_e}{k_i l_i}\right) \tag{3.17}$$

where l_i denotes the axial length of the machine excluding ducts (length of iron) and k_i takes into account the insulation space between the laminations. The factor k_i has a value between 0.87 for 0.014 in. (14 mil) laminations and 0.93 for 0.025 in. (25 mil) laminations.

The assumption that each tooth has uniform cross section throughout its length can only be justified when the diameter of the machine is very large in relation to the slot depth or when the teeth are intentionally designed to have parallel sides as in small random wound machines. Often stator and rotor slots have a considerable taper, which results in either the top or the root of the tooth becoming saturated first, depending upon whether the tooth is associated with the stator or rotor. When tooth taper is appreciable, the simplest approach is to determine the flux density at three points in the tooth; namely at the top, at the center, and at the root of the tooth which must be calculated and the corresponding field intensities found from the iron B–H curves. The net or average field intensity is then computed approximately by Simpson's rule.

In order to simplify the calculations, the assumption is made that the total flux in the tooth remains unaltered throughout all tooth cross sections. The tooth is considered as having a total length (or depth) d_t with corresponding width t_t at the top, a width t_m midway down the tooth, and a width t_r at the root of the tooth. See Figure 3.5. More points along the tooth can be selected, if the shape of the tooth is more complicated.

A design is generally commenced by choosing a value of B for the point of maximum flux density in either the stator or the rotor tooth depending upon which presents the highest flux density. In Figure 3.5, this point is clearly at the tooth root. The flux density at various other parts of the tooth can be obtained by simple ratios. By selecting a number of such values, the corresponding values of B_{g1} can be obtained for

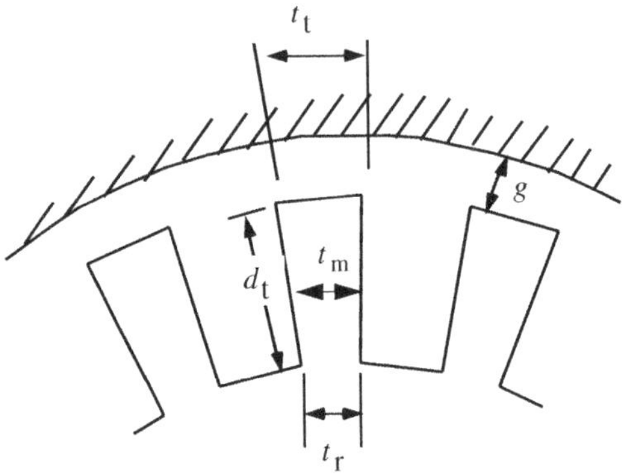

Figure 3.5 Illustrating MMF calculations for tapered teeth.

a particular type of magnetic material. Alternatively, B_{g1} can be selected and the tooth flux density at various points calculated (approximated). It is important to note that knowledge of the peak fundamental component of flux density B_{g1} is approximate at this point, since the permeability varies from tooth to tooth so that the exact air gap flux density distribution is non-sinusoidal. The estimate can be improved, however, when the true variation in flux density has been determined.

Assuming a value for B_{root}, one has at the top and midpoint of the tooth

$$B_{\text{top}} = B_{\text{root}}\frac{t_r}{t_t} \tag{3.18}$$

$$B_{\text{mid}} = B_{\text{root}}\frac{t_r}{\left(\frac{1}{2}\right)(t_t + t_r)} \tag{3.19}$$

Simpson's rule states that an approximation to the integral of a function $y(x)$ over a measure x is

$$\int_0^x y(x)dx = \frac{\Delta x}{3}\{y(0) + 4y(\Delta x) + 2y(2\Delta x) + 4y(3\Delta x)\ldots \\ +4y[(n-1)\Delta x] + y(n\Delta x)\} \tag{3.20}$$

where

$$\Delta x = x/n$$

and n is an even number. In the problem of interest, it is desired to compute

$$H_{\text{t(ave)}} = \frac{1}{d_t}\int_0^{d_t} H_t(l)dl$$

Letting $n = 2$ in equation (3.20), then

$$H_{t(ave)} = \frac{1}{d_t}\left(\frac{d_t}{2}\right)\left(\frac{1}{3}\right)\left[H_t(0) + 4H_t(d_t/2) + H_t(d_t)\right]$$

or

$$H_{t(ave)} = \frac{1}{6}[H_t(0) + 4H_t(d_t/2) + H_t(d_t)] \tag{3.21}$$

If the three values of H are replaced by H_{root} for the value at the tooth bottom, by H_{mid} at the mid-center section, and by H_{top} at the top section of the tooth, Simpson's approximation becomes

$$H_{t(ave)} = \frac{1}{6}H_{top} + \frac{2}{3}H_{mid} + \frac{1}{6}H_{root} \tag{3.22}$$

The total MMF drop in the tooth is then readily computed as

$$\mathcal{F}_{t(ave)} = \left[\frac{1}{6}H_{top} + \frac{2}{3}H_{mid} + \frac{1}{6}H_{root}\right] d_t \tag{3.23}$$

The values of H_{top}, H_{mid}, and H_{root} are determined from the nonlinear B–H curve of the material by entering the curve for the flux densities B_{top}, B_{mid}, and B_{root} and reading H_{top}, H_{mid}, and H_{root} from the curve.

The reluctance associated with the toothed region is

$$\mathcal{R}_t = \frac{\mathcal{F}_{t(ave)}}{\phi_t} \tag{3.24}$$

where

$$\phi_t = B_{top} t_t k_i l_i = B_{root} t_r k_i l_i \tag{3.25}$$

Hence

$$\mathcal{R}_t = \frac{\mathcal{F}_{t(ave)}}{B_{top} t_t k_i l_i} \tag{3.26}$$

A repeated calculation of $\mathcal{R}_t$ over all expected values of B_c results in one of the desired nonlinear elements in the per pole magnetic circuit. A similar calculation can be carried out for the other slotted element, that is, the stator (or rotor). With ordinary cold rolled steel, the peak value of B typically should not normally exceed 1.7 teslas (roughly 110 kilolines/in.^{2}) in the stator or 1.8 teslas (115 kilolines/in.^{2}) in the rotor teeth. For silicon steels, the maximum values are 1.6 and 1.7 teslas, respectively.

3.5 EXAMPLE 1—TOOTH MMF DROP

Consider now the calculation of the MMF drop in the air gap and rotor teeth over one tooth pitch. For purposes of analysis, the dimensions of Figure 3.6 are assumed. The thickness of the laminations is assumed to be 0.014″ ($k_i = 0.87$). The material is

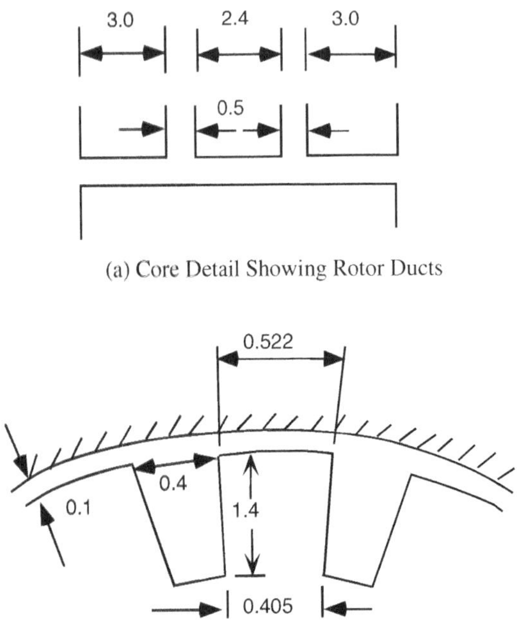

(a) Core Detail Showing Rotor Ducts

(b) Rotor Slot Detail

Figure 3.6 Core and slot dimensions for calculation of air gap and rotor slot MMF, all dimensions in inches.

29 gage, M27 steel, B–H curves of which are given in Figure 1.16. From Figure 3.6 one can deduce that

$$\begin{aligned} \tau_s &= 0.922'' & l_i &= 8.4'' \\ b_o &= 0.4'' & l_o &= 0.5'' \\ d_t &= 1.4'' & g &= 0.1'' \\ t_t &= 0.522'' & t_r &= 0.405'' \end{aligned}$$

The first step in the analysis is to calculate the effective values of length, width, and depth. From the equation corresponding to Case 4 in Section 3.3,

$$l_e \approx 2g + l_i + nl_0 \left[\frac{5}{5 + \frac{l_0}{g}} \right]$$

where l_o is the duct opening and l_i is the length of the iron stack. From this equation one obtains

$$\begin{aligned} l_e &= 0.2 + 8.4 + 2 \cdot 0.5 \left[\frac{5}{5 + \frac{0.5}{0.1}} \right] \\ &= 9.1'' \end{aligned}$$

From equation (3.11)

$$k_c = \frac{\tau_s}{\tau_s - \frac{2b_o}{\pi}\left\{ a\tan\left(\frac{b_o}{2g}\right) - \frac{g}{b_o}\left(\ln\left[1 + \left(\frac{b_o}{2g}\right)^2\right]\right)\right\}}$$

$$= \frac{0.922}{0.922 - \left(\frac{2 \cdot 0.4}{3.1416}\right)\left\{ a\tan\left(\frac{0.4}{2 \cdot 0.1}\right) - \frac{0.1}{0.4}\ln\left[1 + \left(\frac{0.4}{2 \cdot 0.1}\right)^2\right]\right\}}$$

$$= 1.24$$

so that

$$g_e = k_c g = 0.124''$$

Consider now the computation of the MMF drops for a value of 1.86 teslas (120 kilolines/in.^{2}) at the root of the tooth. It should be noted that this value of flux density results in a quite saturated condition and is at the upper end for a conventional design. However, this case was chosen merely as an example of a highly saturated machine. Ultimately it will be found that the voltage required to supply this amount of flux density is greater than the rated value. From the iron *B–H* curves in Figure 1.16, one can find that corresponding to

$$B_{\text{root}} = 1.86 \text{ teslas} \quad \text{T}$$

one has, for low carbon steel

$$H_{\text{root}} = 200 \text{ Oe} \times 0.796 \text{ (A} \cdot \text{t/cm)/Oe} = 160 \text{ A-t/cm}$$

The constituent equation is (Chapter 1, Section 1.20)

$$B_{\text{root}} = (4\pi \cdot 10^{-9})\mu_i H_{\text{root}}$$

when μ_0 is expressed in henries per cm. It is now possible to obtain the relative permeability of the saturated stator iron as

$$\mu_i = \frac{1.86(10)^{-4}}{(4\pi)(10)^{-9}(160)} = 93$$

An estimate for the average flux density in the air gap corresponding to 1.86 teslas at the root of the tooth is consequently, from equation (3.16)

$$B_{g,\text{ave}} = B_{\text{top}}\left(\frac{k_i l_i}{l_e}\right)\left(\frac{t_t}{\tau_s}\right) = B_{\text{root}}\left(\frac{k_i l_i}{l_e}\right)\left(\frac{t_r}{\tau_s}\right)$$

or

$$B_{g,\text{ave}} = B_{\text{root}}\left(\frac{0.87 \times 8.4}{9.1}\right)\left(\frac{0.405}{0.922}\right)$$

$$= 1.86(0.353)$$

$$B_{g,\text{ave}} = 0.656 \text{ T}$$

The flux density in the air gap over the tooth is

$$B_{\mathrm{gt}} = \frac{k_{\mathrm{i}} l_{\mathrm{i}}}{l_{\mathrm{e}}} \frac{t_{\mathrm{r}}}{t_{\mathrm{t}}} B_{\mathrm{root}} = \frac{0.87 \times 8.4}{9.1} \cdot \frac{0.405}{0.522} \cdot 1.86$$

$$= 1.15 \text{ T}$$

The MMF

$$\mathcal{F}_{\mathrm{gt}} = \frac{B_{\mathrm{g,ave}}}{\mu_0}(g_{\mathrm{e}}) = \frac{0.656}{4\pi \cdot 10^{-7}} \cdot 0.124 \left(\frac{2.54}{100}\right)$$

$$= 1644 \text{ A-t}$$

where the factor $2.54/100$ converts inches to meters.

The flux passing through the tooth is,

$$\Phi_{\mathrm{gt}} = B_{\mathrm{top}} l_{\mathrm{e}} t_{\mathrm{t}} = 1.15 \cdot 9.1 \cdot 0.522 \cdot \left(\frac{2.54}{100}\right)^2$$

$$= 0.00352 \text{ webers(Wb)}$$

The reluctance of the gap over the toothed region is

$$\mathcal{R}_{\mathrm{gt}} = \frac{\mathcal{F}_{\mathrm{gt}}}{\Phi_{\mathrm{gt}}} = \frac{1644}{0.00352}$$

$$= 467 \times 10^3 \text{ H}^{-1} \text{inverse henries}(H^{-1})$$

In order to calculate the reluctance of the tooth, it is necessary to first calculate the MMF drop over the tooth. The flux density at the root of the tooth where $B_{\mathrm{root}} = 1.86$ is, from Figure 1.16

$$H_{\mathrm{root}} = 160 \text{ A-t/cm}$$

Since the tooth tapers linearly from the top to the midpoint of the tooth

$$B_{\mathrm{mid}} = B_{\mathrm{root}} \left(\frac{0.405}{\dfrac{0.405 + 0.522}{2}} \right) = 1.65 \text{ T}$$

From the *B–H* curve, one obtains

$$H_{\mathrm{mid}} = 44.5 \text{ A-t/cm}$$

Again, by taking ratios, the flux density at the top of the tooth is

$$B_{\mathrm{top}} = B_{\mathrm{root}} \left(\frac{0.405}{0.522}\right) = 1.44 \text{ T}$$

whereupon

$$H_{\mathrm{top}} = 8 \text{ A-t/cm}$$

The average MMF drop along the tooth is now computed from Simpson's rule (equation 3.22) as

$$\begin{aligned} H_{\mathrm{t(ave)}} &= \frac{H_{\mathrm{top}}}{6} + \frac{2H_{\mathrm{mid}}}{3} + \frac{H_{\mathrm{root}}}{6} \\ &= \frac{1}{6}(8) + \frac{2}{3}(44.5) + \frac{1}{6}(160) \\ &= 57.67 \text{ A-t/cm} \end{aligned}$$

Hence, the average MMF drop down the length of the tooth is

$$\begin{aligned} \mathcal{F}_{\mathrm{t(ave)}} &= H_{\mathrm{t(ave)}} d_{\mathrm{t}} \\ &= (57.67)(1.4)2.54 \\ &= 205.1 \text{ A-t} \end{aligned}$$

so that the reluctance of the rotor tooth is

$$\begin{aligned} \mathcal{R}_{\mathrm{t}} &= \frac{\mathcal{F}_{\mathrm{t(ave)}}}{\Phi_{\mathrm{r}}} = \frac{205.1}{3.52} \times 10^3 \\ &= 58 \times 10^3 \text{ H}^{-1} \end{aligned}$$

Note that this value is about a 12% of the air gap portion of the reluctance. Including the MMF drop in a stator tooth, not considered here, this result would indicate that the MMF drops of the teeth are on the order of 25% that of the air gap. It is clear that the MMF drops in the iron cannot be neglected even when the flux density in the iron reaches a mildly saturated value. The magnetic circuit of the gap, tooth, and slot is given in Figure 3.7.

The effects of saturation are often calculated approximately by computing the flux density and corresponding field intensity at a point 1/3 the distance from the

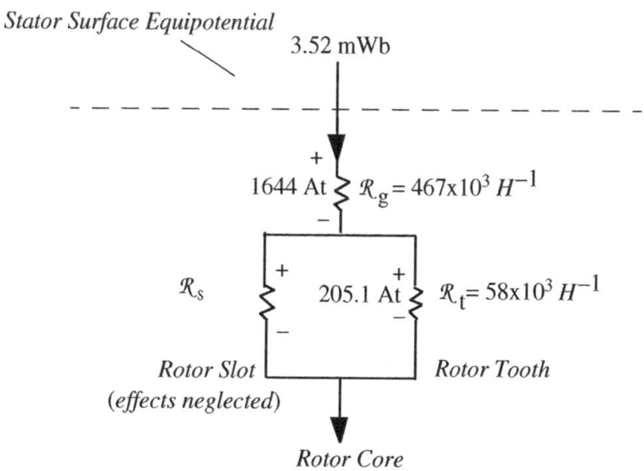

Figure 3.7 Magnetic equivalent circuit for gap, slot, and tooth configuration of Figure 3.6 for $B_m = 1.86\ T$.

narrowest part of the tooth (1/3 the way from the root of the tooth in the example above), given the maximum flux density in the air gap. If one assumes the same value of air gap flux density and the approximate expression for tooth flux, then, for a point 1/3 the way from the root of the tooth of Figure 3.6 and neglecting the effect of finite permeability, either

$$B_{1/3} = \frac{B_{\mathrm{g,ave}}}{\left(\frac{k_{\mathrm{i}} l_{\mathrm{i}}}{l_{\mathrm{e}}}\right)\left[\frac{t_{1/3}}{\tau_{\mathrm{s}}}\right]} \tag{3.27}$$

or

$$B_{1/3} = \left(\frac{t_{\mathrm{m}}}{t_{1/3}}\right) B_{\mathrm{mid}} \tag{3.28}$$

depending upon whether B_{g} or B_{c} has been assumed and where $t_{1/3}$ is the tooth width at the point 1/3 of the way from the root of the tooth. The width $t_{1/3}$ is readily found to have a value

$$\begin{aligned} t_{1/3} &= t_{\mathrm{r}} + \frac{1}{3}(t_{\mathrm{t}} - t_{\mathrm{r}}) \\ &= 0.405 + \frac{1}{3}(0.522 - 0.405) \\ &= 0.444'' \end{aligned}$$

Hence, assuming a value for the flux at the root of the tooth, $B_{\mathrm{root}} = 1.86$ teslas.

$$B_{1/3} = \frac{0.405}{0.444} \cdot 1.86 = 1.7\ \mathrm{T}$$

From the B–H curves of Chapter 1,

$$H_{1/3} = 80 \times 0.796 = 64\ \text{A-t/cm}$$

so that

$$\begin{aligned} \mathcal{F}_{\mathrm{t(ave)}} &= H_{1/3} d_{\mathrm{t}} \\ &= (64)(1.4)2.54 \\ &= 227.6\ \text{A-t} \end{aligned}$$

Note that this value is reasonably close to the value of 205.1 computed previously. While not suitable for high saturation conditions, the estimate improves considerably when the iron is only mildly saturated.

3.6 CALCULATION OF CORE RELUCTANCE

The process of computing tooth permeances as a function of gap flux density over one tooth span has now been completed. The remaining task to solving the complete magnetic circuit is to evaluate the MMF drop in the core or "back iron" portion of the circuit. Figure 3.8 shows the magnetic equivalent circuit for a machine with only six

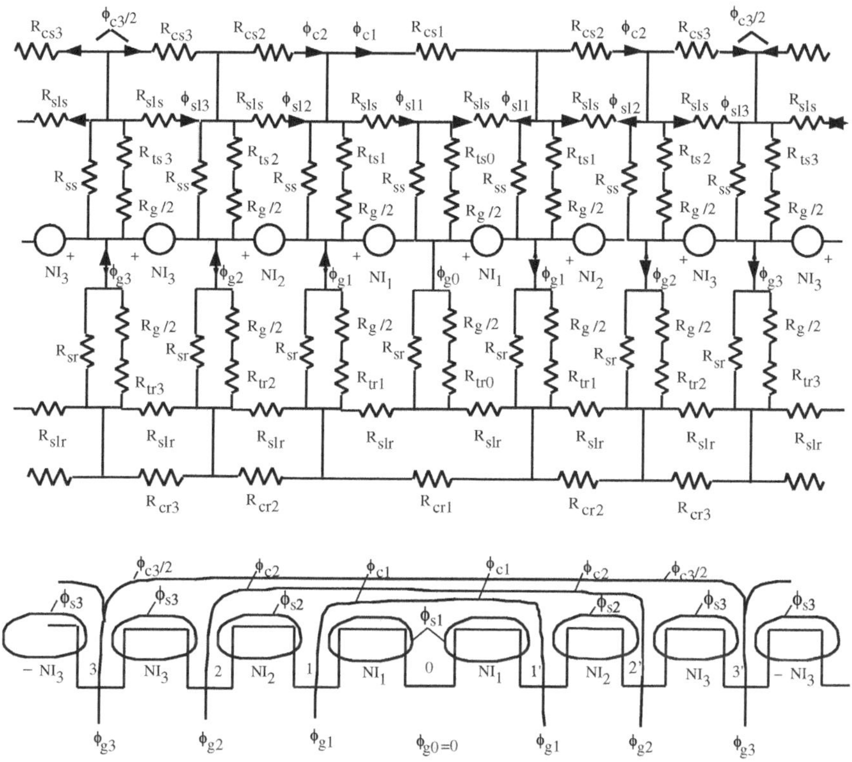

Figure 3.8 Magnetic equivalent circuit for an induction machine with six stator and rotor slots per pole including slot leakage flux.

stator and rotor slots per pole. (In general, the stator and rotor slot numbers should be unequal. The case of unequal slot numbers will be discussed later). Note that a "flux tube" can be identified for each pair of slots and the core portion of the machine supports all such flux tubes in a pole. The flux in the outermost flux tube splits equally between a north and an adjacent south pole (see Figure 3.1). Also note that an even number of slots result in one tooth per pole, that is, essentially supporting no flux at any instant. That is to say, since the MMF distribution is symmetrical, when there are an even number of "steps" in the MMF per half cycle the center "step" MMF must be zero. When an odd number of slots per pole are used, all of the teeth are excited at any instant.

The MMF drop in the core portion can best be computed by using the outermost flux tube (teeth 3 and 3′). However, the problem remains complicated by the fact that the three magnetic circuits represented by the flux tubes corresponding to the gap fluxes, Φ_{g1}, Φ_{g2}, and Φ_{g3} are not independent since the magnetic material in the core area is nonlinear. In addition, leakage flux paths exist (designated in this example by Φ_{sl1}, Φ_{sl2}, and Φ_{sl3}) which close their path through the stator core. Hence, the MMF drop in each of the three core reluctances is a function of the flux density in the core which involves the total flux flowing in all three magnetic circuits taken

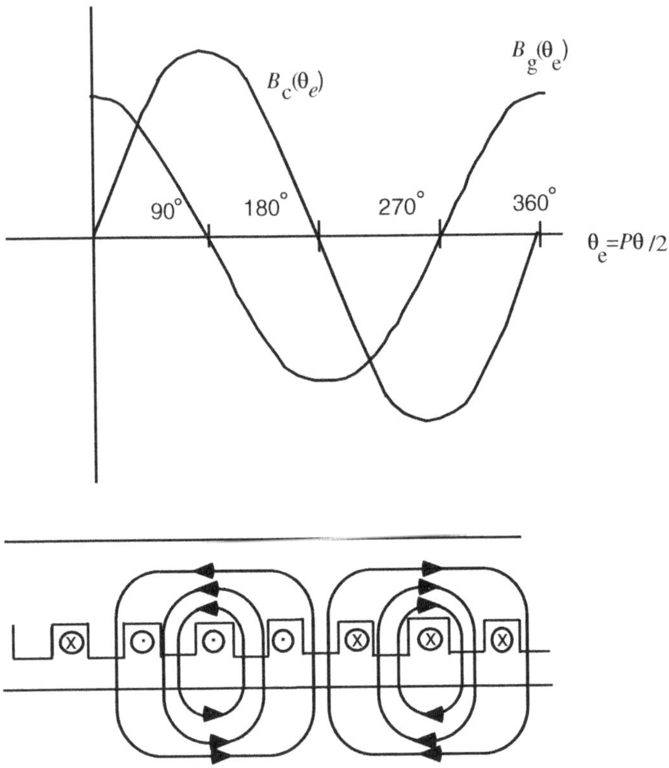

Figure 3.9 Gap and core flux density as a function of electrical angle along the air gap surface.

simultaneously. Nonetheless, the problem can still be solved if it is assumed that the flux density in the gap is known and that it is sinusoidal.

Observe from Figure 3.9 that if the gap flux density is sinusoidal then the flux density in the core area is co-sinusoidal, that is to say that the core flux density is the space integral of the gap flux density. In particular, if P is the number of poles then the spatial distribution of flux density in the air gap can be written as

$$B_g(\theta) = B_{g1} \cos\left(\frac{P\theta}{2}\right) \tag{3.29}$$

where θ is an angular circumferential measure around the air gap. From Figure 3.9, it can be noted that the core flux is zero at the point where the air gap flux density is a maximum ($\theta_e = 0°$). The total flux entering the core as a result of air gap flux from the point $\theta_e = 0°$ to an arbitrary point θ_e is

$$\Phi_c(\theta_e) = \int_0^{\theta=(2/P)\theta_e} B_{g1} \cos\left(\frac{P\theta}{2}\right) r l_e d\theta \tag{3.30}$$

where r denotes the radius from the rotor center line to the cylindrical surface defined by the midpoint of the air gap. In particular, when

$$\frac{P\theta}{2} = \theta_e = \frac{\pi}{2}$$

or

$$\theta = \pi/P$$

the point where the flux in the core reaches a maximum is determined. This point is 90 electrical degrees from the point where the air gap flux density is a maximum. Assuming that the peak fundamental and maximum gap flux are the same, the stator or rotor core flux is computed from equation (3.30) as

$$\Phi_c(90°) = \frac{2}{P} l_e r B_{g1}$$

which can be written as

$$\Phi_c(90°) = \left(\frac{\tau_p}{2} l_e\right)\left(\frac{2}{\pi} B_{g1}\right) \tag{3.31}$$

The quantity $2B_{g1}/\pi$ is the average value of flux density per pole and $\tau_p\, l_e/2$ is the area through which half the air gap flux passes on its way to the stator or rotor core. Half the total flux finds its way in each direction in the core as shown in Figure 3.9. The flux density in the core at the point $\theta = \pi/P$ (90 electrical degrees) is found by taking the flux given by equation (3.31) and dividing by the cross-sectional area of the core. The result is

$$\begin{aligned} B_c(90°) &= \frac{\Phi_c(90°)}{d_c k_i l_i} \\ &= \frac{l_e(\tau_p/2)}{d_c k_i l_i}\left(\frac{2B_{g1}}{\pi}\right) \end{aligned} \tag{3.32}$$

where d_c (d_{cs} or d_{cr}) is the depth of the core (stator or rotor) and k_i and l_i are the appropriate values for either the stator or rotor. The flux in other points of the core can be found in a similar manner.

As noted in Figure 3.8, the total flux in the core results from both the gap flux which closes its path through the core and also a leakage component which does not enter the gap but also closes its path through the core. The flux in any portion of the core is dependent upon the sum of the gap flux plus the leakage flux. Leakage flux calculations are the subject of the next chapter. However, the essential effect of leakage on the core flux must be dealt with here in the main flux calculation. In particular, the leakage flux component which is represented in Figure 3.8 is termed the *slot leakage* flux. It is stated in References [2] and [3] that the total stator leakage reactance can be estimated from the equation

$$X_{ls} = \frac{(P/100)}{1 - \dfrac{P}{100}} X_{ms} \tag{3.33}$$

when the machine is unsaturated and by

$$X_{ls} = \frac{1.5(P/100)}{1 - 1.5\dfrac{P}{100}} X_{ms} \tag{3.34}$$

when the machine is saturated. In equations (3.33) and (3.34), X_m represents the reactance corresponding to the air gap flux (i.e., magnetizing reactance) and X_{ls} denotes the total stator leakage reactance. In general, the slot leakage reactance is only a portion of the total stator leakage reactance. Assuming that the slot leakage makes up 1/2 of the total, one can write that,

$$\Phi_{cs} = \left(\frac{X_{ms} + X_{ls}/2}{X_{ms}}\right)\left(\frac{\Phi_g}{2}\right) \tag{3.35}$$

where Φ_{cs} denotes the flux in the stator core measured at the point of maximum core flux and Φ_g denotes the flux crossing the air gap over one pole pitch. Substituting equation (3.34) into equation (3.35) it can be readily determined that

$$\Phi_{cs} = \frac{1 + K_p}{2K_p}\frac{\Phi_g}{2} \tag{3.36}$$

where Φ_{cs} and Φ_g denote the stator core and gap flux, respectively and

$$K_p = 1 - \frac{1.5P}{100}$$

Hence, to include the effect of slot leakage flux, equation (3.31) must be written as,

$$\Phi_{cs}(90°) = \left(\frac{\tau_p l_e}{2}\right)\left(\frac{2B_{g1}}{\pi}\right)\frac{1 + K_p}{2K_p} \tag{3.37}$$

The flux density in the core at the same point is

$$B_{cs}(90°) = \frac{\tau_p(l_e/2)}{d_{cs}k_{is}l_{is}}\left(\frac{2B_{g1}}{\pi}\right)\frac{1 + K_p}{2K_p} \tag{3.38}$$

The same factor applies to estimating the flux and flux density at other points in the core. It can be similarly shown that the peak flux density in the rotor is

$$B_{cr}(90°) = \frac{\tau_p(l_e/2)}{d_{cr}k_{ir}l_{ir}}\left(\frac{2B_{g1}}{\pi}\right)\frac{3K_p - 1}{2K_p} \tag{3.39}$$

A similar correction factor can be used to account for the effects of rotor leakage flux on the saturation of the rotor core.

The *B–H* curves can now be used in conjunction with Simpson's rule to find the resulting MMF drop in the core similar to the approach taken for tooth saturation. Before taking this step, however, it is useful to pause to consider what happens to the air gap flux as the teeth begin to saturate. Clearly, the teeth with the highest flux density will saturate first so that the flux distribution will begin to assume a "flattened" sine wave with peak value $B_{g,max}$ as shown in Figure 3.10. Note that the amplitude of the fundamental component, B_{g1}, will now be somewhat greater than $B_{g,max}$. Consider now what would happen if a value B_{g1} for teeth 3 and 3′ is assumed and the MMF around the magnetic circuit is computed on this basis. The resulting computation

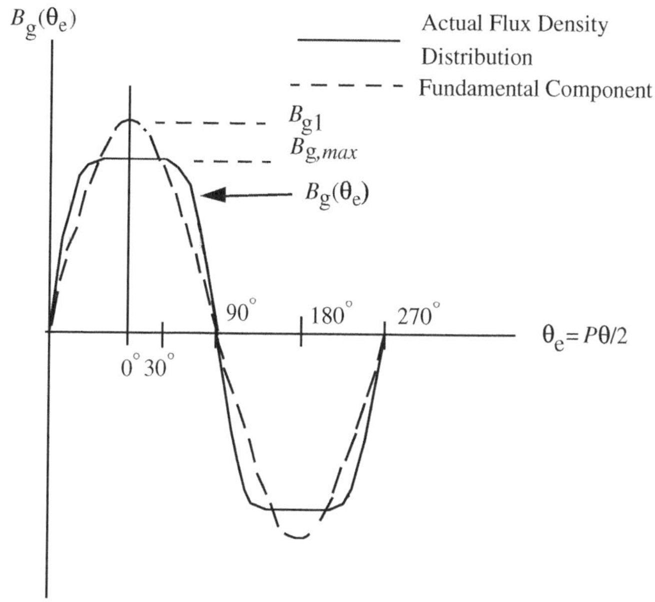

Figure 3.10 Actual and fundamental component of air gap flux density vs. electrical angle around the gap. Effect of slotting neglected.

would clearly be in error on the high side since the flux in the teeth 3 and 3′ of Figure 3.8 would certainly be too high. On the other hand, if a sinusoidal distribution is assumed with amplitude $B_{g,max}$, the MMF drop around the magnetic circuit will be too small since the flux density in the core portion will not be computed accurately and hence the MMF drop in the core will be in error.

This apparent dilemma can be resolved if one pauses to consider that the flattening of the sine wave of air gap flux density is produced primarily by a third harmonic component. Note from Figure 3.10 that the actual waveform and the fundamental component are equal very near 30°. The value is exactly 30° if the fifth, seventh, etc. harmonics are neglected. If it is assumed that the harmonics vary as inverse integers, that is, 1/3 for the third, 1/5 for the fifth, 1/7 for the seventh, etc. a more accurate value of this intersection is about 37°. In practice, values anywhere from 0 to 38.3° might be obtained depending upon the relative content of the third, fifth, and seventh harmonic and their relative polarity (phase shift). It has been shown by Lee [2] that a value of 30° is a reasonable choice for practical machines.

Since the apparent and actual flux densities are equal at a 30° point on the flux density angular distribution, the MMF drops in the teeth centered around this point will be nearly correct if it is assumed an average value of flux density for this tube equals $(\sqrt{3}/2)\ B_{g1}$. Hence, the proper flux tube to use corresponds to slots 2 and 2′ in Figure 3.8. (It should be mentioned that the choice of six slots per pole was intentional. The argument is valid for any number of slot per pole since the MMF pattern can always be rotated so that the 60° point is properly centered over a tooth.) The "ripple" in the flux density wave due to slotting is a type of leakage which will be calculated separately later. It is important to observe that even though the effect of the third harmonic on the computation of tooth MMF has been eliminated, its effect has not been eliminated in the core calculation. Generally, the core flux density is somewhat lower than the tooth flux density so that the extra core flux due to the third harmonic does not result in an appreciable change in core MMF drop. If this is not the case, an iterative solution must be resorted to in order to converge on the proper solution. The proper procedure to follow in such a case is left to the interested reader [4].

The computation of MMF drops in the main magnetic circuit of an induction machine now proceeds as follows:

1. Assuming a fundamental component of flux density amplitude B_{g1}, the corresponding value at a point 30° from the maximum is computed as

$$B_g(30°) = B_{g1}(30°) = \frac{\sqrt{3}}{2}B_{g1} \tag{3.40}$$

2. The corresponding tooth flux density is calculated by means of equation (3.17).
3. The MMF drop in the tooth and gap is computed.
4. Using B_{g1}, one can calculate the flux density at points in both the stator and rotor core where $\theta_e = 30°$, 60°, and 90°. For the stator core, the quantity $B_{cs}(90°)$ is readily derived from equation (3.38). The values of core flux density at 30° and 60° are easily seen to be

$$\begin{aligned} B_{cs}(60°) &= B_{cs}(90°)\cos(30°) \\ &= \frac{\sqrt{3}}{2}B_{cs}(90°) \end{aligned} \tag{3.41}$$

$$\begin{aligned} B_{cs}(30°) &= B_{cs}(90°)\cos(60°) \\ &= 0.5B_{cs}(90°) \end{aligned} \tag{3.42}$$

Similar expressions apply for the rotor core.
5. The corresponding values of field intensity H can be read from the B–H curves of the magnetic material being used. For common steel, the peak core flux density should normally be limited to 1.4 teslas in the stator and rotor cores. For silicon steels, the values should be reduced by approximately 10%.
6. Simpson's rule can now be used to find the average MMF drop over the core span from $\theta_e = 30°$ to $\theta_e = 150°$ for both the stator and rotor. In particular, Simpson's rule for the average value of a function H_{core} over a length l_c using five equidistant points is

$$\begin{aligned} H_{c(ave)} = \left(\frac{1}{l_c}\right)\left(\frac{l_c}{4}\right)\left(\frac{1}{3}\right)\big[&H_c(30°) + 4H_c(60°) + 2H_c(90°) \\ &+ 4H_c(120°) + H_c(150°)\big] \end{aligned} \tag{3.43}$$

From symmetry, one has the following relationships

$$H_c(150°) = H_c(30°)$$
$$H_c(120°) = H_c(60°)$$

so that equation (3.43) reduces to

$$H_{c(ave)} = \frac{1}{6}H_c(30°) + \frac{2}{3}H_c(60°) + \frac{1}{6}H_c(90°) \tag{3.44}$$

The procedure is done for both stator $H_{cs(ave)}$ and rotor core $H_{cr(ave)}$.

7. The mean length of path through the stator core over one pole pitch is computed as

$$l_{cs} = \frac{\pi(D_{is} + 2d_{ss} + d_{cs})}{P} \tag{3.45}$$

where D_{is} is the inner diameter of the stator core, d_{ss} is the depth of one stator slot, and d_{cs} is the radial depth of the stator core as measured from the bottom of a stator slot to the outer radius of the stator core.

For the rotor

$$l_{cr} = \frac{\pi(D_{or} - 2d_{sr} - d_{cr})}{P} \tag{3.46}$$

where d_{sr} is the depth of the rotor slots and D_{or} is the outer diameter of the rotor punching, that is

$$D_{or} = D_{is} - 2g \tag{3.47}$$

8. The MMF drop over the stator and rotor core can now be computed as

$$\mathcal{F}_{cs(ave)} = H_{cs(ave)}\left(\frac{2}{3}l_{cs}\right) \tag{3.48}$$

and

$$\mathcal{F}_{cr(ave)} = H_{cr(ave)}\left(\frac{2}{3}l_{cr}\right) \tag{3.49}$$

9. The MMF's can now be summed to find the impressed MMF per pole at the 30° point, that is

$$2\mathcal{F}_{s1}(30°) = 2\mathcal{F}_g(30°) + 2\mathcal{F}_{ts}(30°) + 2\mathcal{F}_{tr}(30°) + \mathcal{F}_{cs(ave)} + \mathcal{F}_{cr(ave)} \tag{3.50}$$

10. The actual value of MMF corresponding to the peak of an assumed sine wave of flux density is

$$2\mathcal{F}_{s1}(0°) = \frac{2}{0.866}\mathcal{F}_{s1}(30°) \quad (= 2\mathcal{F}_{s1} \text{ in Eq.3.1}) \tag{3.51}$$

3.7 EXAMPLE 2—MMF DROP OVER MAIN MAGNETIC CIRCUIT

A certain 250 HP, 8 pole, 2400 V, 60 Hz induction machine has the following constants and dimensions (in inches):

Stator OD = $D_{os} = 31.5''$	Gross core length = $9.25''$
Stator ID = $D_{is} = 24.08''$	Number of opposing stator
Rotor OD = $D_{or} = 24.0''$	and rotor ducts = 2
Rotor ID = $D_{ir} = 19''$	Width of stator and rotor ducts = $0.375''$
No. of stator slots = $S_1 = 120$	Lamination thickness = $0.014''$
No. of rotor slots = $S_2 = 97$	Type of steel = 29 Gage, M27 Steel

From this data one can deduce the following information

Gap = $g = 0.04''$
Length of stator and rotor stack = $l_{is} = l_{ir} = 8.5''$
Stacking factor (stator and rotor) = $k_{is} = k_{ir} = 0.93$
Pole pitch measured at the air gap = $\tau_p = 9.456''$
Stator slot pitch = $\tau_s = 0.630''$
Rotor slot pitch = $\tau_r = 0.733''$

Details of the stator and rotor slot shapes are given in Figure 3.11. From this figure and the above information, one can establish the following.

For the stator:

Depth of slot = $d_{ss} = 2.2''$
Depth of core = $d_{cs} = 1.51''$
Tooth width at top = $t_{ts} = 0.256''$
Tooth width at bottom = $t_{bs} = 0.369''$
Slot opening = $b_{os} = 0.374''$

For the rotor:

Depth of slot = $d_{sr} = 0.67''$
Depth of core = $d_{cr} = 2.08''$
Tooth width at top = $t_{tr} = 0.392''$
Tooth width at bottom = $t_{br} = 0.348''$
Slot opening = $b_{or} = 0.09''$

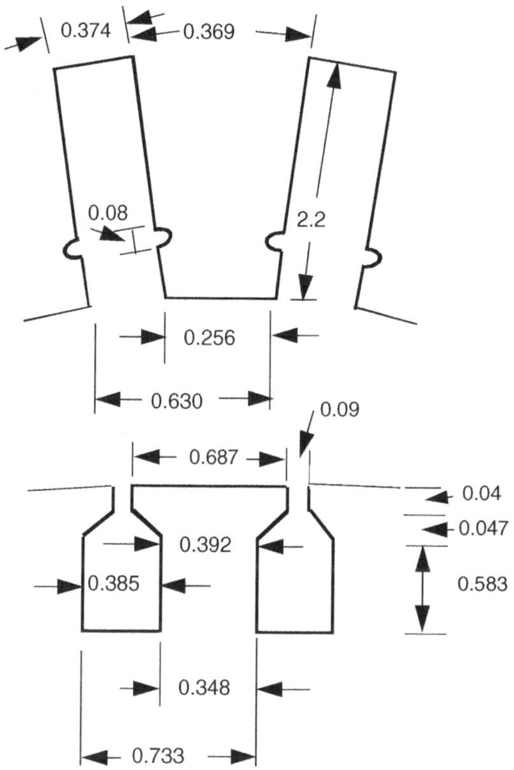

Figure 3.11 Details of the stator and rotor slot shapes for 250 HP induction machine.

Finally, the effective length of the stator (and rotor in this case) is obtained from Case 3 of Section 3.3. The relevant equation is, repeating

$$l_e = l_{is} + 2g + nl_0 \left(\frac{5}{5 + \frac{l_0}{g}} \right)^2$$

where $l_{0s} = l_{0r} = l_0$. From which

$$\begin{aligned} l_e &= 8.5 + 2(0.04) + (2)(0.375) \left(\frac{5}{5 + \frac{(0.375)}{0.04}} \right)^2 \\ &= 8.67'' \end{aligned}$$

One must now determine the MMF per pole necessary to produce a fundamental component of air gap flux density of 0.775 teslas (50,000 lines/in.^{2}). In this case the flux which travels straight down the slot will be neglected.

MMF and Reluctance of Air Gap

From equation (3.11), Carter's coefficient for the stator slots is

$$k_{cs} = \frac{0.63}{0.63 - 2\left(\frac{0.374}{\pi}\right)\left\{\text{a}\tan\left(\frac{0.374}{2(0.04)}\right) - \frac{0.04}{0.374}\text{In}\left[1 + \left(\frac{0.374}{2(0.04)}\right)^2\right]\right\}}$$

$$= 1.633$$

Carter's coefficient for the rotor slots is

$$k_{cr} = \frac{0.777}{0.777 - 0.0573(0.844 - 0.444(0.818))}$$

$$= 1.037$$

The effective air gap is therefore

$$g_e = k_{cs}k_{cr}g$$

$$= (1.633)(1.037)(0.04)$$

$$= 0.0677''$$

The amplitude of flux density at the point 30° away from the maximum is

$$B_{g1}(30°) = \frac{\sqrt{3}}{2}B_{g1}$$

$$= (0.866)(0.775) = 0.671 \text{ T}$$

The flux entering the stator tooth at this point is

$$\Phi_{ts} = B_{g1}(30°)\tau_s l_s$$

$$= 0.671(0.63)(8.67)\left(\frac{2.54}{100}\right)^2$$

$$= 2.37 \text{ mWb (milli-webers)}$$

The MMF necessary to overcome the air gap reluctance is therefore

$$\mathcal{F}_g(30°) = \frac{B_{g1}(30°)}{\mu_0}g_e$$

$$\mathcal{F}_g(30°) = \frac{(0.67)(0.0677)}{(4\pi 10^{-7})}\frac{2.54}{100}$$

$$= 918.5 \text{ A-t}$$

The air gap reluctance referred to one stator tooth is

$$\mathcal{R}_{gs} = \frac{g_e}{\mu_0 l_e \tau_s} = \frac{0.0677}{(\mu_0)(8.67)(0.63)}\left(\frac{100}{2.54}\right) = 388.4 \times 10^3 \text{ H}^{-1}$$

or referred to one rotor tooth is

$$\mathcal{R}_{gr} = \frac{g_e}{\mu_0 l_e \tau_r} = \frac{0.0677}{(\mu_0)(8.67)(0.777)}\left(\frac{100}{2.54}\right) = 314.8 \times 10^3 \text{ H}^{-1}$$

MMF and Reluctance of Stator Tooth and Slot

The tooth width midway down the stator tooth is

$$\begin{aligned} t_{ms} &= t_{ts} + \frac{1}{2}(t_{bs} - t_{ts}) \\ &= 0.256 + \frac{1}{2}(0.369 - 0.256) \\ &= 0.3125'' \end{aligned}$$

The flux density at the top of the stator tooth is

$$\begin{aligned} B_{top,s} &= B_{g1}(30°)\frac{\tau_s l_e}{t_{ts} k_{is} l_{is}} \\ &= \frac{(0.67)(0.630)(8.67)}{(0.2572)(0.93)(8.5)} \\ &= 1.81\ \text{T} \end{aligned}$$

Note that this result indicates that the teeth are highly saturated. The "dimples" in the stator teeth to accommodate the slot wedge will probably not appreciably affect the answer under such conditions. The possibility of this constriction producing premature saturation under reduced flux conditions should normally be examined carefully.

Since the flux density at the top of the tooth is known, the corresponding flux densities at the midpoint and at the root of the tooth are found from the following ratios

$$\begin{aligned} B_{mid,s} &= \frac{B_{top,s} t_{ts}}{t_{ms}} \\ &= \frac{(1.81)(0.256)}{(0.3125)} \\ &= 1.48\ \text{T} \end{aligned}$$

$$\begin{aligned} B_{root,s} &= \frac{B_{top,s} t_{ts}}{t_{bs}} \\ &= \frac{(1.81)(0.256)}{(0.369)} \\ &= 1.26\ \text{T} \end{aligned}$$

From the *B–H* characteristics of 29 Gauge, M27 steel, shown in Figure 1.16, the corresponding values of field intensity are determined to be

$$\begin{aligned} H_{top,s} &= 222\ \text{A-t/in.} \\ H_{mid,s} &= 21\ \text{A-t/in.} \\ H_{root,s} &= 6.9\ \text{A-t/in.} \end{aligned}$$

Hence the average field intensity along the stator tooth is

$$\begin{aligned} H_{ts(ave)} &= \frac{1}{6}H_{top,s} + \frac{2}{3}H_{mid,s} + \frac{1}{6}H_{root,s} \\ &= 52.15 \text{ A-t/in.} \end{aligned}$$

and the corresponding average MMF drop in the stator tooth is

$$\begin{aligned} \mathcal{F}_{ts(ave)} &= H_{ts(ave)} d_{ss} \\ &= (52.15)(2.2) \\ &= 114.7 \text{ A-t} \end{aligned}$$

The reluctance of the stator tooth is computed as

$$\mathcal{R}_{ts} = \frac{\mathcal{F}_{ts(ave)}}{\Phi_{ts}} = \frac{114.7}{0.00236}$$

$$\mathcal{R}_{ts} = 48.5 \times 10^3 \text{ H}^{-1}$$

It could be noted here that since the stator tooth is highly saturated it would be wise to recompute the tooth reluctance using more than three samples of the tooth MMF using the general expression, equation (3.20). This can easily be done in practice with a suitable computer algorithm.

MMF and Reluctance of Rotor Tooth and Slot

For simplicity, the rotor slot will be considered as an open slot having the constant width $b_o = 0.385$. The error caused by this assumption will be negligible, since it is expected that the flux density in the rotor tooth will be smallest at this point. This assumption should be checked later and if necessary a correction made for the MMF drop down the overhang section.

The tooth width at the midpoint of the tooth is

$$\begin{aligned} t_{mr} &= t_{tr} - \frac{1}{2}(t_{tr} - t_{rr}) \\ &= 0.392 - \frac{1}{2}(0.392 - 0.348) \\ &= 0.370'' \end{aligned}$$

Using the uncorrected value of air gap flux density, the flux density at the top of the rotor tooth is

$$\begin{aligned} B_{top,r} &= B_{g1}(30^\circ)\frac{\tau_r l_e}{t_{tr} k_{ir} l_{ir}} \\ &= (0.67)\frac{(0.777)(8.67)}{(0.392)(0.93)(8.5)} \\ &= 1.46 \text{ T} \end{aligned}$$

From this result

$$\begin{aligned} B_{\text{mid,r}} &= B_{\text{top,r}}\frac{t}{t_{\text{mr}}} \\ &= 1.46\frac{(0.392)}{0.370} \\ &= 1.55\ \text{T} \end{aligned}$$

$$\begin{aligned} B_{\text{root,r}} &= B_{\text{top,r}}\frac{t_{\text{tr}}}{t_{\text{rr}}} \\ &= 1.46\frac{(0.392)}{0.348} \\ &= 1.64\ \text{T}. \end{aligned}$$

Using the B–H curves for M27 steel the corresponding field intensities are

$$\begin{aligned} H_{\text{top,r}} &= 10.2\ \text{A-t/in.} \\ H_{\text{mid,r}} &= 12.4\ \text{A-t/in.} \\ H_{\text{root,r}} &= 50\ \text{A-t/in.} \end{aligned}$$

Hence,

$$\begin{aligned} H_{\text{tr(ave)}} &= \frac{1}{6}H_{\text{top,r}} + \frac{2}{3}H_{\text{mid,r}} + \frac{1}{6}H_{\text{root,r}} \\ &= 18.3\ \text{A-t/in.} \end{aligned}$$

Note that $H_{\text{top,r}}$ is relatively small so that its contribution to the average MMF drop is minor. For example, if H_{tr} is half as large (5.1 A-t/in.) then the corresponding $H_{\text{tr(ave)}}$ is 17.45 A-t/in., a difference of only 5%. The assumption that the rotor slot overhang can be neglected is, therefore, justified.

The MMF drop in the rotor teeth is computed from

$$\begin{aligned} \mathcal{F}_{\text{tr(ave)}} &= H_{\text{tr(ave)}}d_{\text{or}} \\ &= (18.3)(0.67) \\ &= 12.26\ \text{A-t} \end{aligned}$$

The flux in the rotor tooth is

$$\begin{aligned} \Phi_{\text{tr}} &= B_{\text{g1}}(30^\circ)\tau_{\text{r}}l_{\text{e}} \\ &= (0.671)(0.777)(8.67)\left(\frac{2.54}{100}\right)^2 \\ &= 2.92\ \text{mWb} \end{aligned}$$

The reluctance of the rotor tooth is

$$\begin{aligned} \mathcal{R}_{\text{tr}} &= \frac{\mathcal{F}_{\text{tr(ave)}}}{\Phi_{\text{tr}}} \\ &= \frac{12.26}{0.00292} \\ &= 4.20 \times 10^3\ \text{H}^{-1} \end{aligned}$$

The MMF drop across the tooth and half the gap is assumed to be equal to the slot MMF or

$$\begin{aligned}\mathcal{F}_{\mathrm{sr}} &= \mathcal{F}_{\mathrm{tr(ave)}} + \frac{1}{2}\mathcal{F}_{\mathrm{g}}(30°) \\ &= 12.26 + \frac{918.5}{2} \\ &= 471.5 \text{ A-t}\end{aligned}$$

MMF and Reluctance of Stator Core

The maximum flux in the stator core is obtained from equation (3.31) as

$$\begin{aligned}\Phi_{\mathrm{cs}} &= \left(\frac{2}{\pi}B_{\mathrm{g1}}\right)\left(\frac{\tau_{\mathrm{p}}}{2}l_{\mathrm{e}}\right) \\ &= \left(\frac{2}{\pi}0.775\right)\left(\frac{9.456}{2}\right)(8.67)\left(\frac{2.54}{100}\right)^2 \\ &= 13.0 \text{ mWb}\end{aligned}$$

The peak fundamental component of flux density in the core is therefore, from equation (3.32),

$$\begin{aligned}B_{\mathrm{cs}}(90°) &= \frac{\Phi_{cs}}{d_{\mathrm{cs}}k_{\mathrm{is}}l_{\mathrm{is}}} \\ &= \frac{0.013}{(1.76)(0.93)(8.5)}\left(\frac{100}{2.54}\right)^2 = 1.69 \text{ T}\end{aligned}$$

The corresponding values of core flux density at points 30° and 60° from the maximum are

$$\begin{aligned}B_{\mathrm{cs}}(60°) &= \frac{\sqrt{3}}{2}B_{\mathrm{cs}}(90°) \\ &= 1.47 \text{ T} \\ B_{\mathrm{cs}}(30°) &= \frac{1}{2}B_{\mathrm{cs}}(90°) \\ &= 0.85 \text{ T}\end{aligned}$$

From the *B–H* curves, the field intensities at the corresponding three points in the stator core are

$$\begin{aligned}H_{\mathrm{cs}}(90°) &= 73 \text{ A-t/in.} \\ H_{\mathrm{cs}}(60°) &= 10.2 \text{ A-t/in.} \\ H_{\mathrm{cs}}(30°) &= 1.5 \text{ A-t/in.}\end{aligned}$$

The average value of *H* over the stator core portion of the flux path is

$$\begin{aligned}H_{\mathrm{cs(ave)}} &= \frac{1}{6}H_{\mathrm{cs}}(90°) + \frac{2}{3}H_{\mathrm{cs}}(60°) + \frac{1}{6}H_{\mathrm{cs}}(30°) \\ &= 19.2 \text{ A-t/in.}\end{aligned}$$

The length of one pole pitch at the center line of the stator core can be computed from the equation

$$l_{cs} = \frac{\pi(D_{os} - d_{cs})}{P}$$
$$= \frac{\pi}{8}(32 - 1.51)$$
$$= 11.78''$$

The MMF drop over the stator core is therefore

$$\mathcal{F}_{cs(ave)} = H_{cs(ave)}\left(\frac{2}{3}l_{cs}\right)$$
$$= (19.2)\left(\frac{2}{3}\right)(11.78)$$
$$= 150.88 \text{ A-t}$$

The effective core reluctance of the flux tube corresponding to the tooth and slot in question must be equal

$$\mathcal{R}_{cs} = \frac{\mathcal{F}_{cs(ave)}}{\Phi_{cs}}$$
$$= \frac{150.88}{0.013}$$
$$= 11.56 \times 10^3 \text{ H}^{-1}$$

MMF and Reluctance of Rotor Core

Since leakage flux has been neglected, the flux per pole in the rotor core is the same as in the stator core, that is

$$\Phi_{cr} = 13.0 \text{ mWb}$$

The peak fundamental component of flux density in the core is

$$B_{cr}(90°) = \frac{\Phi_{cr}}{k_{ir} l_{ir} d_{cr}}$$
$$= \frac{0.013}{(0.93)(8.5)(2.58)}\left(\frac{100}{2.54}\right)^2$$
$$= 0.794 \text{ T}$$

Hence

$$B_{cr}(60°) = \frac{\sqrt{3}}{2} B_{cr}(90°)$$
$$= 0.690 \text{ T}$$
$$B_{cr}(30°) = \frac{1}{2} B_{cr}(90°)$$
$$= 0.397 \text{ T}$$

From the material *B–H* curves

$$H_{cr}(90°) = 1.4 \text{ A-t/in.}$$
$$H_{cr}(60°) = 1.0 \text{ A-t/in.}$$
$$H_{cr}(30°) = 0.7 \text{ A-t/in.}$$

from which one can obtain

$$H_{cr(ave)} = \frac{1}{6} H_{cr}(90°) + \frac{2}{3} H_{cr}(60°) + \frac{1}{6} H_{cr}(30°)$$
$$= 1.02 \text{ A-t/in.}$$

The length of one pole pitch at the center of the rotor core cross section is

$$l_{cr} = (D_{ir} + d_{cr}) \frac{\pi}{P}$$
$$= (19 + 2.08) \frac{\pi}{8}$$
$$= 8.28''$$

Hence

$$\mathcal{F}_{cr(ave)} = H_{cr(ave)} \left(\frac{2}{3} l_{cr} \right)$$
$$= (1.02) \left(\frac{2}{3} \right) (8.28) = 5.61 \text{ A-t}$$

and

$$\mathcal{R}_{cr} = \frac{\mathcal{F}_{cr(ave)}}{\Phi_{cr}}$$
$$= \left(\frac{33.2}{0.0131} \right) = 430.0 \text{ H}^{-1}$$

The total MMF drop around the magnetic circuits is obtained by summing up the individual MMF drop around the entire magnetic circuit comprised of two air gaps, two stator and rotor teeth, and the stator core and the rotor core. The required MMF per pole at the 30° point needed to produce the specified air gap flux density is

$$2\mathcal{F}_p(30°) = 2\mathcal{F}_g(30°) + 2\mathcal{F}_{tr} + 2\mathcal{F}_{ts} + \mathcal{F}_{cr} + \mathcal{F}_{cs}$$
$$= 2(918.5) + 2(12.26) + 2(114.73) + 5.61 + 150.88$$
$$\mathcal{F}_p(30°) = 1124 \text{ A-t}$$

The corresponding value of MMF per pole at the maximum value of air gap flux density of approximately 0.775 tesla (50 kilolines/in.^{2}) is

$$\mathcal{F}_{\mathrm{p}}(0^\circ) = \mathcal{F}_{\mathrm{p1}} = \frac{2}{\sqrt{3}}\mathcal{F}_{\mathrm{p}}(30^\circ)$$

$$= 1298 \text{ A-t}$$

3.8 MAGNETIC EQUIVALENT CIRCUIT

Figure 3.12 shows an equivalent circuit of the per pole magnetic circuit for the 250 HP example of Section 3.7. Note that since the slot pitches of the stator and rotor teeth are not equal, the reluctance for one-half the air gap differs on the two sides of the gap. Conceptual problems seem to arise since a different amount of flux "flows" in the stator and rotor sides of the equivalent circuit, since the width of the stator and rotor teeth are typically not identical.

This conceptual difficulty can be eliminated if the actual rotor slots and teeth are replaced by "equivalent" slots and teeth which produce the same MMF drop as

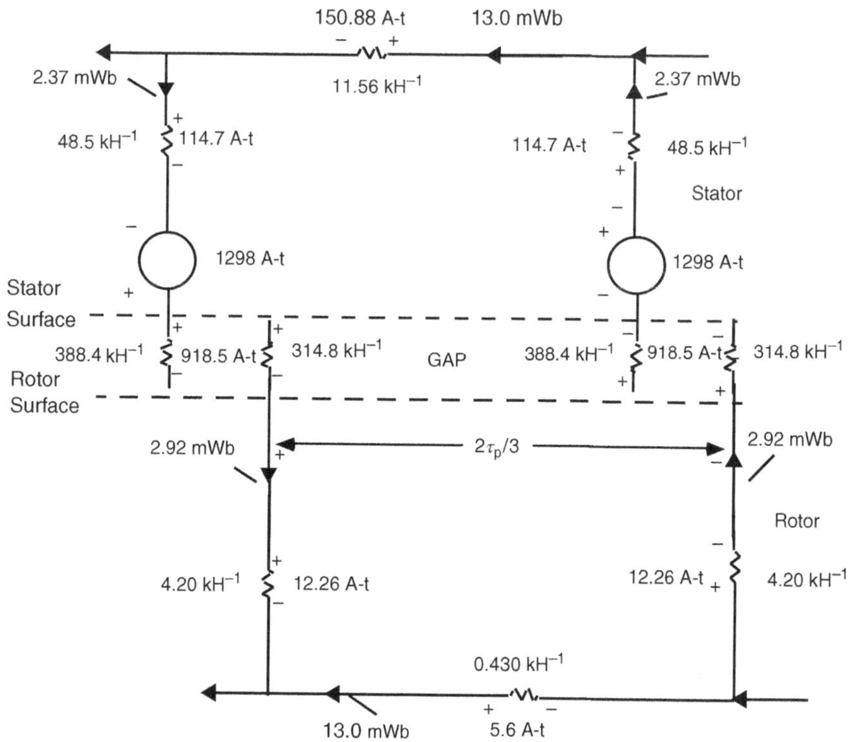

Figure 3.12 Magnetic equivalent circuit of one pole of 250 HP induction machine at no load for a peak fundamental flux density of 0.775 tesla, leakage fluxes ignored.

the actual rotor teeth but carry the same flux as the stator teeth. However, this degree of complexity for the purpose of achieving a simpler equivalent circuit is generally not warranted.

The concept of an equivalent circuit is potentially as powerful a concept in magnetic circuit analysis of electric machines as it traditionally has been in electric circuit analysis. However, the approach has not been widely adopted. Rather, hybrid solutions such as that outlined in Section 3.6 are more popular primarily due to the long history of this subject and the unavailability, until recently, of powerful digital computer solution algorithms. Clearly, the desired answer is best obtained if one simply establishes the reluctance of the four nonlinear elements of Figure 3.12 (stator core, rotor core, stator tooth, and rotor tooth) as a function of the flux through the reluctance and then turn the work over to a standard nonlinear algebraic equation solver. Although there is almost an urgent need to keep the complexity of the computations at a minimum with traditional methods, this is clearly not the case if an equivalent circuit solution on a digital computer is employed. Indeed, circuits of nearly any complexity can be set up with relative ease and problems encountered with indentations or overhangs in the teeth and core (so studiously avoided in Section 3.6) can be handled without much difficulty.

3.9 FLUX DISTRIBUTION IN HIGHLY SATURATED MACHINES

In general, the saturable portion of the magnetic circuit of an induction machine consists of the stator teeth, rotor teeth, stator core, and rotor core. Depending upon the design, any one of these elements or any combination of them can be placed in deep saturation. When a sinusoidal MMF is applied, the resultant air gap flux density may be either *flat-topped* as shown in Figure 3.10 or *peaked* depending upon whether the teeth or cores are more highly saturated. It is rather evident that saturations of the stator and rotor teeth are the cause of the flattening of the air gap flux density wave. Peaked waves, on the other hand can occur when the core saturates. If the core is saturated, the distribution of flux density along the circumference of the core is essentially constant. Since the flux in the cores is the spatial integral of the flux in the air gap, the derivative of the square wave produces a peaked (approaching a "pulse") flux density wave in the air gap of the machine. In most practical machines, however, the teeth are more saturated than the cores so that the air gap flux density wave is nearly always flat topped. In this case, the MMF drop due to the harmonic flux in the core can be neglected.

When the machine is highly saturated, the easiest way to calculate core ampere-turns is illustrated in Figure 3.13. If the flux density waveform in the core remains sinusoidal, the curve can be approximated by three straight lines as shown with terminal points 0°, 45°, 70°, and 90°. In each of the three sections, the flux density variation is assumed to be linearly related to the distance along the core. As in calculating the ampere-turns of a tapered tooth, the ampere-turns drop for each section can be determined by using Simpson's rule. Alternatively, one can determine the effective flux

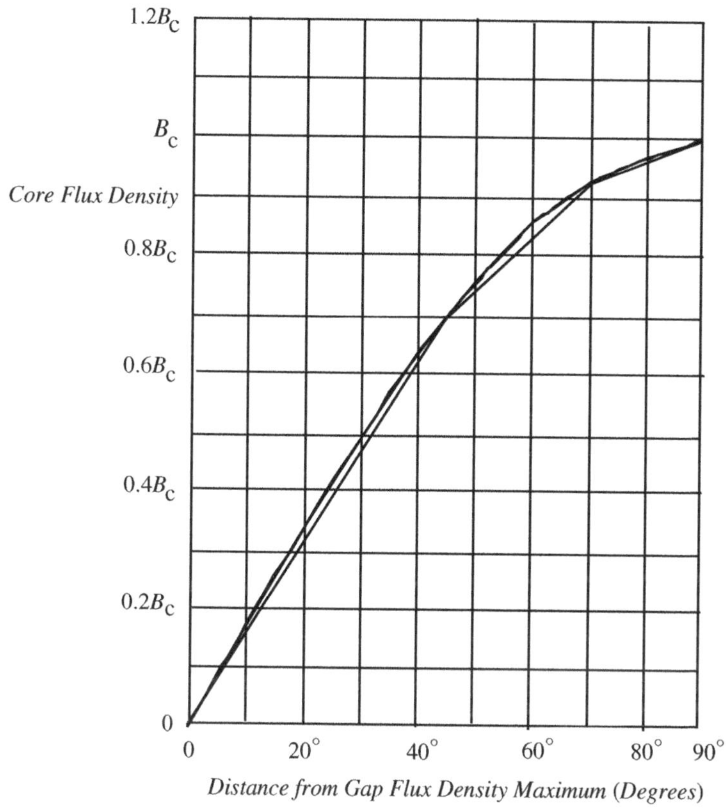

Figure 3.13 Stator or rotor core flux density distribution showing three segment linear approximation.

density one-third from the high-density end of the segment and then use the appropriate DC magnetization curve for the steel at this point to determine the equivalent H value. The three effective core densities in Figure 3.13 are B_c sin 30° = 0.5 B_c, $B_c$sin 61.67° = 0.88B_c, and $B_c$sin 83.3° = 0.993B_c, where B_c is an assumed maximum core flux density which is not sufficiently large so as to deeply saturate the core and make the sinusoidal distribution of core flux density assumption invalid. If H_{30} denotes the field intensity at the 30° point (1/3 of the way from the flux density at the 45° point) and $H_{61.7}$ and $H_{83.3}$ are the corresponding magnetic field intensities for the other two straight line segments, then the core ampere-turn drop over the interval from 0° to 45° is

$$\mathcal{F}_{c(0-45)} = \frac{1}{2}\left(\frac{l_c}{2}\right) H_{30} \tag{3.52}$$

Over the interval from 0° to 70°

$$\mathcal{F}_{c(0-70)} = \frac{1}{2}\left(\frac{l_c}{2}\right) H_{30} + \frac{5}{18}\left(\frac{l_c}{2}\right) H_{61.7} \tag{3.53}$$

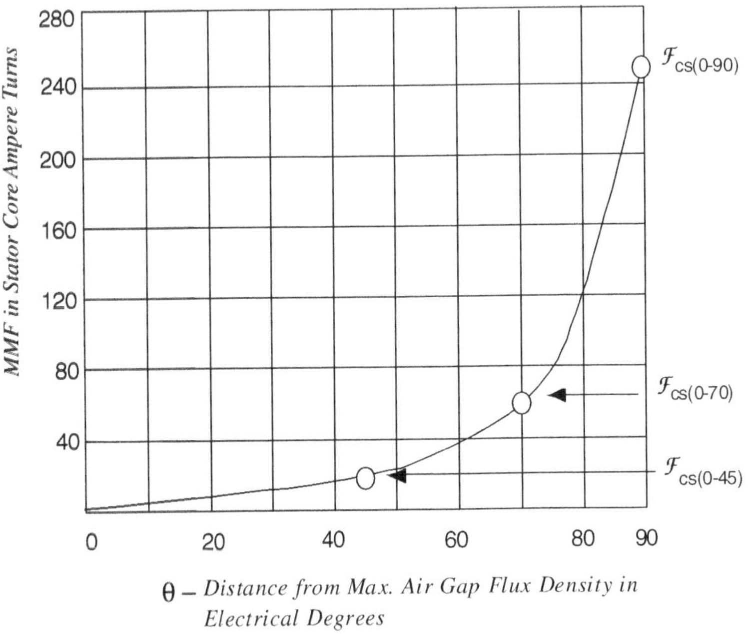

Figure 3.14 Typical plot of core ampere turns vs. angular position relative to maximum air gap flux density.

and over the interval from 0° to 90°

$$\mathcal{F}_{c(0-90)} = \frac{1}{2}\left(\frac{l_c}{2}\right)H_{30} + \frac{5}{18}\left(\frac{l_c}{2}\right)H_{61.7} + \frac{2}{9}\left(\frac{l_c}{2}\right)H_{83.3} \tag{3.54}$$

where l_c is the total length of the path of one pole pitch as measured at the midpoint of the core. The same procedure can be used to determine both the stator and the rotor core drops. A typical curve for the stator core MMF drop is plotted in Figure 3.14. It should be mentioned that this curve corresponds to Example #2 of Section 3.7, wherein these values are $\mathcal{F}_{cs(0-45)} = 9.72, \mathcal{F}_{cs(0-70)} = 61.7$, and $\mathcal{F}_{cs(0-90)} = 247.6$. A similar curve can be obtained for the rotor core MMF drop.

The MMF producing the air gap flux density at any location θ between 0° and 90° is then, from Ampere's law,

$$\mathcal{F}_{gt}(\theta) = \mathcal{F}_{s1}\cos\theta - \mathcal{F}_{cs(\theta-90°)}(\theta) - \mathcal{F}_{cr(\theta-90°)}(\theta) \tag{3.55}$$

where $\mathcal{F}_{cs}(\theta)$ and $\mathcal{F}_{cr}(\theta)$ are obtained from Figure 3.14, $\mathcal{F}_{gt}(\theta)$ is the total MMF drop in the stator tooth, rotor tooth, and air gap at the point corresponding to θ and $\mathcal{F}_{s1}$ is defined by equation (3.1).

In general, while the flux density in the stator and rotor cores can be considered as nearly sinusoidal, the saturation of the teeth make the air gap flux usually flat-topped (unless the core is more saturated rather than the teeth in which case the gap flux density is peaked). Including only the possibility of the first, third, fifth, and seventh harmonics, the flux density in the gap can be expressed as

$$B_g(\theta_e) = B_{g1}\cos\theta_e + B_{g3}\cos 3\theta_e + B_{g5}\cos 5\theta_e + B_{g7}\cos 7\theta_e \tag{3.56}$$

When $\theta_e = 0°$, equation (3.56) becomes

$$B_g(0) = B_{g1} + B_{g3} + B_{g5} + B_{g7}$$

When $\theta_e = 30°$,

$$B_g(30°) = \frac{\sqrt{3}}{2}B_{g1} + 0.B_{g3} - \frac{\sqrt{3}}{2}B_{g5} - \frac{\sqrt{3}}{2}B_{g7}$$

and when $\theta_e = 60°$,

$$B_g(60°) = \frac{1}{2}B_{g1} - B_{g3} + \frac{1}{2}B_{g5} + \frac{1}{2}B_{g7}$$

These three equations can now be solved for B_{g1} as

$$B_{g1} = \frac{1}{3}\left[B_g(0) + \sqrt{3}B_g(30°) + B_g(60°)\right] \tag{3.57}$$

An iterative process can now be used to determine the solution.

- First a value of B_{mid} is selected for the stator or rotor tooth. Choose one that is expected to be the most heavily saturated. The corresponding value of $B_{g,ave}$ can be determined from equations (3.17) and (3.19). As a first guess, one could select for example $B_{g1} = B_{g,ave}$.
- The corresponding values of B_{cs} and B_{cr} are obtained from equations (3.38) and (3.39) assuming sinusoidal air gap flux.
- Since the harmonics in the air gap are assumed to not affect the MMF drop in the cores, the MMF drops in the stator and rotor cores are thereby fixed and curves similar to Figure 3.14 can be established.
- The MMF per pole is then calculated from equation (3.55) by adding the MMF necessary to carry the fundamental component of flux in the cores plus the MMF needed to drive the gap flux corresponding to $B_{g,max}$ through the gap plus the stator and rotor teeth, whereupon,

$$\mathcal{F}_{s1} = \mathcal{F}_{gt}(0) + \mathcal{F}_{cs(0-90)} + \mathcal{F}_{cr(0-90)} \tag{3.58}$$

- The flux densities at the 30 and 60 degree points are then determined by first calculating the MMF available at these points to establish the air gap flux. For example,

$$\mathcal{F}_{gt}(30°) = \mathcal{F}_{s1}\cos(30°) - \mathcal{F}_{cs(30-90)} - \mathcal{F}_{cr(30-90)} \tag{3.59}$$

  where $\mathcal{F}_{cs}$ and $\mathcal{F}_{cr}$ are the core MMF drops obtained from curves similar Figure 3.14 as described in the text following equation (3.55).
- The corresponding air gap flux densities $B_g(30°)$ and $B_g(60°)$ are then calculated from $\mathcal{F}_{gt}(30°)$ and $\mathcal{F}_{gt}(60°)$.

- Equation (3.57) is then solved and compared with the guessed value of B_{g1}. If the result does not match, a new factor relating $B_{g,max}$ and B_{g1} is guessed, for example, $B_{g,max} = 0.85\ B_{g1}$. The iteration continues until a match is obtained within 1%. Usually, only three or four iterations will do sufficiently well.
- Having obtained the correct value of $B_g(\theta_e)$, the saturation harmonics at no load can now be computed as follows,

$$B_{g3} = -\frac{1}{3}\left[2B_g(60°) - B_g(0°)\right] \tag{3.60}$$

$$B_{g5} = \frac{1}{5}\left[2B_g(72°) - 2B_g(36°) + B_g(0°)\right] \tag{3.61}$$

$$B_{g7} = -\frac{1}{7}\left[2B_g(77.1°) - 2B_g(51.4°) + 2B_g(25.7°) - B_g(0)\right] \tag{3.62}$$

If the saturation harmonics are sufficiently large, they could affect the core flux density making the initial assumption of a sinusoidally distributed core flux density invalid. In this case, a new curve of core flux versus position along the stator and/or rotor core periphery, similar to Figure 3.13 can be determined. This new curve will have a somewhat more "peaked" rather than a "flattened" shape compared to a sine wave. That is, the peak core flux density will be slightly greater than the initial value assumed at the beginning of the calculation procedure. The process of calculating the core MMF drop can now be repeated with the new function of core flux density and iteration used until convergence is obtained.

While the determination of these "saturation harmonics" are not necessary to achieve the fundamental task, that of calculating the magnetizing reactance, they are important contributors to stray losses a subject which will be discussed in Chapter 5.

3.10 CALCULATION OF MAGNETIZING REACTANCE

Although a solution for the main magnetic circuit of the induction machine has essentially been completed, it is an interesting fact that the exact distribution of windings in the stator slots has yet to be defined and therefore it is not yet possible to specify the voltage and current supplied to the machine. Assume now, however, that the winding distribution in the stator slots have been specified so that the fundamental component of impressed MMF is known. That is, at the center line for each of the P poles of the machine, equation (3.1) prevails for a three-phase winding, and

$$\mathcal{F}_{s1} = \left(\frac{3}{2}\right)\left(\frac{4}{\pi}\right)\left(\frac{k_1 N_t}{CP}\right) I_s \tag{3.63}$$

where

$$k_1 = k_{p1} k_{d1} k_{\chi 1} k_{s1} \tag{3.64}$$

Since the MMF $\mathcal{F}_{s1}$ is assumed known and the winding distribution fixes the pitch, distribution, and skew factors as well as the total turns, poles, and circuits, the corresponding maximum current per phase is readily computed.

The remaining task is to relate the air gap flux density in the gap to the flux linking the winding and thereby to the terminal voltage. In general, the MMF can be considered as sinusoidal so that

$$\mathcal{F}_s(\theta) = \mathcal{F}_{s1} \cos\left(\frac{P\theta}{2}\right) \tag{3.65}$$

where $\theta_e = (P/2)\theta$ is the circumferential angle in electrical degrees measured along the stator periphery and represents the distance away from the maximum defined by Figure 3.10.

Recall from Chapter 2 that the net stator MMF distribution is made up of three MMF distributions, one for each of the three stator phases a, b, and c which are physically displaced around the air gap by 120 electrical degrees. When harmonics are neglected, these three-phase MMFs are also sinusoidal. For purposes of convenience, let the magnetic axis of phase a coincide with the positive peak value of MMF $\mathcal{F}_p$. Then the phase *a* current is also a maximum and the corresponding MMF distribution of phase *a* is, from equation (2.58)

$$\mathcal{F}_a(\theta) = \frac{4}{\pi}\left(\frac{k_1 N_t}{CP}\right) I_s \cos\left(\frac{P\theta}{2}\right) \tag{3.66}$$

where I_s/C is the amplitude of the current in each of the C circuits of phase *a*.

It has also been established that if the MMF impressed by all three-phase currents is given by equation (3.63) then the flux density has the identical variation in the air gap or,

$$B_g(\theta) = B_{g1} \cos\left(\frac{P\theta}{2}\right) \tag{3.67}$$

The phase a turns linked by the flux density $B_g(\theta)$ over one pole of the machine inherently takes into account the fact that the machine is excited by three-phase currents and therefore includes the mutual coupling between phases. The calculation must also take into account the partial linkages caused by the sinusoidal winding distribution. For this purpose, the magnetizing inductance is most easily calculated by first computing the magnetic energy stored in the gap of one pole corresponding to the air gap flux linking one phase (flux linkage per pole per phase). In terms of the gap magnetic flux density and magnetic field of phase a, per pole

$$W_{mp} = \frac{1}{2} \int_{\text{one pole}} \boldsymbol{B}_g \bullet \boldsymbol{H}_a dV \tag{3.68}$$

Upon specifying that $\boldsymbol{B}_g$ and $\boldsymbol{H}_a$ are radially directed, taking the dot product and noting that the field intensity of one of the three phases is

$$H_a = \left(\frac{2}{3}\right) \frac{\mathcal{F}_{s1}}{g} \cos\left(\frac{P\theta}{2}\right)$$

then, since B_g and H_a are only functions of θ, the integral reduces to

$$W_{\mathrm{mp}} = \left(\frac{1}{2}\right) B_{\mathrm{g1}} \left(\frac{4}{\pi}\right) \left(\frac{k_1 N_{\mathrm{t}} I_{\mathrm{s}}}{CP}\right) \left(\frac{D_{\mathrm{is}}}{2}\right) l_{\mathrm{e}} \int_{-\frac{\pi}{P}}^{\frac{\pi}{P}} \cos^2\left(\frac{P\theta}{2}\right) d\theta \tag{3.69}$$

Equation (3.69) works out to,

$$W_{\mathrm{mp}} = B_{\mathrm{g1}} \frac{k_1 N_{\mathrm{t}} I_{\mathrm{s}}}{CP^2} D_{\mathrm{is}} l_{\mathrm{e}} \tag{3.70}$$

which can also be written as

$$W_{\mathrm{mp}} = \frac{1}{2} \left(\frac{k_1 N_{\mathrm{t}}}{P}\right) \left(\frac{I_{\mathrm{s}}}{C}\right) \left(\frac{2}{\pi} B_{\mathrm{g1}}\right) (\tau_{\mathrm{p}} l_{\mathrm{e}}) \tag{3.71}$$

where the pole pitch

$$\tau_{\mathrm{p}} = \frac{\pi D_{\mathrm{is}}}{P}$$

The first two product terms in equation (3.71) can be recognized as the MMF per pole while the second two terms correspond to the average value of flux density per pole times the cross-sectional area of one pole through which this flux density passes, that is, the flux per pole.

Since I_s/C is equivalent to the current in the set of coils making up one of the poles, the magnetic energy stored per pole can also be written as,

$$W_{\mathrm{mp}} = \frac{1}{2} \lambda_{\mathrm{p}} \left(\frac{I_{\mathrm{s}}}{C}\right) \tag{3.72}$$

Thus, solving for the flux linkage per pole

$$\lambda_{\mathrm{p}} = \left(\frac{k_1 N_{\mathrm{t}}}{P}\right) \left(\frac{2}{\pi} B_{\mathrm{g1}}\right) (\tau_{\mathrm{p}} l_{\mathrm{e}}) \tag{3.73}$$

For balanced windings, the flux linking each of the individual poles is identical. The total flux linkages are found by multiplying the flux linkages per pole by the number of poles connected in series, that is

$$\lambda_{\mathrm{ms}} = \left(\frac{P}{C}\right) \lambda_{\mathrm{p}} = k_1 \left(\frac{N_{\mathrm{t}}}{C}\right) \left(\frac{2}{\pi} B_{\mathrm{g1}}\right) (\tau_{\mathrm{p}} l_{\mathrm{e}}) \tag{3.74}$$

$$= k_1 N_{\mathrm{s}} \left(\frac{2}{\pi} B_{\mathrm{g1}}\right) (\tau_{\mathrm{p}} l_{\mathrm{e}}) \tag{3.75}$$

where N_s is now defied as the number of series connected turns per phase.

The magnetizing inductance per phase is, by definition, found by taking the ratio of the total air gap flux linkages per winding phase to the current in the phase. Hence, by use of equations (3.75) and (3.63), the magnetizing inductance is, in terms N_s, the number of series connected turns

$$L_{ms} = \frac{\lambda_{ms}}{I_s} = \left(\frac{3}{2}\right)\left(\frac{4}{\pi}\right)\frac{k_1^2 N_s^2}{P}(\tau_p l_e)\left(\frac{\frac{2B_{g1}}{\pi}}{\mathcal{F}_{s1}}\right) \tag{3.76}$$

Note that the "3/2" and "4/π" terms are consistent with the definition of the MMF of a three-phase system, (equation 2.73). The quantity "$2B_{g1}/\pi$" corresponds to the average value of a sinusoidal distribution of flux density over one pole having amplitude B_{g1}.

For purposes of analysis, it is frequently convenient to define another equivalent gap g_e which now takes into account the MMF drops in the iron portion of the magnetic circuit as well as the slot openings. If all of the MMF drop is now assumed to occur over this equivalent gap length, then from Ampere's law

$$\frac{B_{g1}}{\mu_0} g_e = \mathcal{F}_{s1} \tag{3.77}$$

Hence,

$$\frac{B_{g1}}{\mathcal{F}_{s1}} = \frac{\mu_0}{g_e} \tag{3.78}$$

where now

$$g_e = k_{sat} k_c g \tag{3.79}$$

Substituting this result into equation (3.76)

$$L_{ms} = \left(\frac{3}{2}\right)\left(\frac{8}{\pi^2}\right)\frac{k_1^2 N_s^2}{P}\mu_0 \frac{\tau_p l_e}{g_e} \tag{3.80}$$

Alternatively

$$L_{ms} = \left(\frac{12}{\pi^2}\right)\frac{k_1^2 N_s^2}{P}\mathcal{P}_p \tag{3.81}$$

where $\mathcal{P}_p$ is the effective permeance of one pole given by

$$\mathcal{P}_p = \mu_0 \frac{\tau_p l_e}{g_e} \tag{3.82}$$

Equation (3.80) has many alternative forms. In terms of the diameter D_{is}, one can also write L_{ms} as

$$L_{ms} = \frac{12}{\pi}\left(\frac{k_1 N_s}{P}\right)^2 \mu_0 \frac{D_{is} l_e}{g_e} \tag{3.83}$$

The fundamental component (effective number) of series connected turns is

$$N_{se} = \frac{4}{\pi} k_1 \frac{N_t}{C} = \frac{4}{\pi} k_1 N_s \tag{3.84}$$

then

$$L_{ms} = \frac{3\pi}{4} \left(\frac{N_{se}}{P} \right)^2 \mu_0 \frac{D_{is} l_e}{g_e} \tag{3.85}$$

Or, finally, in terms of the radius r measured along the surface of the gap,

$$L_{ms} = \frac{3}{2} \left(\frac{N_{se}}{P} \right)^2 \mu_0 (\pi) \frac{r l_e}{g_e} \tag{3.86}$$

This equation is the form typically found during the derivation of machine d–q parameters in texts concerned with the terminal behavior of an AC machine [5]. It is useful to pause here to note that the magnetizing inductance is independent of the number of slots and inversely proportional to the square of the number of poles. Hence, the magnetizing inductance decreases rapidly with pole number, making the magnetizing current of an induction machine become excessive when the pole number becomes too large.

3.11 EXAMPLE 3—CALCULATION OF MAGNETIZING INDUCTANCE

The 250 HP induction machine in Example 2 has the following additional information supplied for the stator winding

Phase connection	Y
Circuits	1
Coil pitch in slots	12
Coil sides per phase belt	5
Coil sides/slot	2
Conductors/coil side	6

From the previous example, it is known that the total number of stator slots is 120 and that there are 8 poles. Hence the number of stator slots per pole is

$$\text{Stator Slots/Pole} = S_1/P = 120/8 = 15$$

The pitch of the winding is defined as

$$Pitch = W/\tau_p$$

However, this ratio is more easily computed by simply taking the ratio of slots corresponding to the coil pitch and pole pitch, that is,

$$Pitch = Slots\ per\ Coil\ Pitch/Slots\ per\ Pole\ Pitch = 12/15 = 0.8$$

From this result, one obtains from equation (2.19), the pitch factor for the fundamental component of MMF

$$k_{p1} = \sin\left(\frac{\pi}{2}\frac{W}{\tau_p}\right)$$
$$= \sin(0.8\pi)/2 = 0.951$$

Since five of the fifteen slots per pole are occupied by coil sides of the same phase winding ($q = 5$), the phase belt in per unit of a pole pitch Z/τ_p can also be computed by taking the appropriate ratio of slots,

$$Z/\tau_p = (\textit{Slots per Phase belt})/(\textit{Slots per Pole}) = 5/15 = 1/3$$

Hence, from equation (2.29) or Table 2.1 the distribution factor is

$$k_{d1} = \frac{1}{q}\frac{\sin\left(\frac{Z}{\tau_p}(\pi/2)\right)}{\sin\left(\frac{Z}{\tau_p}\frac{\pi}{2q}\right)} = \frac{1}{5}\frac{\sin(\pi/6)}{\sin(\pi/30)}$$
$$= \left(\frac{1}{5}\right)\frac{0.5}{0.1045}$$
$$= 0.9567$$

The slot opening factor is determined from

$$k_{\chi 1} = \frac{\sin(\chi/2)}{\chi/2}$$

where

$$\chi = \frac{b_{os}}{\tau_p}180 = \frac{0.374}{9.456}180 = 7.12^\circ$$

whereupon

$$k_{\chi 1} = \frac{\sin(3.56^\circ)}{3.56 \cdot \frac{\pi}{180}} = 0.9993$$

The effects of the slot opening on the fundamental component of MMF is clearly negligible except for open slot machines with a small number of slots per pole per phase and is frequently neglected in calculations of this type.

Since the skew of the stator winding is not supplied it is assumed unskewed. Therefore,

$$k_{s1} = 1.0$$

The overall winding factor for the stator winding of the machine is

$$k_1 = k_{p1}\, k_{d1}\, k_{x1}\, k_{s1}$$
$$= (0.951)\,(0.9567)(0.9993)(1.0) = 0.91$$

The total number of series connected turns per phase can be found from the following.

$$\begin{aligned}\text{Turns/Phase/Circuit} &= \text{(Slots/Phasebelt/Phase)(Turns/Coilside)(Coilsides/Slot)}\\ &\quad \times\text{(Poles)/(Circuits)}\\ &= (5)(3)(2)(8)/(1)\\ N_t/C = N_s &= 240\end{aligned}$$

Recall from Example 2 that the MMF needed to obtain an air gap flux density of 0.775 Tesla is 1495 A-t. The corresponding maximum current in each phase corresponding to this value is, from equation (3.63)

$$\begin{aligned}I_s &= \frac{\mathcal{F}_{p1}}{\left(\frac{3}{2}\right)\left(\frac{4}{\pi}\right)\left(\frac{k_1 N_s}{P}\right)}\\ &= \frac{1495}{\left(\frac{3}{2}\right)\left(\frac{4}{\pi}\right)\left(\frac{(0.91)(240)}{8}\right)} \qquad (3.87)\\ &= 28.67 \quad \text{A pk}\end{aligned}$$

The magnetizing inductance per phase can be found conveniently from equation (3.76) as

$$\begin{aligned}L_{ms} &= \left(\frac{3}{2}\right)\left(\frac{4}{\pi}\right)\frac{k_1^2 N_s^2}{P}\frac{\left(\frac{2}{\pi}B_{g1}\right)(\tau_p l_e)}{\mathcal{F}_{s1}}\\ &= \left(\frac{3}{2}\right)\left(\frac{4}{\pi}\right)\frac{(0.91)^2(240)^2}{(8)}\frac{\left(\frac{2}{\pi}(0.775)(9.456)(8.67)\right)}{1495} x \left(\frac{2.54}{100}\right)^2\\ &= 0.199 \text{ H}\end{aligned}$$

It was mentioned earlier that the actual gap g can be replaced by an equivalent gap which incorporates both the fringing of the flux due to the stator and rotor teeth and ducts and also allows for the saturation of the teeth and core. For the sample problem, this value can be obtained from equation (3.78) as

$$\begin{aligned}g_e &= \frac{\mu_0 \mathcal{F}_{s1}}{B_{g1}}\\ &= \frac{(\mu_0)(1495)}{0.775}\left(\frac{100}{2.54}\right)\\ &= 0.0954\end{aligned}$$

Note that this equivalent gap is more than twice as large as the actual gap of 0.04 in.

It should now be recalled that the computation of magnetizing inductance was launched by assuming an arbitrary value of air gap flux density of 0.775 tesla (50,000 lines/in.^{2}). In practice, this value is near the upper limit of air gap flux density for this machine and it is of interest to investigate the corresponding value of voltage drop in the per phase equivalent circuit. If the machine is excited by 60 Hz, the voltage at the air gap can be readily calculated as

$$\begin{aligned} V_{\mathrm{m}} &= \omega_{\mathrm{e}} L_{\mathrm{ms}} I_{\mathrm{m}} \\ &= (377)(0.199)(28.67) \\ &= 2151 \text{ V peak} \end{aligned}$$

Neglecting leakage inductance, the corresponding RMS line-to-line voltage is approximately

$$V_{\mathrm{l-l}} \approx \frac{\sqrt{3}}{\sqrt{2}} V_{\mathrm{m}} = 2630 \text{ V} \cdot \text{rms}$$

Hence, the particular case that has been studied corresponds to a 10% overexcited condition since the rated voltage of the machine was stated to be 2400 V.

Often, an accurate calculation of the magnetic circuit is desired only for the rated voltage condition. In this case, the approximate flux density in the gap can be calculated from equation (3.74) as

$$B_{g1} = \frac{\frac{\pi}{2}\lambda_{\mathrm{p}}}{\left(\frac{k_1 N_{\mathrm{t}}}{P}\right)(\tau_{\mathrm{p}} l_{\mathrm{e}})} \tag{3.88}$$

where

$$\lambda_{\mathrm{p}} = \left(\frac{C}{P}\right)\left(\frac{V_{\mathrm{l-l}}}{\omega_{\mathrm{e}}}\right)\frac{\sqrt{2}}{\sqrt{3}} \tag{3.89}$$

Hence

$$B_{g1} = \frac{\frac{\pi}{2}\sqrt{\frac{2}{3}}\frac{V_{\mathrm{l-l}}}{\omega_{\mathrm{e}}}}{(k_1 N_{\mathrm{s}})(\tau_{\mathrm{p}} l_{\mathrm{e}})} \tag{3.90}$$

In the case of the example 250 HP machine, the gap flux density at rated line-to-line voltage is

$$\begin{aligned} B_{g1} &= \frac{\frac{\frac{\pi}{2}\sqrt{\frac{2}{3}}2400}{377}}{(0.91)(240)(9.456)(8.67)} x \left(\frac{100}{2.54}\right)^2 \\ &= 0.707 \text{ tesla} \end{aligned}$$

rather than the value of 0.775 tesla assumed in Section 3.7.

3.12 CONCLUSION

This chapter has served to present one of the two key elements in the design of an electrical machine, namely the determination of the magnetizing flux paths in the machine. The magnetizing flux, interacting with the induced rotor current, determines, in essence, the developed torque in an induction machine and also plays a key role in the torque produced within any type of electrical machine.

REFERENCES

[1] M. Liwschitz-Garik and C. C. Whipple, *A-C Machines*, 2nd edition, D.Van Nostrand Co. Inc., Princeton, 1961.
[2] C. H. Lee, "Saturation Harmonics of Polyphase Induction Machines," *AIEE Transactions*, October 1961, pp. 597–603.
[3] B. J. Chalmers and R. Dodgson, "Waveshapes of flux density in polyphase induction motors under saturated conditions," *IEEE Transactions on Power Apparatus and Systems*, vol. PAS-90, no. 2, March/April 1971, pp. 564–569.
[4] C. G. Veinott, *Theory and Design of Small Induction Motors*, McGraw-Hill, New York, 1959.
[5] D. W. Novotny and T. A. Lipo, *Vector Control and Dynamics of AC Drives*, Oxford University Press, New York, 1996.

CHAPTER 4

USE OF MAGNETIC CIRCUITS IN LEAKAGE REACTANCE CALCULATIONS

IT IS IMPORTANT TO RECALL THAT torque production in most electrical machines involves the interaction of the air gap flux with the rotor MMF. The theoretical underpinnings to calculate the air gap flux and associated air gap inductance for practical machine geometries have now been obtained. To a first order of approximation, the air gap flux is a function of the applied voltage and, hence, does not vary widely over a change in mechanical load. The second quantity, namely the rotor MMF, varies with the rotor (and therefore usually stator) current and hence has a much more dominant role in torque production. The stator and rotor currents, are, in turn, primarily a function of the leakage reactances of the machine. In particular, key performance characteristics such as breakdown torque, starting torque, and inrush current are nearly independent of magnetizing inductance but depend critically upon the leakage inductances. In addition, the key electromagnetic time constants are dependent almost solely upon the leakage inductances (together with the resistances) of the machine. Since the leakage inductance arises from many factors, it is a more complicated issue than magnetizing inductance. This complex but fascinating aspect of motor design is the subject of this chapter.

4.1 COMPONENTS OF LEAKAGE FLUX IN INDUCTION MACHINES

In general, five principle components of leakage flux can be identified for induction machines.

1. Slot Leakage Flux. Slot leakage flux is shown in Figure 4.1. Note that this component of flux crosses the slot from one tooth to the next, linking with that portion of the conductors below it by returning through the core portion of the punching.

2. End-Winding Leakage Flux. The overhang or end-winding portion of a winding produces a distinctly different component of leakage flux whose magnetic

Introduction to AC Machine Design, First Edition. Thomas A. Lipo.

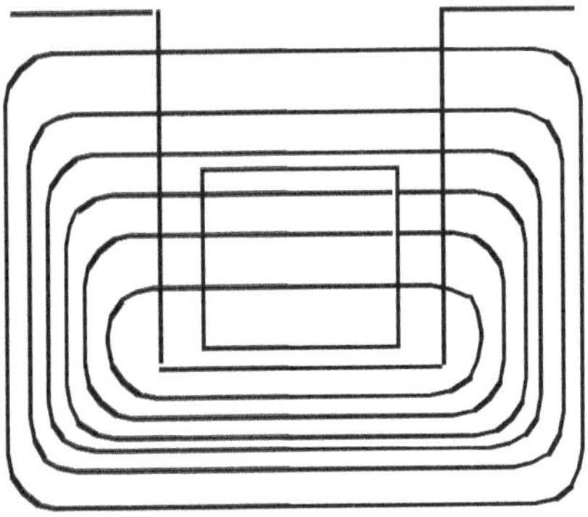

Figure 4.1 Slot leakage flux in an isolated slot.

circuit is almost entirely in air. Nonetheless, this component forms an appreciable portion of the leakage inductance since 55–75% of the copper of a two pole stator winding is located in the end-winding region. Corresponding values for 4-, 6-, and 8-pole machines are 49–68%, 43–60%, and 39–52%, respectively. The complexity of the flux paths associated with practical winding arrangements calls for very complex mathematical models. As a result, empirical correction factors are sometimes employed for better accuracy.

3. Harmonic or Belt Leakage Flux. This flux is due to the fact that the primary and secondary MMF distributions are not, in general, the same. Since the corresponding harmonic fluxes do not couple both stator and rotor equally, the net result is that they produce a form of leakage flux.

4. Zigzag Leakage Flux. This component of leakage flux passes across the air gap from one tooth to another in a zigzag fashion as shown in Figure 4.2. It can be observed that the magnitude of this flux depends upon the length of the air gap and the relative instantaneous positions of the tooth tips.

5. Skew Leakage Flux. The skew leakage flux is present only when the stator or rotor windings are skewed. Skewing, in essence, results in a decrease in the amount of air gap flux coupling both the stator and rotor windings for a given amplitude of air gap flux. The result has the same effect as an increase in stator leakage flux and therefore is regarded as a type of leakage.

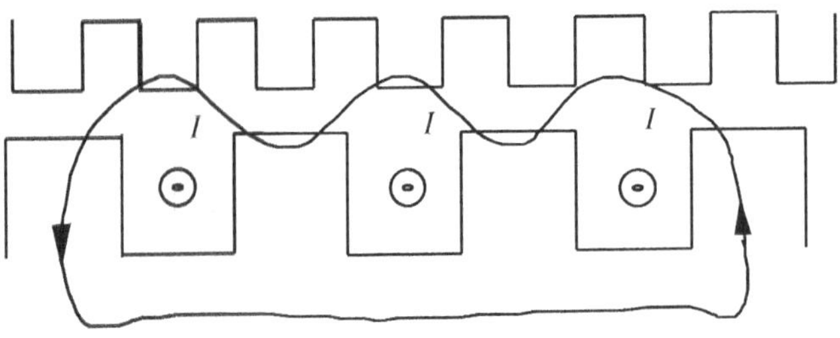

Figure 4.2 Zigzag leakage over one-phase belt.

4.2 SPECIFIC PERMEANCE

The leakage flux distribution in a practical machine is clearly a three-dimensional problem. Nonetheless, by taking a cross section of the machine sufficiently remote from the ends, it is possible to reduce the problem to a two-dimensional one. The flux which is forced across a slot is driven by the current which flows in the slot itself. By Ampere's law, the MMF supplied by the current in the slot is consumed by an MMF drop in the iron plus the drop in air across the slot. Since the slot width is extremely large, many times larger than the air gap of the machine, the percent MMF drop in the iron is a small percentage of the drop even if the tooth is entering saturation. Hence, to a very good approximation, one can frequently neglect the iron drop in a rough calculation. The permeance of the flux paths can now be found by summing up all tubes of flux of cross-sectional area $l_e dx$, where l_e is the effective length of the machine for stator slot leakage flux and includes the "bowing out" of the slot leakage fluxes near stator-based air ducts and at the end of the machine.

As an illustration of the computation of slot leakage flux, consider a flux line passing through an arbitrary point in the conductor as shown in Figure 4.3. Neglecting the MMF drop in the iron part of the path, Ampere's law states

$$\int_{1}^{2} \boldsymbol{H} \bullet \boldsymbol{dl} = n(x)I \tag{4.1}$$

where the points 1 and 2 are marked in Figure 4.3 and $n(x)$ is the partial number of conductors enclosed by the flux line. If H is constant along the path 1–2

$$H_y = n(x)I \tag{4.2}$$

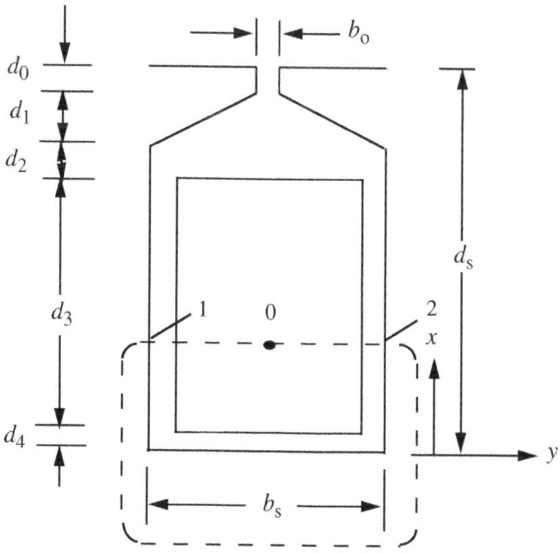

Figure 4.3 Determination of slot leakage permeance.

or

$$\frac{B_y y}{\mu_0} = n(x) I \tag{4.3}$$

The differential flux passing through a differential area $l_e dx$ centered about the path 1–2 is

$$d\Phi = B_y l_e dx = \mu_0 \frac{n(x) I l_e dx}{y} \tag{4.4}$$

The flux linkages enclosed by the flux $d\Phi$ is

$$d\lambda = n(x)\, d\Phi = \mu_0 n(x)^2 \frac{I l_e dx}{y} \tag{4.5}$$

$$d\lambda = \mu_0 n_s^2 I \left[\frac{n(x)}{n_s}\right]^2 \frac{l_e dx}{y} \tag{4.6}$$

where n_s is the total number of conductors or "turns" in the slot. The differential inductance associated with the flux path is

$$dL = \frac{d\lambda}{I} = \mu_0 n_s^2 \left[\frac{n(x)}{n_s}\right]^2 \frac{l_e dx}{y} \tag{4.7}$$

Since by definition

$$dL = n_s^2 d\mathcal{P} \tag{4.8}$$

The effective differential permeance over the path 1–2 is

$$d\mathcal{P} = \mu_0 \left[\frac{n(x)}{n_s}\right]^2 \frac{l_e dx}{y} \tag{4.9}$$

It can be observed that throughout this analysis the effective length l_e has been carried along simply as a constant. For this reason, it is convenient to define a two-dimensional permeance, or effective *specific permeance*, obtained by dividing equation (4.9) by the effective length, or

$$dp = \mu_0 \left[\frac{n(x)}{n_s}\right]^2 \frac{dx}{y} \tag{4.10}$$

The total specific permeance is found by integrating equation (4.10) over the slot depth in question or

$$p = \mu_0 \int \left[\frac{n(x)}{n_s}\right]^2 \frac{dx}{y} \tag{4.11}$$

Note that when $0 < x < d_4$ the current enclosed by a tentative flux path is zero so that the flux below the conductor in the slot is exactly zero. Alternatively when $d_4 + d_3 < x < d_s$, all of the turns are enclosed so that $n(x) = n_s$ and equation (4.10) reduces to the form

$$dp = \mu_0 \frac{dx}{y} \tag{4.12}$$

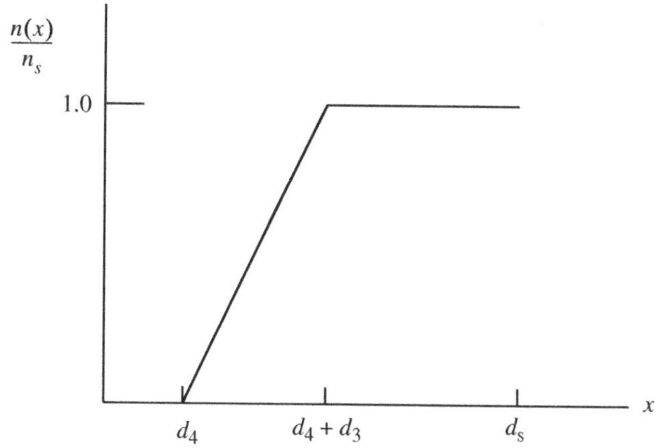

Figure 4.4 Plot of per unit turns vs. distance from bottom of slot corresponding to slot of Figure 4.3.

A plot of the per unit turns as a function of the distance from the bottom of the slot of Figure 4.3 is plotted in Figure 4.4. Recall that this quantity can be viewed as a normalized turns function similar to the winding function defined in Chapter 1.

4.3 SLOT LEAKAGE PERMEANCE CALCULATIONS

First, consider the simple slot shape with semi-closed slot opening shown in Figure 4.5. This shape is probably the most popular configuration for induction machine stator slots. Note that for $d_4 < x < d_3 + d_4$, the typical tube of flux links with a portion

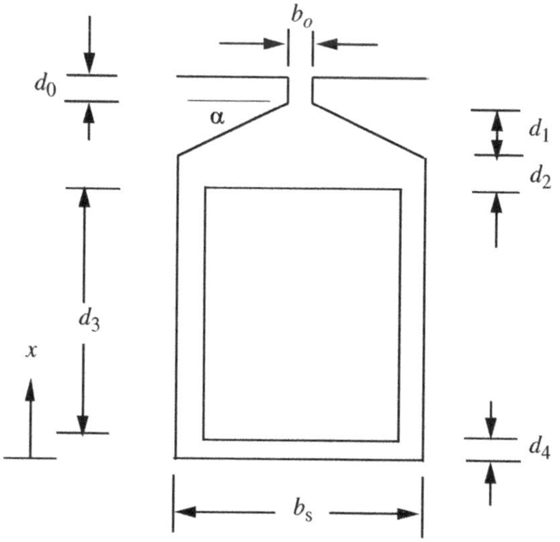

Figure 4.5 Semi-closed slot with parallel sides.

$(x - d_4)/d_3$ of the total conductors. Hence, by equation (4.11) the specific permeance for this portion of the slot is

$$p_3 = \mu_0 \int_{d_4}^{(d_3+d_4)} \left(\frac{x - d_4}{d_3}\right)^2 \frac{dx}{b_s} \tag{4.13}$$

which reduces to

$$p_3 = \mu_0 \frac{d_3}{3b_s} \tag{4.14}$$

For the paths within the depths d_2 and d_0, the specific permeances are clearly

$$p_2 = \mu_0 \frac{d_2}{b_s} \tag{4.15}$$

$$p_0 = \mu_0 \frac{d_0}{b_o} \tag{4.16}$$

For a flux path within the depth d_1 the flux lines clearly do not go straight across the slot since the flux must leave the surface of the slot at right angles. However, the error is not great if this feature is neglected. The length of a typical flux line going across the slot within this region is then given by

$$y = b_s - \frac{(b_s - b_o)(x - d_2 - d_3 - d_4)}{d_1} \tag{4.17}$$

and thus

$$p_1 = \int_{d_4+d_3+d_2}^{d_4+d_3+d_2+d_1} \frac{\mu_0 dx}{b_s - (b_s - b_o)[(x - d_2 - d_3 - d_4)/d_1]} \tag{4.18}$$

Equation (4.18) reduces to the expression

$$p_1 = \frac{\mu_0 d_1}{b_s - b_o} \ln\left(\frac{b_s}{b_o}\right) \tag{4.19}$$

If desired, a more exact solution can be obtained by assuming that the flux lines over the slanted portion of the slot form arcs of a circle. The solution proceeds very similar to Section 3.2. In this case, the more accurate solution is

$$p_1 = \frac{\mu_0}{\pi - 2\alpha} \ln\left(\frac{b_s}{b_o}\right) \tag{4.20}$$

where α is the slant angle noted in Figure 4.5.

Since the four leakage paths are in parallel, the specific permeance for the whole slot is simply the sum of the four individual permeances so that

$$p_1 = \mu_0 \left[\frac{d_3}{3b_s} + \frac{d_2}{b_s} + \frac{1}{\pi - 2\alpha} \ln\left(\frac{b_s}{b_o}\right)\right] + \frac{d_0}{b_0} \tag{4.21}$$

It is important to mention that it has been assumed that a uniform distribution of current exists over the slot. In the case of a stator winding, this is nearly always

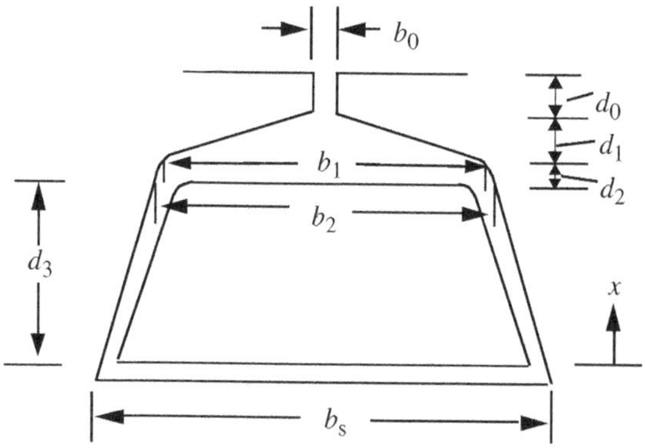

Figure 4.6 Tapered semi-closed "coffin-shaped" slots.

the case since the stator coils are invariably many turns of wire so that the current in the slot is forced to be evenly distributed. However, for rotor slots only one conductor occupies the slot and eddy currents may reduce the specific permeance to a smaller value depending upon the depth of the slot and the frequency of the flux reversals within the slot. A further discussion of this phenomena, called the *deep bar effect*, will be deferred to Chapter 5.

As a second, more difficult, example of specific permeance calculation, consider the tapered or "coffin-shaped" slot shown in Figure 4.6. This shape is widely used as a rotor slot configuration. From previous work, the specific permeance over the portions d_2, d_1, and d_0 have the value

$$p_{012} = \mu_0 \left[\frac{d_2}{b_2 - b_1} \ln \left(\frac{b_2}{b_1} \right) + \frac{1}{\pi - 2\alpha} \ln \left(\frac{b_1}{b_o} \right) + \frac{d_0}{b_o} \right] \tag{4.22}$$

Over the portion d_3 the length of a tube of flux at a distance x from the bottom is

$$y = b_s - x \left(\frac{b_s - b_2}{d_3} \right) \tag{4.23}$$

The area of the bar encircled by a flux line at a distance x from the bottom is

$$\begin{aligned} A(x) &= \int_0^x y \, dx \\ &= \int_0^x \left[b_s - \frac{(b_s - b_2)}{d_3} \right] dx \\ &= x \left[b_s - \frac{b_s - b_2}{d_3} \frac{(b_s - y)}{(b_s - b_2)} \frac{d_3}{2} \right] \\ &= \frac{x(y + b_s)}{2} \end{aligned} \tag{4.24}$$

The total area occupied by the conductor is

$$\begin{aligned} A_c &= A(x)\Big|_{x=d_3, y=b_2} \\ &= d_3 \frac{(b_s + b_2)}{2} \end{aligned} \tag{4.25}$$

Hence the fraction of the total turns enclosed by the flux line is

$$\begin{aligned} \frac{n(x)}{n_s} &= \frac{A(x)}{A_c} = \frac{[x(y + b_s)]/2}{[d_3(b_s + b_2)]/2} \\ &= \frac{y + b_s}{b_2 + b_s}\left(\frac{x}{d_3}\right) \end{aligned} \tag{4.26}$$

The specific permeance of the conductor-occupied region is, from equation (4.11),

$$p_3 = \mu_0 \int_0^{d_3} \left(\frac{y + b_s}{b_2 + b_s}\right)^2 \left(\frac{x}{d_3}\right)^2 \frac{dx}{y} \tag{4.27}$$

which yields ultimately

$$p_3 = \mu_0 \frac{d_3}{b_s} \left[\frac{\beta^2 - \dfrac{\beta^4}{4} - \ln \beta - \dfrac{3}{4}}{(1 - \beta)(1 - \beta^2)^2} \right] \tag{4.28}$$

where

$$\beta = \frac{b_2}{b_s}$$

The total permeance of the coffin-shaped slot is simply the sum of p_{012} and p_3 defined respectively by equations (4.22) and (4.28).

As a final example consider the circular slot of Figure 4.7. Again this is a widely used configuration for rotor slots particularly in larger machines where copper rods

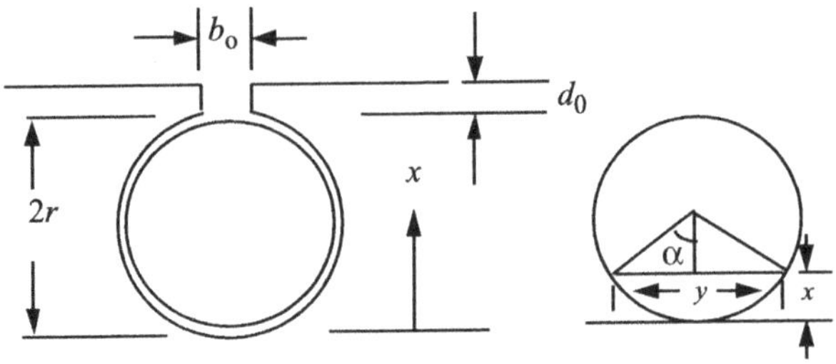

Figure 4.7 Circular slot.

are inserted in the slots and soldered together at each end of the rotor to form a squirrel cage. The specific permeance of the slot opening is, almost by inspection

$$p_0 = \frac{\mu_0 d_0}{b_o}$$

Over the circular portion, the length of an infinitesimal tube of flux at a distance x from the bottom of the slot can be expressed as

$$y = 2r \sin \alpha \tag{4.29}$$

where

$$\alpha = \cos^{-1}\left(\frac{r-x}{r}\right) \tag{4.30}$$

or,

$$x = r - r\cos\alpha$$

so that

$$dx = r \sin \alpha d\alpha \tag{4.31}$$

The area of the segment of the circle below the tube of flux is

$$\begin{aligned} A(x) &= \int_0^x y\,dx \\ &= \int_0^\alpha 2r^2 \sin^2 \alpha d\alpha \\ &= \frac{r^2}{2}(2\alpha - \sin 2\alpha) \end{aligned} \tag{4.32}$$

Hence, the portion of conductors producing flux in the infinitesimal strip is

$$\begin{aligned} \frac{n(x)}{n_s} = \frac{A(x)}{A_c} &= \frac{\frac{r^2}{2}(2\alpha - \sin 2\alpha)}{\pi r^2} \\ &= \frac{\alpha - \frac{1}{2}\sin 2\alpha}{\pi} \end{aligned} \tag{4.33}$$

The specific permeance of the circular part of the slot is

$$\begin{aligned} p_c &= \mu_0 \int_0^\pi \left(\frac{\alpha - \frac{1}{2}\sin 2\alpha}{\pi}\right)^2 \frac{r \sin \alpha}{2r \sin \alpha} d\alpha \\ &= \frac{\mu_0}{2\pi^2} \int_0^\pi \left(\alpha - \frac{1}{2}\sin 2\alpha\right)^2 d\alpha \end{aligned}$$

which reduces to

$$p_c = \mu_0 \left(\frac{\pi}{6} + \frac{5}{16\pi} \right) \tag{4.34}$$

so that, for the entire slot

$$p_s = \mu_0 \left(0.623 + \frac{d_0}{b_o} \right) \tag{4.35}$$

4.4 SLOT LEAKAGE INDUCTANCE OF A SINGLE-LAYER WINDING

In many machines, only one coil side occupies each of the stator slots. In such cases the slot leakage inductance associated with only one coil side is clearly

$$L_{slot} = n_s^2 l_e p_s \tag{4.36}$$

where the specific permeance p_s has been calculated for the slot shape in question. Recall that the quantity n_s is the number of series-connected conductors in the slot. Assuming that the machine is connected as a three-phase system and has S total slots, then the leakage inductance associated with one stator phase belt of a three-phase machine is

$$\begin{aligned} L_{phase\ belt} &= qL_{slot} = \left(\frac{S}{3P} \right) L_{slot} \\ &= \frac{S}{3P} n_s^2 l_e p_s \end{aligned} \tag{4.37}$$

The leakage inductance of one circuit is the leakage inductance per phase belt times the number of phase belts connected in series

$$\begin{aligned} L_{circuit} &= \frac{P}{C} L_{phase\ belt} \\ &= \frac{S}{3C} n_s^2 l_e p_s \end{aligned} \tag{4.38}$$

The inductance per phase is the parallel combination of C circuit inductances

$$L_{phase} = L_{circuit}/C = \frac{S}{3C^2} n_s^2 l_e p_s \tag{4.39}$$

The total number of series-connected stator turns N_s is related to the number of series connected conductors per slot by

$$\begin{aligned} \text{Series-connected turns/phase/circuit} &= N_s = N_t/C \\ &= \frac{\text{(Turns/coil side)(coil sides/slot)(slots)}}{\text{(phases)(circuits)}} \end{aligned}$$

or, for a three-phase machine with c coil sides per slot

$$\begin{aligned} N_s &= \frac{(n_c/2)(c)S}{3 \cdot C} \\ &= \frac{n_c c S}{6C} \end{aligned} \tag{4.40}$$

Since the number of series-connected conductors per slot $n_s = n_c c$

$$n_s = \frac{6CN_s}{S} \tag{4.41}$$

Hence, the slot inductance per phase can be written in the form

$$\begin{aligned} L_{phase} &= \frac{S}{3}\frac{36N_s^2}{S^2} l_e p_s \\ &= 12N_s^2 l_e \frac{p_s}{S} \end{aligned} \tag{4.42}$$

When the number of phases differs from three, it is immediately apparent that

$$L_{phase} = 4N_s^2 l_e \frac{m p_s}{S} \tag{4.43}$$

where m denotes the number of phases. While expressly written for the stator slot leakage the solution for a wound rotor winding is clearly similar. It is important to take note that the slot leakage inductance is not inversely proportional to the number of poles as was the case for the magnetizing inductance (equation 3.83). On the other hand, it is inversely proportional to the number of slots so that the leakage inductance can by minimized by increasing the number of slots to as large a value as possible.

4.5 SLOT LEAKAGE PERMEANCE OF TWO-LAYER WINDINGS

It has already been noted that two-layer windings are very extensively utilized in both induction and synchronous machines. Figure 4.8 shows a typical parallel-sided slot indicating T as the top coil side and B as the bottom coil side. Since the return portion of coil T will be in the bottom of another slot and vice versa for B, the total inductance of the coils will be made up of the inductances of one top and one bottom side slot inductance. In cases where coils T and B are members of the same phase, the total inductance will also include a term which is twice the mutual inductance between the top and bottom coil sides. If coils T and B are not members of the same phase then they are mutually coupled and a mutual inductance term exists.

From previous work, it can be readily established that the specific slot permeance of the top coil side is

$$p_T = \mu_0 \left[\frac{d_3}{3b_s} + \frac{d_2}{b_s} + \frac{d_1}{b_s - b_o} \ln\left(\frac{b_s}{b_o}\right) + \frac{d_0}{b_o} \right] \tag{4.44}$$

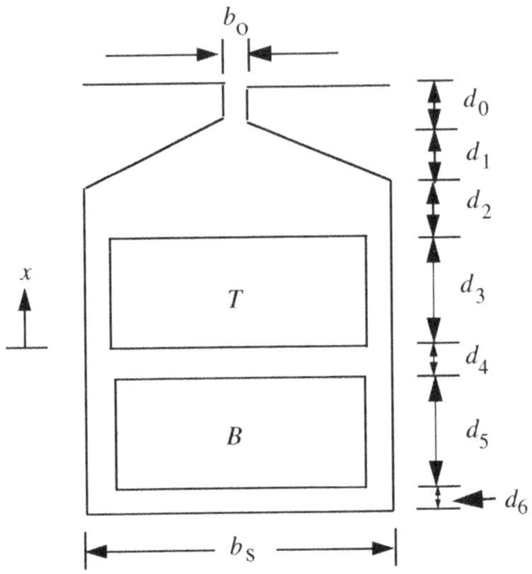

Figure 4.8 Semi-closed slot with double-layer winding.

The specific slot permeance of the bottom coil side is

$$p_B = \mu_0 \left[\frac{d_5}{3b_s} + \frac{d_2 + d_3 + d_4}{b_s} + \frac{d_1}{b_s - b_o} \ln\left(\frac{b_s}{b_o}\right) + \frac{d_0}{b_o} \right] \tag{4.45}$$

The presence of two coil sides in the slot also results in a mutual inductance between the coils. The specific slot permeance corresponding to this mutual inductance is now to be calculated. The effect of the flux produced by the current in the top coil side linking with the bottom coil side is first determined. The flux in an infinitesimal strip of height dx at a distance x from the bottom of the top coil side is given by

$$\begin{aligned} d\Phi &= \mathcal{F}(x)\, l_e dp\,(x) \\ &= \left(I n_c \frac{x}{d_3} \right) \frac{\mu_0 l_e}{b_s} dx \end{aligned} \tag{4.46}$$

where n_c denotes the series-connected *conductors per coil side* or alternatively *the number of turns per coil and* I is, here, the current in each conductor. This flux clearly links with all the conductors in the bottom coil, therefore the flux linkages associated with this flux is

$$d\lambda_{3m} = \mu_0 n_c^2 I \frac{l_e}{b_s} \frac{x}{d_3} dx$$

The total flux linkages corresponding to the mutual flux is

$$\begin{aligned} \lambda_{3m} &= \int_0^{d_3} d\lambda_{3m} \\ &= \mu_0 n_c^2 I l_e \frac{d_3}{2b_s} \end{aligned} \tag{4.47}$$

The specific permeance associated with the conductor portion for the mutual flux is therefore

$$p_{3\mathrm{m}} = \mu_0 \frac{d_3}{2b_s} \tag{4.48}$$

In the non-conductor portion, the flux is produced by all the conductors in the top coil side and the flux so produced links with all the conductors in the bottom coil side. Therefore, previously derived expressions can be used for the regions involving d_0, d_1, and d_2. The relevant specific permeances are

$$p_{0\mathrm{m}} = \mu_0 \frac{d_0}{b_\mathrm{o}}$$

$$p_{1\mathrm{m}} = \mu_0 \frac{d_1}{b_\mathrm{s} - b_\mathrm{o}} \ln\left(\frac{b_\mathrm{s}}{b_\mathrm{o}}\right)$$

$$p_{2\mathrm{m}} = \mu_0 \frac{d_2}{b_\mathrm{s}}$$

Therefore, the total specific slot permeance corresponding to mutual flux linkages is

$$\begin{aligned} p_{\mathrm{TB}} &= p_{0\mathrm{m}} + p_{1\mathrm{m}} + p_{2\mathrm{m}} + p_{3\mathrm{m}} \\ &= \mu_0 \left[\frac{d_\mathrm{o}}{b_\mathrm{o}} + \frac{d_1}{b_\mathrm{s} - b_\mathrm{o}} \ln\left(\frac{b_\mathrm{s}}{b_\mathrm{o}}\right) + \frac{d_2}{b_\mathrm{s}} + \frac{d_3}{2b_\mathrm{s}}\right] \end{aligned} \tag{4.49}$$

Since the reciprocity holds even for the case of a nonlinear magnetic circuit, the flux linking coil "T" due to the current in coil "B" will be identical to the reverse case that has been calculated. Clearly, the corresponding permeance terms are equal so that

$$p_{\mathrm{TB}} = p_{\mathrm{BT}}$$

It is an interesting exercise to verify this result by assuming a differential flux in coil B and finding the flux linkages in coil T in the manner outlined above.

4.6 SLOT LEAKAGE INDUCTANCES OF A DOUBLE-CAGE WINDING

One frequent design modification of an induction motor involves the use of a double rotor cage construction as shown in Figure 4.9. By proper design of the cage, the relative distribution of currents between the top and the bottom bar can be controlled, thereby reducing starting currents without appreciably sacrificing behavior near rated speed. A more thorough discussion of this phenomenon will be given in Chapter 5. In general, as the frequency of flux exciting the slot increases, the current in the bar is forced to flow primarily in the top bar. Hence, the parameters associated with this bar fix the characteristics of the rotor circuit during the starting condition when the rotor slip frequency is high (i.e., equal to the excitation frequency). When the motor is operating near its rated speed condition, the slip frequency is low so that the total

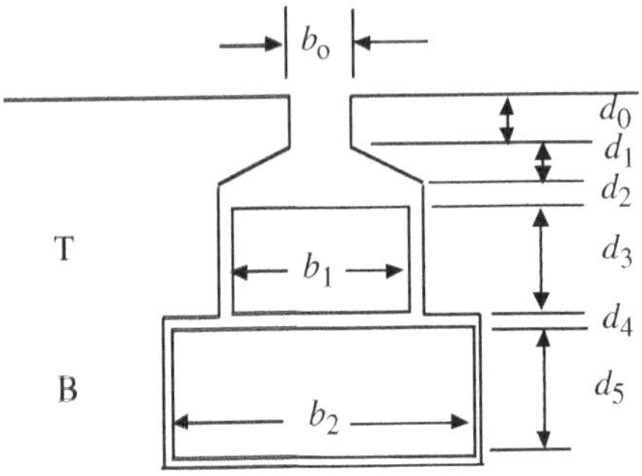

Figure 4.9 Double-cage rotor bar construction.

cross-sectional area associated with the two bars fixes the characteristics of the rotor circuit. Since it is desirable that the starting current remains low, a small top bar of relatively small cross section is implied as indicated in Figure 4.9.

The permeances of the top and bottom bars of Figure 4.9 can be easily deduced from equations (4.44), (4.45), and (4.49) by analogy. For a double rotor cage with rectangular bars in which b_1 and b_2 span the entire slot width

$$p_T = \mu_0 \left[\frac{d_3}{3b_1} + \frac{d_2}{b_1} + \frac{d_1}{b_1 - b_o} \ln\left(\frac{b_1}{b_o}\right) + \frac{d_0}{b_o} \right] \tag{4.50}$$

$$p_B = \mu_0 \left[\frac{d_5}{3b_2} + \frac{d_2 + d_3}{b_1} + \frac{d_4}{b_2} + \frac{d_1}{b_1 - b_o} \ln\left(\frac{b_1}{b_o}\right) + \frac{d_0}{b_o} \right] \tag{4.51}$$

$$p_{TB} = \mu_0 \left[\frac{d_3}{2b_1} + \frac{d_2}{b_1} + \frac{d_1}{b_1 - b_o} \ln\left(\frac{b_1}{b_o}\right) + \frac{d_0}{b_o} \right] \tag{4.52}$$

The leakage fluxes which link the top and bottom bars are clearly

$$\lambda_{lT} = L_{lT} i_T + L_{lTB} i_B \tag{4.53}$$

$$\lambda_{lB} = L_{lTB} i_T + L_{lB} i_B \tag{4.54}$$

where $L_{lT} = l_{er}\, p_T$, etc., and l_{er} is the rotor equivalent length. Equations (4.53) and (4.54) suggest the equivalent circuit of one rotor bar shown in Figure 4.10. The

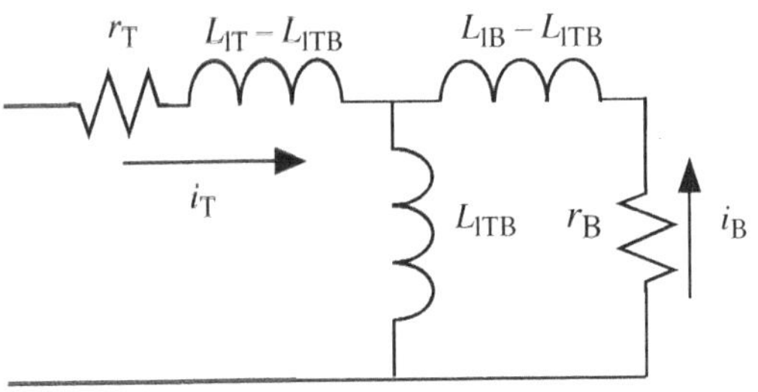

Figure 4.10 Equivalent circuit for double bar arrangement.

associated bar resistances are also shown. Without additional frequency-dependent effects, the resistances of the top and bottom bars, r_{T} and r_{B}, are calculated in the conventional manner. That is, for example

$$r_{\mathrm{T}} = \rho \frac{l_{\mathrm{br}}}{d_3 b_1} \tag{4.55}$$

where ρ is the resistivity of the conductor in Ohm-m and l_{br} is the length of the rotor bar. It is interesting to note that this equivalent circuit results in negative inductance elements. For example

$$L_{\mathrm{lT}} - L_{\mathrm{lTB}} = -\mu_0 l_{\mathrm{br}} \left[\frac{d_3}{6b_1} \right]$$

This anomaly should not be of concern here since the resulting loop equations which describe the flux linkages of the two circuits ultimately generate positive inductance coefficients.

4.7 SLOT LEAKAGE INDUCTANCE OF A DOUBLE-LAYER WINDING

In cases where the pitch of the winding is unity, each slot will be occupied only by coil sides associated with the same phase. The corresponding leakage inductance for one slot or, alternatively, for two coil sides will be given by

$$L_{\mathrm{slot}} = n_{\mathrm{c}}^2 l_{\mathrm{e}} (p_{\mathrm{T}} + p_{\mathrm{B}} + 2p_{\mathrm{TB}}) \tag{4.56}$$

From equation (4.39), the corresponding leakage inductance per phase for a P-pole, C-circuit stator winding is

$$L_{\mathrm{phase}} = \frac{S}{3C^2} n_{\mathrm{c}}^2 l_{\mathrm{e}} (p_{\mathrm{T}} + p_{\mathrm{B}} + 2p_{\mathrm{TB}}) \tag{4.57}$$

It can be noted that due to the two-layer winding, the number of conductors per coil side n_{c} is one-half the number of conductors per slot n_{s}. It is useful to continue to express the slot inductance in terms of the number of conductors per slot rather than per coil. Equation (4.57) can also be written in the form

$$L_{\mathrm{phase}} = \frac{S}{3C^2} n_{\mathrm{c}}^2 l_{\mathrm{e}} \left(\frac{p_{\mathrm{T}} + p_{\mathrm{B}} + 2p_{\mathrm{TB}}}{4} \right) \tag{4.58}$$

The quantity

$$(p_{\mathrm{T}} + p_{\mathrm{B}} + 2p_{\mathrm{TB}})/4$$

can be considered as the effective specific permeance per slot. If one defines

$$p_{\mathrm{s}} = \frac{1}{4} (p_{\mathrm{T}} + p_{\mathrm{B}} + 2p_{\mathrm{TB}})$$

then

$$L_{\text{phase}} = \frac{S}{3C^2} n_s^2 l_e p_s \tag{4.59}$$

It can be again shown that the conductors per slot are related to the series-connected number of turns per phase by

$$N_s = \frac{n_s S}{6C}$$

so that

$$L_{\text{phase}} = 12 N_s^2 l_e \frac{p_s}{S} \tag{4.60}$$

which has the same form as equation (4.42).

It should be recalled that the above expression for slot leakage inductance is valid only when the pitch is unity. Consider now the more practical case in which the pitch is less than one. Figure 4.11 shows a typical situation for a machine with a 60° phase belt in which the pitch lies between 2/3 and unity. Note that just as many coil sides are on the bottom and on the top of the slot as was the case where the pitch is unity. Hence, that portion of the slot leakage inductance per phase associated with the coil sides in the bottom of the slots can be written by analogy with equation (4.39) as

$$\begin{aligned} L_{lB} &= \frac{S}{3C^2} n_c^2 l_e p_B \\ &= \frac{S}{3C^2} n_s^2 l_e \left(\frac{p_B}{4}\right) \end{aligned} \tag{4.61}$$

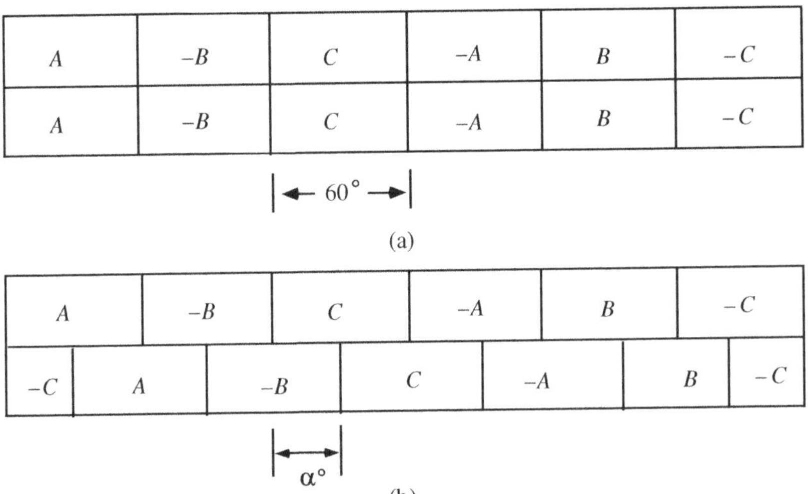

Figure 4.11 Winding distribution for (a) full-pitch; (b) fractional-pitch winding, $2/3 < p < 1$, where $p = 1 - \alpha/180°$.

That portion of the leakage inductance associated with coil sides in the top of the slots is

$$L_{lT} = \frac{S}{3C^2} n_c^2 l_e \left(\frac{p_T}{4}\right) \tag{4.62}$$

Note that the mutual inductance from coil sides of the same phase in the top and bottom of the slots has not decreased. Clearly, when the pitch p is unity the result will be the same as equation (4.49), and when $p = 2/3$ this term will be zero. Hence when $2/3 < p < 1$, the leakage inductance due to mutual coupling is

$$L_{lTB} = \frac{S}{3C^2} n_s^2 l_e \left[\left(\frac{3p-2}{4}\right) p_{TB}\right] \tag{4.63}$$

$$= 12N_s^2 \frac{l_e}{S} \left[\left(\frac{3p-2}{4}\right) p_{TB}\right]$$
$$= L_{lM}(3p-2) \tag{4.64}$$

where

$$L_{lM} = 12N_s^2 \frac{l_e}{S} \left(\frac{p_{TB}}{4}\right) \tag{4.65}$$

Since the stator is symmetrically wound, the "self" component of leakage flux for all three phases is

$$L_{sls} = \left[L_{lT} + L_{lB} + 2L_{lTB}\right]$$
$$= \frac{S}{3C^2} n_s^2 l_e \left[\frac{p_T}{4} + \frac{p_B}{4} + \frac{p_{TB}}{2}(3p-2)\right] \tag{4.66}$$
$$= 12N_s^2 \frac{l_e}{S} \left[\frac{p_T}{4} + \frac{p_B}{4} + \frac{p_{TB}}{2}(3p-2)\right]$$
$$= 12N_s^2 \frac{l_e}{S} p_s \tag{4.67}$$

where

$$p_s = \frac{p_T}{4} + \frac{p_B}{4} + \frac{p_{TB}}{2}(3p-2) \tag{4.68}$$

Note that equation (4.67) is an extended definition of slot leakage corresponding to equation (4.43) which now includes the effect of a non-unity pitch.

When the pitch p is not equal to unity, it is apparent that there now exists a mutual coupling term between the three phases due to slot flux. This term is zero when $p = 1$ and reaches a maximum when $p = 2/3$. By analogy with previous work, it is not difficult to verify that the proper expression for the mutual component of leakage inductance between any two phases is

$$L_{\text{slm}} = -\frac{S}{3C^2} n_s^2 l_e p_{\text{TB}} \frac{(3-3p)}{4}$$
$$= -12N_s^2 \frac{l_e}{S} p_{\text{TB}} \frac{(3-3p)}{4} \tag{4.69}$$
$$= -L_{\text{lM}}(3-3p) \tag{4.70}$$

Note the presence of the negative sign which arises because the currents in the two coil sides are in opposition.

When $1/3 < p < 2/3$ it can be verified from Figure 4.11 that the inductance L_{lTB} remains at zero while the mutual inductance increases to a positive maximum. When $0 < p < 1/3$ the inductance L_{lTB} decreases to a negative maximum while the mutual inductance decreases to zero.

It is clear that, in general, the self and mutual leakage components of slot flux can be expressed by

$$L_{\text{sls}} = L_{\text{lT}} + L_{\text{lB}} + 2k_s(p)L_{\text{lM}} \tag{4.71}$$

$$L_{\text{slm}} = k_m(p)L_{\text{lM}} \tag{4.72}$$

where L_{lT} and L_{lB} are calculated for the case of unity pitch, that is

$$L_{\text{lT}} = 3N_s^2 l_e \frac{p_{\text{T}}}{S} \tag{4.73}$$

$$L_{\text{lB}} = 3N_s^2 l_e \frac{p_{\text{B}}}{S} \tag{4.74}$$

and

$$L_{\text{lM}} = 3N_s^2 l_e \frac{p_{\text{TB}}}{S} \tag{4.75}$$

The quantities k_s and k_m are called the coil pitch slot factors. When $2/3 < p < 1$ the slot factors are given by

$$k_s = 3p - 2 \tag{4.76}$$

$$k_m = 3p - 3 \tag{4.77}$$

When $1/3 < p < 2/3$

$$k_s = 0 \tag{4.78}$$

$$k_m = 3(1-2p) \tag{4.79}$$

When $0 < p < 1/3$

$$k_s = 3p - 1 \tag{4.80}$$

$$k_m = 3p \tag{4.81}$$

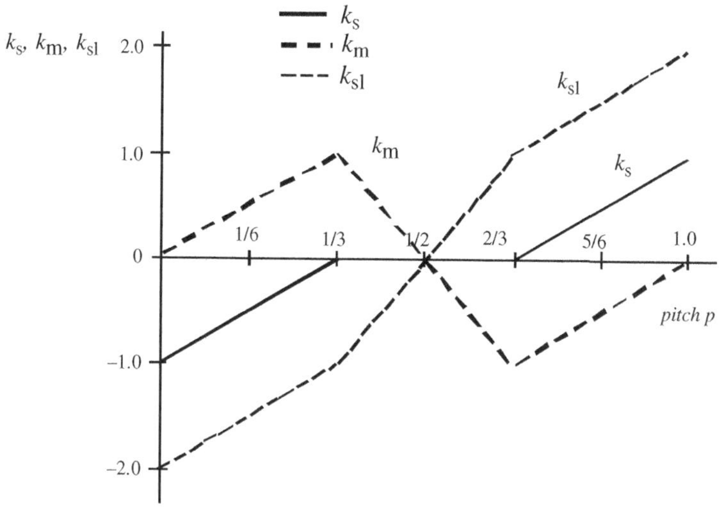

Figure 4.12 Slot factors for self and mutual components of slot leakage for leakage flux versus winding pitch for 60° phase belts.

The slot factors k_s and k_m are plotted versus the pitch p in Figure 4.12. It is a useful exercise to calculate the slot factors for machines with other phase belts, for example with 120° phase belts.

The fact that the three phases have been shown to be mutually coupled suggests that the proper expressions for the slot leakage flux linkages of the three-phase windings are

$$\lambda_{sla} = L_{sls} i_a + L_{slm} i_b + L_{slm} i_c \tag{4.82}$$

$$\lambda_{slb} = L_{slm} i_a + L_{sls} i_b + L_{slm} i_c \tag{4.83}$$

$$\lambda_{slc} = L_{slm} i_a + L_{slm} i_b + L_{sls} i_c \tag{4.84}$$

Note that this type of coupling is not normally present in the usual per phase induction machine equivalent circuit. Hence, it is important to resolve this apparent inconsistency. This difficulty is readily put to rest if it is recalled that when a machine is connected as a three-phase machine without a neutral return,

$$i_a + i_b + i_c = 0$$

Substituting this expression in equations (4.82), (4.83), and (4.84)

$$\lambda_{sla} = (L_{sls} - L_{slm}) i_a \tag{4.85}$$

$$\lambda_{slb} = (L_{sls} - L_{slm}) i_b \tag{4.86}$$

$$\lambda_{slc} = (L_{sls} - L_{slm}) i_c \tag{4.87}$$

Hence, the "slot mutual coupling" has been easily removed. From equations (4.71) and (4.72)

$$L_{sls} - L_{slm} = L_{lT} + L_{lB} + \left[2k_s(p) - k_m(p)\right] L_{lM} \tag{4.88}$$

or simply

$$L_{\text{slot}} = L_{\text{lT}} + L_{\text{lB}} + k_{\text{sl}}(p)\, L_{\text{lM}} \tag{4.89}$$

where

$$L_{\text{slot}} \triangleq L_{\text{sls}} - L_{\text{slm}}$$

$$k_{\text{sl}}(p) \triangleq 2k_{\text{s}}(p) - k_{\text{m}}(p)$$

References to equations (4.76) to (4.81) verifies that for $2/3 < p < 1$

$$k_{\text{sl}}(p) = 3p - 1 \tag{4.90}$$

for $1/3 < p < 2/3$

$$k_{\text{sl}}(p) = 3(2p - 1) \tag{4.91}$$

for $0 < p < 1/3$

$$k_{\text{sl}}(p) = 3p - 2 \tag{4.92}$$

The function $k_{\text{sl}}(p)$ is also plotted in Figure 4.12.

4.8 END-WINDING LEAKAGE INDUCTANCE

4.8.1 Method of Images

As already mentioned, the exact calculation of the end-winding leakage is very difficult since the effect of adjacent coils and adjacent phases on each other, as well as the effect of coupling of the rotor and stator circuits, must be considered. Probably the most rigorous solution in the literature is that of Alger [1]. The derivation is far too complicated to be repeated here. However, considerable insight into the solution of the problem can be gained from the method of images [2].

Consider Figure 4.13 which shows an infinite line current I at a distance d from the face of a semi-infinite magnetic slab of relative permeability μ_{r} and finite conductivity. If an AC current flows in the line, a surface current will be induced on the surface of the magnetic slab. From Section 1.10, the tangential components of H across the boundary must follow

$$H_{\text{ta}} - H_{\text{ti}} = K_{\text{s}} \tag{4.93}$$

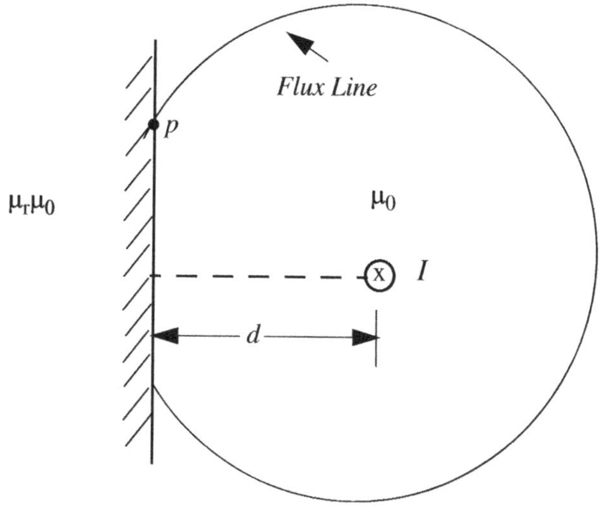

Figure 4.13 Line current parallel to the face of a semi-infinite magnetic slab representing the lamination stack.

where H_{ta} and H_{ti} denote the tangential components on the air and iron side of the surface, respectively. Assuming that a portion of the surface current flows on the iron surface and the remainder on the air surface

$$H_{ta} - H_{ti} = K_{sa} + K_{si} \tag{4.94}$$

On the surface facing the air, the current K_{sa} sets up a resulting magnetic field which is perpendicular to the surface. Similarly, K_{si} produces a magnetic field which is normal to the surface on the iron side. From Section 1.10, the normal components of B must be continuous across this boundary so that at any point on the surface

$$B_{na} = B_{ni} \tag{4.95}$$

The values of flux density must be proportional to the current source so that

$$B_{na} \propto \mu_0(K_{sa})$$

and

$$B_{ni} \propto \mu_0\mu_r(K_{si})$$

where $\propto$ denotes "proportional to". Taking the ratio of the two flux density equations

$$1 = \frac{K_{sa}}{\mu_r K_{si}}$$

so that

$$K_{si} = \frac{K_{sa}}{\mu_r}$$

Equation (4.94) becomes

$$H_{ta} - H_{ti} = K_{sa}\left(1 + \frac{1}{\mu_r}\right) \tag{4.96}$$

or

$$\frac{B_{\text{ta}}}{\mu_0} - \frac{B_{\text{ti}}}{\mu_0 \mu_{\text{r}}} = K_{\text{sa}} \left(\frac{1 + \mu_{\text{r}}}{\mu_{\text{r}}} \right) \tag{4.97}$$

Discontinuities cannot occur in either the normal or tangential components of flux density across any boundary so that finally

$$K_{\text{sa}} = \left(\frac{\mu_{\text{r}} - 1}{\mu_{\text{r}} + 1} \right) \frac{B_{\text{ta}}}{\mu_0} \tag{4.98}$$

Consider now replacing the problem with an equivalent one. If the magnetic slab is removed, the current I will no longer induce a current since it is at an air–air interface. However, the problem solution will be the same if one supposes that there exists an image current at the distance d to the left of the boundary having the value

$$I_{\text{i}} = \left(\frac{\mu_{\text{r}} - 1}{\mu + 1} \right) I \tag{4.99}$$

which will, it should be evident, produce the same tangential component of flux density as equation (4.98).

The concept of an image current can now be readily extended to include currents of finite length oriented parallel or perpendicular to the magnetic surface as well as other shapes such as a circle as shown in Figure 4.14.

In practical cases where the iron is not infinitely permeable the current embedded in the iron, that is the slots, must be considered. Consider the addition of a conductor to a problem involving a current element I in air of semi-infinite extent as shown in Figure 4.15a. The current element in air produces an image of $I(\mu_{\text{r}} - 1)/(\mu_{\text{r}} + 1)$ as indicated in Figure 4.15b. If a hypothetical slot-based current of semi-infinite extent on the iron side is added to the problem as in Figure 4.15c, the net result will be a

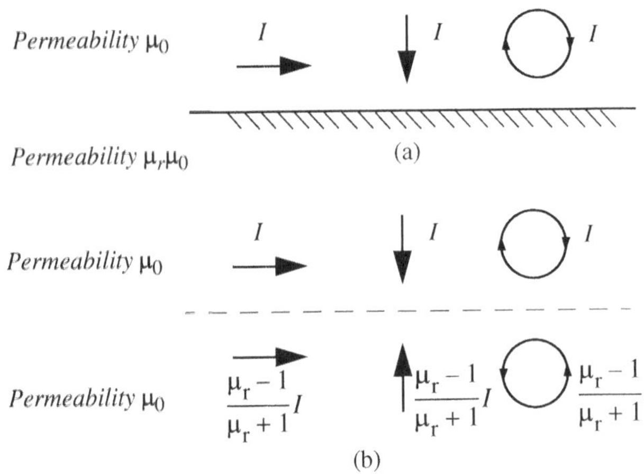

Figure 4.14 Virtual current images replacing a semi-infinite magnetic slab: (a) original problem; (b) equivalent problem.

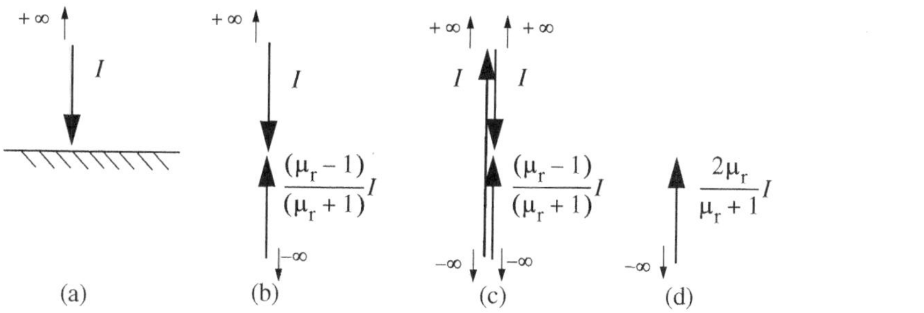

Figure 4.15 (a) Current of semi-infinite length impinging on a magnetic slab of semi-infinite extent, (b) equivalent problem using the method of images, (c) addition of a slot current of ±∞ extent, (d) equivalent problem to (c) representing image equivalent of a slot current.

complete cancellation of the current element in air and a net current on the slot side of the boundary of value

$$I + \frac{\mu_r - 1}{\mu_r + 1}I = \frac{2\mu_r}{\mu_r + 1}I \tag{4.100}$$

The solution of the problems Figures 4.15c and 4.15d are clearly the same so that Figure 4.15d correctly represents the contribution of the slot current to the complete solution. It is thus seen that the iron has the effect of adding a current $I(\mu_r - 1)/(\mu_r + 1)$ in the same place as the actual current I in the slot when the problem is replaced by an equivalent current totally in air. $I(\mu_r - 1)/(\mu_r + 1)$.

4.8.2 End-Winding Leakage Inductance of Random-Wound Coils

It is now possible to examine the solution of a practical end winding for a random-wound machine. Since random-wound coil bundles are flexible, the shape of the end winding tends to approximate a rectangular shape as shown in Figure 4.16a. When the

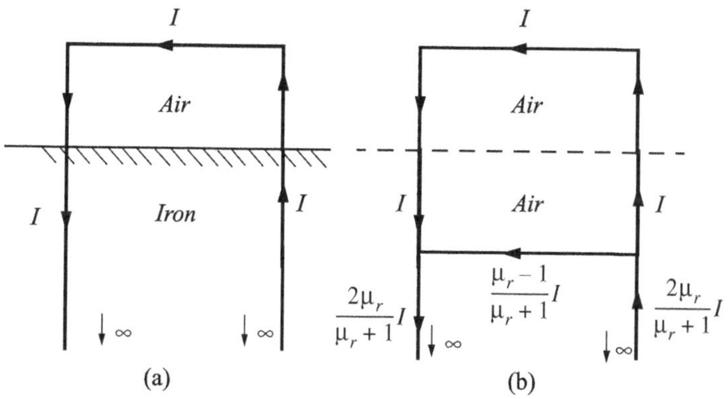

Figure 4.16 (a) Idealized model of one coil of a random wound winding; (b) equivalent model using method of images.

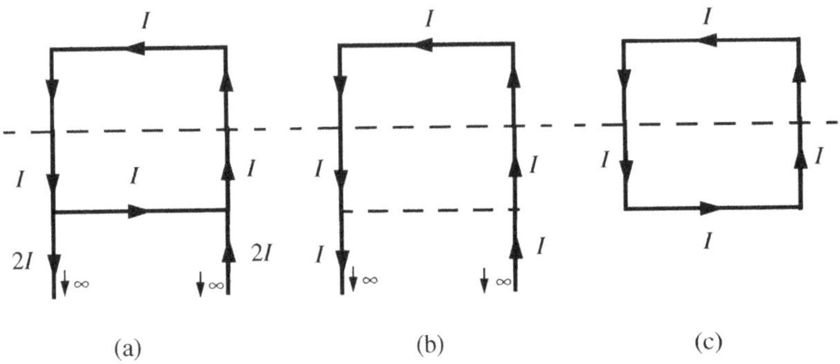

Figure 4.17 Three idealized cases: (a) $\mu_r = \infty$ (perfect iron); (b) $\mu_r = 1$ (air); (c) $\mu_r = 0$ (perfect conductor).

slot current is included in the solution, the equivalent image-based problem becomes that of Figure 4.16b. In principle, the solution can now proceed by setting the relative permeability to a sufficiently large number so that the quantities $2I\mu_r/(\mu_r + 1)$ and $I(\mu_r - 1)/(\mu_r + 1)$ approach $2I$ and I, respectively. The problem then reduces to that of Figure 4.17a. However, measurements of the end-winding leakage have demonstrated that the conductivity of the iron laminations have a very dramatic effect since the eddy currents induced on the surface of the laminations act to "shadow" the images. As the conductivity of the material increases, the effective permeability reduces accordingly. It has been shown from careful measurements of a large machine that more accurate results are obtained when the effective permeability is somewhere between zero (infinitely conducting sheet) and one (equivalent to air) [3,4]. The problem takes on the form of either Figure 4.17b or 4.17c.

4.8.3 End-Winding Leakage Inductance of a Coil with Stator Iron Treated as a Perfect Conductor

Consider first the solution of Figure 4.17c redrawn as Figure 4.18 [4,7]. The shape of the equivalent coil consists of four straight lines. The rectangle ABCD represents the inner area of one end-winding coil while the rectangle ADFE represents the image of the coil due to the presence of the stator iron stack. The inductance of one end-winding coil must be computed as the flux through ABCD and not ADFE. However,

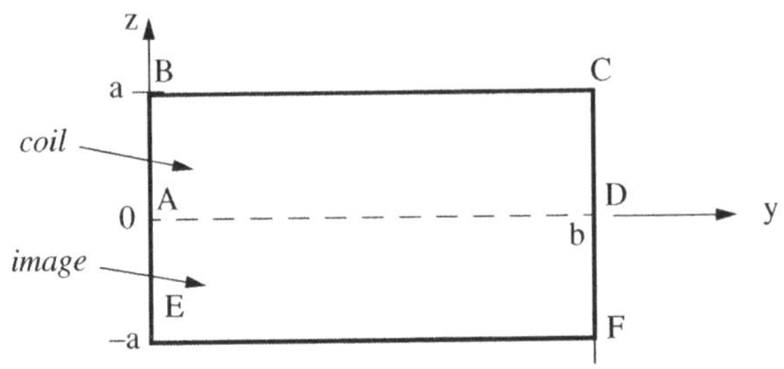

Figure 4.18 Rectangular loop representing both end-winding portions of one coil.

the corresponding winding coil end on the other side has the same equivalent representation. The inductance of that coil should be computed by finding the flux linking ADFEA and not ABCDA. It should therefore then be apparent that the total end-winding inductance is therefore found by simply taking the flux linking the entire area ABCDFEA.

The flux through the loop ABCDFEA due to the wire segment BE can be obtained by using the result for the B field due to the length of a finite length found in Chapter 1, that is from equation (1.12)

$$\mathbf{B}(y,z) = \frac{\mu_0 I}{4\pi y}\left[\frac{z+a}{\sqrt{y^2+(z+a)^2}} - \frac{z-a}{\sqrt{y^2+(z-a)^2}}\right]\boldsymbol{u}_x \tag{4.101}$$

Thus, by symmetry, the total flux through the loop due to current over the segments BE and CF is

$$\Phi_{\mathrm{BE}} = \int_{-a}^{a}\int_{\varepsilon}^{b} B(y,z)\,dydz \tag{4.102}$$

where ε represents the outer radius of the winding bundle.

If the expression for Φ_{BE} is written explicitly

$$\Phi_{\mathrm{BE}} = \frac{\mu_0 I}{4\pi}\int_{-a}^{a}\int_{\varepsilon}^{b}\frac{1}{y}\left(\frac{z+a}{\sqrt{y^2+(z+a)^2}} - \frac{z-a}{\sqrt{y^2+(z-a)^{-2}}}\right)dydz \tag{4.103}$$

From Dwight, Entry 221.01 [6], and evaluating the inner integral, the expression reduces to [7]

$$\Phi_{\mathrm{BE}} - \frac{\mu_0 I}{4\pi}\int_{-a}^{a}\left\{-\ln\left[\frac{a+z}{b} + \sqrt{\left(\frac{a+z}{b}\right)^2+1}\right] - \ln\left[\frac{a-z}{b} + \sqrt{\left(\frac{a-z}{b}\right)^2+1}\right]\right.$$
$$\left. + \ln\left[\frac{a+z}{\varepsilon} + \sqrt{\left(\frac{a+z}{\varepsilon}\right)^2+1}\right] + \ln\left[\frac{a-z}{\varepsilon} + \sqrt{\left(\frac{a-z}{\varepsilon}\right)^2+1}\right]\right\}dz \tag{4.104}$$

This expression can also be integrated when the quantities $(a \pm z)/d$ and $(a \pm b)/\varepsilon$ are replaced by a dummy variable x. Using Dwight Entry 525 [6] the integral becomes

$$\Phi_{\mathrm{BE}} = \frac{\mu_0 I}{2\pi}\left\{-2a\ln\left[\frac{2a}{b} + \sqrt{\left(\frac{2a}{b}\right)^2+1}\right] + \sqrt{(2a)^2+b^2}\right.$$
$$\left. +2a\ln\left[\frac{2a}{\varepsilon} + \sqrt{\left(\frac{2a}{\varepsilon}\right)^2+1}\right] - 2a - b\right\} \tag{4.105}$$

where it has been assumed that ε is negligibly small compared to b (but not zero).

The solution for the flux linking the loop due to the current in the portion BC can be readily found from equation (4.105) by simply interchanging the quantities $2a$ and b. The result is

$$\Phi_{\text{CB}} = \frac{\mu_0 I}{2\pi}\left\{-b\ln\left[\frac{b}{2a}+\sqrt{\left(\frac{b}{2a}\right)^2+1}\right]+\sqrt{(2a)^2+b^2}\right.$$
$$\left.+b\ln\left[\frac{b}{\varepsilon}+\sqrt{\left(\frac{b}{\varepsilon}\right)^2+1}\right]-2a-b\right\} \tag{4.106}$$

By symmetry $\Phi_{\text{FC}} = \Phi_{\text{BE}}$ and $\Phi_{\text{EF}} = \Phi_{\text{CB}}$. The total flux linking the loop is then simply

$$\Phi_{\text{total}} = 2\Phi_{\text{BE}} + 2\Phi_{\text{CB}} \tag{4.107}$$

Thus the inductance of the end winding of a one-turn coil of axial length a and circumferential span b is

$$L_{\text{ew1}} = \frac{\Phi_{\text{total}}}{I} = \frac{\mu_0}{\pi}\left\{-2a\ln\left[\frac{2a}{b}+\sqrt{\left(\frac{2a}{b}\right)^2+1}\right]-b\ln\left[\frac{b}{2a}+\sqrt{\left(\frac{b}{2a}\right)^2+1}\right]\right.$$
$$\left.+2a\ln\left(\frac{4a}{\varepsilon}\right)+b\ln\left(\frac{2b}{\varepsilon}\right)+2\sqrt{4a^2+b^2}-2b-4a\right\}\quad a,b \gg \varepsilon \tag{4.108}$$

The result includes the effect of the end-winding portions on both ends of the machine.

4.8.4 End-Winding Leakage Inductance of a Coil with Stator Iron Treated as Air

Consider now the case where the stator iron appears effectively as air [5]. In this case shown in Figure 4.19, the image of the coils extends into the iron with infinite extent. Letting $-a$ approach $-\infty$, equation (4.101) becomes.

$$\boldsymbol{B}(\boldsymbol{Y},\boldsymbol{Z}) = \frac{\mu_0 I}{4\pi y}\left[\frac{z+a}{\sqrt{y^2+(z+a)^2}}\right]\boldsymbol{u}_x \tag{4.109}$$

The flux linking the winding coil due to current in wire BE can now be expressed as

$$\Phi_{\text{BE}} = \int_0^a\int_\varepsilon^b B(y,z)\,dydz \tag{4.110}$$

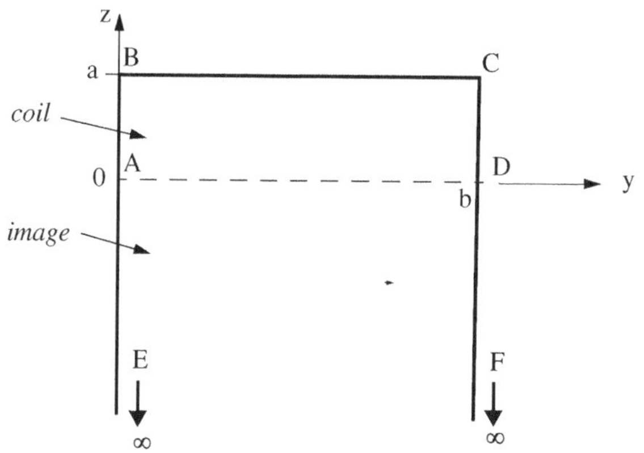

Figure 4.19 Loop representing end-winding portions of one coil and treating the stator iron as air.

or, explicitly

$$\Phi_{BE} = \frac{\mu_0 I}{4\pi} \int_0^a \int_\varepsilon^b \frac{1}{y} \left(\frac{z+a}{\sqrt{y^2+(z+a)^2}} - 1 \right) dydz \tag{4.111}$$

Upon performing the integration

$$\Phi_{BE} = \frac{\mu_0 I}{4\pi} \left\{ -\sqrt{a^2+b^2} + \sqrt{4a^2+b^2} - a + a\ln\left[\frac{2\left(a+\sqrt{a^2+b^2}\right)}{b}\right] \right.$$
$$\left. -2a\ln\left[\frac{2a+\sqrt{4a^2+b^2}}{2b}\right] + a\ln\frac{\varepsilon}{b} + a\ln\left[\frac{4a}{\varepsilon}\right] \right\} \tag{4.112}$$

The contribution of the portion CB is one-half the value of the previous case,

$$\Phi_{CB} = \frac{\mu_0 I}{4\pi} \left\{ -b\ln\left[\frac{b}{2a} + \sqrt{\left(\frac{b}{2a}\right)^2 + 1}\right] + \sqrt{(2a)^2+b^2} \right.$$
$$\left. +b\ln\left[\frac{b}{\varepsilon} + \sqrt{\left(\frac{b}{\varepsilon}\right)^2 + 1}\right] - 2a - b \right\} \tag{4.113}$$

The total flux linking the winding is

$$\Phi_{total} = 2\Phi_{BE} + \Phi_{CB} \tag{4.114}$$

Including both ends of the machine, the end-winding inductance then becomes, finally

$$L_{ew1} = \frac{2\Phi_{total}}{I} \tag{4.115}$$

$$L_{\text{ew1}} = \frac{\mu_0}{2\pi}\left\{-2\sqrt{a^2+b^2}+3\sqrt{4a^2+b^2}-4a-b+2a\ln\left[\frac{2\left(a+\sqrt{a^2+b^2}\right)}{b}\right]\right.$$
$$-4a\ln\left[\frac{2a+\sqrt{4a^2+b^2}}{2b}\right]+2a\ln\frac{\varepsilon}{b}+2a\ln\left[\frac{4a}{\varepsilon}\right] \tag{4.116}$$
$$\left.-b\ln\left[\frac{b}{2a}+\sqrt{\left(\frac{b}{2a}\right)^2+1}\right]+b\ln\left[\frac{b}{\varepsilon}+\sqrt{\left(\frac{b}{\varepsilon}\right)^2+1}\right]\right\}$$

The results of equations (4.108) and (4.116) can now be used to determine the inductance of a complete phase. The choice of which of these two equations to choose must be left to the reader. However, careful measurements of the end-winding leakage in Reference [3] indicates that a value for the permeability of iron lamination roughly half way between $\mu_r = 0$ and $\mu_r = 1$ to be reasonable. Hence, a somewhat better estimate for this leakage inductance suggests the average value of equations (4.108) and (4.116). With a single turn with the end turn in the air, the relative permeance of the inductances of equations (4.108) and (4.116) is simply $L_{\text{ew1}}/\mu_0 = \mathcal{P}_{\text{ew1}}$. The permeance is plotted in Figure 4.20 for the cases where the iron surface acts as a perfect conductor, air, and the average of the two.

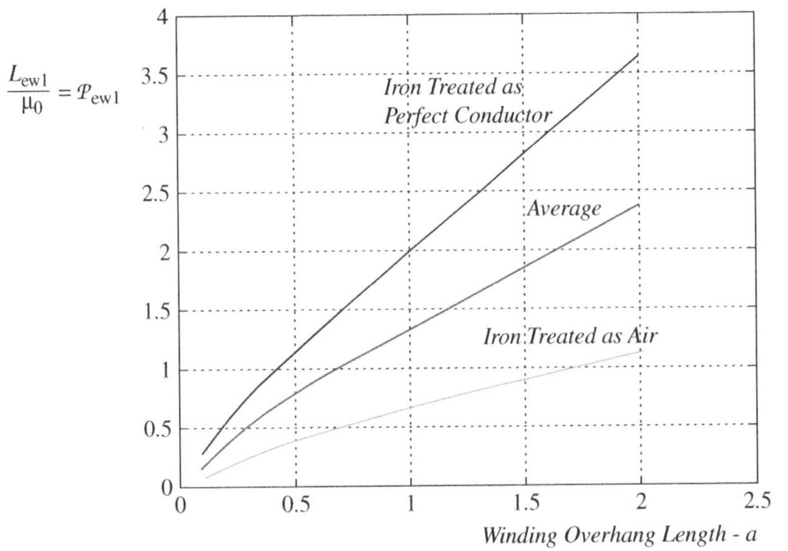

Figure 4.20 Permeance for a rectangular end widing where the winding pitch $b = 1$, coil diameter $\varepsilon = 0.1$ as a function of overhang length a.

4.8.5 End-Winding Leakage Inductance per Phase

If the coil has n_s turns instead of one turn, the end-winding inductance of one coil will be

$$L_{ew,coil} = n_s^2(L_{ew1} + L_{internal}) \tag{4.117}$$

where, from equation (1.138)

$$L_{internal} = \frac{\mu_0}{8\pi}(4a + 2b) \tag{4.118}$$

Assuming q coils in a phase belt and since a phase belt includes only half coils, the inductance of the q half coils will be

$$L_{ew,phase\ belt} = q(k_{d1}k_{p1}n_s)^2\frac{(L_{ew1} + L_{internal})}{2} \tag{4.119}$$

It has been shown earlier that for a three-phase machine, (equation (4.41))

$$n_s = \frac{6CN_s}{S} \tag{4.120}$$

and, with three phases

$$q = \frac{S}{3P}$$

Thus

$$L_{ew,phase\ belt} = \frac{S}{3P}\left(k_{d1}k_{p1}\frac{6CN_s}{S}\right)^2\frac{(L_{ew1} + L_{internal})}{2} \tag{4.121}$$

The overall inductance is, from equations (4.38) and (4.39)

$$L_{ew,\ circuit} = \frac{P}{C}L_{ew,\ phase\ belt} \tag{4.122}$$

$$= \left(\frac{P}{C}\right)\left(\frac{S}{3P}\right)\left(\frac{6C}{S}\right)^2(k_{d1}k_{p1}N_s)^2\frac{(L_{ew1} + L_{internal})}{2} \tag{4.123}$$

Thus the total end-winding leakage per phase can be expressed in terms of the total number of series-connected turns per phase N_s as

$$L_{ew,phase} = \frac{6}{S}k_{d1}^2k_{p1}^2N_s^2(L_{ew1} + L_{internal}) \tag{4.124}$$

4.8.6 End-Winding Leakage of Form-Wound Coils

The following formulas are taken from Liwschitz-Garik [5] which are the results partly of theoretical derivations and partly of experience. Refer to Figure 4.21 which illustrates an idealized portion of the end-winding region for the case of form-wound coils. Figure 4.21 shows two coils of a phase belt having q slots per pole per phase. Again, when the coils are now spread over q slots and then pitched

$$L_{ew} = q(k_{d1}k_{p1}n_s)^2 p_{ew}l_{ew} \tag{4.125}$$

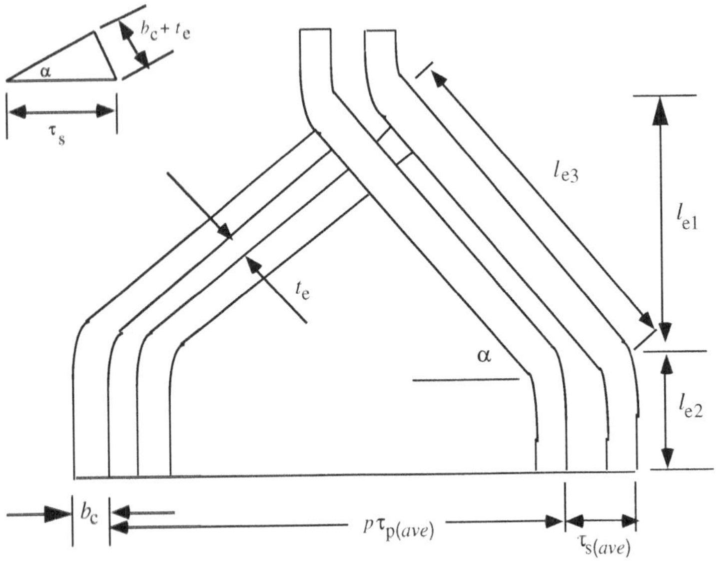

Figure 4.21 End-winding configuration for form-wound coils.

Here the inductance of equation (4.125) expresses the end-winding leakage for one entire phase belt comprised of q slots and q coils (two-layer winding). The corresponding inductance *per slot* is

$$\begin{aligned} L_{ew/slot} &= L_{ew}/q \\ &= k_{p1}^2 k_{d1}^2 n_s^2 p_{ew} l_{ew} \end{aligned} \tag{4.126}$$

For machines of the type shown in Figure 4.21 (i.e., machines with form-wound stators), a suggested estimate of the specific permeance for the end winding is [5]

$$p_{ew} = \mu_0(1.2) \tag{4.127}$$

$$l_{ew} = 2(l_{e2} + l_{e1}/2) \tag{4.128}$$

where the factor 2 takes into account the fact that the coil has sides projecting from either side of the machine, so that with l_{e1} and l_{e2} in inches and $\mu_0 = 3.192 \times 10^{-8}$ henries per inch

$$L_{ew/slot} = \mu_0 k_{p1}^2 k_{d1}^2 n_s^2 (2.4)[l_{e2} + l_{e1}/2] \tag{4.129}$$

where in this result l_{e1} and l_{e2} are in inches and $\mu_0 = 3.192 \times 10^{-8}$ henries per inch.

The significance of l_{e1} and l_{e2} is seen from Figure 4.21. The method used in deriving equation (4.128) consists in treating the end-winding leakage as a revolving field in air. The presence of the iron is neglected in the derivation. Furthermore, it is assumed that the fluxes are confined to radial planes and are arcs of circles. Note that the pitch of the winding starts off at $\tau_{p(ave)}$ at the surface of the end lamination of the stack but drops off to zero by the time it reaches the end of the coil. Hence, an effective length of $l_{e1}/2$ is used in the formula. Clearly, the end-winding field is not truly a two-dimensional one but also spreads out axially. This effect was investigated experimentally by making approximate plots of the three-dimensional field. The effect of axial flux, neglected in the idealized analysis, is accounted for by the constant 2.4. For purposes of comparison, the permeance value of 2.4 occurs at a normalized overhang length of $a = 2$ when using the average value plotted in Figure 4.20.

Note also the quantity t_e in Figure 4.21 which represents the air space between two insulated coil sides. In general, this quantity is controlled in order to provide insulation between adjacent coils. If t_e is fixed, it can be shown that l_{e1} is related to t_e by

$$l_{e1} = \frac{p\tau_{p1}(b_c + t_e)}{2\sqrt{\tau_{s1}^2 - (b_c + t_e)^2}} \tag{4.130}$$

where τ_{s1} and $p\tau_{p1}$ are the slot and pole pitches measured at the middle rather than at the surface of the slot. The quantity b_c is the breadth of the coil in the slot as opposed to the breadth of the slot itself. The value of l_{e2} also depends upon the voltage and ranges from 0.25 to 4 in. for a 13,200 V machine.

From equation (4.39), the total end-winding leakage per phase is

$$L_{ew} = \frac{S}{3C^2} L_{ew/slot} \tag{4.131}$$

For a three-phase machine, from equation (4.41),

$$n_s = \frac{6CN_s}{S}$$

The total end-winding leakage per phase can now be expressed in terms of the total number of series-connected turns per phase N_s as

$$L_{ew} = \mu_0 \frac{S}{3C^2}\left(\frac{6CN_s}{S}\right)^2 k_{p1}^2 k_{d1}^2 (2.4)(l_{e2} + l_{e1}/2) \tag{4.132}$$

$$= 12\mu_0 \frac{N_s^2}{S} k_{p1}^2 k_{d1}^2 (2.4)(l_{e2} + l_{e1}/2) \tag{4.133}$$

A second alternative to calculating the end-winding leakage inductance of a form-wound coil is to use an approximation based on the value calculated in Sections 4.8.3 and 4.8.4. Assuming that the area linked by the leakage flux is the same in both cases, by simple geometry it can be established that if the length a in Figure 4.18 is set equal to $l_{e2} + l_{e1}/2$ in Figure 4.21 and b set to the pole pitch $\tau_{p(ave)}$; equation (4.124) can be used directly. While necessarily a crude approximation, the estimation is sufficiently good for an estimate. It should be mentioned that more accurate calculations of end-winding leakage inductance are available utilizing Neumann Integrals [3, 7]. The work involved, while intellectually satisfying, is rarely warranted given the uncertainty of the effect of eddy currents induced in the outer stator laminations.

4.8.7 Squirrel-Cage End-Winding Inductance

Figure 4.22 illustrates the end-winding region for a squirrel cage machine. An expression for the squirrel-cage end winding leakage was presented in Reference [8].

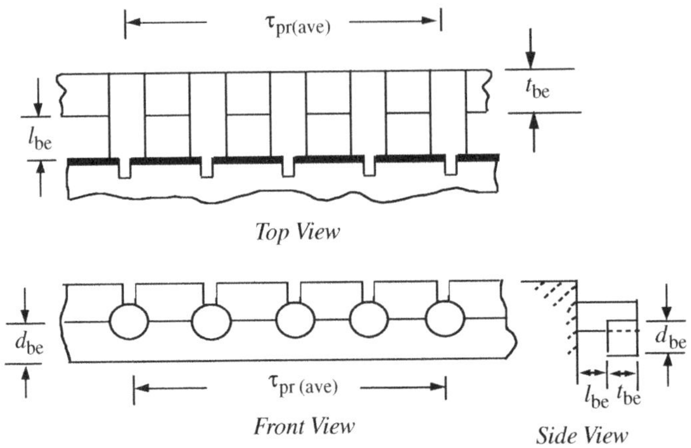

Figure 4.22 End-winding configuration for squirrel-cage rotor construction.

The solution is based on the use of Grover's formula for the self-inductance of a circular ring of mean radius $D_b/2$ with a square cross section [9]. Using the notation of Figure 4.22, and assuming an equivalent square with cross section $d_{be}t_{be}$ the result for a portion $\pi D_b/S_2$ of the end ring is, with length in centimeters

$$L_{er} = \mu_0 \frac{D_b}{2S_2} \left\{ \frac{1}{2} \left[1 + \frac{1}{6} \left(\frac{d_{be}t_{be}}{D_b^2} \right) \right] \ln \left[8 \frac{D_b^2}{d_{be}t_{be}} \right] - 0.8434 + 0.2041 \left(\frac{d_{be}t_{be}}{D_b^2} \right) \right\} \text{H} \tag{4.134}$$

where $\mu_0 = 4\pi \times 10^{-9}$. Here the inductance is expressed in terms of the rotor end ring current not the bar current. The expression does not include the effect of the rotor bar extension l_{be} which must also ultimately be included. The result is valid for small cross sections, $d_{be}/D_b < 0.2$ but values for cross sections greater than 0.2 are provided in Table 21 of Reference [9].

4.9 STATOR HARMONIC OR BELT LEAKAGE

Up to this point the harmonic components of flux in the air gap have been neglected. However, it is clear that since the stator windings consist of coils placed in a finite number of slots, the shape of the flux wave is anything but sinusoidal. Clearly, these unwanted flux components must be accounted for in some manner and since they accomplish no useful work they may be considered as leakage components. In the analysis of these leakage components, it is useful to distinguish between the effects of the harmonic components of MMF which arise because of the discrete number of stator coils and the effects of the stator slotting in which these coils are placed. The first effect is said to produce *belt leakage inductance* while the second component, to be considered in the next Section 4.10 contributes to *zig zag leakage inductance*. First, consider the stator MMF produced by the stator windings. It has already been shown in Chapter 2 that any practical winding produces in the air gap, besides the main (fundamental) component of MMF, space harmonics which travel with different speeds with respect to the stator. The main flux wave induces a voltage across the air gap which along determines the useful power transferred to the rotor. However, the space harmonics also induce voltages in the stator winding. These voltages are of different magnitudes but they are all of the same frequency since they all result from a fundamental component of stator current. Also, they are all in phase which means that they all are purely reactive if they are not "shorted out" by opposing rotor currents.

From Chapter 2, equation (2.73), the hth harmonic MMF in the gap corresponding to a P-pole, C-circuit winding is

$$\mathcal{F}_{\text{pn}} = \frac{3}{2}\frac{4}{\pi}\left(\frac{N_{\text{t}}}{CP}\right)\left(\frac{k_{\text{h}}}{h}\right)I_{\text{m}} \tag{4.135}$$

where

$$k_{\text{h}} = k_{\text{ph}}k_{\text{dh}}k_{\text{sh}}k_{\chi\text{h}} \tag{4.136}$$

The flux density in the air gap resulting from this impressed MMF is

$$B_{\text{gh}} = \frac{\mu_0 \mathcal{F}_{\text{ph}}}{g_{\text{e}}} \tag{4.137}$$

where g_{e} is the effective gap including the effects of saturation and is defined by equation (3.78).

The flux linkages coupling one pole of the hP pole MMF harmonic is found by determining the flux linking its corresponding fractional number of turns and then summing by integration, that is

$$\begin{aligned}\lambda_{\text{ph}} &= \int_0^{(2\pi)/(h\times P)} N_{\text{ah}}(h\phi)B_{\text{g}}(h\phi)\, l_{\text{e}} r d\phi \\ &= \frac{4}{\pi}\left(\frac{k_{\text{h}}N_{\text{t}}}{hP}\right)\int_0^{(2\pi)/(h\times P)} B_{\text{gh}}\cos^2\left(\frac{hP\phi}{2}\right) l_{\text{e}} r d\phi\end{aligned}$$

Recall from Chapter 2 that the winding function N_{ah} is simply the same as the MMF $\mathcal{F}_{\text{ph}}$ per unit current. Upon integrating

$$\lambda_{\text{ph}} = \left(\frac{N_{\text{t}}}{P}\right)\left(\frac{k_{\text{h}}}{h^2}\right)\left(\frac{2B_{\text{gh}}}{\pi}\right)\tau_{\text{p}} l_{\text{e}} \tag{4.138}$$

The total series-connected flux linkages of all hP poles is

$$\begin{aligned}\lambda_{\text{h}} &= \frac{hP}{C}\lambda_{\text{ph}} \\ &= \left(\frac{N_{\text{t}}}{C}\right)\left(\frac{k_{\text{h}}}{h}\right)\left(\frac{2B_{\text{gh}}}{\pi}\right)\tau_{\text{p}} l_{\text{e}}\end{aligned} \tag{4.139}$$

The inductance for the hth harmonic of the MMF is found by substituting equations (4.135) and (4.137) into equation (4.139) and then dividing the result by the current in the stator winding I_m. The result is

$$L_h = \frac{3}{2}\left(\frac{8}{\pi^2}\right)\left(\frac{N_t^2}{C^2 P}\right)\left(\frac{k_h^2}{h^2}\right)\frac{\mu_0 \tau_p l_e}{g_e} \tag{4.140}$$

$$= \frac{3}{2}\left(\frac{8}{\pi^2}\right)\left(\frac{N_s^2}{P}\right)\left(\frac{k_h}{h}\right)^2\frac{\mu_0 \tau_p l_e}{g_e} \tag{4.141}$$

Since the same current "flows through" all of the harmonic winding components, the inductances corresponding to each of the harmonics are in series. The total leakage inductance resulting from all of the harmonics of MMF is

$$L_{lk} = \frac{3}{2}\left(\frac{8}{\pi^2}\right)\left(\frac{N_s^2}{P}\right)\mu_0\frac{\tau_p l_e}{g_e}\left[\sum_{h=2}^{\infty}\left(\frac{k_h}{h}\right)^2\right] \tag{4.142}$$

Comparison of this result with equation (3.80) indicates that equation (4.142) can be written in the compact form

$$L_{lk} = L_{ms}\left[\frac{1}{k_1^2}\sum_{h=2}^{\infty}\left(\frac{k_h}{h}\right)^2\right] \tag{4.143}$$

A plot of the term $\sum\left(\frac{k_h}{h}\right)^2$ for the commonly used three-phase, 60° phase belt winding is given in Figure 4.23 for different numbers of slots per pole per-phase q as

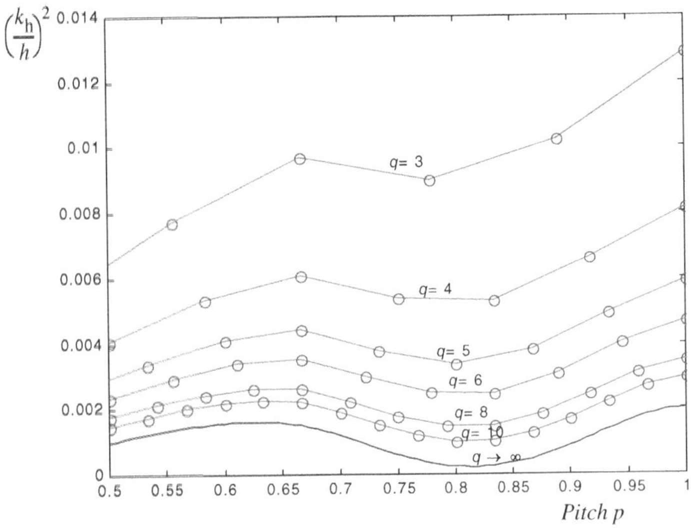

Figure 4.23 Belt leakage coefficient for integral slot windings and 60° phase belts. Note: for $q = 2$ and $p = 0.667, 0.833$, and 1.0, $\sum (k_h/h)^2 = 0.0199$, 0.0205 and 0.0265.

the parameter. It can be seen from this figure that the harmonic leakage depends upon the number of slots per pole per phase and also on the pitch. The harmonic leakage decreases approximately with the square of q.

It is important to mention that the harmonic leakage is important only for wound rotor machines. When the induction machine is equipped with a squirrel cage the harmonics in the air gap produced by the non-fundamental components of MMF will each induce a component of rotor current which will tend to "short" the harmonic inductance. Although belt harmonic fluxes still remain in the gap, they are now very small, being the summation of the harmonic stator fluxes and the opposing harmonic rotor fluxes. The problem is also much more complicated since the rotor resistance as well as reactance now enters the picture. The losses resulting from currents which flow in a squirrel-cage rotor due to the harmonic stator fluxes are called belt harmonic losses and will be treated in Chapter 5. Although the losses produced by the belt harmonics are important, the belt leakage inductance is normally neglected for squirrel-cage machines.

4.10 ZIGZAG LEAKAGE INDUCTANCE

Consider a symmetrical induction machine having n_s conductors per stator slot and n_r conductors per rotor slot. Assume that the rotor is arranged in such a position that the center line of one rotor tooth lines up with the center line of a stator tooth. Figure 4.24a indicates such a position. In general, the stator and rotor ampere-turns tend to cancel much the same as a transformer. The difference between stator and rotor ampere-turns sets up the air gap magnetic field which is rotating around the gap. If this difference component is discarded (the zigzag leakage for this component can

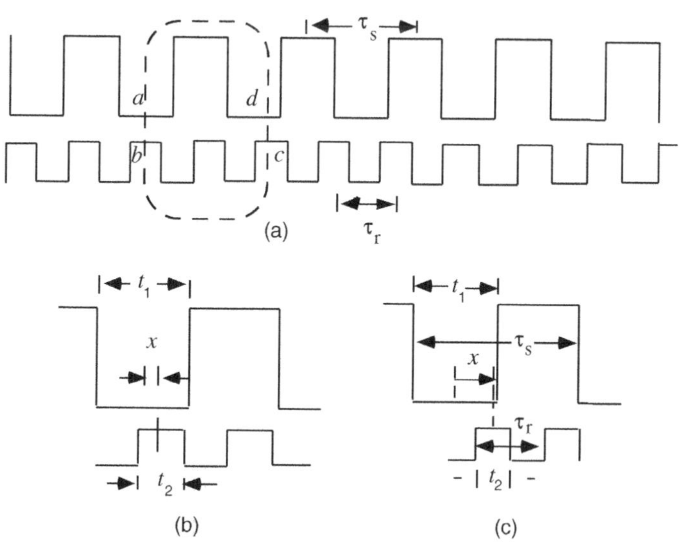

Figure 4.24 Illustrating calculation of zigzag leakage inductance.

be shown to be zero), then at this point the remaining stator and rotor ampere-turns are equal and opposite. Hence the MMF per unit length of the air gap induced by this component of stator current is the same as that induced by the rotor current. It follows that

$$\frac{n_s I_s}{\tau_s} = \frac{n_r I_r}{\tau_r} \tag{4.144}$$

Consider now line integral which encircles one stator and two rotor slots which follow the center line of the rotor teeth as it crosses the air gap. The path taken corresponds to path *abcda* in Figure 4.24a. From Ampere's law

$$\mathcal{F}_{ab} + \mathcal{F}_{cd} = n_s I_s - 2 n_r I_r \tag{4.145}$$

From equation (4.144) this reduces to

$$\mathcal{F}_{ab} + \mathcal{F}_{cd} = n_s I_s \left(1 - \frac{2\tau_r}{\tau_s}\right) \tag{4.146}$$

Since the MMF is assumed to be sinusoidally varying, at least two of the stator and rotor teeth are clearly at (or near) zero MMF. If it is assumed that the tooth corresponding to point a is a zero-potential point then so also is point b if equation (4.144) is assumed. Hence, one can write that

$$\mathcal{F}_{ab} = 0 \tag{4.147}$$

$$\mathcal{F}_{cd} = n_s I_s \left(\frac{\tau_s - 2\tau_r}{\tau_s}\right) \tag{4.148}$$

Note, however, that $2\tau_r - \tau_s$ is the distance between the center lines of the stator and rotor teeth adjacent to the two teeth that are in alignment. If one defines

$$2\tau_r - \tau_s = x \tag{4.149}$$

then

$$\mathcal{F}_{cd} = n_s I_s \frac{x}{\tau_s} \tag{4.150}$$

where x is measured from the center of the stator tooth in question. This is actually a general result which states that the magnetic potential difference between any two opposing teeth is equal to the MMF per stator slot times the ratio of the distance between the two tooth center lines to one stator slot pitch.

In general, as the rotor tooth moves through one stator slot pitch there are three regions to be considered (1) where the rotor tooth is entirely opposite the stator tooth, (Figure 4.24b) and (2) partly opposite the stator slot, (Figure 4.24c), and (3) where the rotor tooth is entirely opposite to the stator slot.

In the first region, that is where $0 < x < (t_1 - t_2)/2$, the MMF between the two teeth are, according to equation (4.150)

$$\mathcal{F}(x) = n_s I_s \frac{x}{\tau_s} \tag{4.151}$$

and the permeance is

$$\mathcal{P}(x) = \mu_0 \frac{l_e t_2}{g_e} \tag{4.152}$$

Hence, the flux corresponding to zigzag leakage is

$$\Phi(x) = \mathcal{F}(x)\mathcal{P}(x) \tag{4.153}$$

or

$$\Phi(x) = \mu_0 n_s I_s \frac{l_e t_2}{g_e} \frac{x}{\tau_s} \tag{4.154}$$

In the second region, $(t_1 - t_2)/2 < x < (t_1 + t_2)/2$. The MMF between the two teeth remains defined by equation (4.150). The permeance is

$$\mathcal{P}(x) = \frac{\mu_0 l_e}{g_e}\left(\frac{t_1 + t_2}{2} - x\right) \tag{4.155}$$

so that

$$\Phi(x) = \mu_0 n_s I_s \frac{l_e}{g_e} \frac{x}{\tau_s}\left(\frac{t_1 + t_2}{2} - x\right) \tag{4.156}$$

In the third region, $(t_1 + t_2)/2 < x < \tau_s/2$. The MMF is again given by equation (4.150). The permeance in region three is assumed to be so small that it can be assumed as zero. The flux which flows between stator and rotor teeth due to zigzag leakage is therefore

$$\Phi(x) = \mathcal{F}(x)\mathcal{P}(x) = 0$$

A plot of the rotor tooth MMF with respect to the stator tooth MMF, the corresponding permeance per tooth, and the resulting flux are given in Figure 4.25. It is clear that this type of behavior can be represented exactly only with an extremely complicated equivalent circuit. However, the exact details concerning the instantaneous behavior of this leakage mechanism is not of interest here but only in its gross overall effect. Hence, it is permissible to replace the actual situation by an equivalent one which produces the same average energy storage. In effect, one may define an equivalent inductance for zigzag leakage as

$$L_{zz/s} = \frac{W_{m,ave}}{\frac{1}{2} I_s^2} \tag{4.157}$$

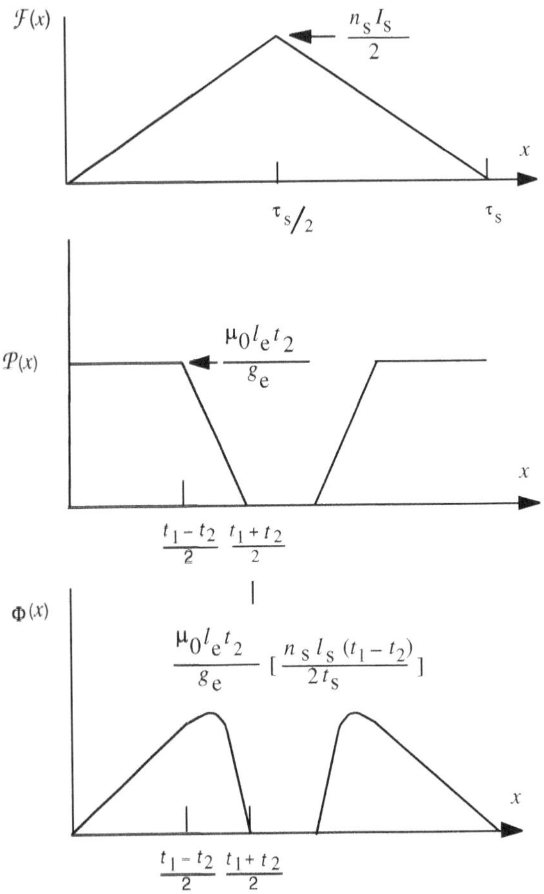

Figure 4.25 Illustrating zigzag leakage permeance variation.

where $W_{m,ave}$ is the average energy stored in the stator–rotor tooth region during the motion from $x = 0$ to $x = \tau_s / 2$. That is

$$W_{m,ave} = \frac{2}{\tau_s} \int_0^{(t_1-t_2)/2} (n_s I_s)^2 \left(\frac{x}{\tau_s}\right)^2 \mu_0 \frac{l_e t_2}{g_e} dx + \frac{2}{\tau_s} \int_{(t_1-t_2)/2}^{(t_1+t_2)/2} (n_s I_s)^2 \left(\frac{x}{\tau_s}\right)^2 \left(\mu_0 \frac{l_e}{g_e}\right) \left(\frac{t_1+t_2}{2} - x\right) dx \tag{4.158}$$

Equation (4.158) works out to

$$W_{m,ave} = \frac{\mu_0 l_e (n_s I_s)^2 t_1 t_2 (t_1^2 + t_2^2)}{12 g_e \tau_s^3} \tag{4.159}$$

so that the zigzag leakage inductance per slot is

$$L_{zz/s} = \frac{\mu_0 l_e n_s^2 t_1 t_2 (t_1^2 + t_2^2)}{6 g_e \tau_s^3} \tag{4.160}$$

The corresponding specific permeance is clearly

$$p_{zz} = \frac{\mu_0 t_1 t_2 (t_1^2 + t_2^2)}{6 g_e \tau_s^3} \tag{4.161}$$

Note that this result is independent of the value of current in the slots surrounding the stator and rotor teeth in question. Hence, the result is valid for any stator tooth and slot. The resulting zigzag inductance per phase is found in a manner analogous to equations (4.39) to (4.42). For a three-phase machine, the result is

$$L_{lzz} = \frac{12 N_s^2}{S_1} l_e p_{zz} \tag{4.162}$$

At this point equation (4.162) is valid only for one-layer windings. When the machine is wound with two layers, two types of slots exist, one type in which the two coil sides are members of the same phase, and a second type in which the coil sides in the slot are members of different phases. Since the zigzag inductance is assumed to not vary with time, choose for purposes of analysis the instant of time in which the current in phase a is a maximum I_m. If the machine is connected without a neutral return, then the current in the other two phases split equally so that the current in the b and c phases is $-I_m/2$. Referring to Figure 4.11, the number of slots having coils sides carrying current I_m in both the top and bottom coils are $(3p-2)S_1/3$, where p is the pitch. Note that all of these coils are associated with phase a. The component of zigzag leakage inductance associated with these coils is

$$L_{zz1} = (3p-2)\frac{S_1}{3C^2} L_{zz/s} \tag{4.163}$$

The number of slots having coil sides carrying both current I_m and $-I_m/2$ are $(3-3p)S_1/3$. With due respect to sign the ampere-turns acting in these slots are

$$\frac{n_s}{2} I_m - \frac{n_s}{2}\left(\frac{-I_m}{2}\right) = \frac{3}{4}(n_s I_m) \tag{4.164}$$

so that the average energy stored in the corresponding rotor and stator teeth is 9/16 of the energy stored in the teeth with coil sides both carrying I_m, and thus the zigzag inductance for these teeth is also 9/16 as large. The zigzag inductance of this component is

$$L_{zz2} = (3-3p)\left(\frac{S_1}{3C^2}\right)\left(\frac{9}{16}\right) L_{zz/s}$$

The total zigzag inductance per phase for 2 layer windings is therefore

$$L_{1zz} = (3p-2)\left(\frac{S_1}{3C^2}\right)L_{zz/s} + (3-3p)\left(\frac{S_1}{3C^2}\right)\left(\frac{9}{16}\right)L_{zz/s} \tag{4.165}$$

which works out to

$$L_{1zz} = \frac{S_1}{3C^2}\frac{(21p-5)}{16}L_{zz/s} \tag{4.166}$$

$$= \frac{3N_s^2}{S_1}l_e\frac{(21p-5)}{4}p_{zz} \qquad 2/3 < p < 1 \tag{4.167}$$

Note that the mutual coupling effect considered in Section 4.7 has been included as part of this analysis. Equation (4.166) is valid only for pitches greater or equal to 2/3. When $1/3 < p \leq 2/3$, the zigzag leakage inductance remains fixed at the value

$$L_{1zz} = \left(\frac{27}{4}\right)N_s^2\frac{l_e}{S_1}p_{zz} \qquad 1/3 < p < 2/3 \tag{4.168}$$

The zigzag leakage inductance clearly decreases to zero for p less than 1/3 but this is more of academic interest. For the record

$$L_{1zz} = \left(\frac{81p}{4}\right)N_s^2\frac{l_e}{S_1}p_{zz} \qquad 0 < p < 1/3 \tag{4.169}$$

It is important to mention that this analysis has been carried out assuming open stator and rotor slots. In cases where the stator and/or rotor slots are semi-closed, the tooth widths at the surface of the teeth must be used in equation (4.160).

The last component of leakage, namely skew leakage, will be deferred until a more detailed discussion of the rotor circuits is completed.

4.11 EXAMPLE 4—CALCULATION OF LEAKAGE INDUCTANCES

Again the example of the 250 HP machine treated in Examples 2 and 3 will be used. In addition to the data already given, the following information is provided.

Stator conductors:	Bare	$0.129 \times 0.204''$
	Insulated	$0.146 \times 0.220''$
Stator slot insulation:	Width	$0.145''$
	Depth	$0.240''$
Bars in parallel per turn:		1
Distance between coil sides:		$1/16''$
Winding pitch:		$12/15 = 0.8$
Rotor conductors:		$3/8 \times 9/16''$ bar
Rotor bar extension:		$1\ 5/8''$
Stator coil extension (straight portion):		$1\ 1/4''$
Rotor skew:		1 stator slot pitch

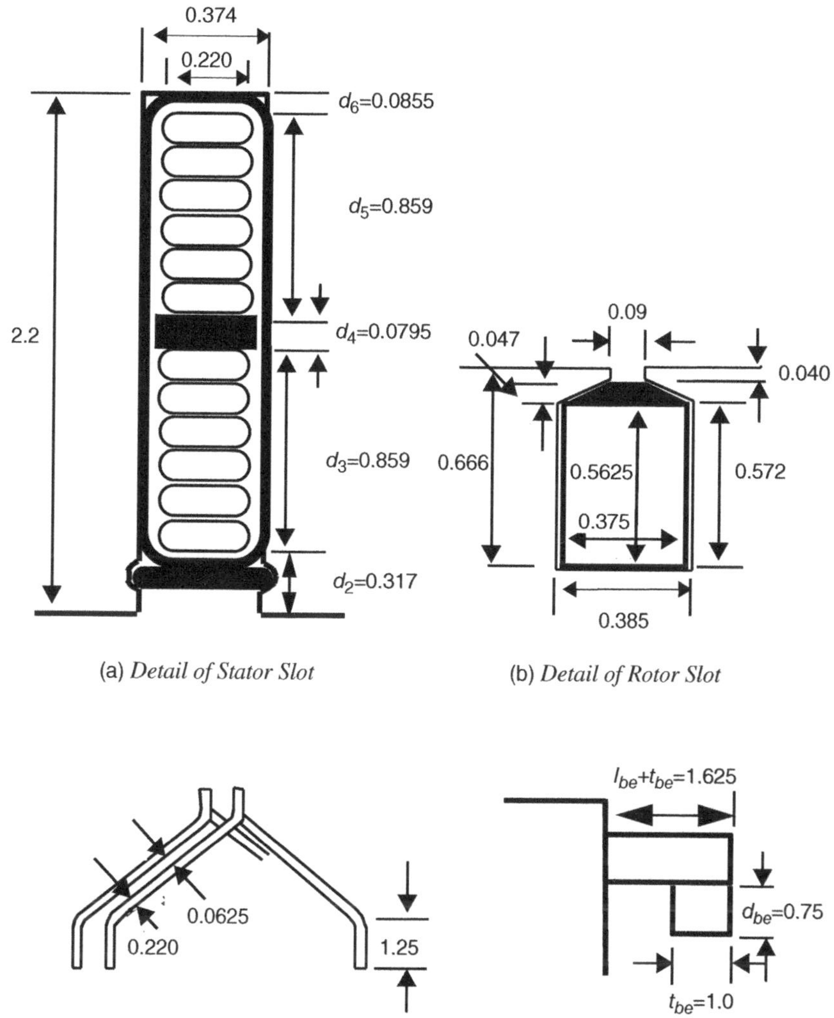

(a) *Detail of Stator Slot* (b) *Detail of Rotor Slot*

(c) *Detail of Stator End Winding* (d) *Detail of Rotor End Winding*

Figure 4.26 Detailed slot and end-winding configurations for 250 HP machine.

Using the above data, the arrangement of conductors in a typical stator and rotor slot can be surmised. Details of stator and rotor slots after winding are shown in Figure 4.26. Note that the stator slot has three types of insulation: (1) insulation between conductors, (2) insulation between coil sides, and (3) insulation to ground. The stator is usually heavily insulated since the potential to ground is generally high (2400 V in this case). The rotor is much less insulated since there is effectively only one turn per slot so that the potential to ground is small. For smaller machines, the insulation material used to isolate the rotor bars from the rotor laminations may be simply an oxide build up intentionally given to either the rotor bars or slots during a heat treatment process. In general, the total rotor copper cross section is selected to be in the range of 60–80% of the stator copper cross section for a squirrel-cage machine and 85–90% for a wound rotor machine. The limiting factor is heating of the bars which limits the current density during starting to values typically on the order of 45,000 A/in.^{2} (70 A/mm^2). The corresponding values for the stator are in the range of 30–35,000 A/in.^{2} (46–54 A/mm^2), increasing with machine size. Figure 4.26 also shows details of the stator and rotor end-winding regions.

Stator Slot Leakage Inductance

From Figure 4.26, note that the clearance from the insulated conductors to the sides of the slot is

$$(0.374 - 0.220)/2 = 0.077''$$

It will be assumed that the same tolerance is allowed at the bottom of the slot. The distance d_6 as defined by Figure 4.8 is equal to this distance plus one insulation thickness for the insulated stator conductor

$$d_6 = 0.077 + (0.146 - 0.129)/2 = 0.0855''$$

The distance d_5 measured from the start of the conductor of the bottommost conductor to the end of the conductor in the topmost conductor is

$$\begin{aligned} d_5 &= 5(0.146) + 0.129 \\ &= 0.859'' \end{aligned}$$

The distance d_4 between coil sides is equal to the thickness of the coil side insulator plus twice the insulation thickness for one conductor

$$\begin{aligned} d_4 &= 0.0625 + (0.146 - 0.129) \\ &= 0.0795'' \end{aligned}$$

The heights d_0 and d_1 are clearly zero for the case of an open slot. The height of the slot above the top coil side conductors is

$$\begin{aligned} d_2 &= d_s - d_6 - 2d_5 - d_4 \\ &= 2.2 - 0.0855 - 2(0.859) - 0.0795 \\ &= 0.317 \end{aligned}$$

The specific permeances p_{T}, p_{B}, p_{TB} are

$$
\begin{aligned}
p_{\mathrm{T}} &= \mu_0 \left[\frac{d_3}{3b_{\mathrm{s}}} + \frac{d_2}{b_{\mathrm{s}}}\right] \\
&= \mu_0 \left[\frac{0.859}{(3)(0.374)} + \frac{0.317}{0.374}\right] \\
&= 2.027 \times 10^{-6}\ \mathrm{H/m} \\
p_{\mathrm{B}} &= \mu_0 \left[\frac{d_5}{3b_{\mathrm{s}}} + \frac{d_2 + d_3 + d_4}{b_{\mathrm{s}}}\right] \\
&= \mu_0 \left[\frac{0.859}{(3)(0.374)} + \frac{0.317 + 0.859 + 0.0795}{0.374}\right] \\
&= 5.18 \times 10^{-16}\ \mathrm{H/m} \\
p_{\mathrm{TB}} &= \mu_0 \left[\frac{d_2}{b_{\mathrm{s}}} + \frac{d_3}{2b_{\mathrm{s}}}\right] \\
&= \mu_0 \left[\frac{0.317}{0.374} + \frac{0.859}{2(0.374)}\right] \\
&= 2.508 \times 10^{-6}\ \mathrm{H/m}
\end{aligned}
$$

Since only ratios are needed to calculate specific permeance, conversion of lengths to SI units is not necessary.

In Example 2 of Chapter 3, the effective length of the stator has been obtained from Case 3 of Section 3.3. The relevant equation is, repeating

$$
l_{\mathrm{es}} = l_{\mathrm{is}} + 2g + nl_{\mathrm{o}}\left(\frac{5}{5 + \dfrac{2l_{\mathrm{o}}}{g}}\right)^2
$$

from which

$$
\begin{aligned}
l_{\mathrm{es}} &= 8.5 + 2(0.04) + (2)(0.375)\left(\frac{5}{5 + 2\dfrac{(0.375)}{0.04}}\right)^2 \\
&= 8.5 + 0.08 + 0.09 = 8.67''
\end{aligned}
$$

The slot leakage inductance per phase for the top, bottom, and the mutual coupling between top and bottom coils can be obtained from equations (4.73), (4.74), and

(4.75) with N_s as calculated in Example 3 of Chapter 3. For the leakage inductance of the top portion of the coils of one phase

$$L_{lT} = \frac{3N_s^2 l_e}{S_1} p_T$$

$$= \frac{(3)(240)^2 \left(\frac{8.67}{39.37}\right)}{120} 2.027 \times 10^{-6}$$

$$= 0.643 \text{ mH}$$

For the bottom portion of the coils,

$$L_{lB} = \frac{3N_s^2 l_e}{S_1} p_B$$

$$= \frac{(3)(240)^2 \frac{(8.67)}{39.37}}{120} 5.18 \times 10^{-6}$$

$$= 1.643 \text{ mH}$$

The mutual coupling between top and bottom coils for unity pitch is

$$L_{lTB} = \frac{3N_s^2 l_e}{S_1} p_{TB}$$

$$= \frac{3(240)^2 \frac{(8.67)}{39.37}}{120} 2.508 \times 10^{-6}$$

$$= 0.795 \text{ mH}$$

From Example 3 it was determined that the pitch of the stator coils is 0.8. From equation (4.90), the slot factor for mutual coupling is

$$k_{sl} = 3p - 1$$

$$= (3)(0.8) - 1$$

$$= 1.4$$

The total slot leakage inductance per phase is, from equation (4.89), therefore

$$L_{lsl} = L_{lT} + L_{lB} + k_{sl}(p)L_{lM}$$

$$= 0.643 + 1.643 + (1.4)(0.795)$$

$$= 3.400 \text{ mH}$$

Stator End-Winding Leakage Inductance

From the additional information that has been given, the length l_{e2} (Figure 4.21) is 1.25″. The spacing between adjacent coil sides in the slot is specified as 0.0625″. If this minimum spacing is maintained in the end-winding region, then the quantity t_e in Figure 4.21 is also 0.0625″. Hence from equation (4.130),

$$l_{e1} = \frac{p\tau_{p1}(b_c + t_e)}{2\sqrt{\tau_{s1}^2 - (b_c + t_e)^2}}$$

The pole pitch at the midpoint of the stator slot is

$$\tau_{p1} = \frac{\pi}{P}(D_{is} + d_s)$$
$$= \frac{\pi}{8}(24.08 + 2.2) = 10.32''$$

The slot pitch at the midpoint of the stator slot is

$$\tau_{s1} = \frac{P\tau_{p1}}{S_1}$$
$$= \frac{8(10.32)}{120} = 0.688''$$

The length of the end-winding extension over the diagonal region is therefore

$$l_{e1} = \frac{(0.8)(10.32)(0.22 + 0.0625)}{2\sqrt{0.688^2 - (0.22 + 0.0625)^2}}$$
$$= 1.859''$$

The stator end-winding leakage inductance per phase is found from equation (4.133) as

$$L_{lew} = 12\mu_0 \frac{N_s^2}{S_1} k_{p1}^2 k_{d1}^2 (2.4)\left(l_{e2} + \frac{l_{e1}}{2}\right)$$
$$= \frac{(12)(\mu_0)}{120}(240)^2(0.91)^2(2.4)\left(\frac{1.25 + 1.859/2}{39.37}\right)$$
$$= 0.796 \text{ mH}$$

Belt Leakage Inductance

The belt leakage inductance is essentially zero since the machine is equipped with a squirrel-cage rotor.

Zigzag Leakage Inductance

The specific permeance corresponding to zigzag leakage flux is given by equation (4.161) as

$$p_{zz} = \frac{\mu_0 t_{ts} t_{tr}(t_{ts}^2 + t_{tr}^2)}{6 g_e \tau_s^3}$$

Referring to Figure 3.5, it can be noted that the rotor tooth width could taper quickly as one progresses down the tooth. When the machine is not saturated the tooth width at the gap clearly applies while for deep saturation the tooth width at the root may be preferable. It was learned in Example 2 of Chapter 3, however that the rotor teeth will not be very saturated for the assumed no-load gap flux density of 0.775 teslas. The tooth width at the air gap will therefore be assumed here.

$$p_{zz} = \frac{(4\pi \times 10^{-7})(0.256)(0.687)(0.256^2 + 0.687^2)}{(6)(0.0677)(0.630)^3}$$
$$= 1.170 \times 10^{-6}\ \mathrm{H/m}$$

Since the pitch is between unity and two-thirds, equation (4.167) applies and

$$L_{1zz} = \frac{3N_s^2}{S_1} l_{es} \frac{(21p-5)}{4} p_{zz}$$
$$= \frac{(3)(240^2)}{120}\left(\frac{8.67}{39.37}\right)\frac{[21(0.8)-5]}{4}(1.170 \times 10^{-6})$$
$$= 1.094\ \mathrm{mH}$$

Rotor Slot Leakage Inductance per Bar

As yet the rotor leakage inductances cannot be computed for an entire phase. The leakage associated with one rotor bar, however, can be computed from equations (4.19) and (4.21) as

$$p_{sl} = \mu_0 \left[\frac{d_{3r}}{3b_{sr}} + \frac{d_{1r}}{(b_{sr} - b_{or})}\ln\left(\frac{b_{sr}}{b_{or}}\right) + \frac{d_{or}}{b_{or}}\right]$$
$$= (\mu_0)\left[\frac{0.5625}{(3)(0.385)} + \frac{0.047}{(0.385 - 0.09)}\log_e\left(\frac{0.385}{0.09}\right) + \frac{0.040}{0.09}\right]$$
$$= 1.461 \times 10^{-6}\ \mathrm{H/m}$$

From equation (4.36), the slot leakage per bar is

$$L_b = n_s^2 l_{er} p_{sl}$$
$$= (1)^2\left(\frac{8.68}{39.37}\right)1.461 \times 10^{-6}$$
$$= 0.32\ \mu\mathrm{H/bar}$$

Note that the quantity l_{er} takes into account the slight additional length of the rotor bar due to skew, that is,

$$l_{er} = \frac{l_{es}}{\cos\left(\frac{2\pi}{S_1}\right)} = \frac{8.67}{\cos\left(\frac{2\pi}{120}\right)} = 8.68''$$

Rotor End-Winding Leakage Inductance per Ring Segment

The end winding leakage inductance per segment of the end ring is obtained from Eq. (4.134) as

$$L_e = \mu_0 \frac{D_b}{2S_2}\left\{\frac{1}{2}\left[1+\frac{1}{6}\left(\frac{d_{be}t_{be}}{D_b^2}\right)\right]\ln\left[8\frac{D_b^2}{d_{be}t_{be}}\right] - 0.8434 + 0.2041\left(\frac{d_{be}t_{be}}{D_b^2}\right)\right\}$$

From Figure 4.26 the rotor diameter measured at the middle of the end ring is

$$D_b = [D_{or} - 2d_{sr} - d_{be}] = [24 - 2(0.67) - 0.75] = 21.91'' = 55.65 \text{ cm}$$

$$L_e = (4\pi 10^{-9})\left(\frac{55.65}{2\times 97}\right)\left(\left(\frac{1}{2}\right)\left(1+\frac{1}{6}\left(\frac{1\times 0.75}{21.91^2}\right)\right)\right.$$
$$\left.\ln\left(8\times\frac{21.91^2}{1\times 0.75}\right) - 0.8434 + 0.2041\frac{1\times 0.75}{21.91^2}\right)$$
$$= 0.01251 \quad \mu\text{H/end ring segment}$$

The next task in the calculation of machine parameters is to relate the inductance per bar just calculated to an expression which involves the overall rotor leakage inductance per phase. This is the topic of Section 4.12.

4.12 EFFECTIVE RESISTANCE AND INDUCTANCE PER PHASE OF SQUIRREL-CAGE ROTOR

It has been shown that if the stator is excited with a sinusoidal set of stator voltages, then the resulting currents produce an MMF, the fundamental component of which rotates about the air gap at synchronous speed. This fundamental component of MMF produces a useful, torque-producing air gap flux while the higher harmonics have been shown to result essentially in leakage components. The useful air gap flux density produced by this MMF has already been calculated in Chapter 3. It is now time to turn attention to the induced rotor currents which clearly also exist since the rotating stator air gap flux density certainly also links rotor as well as stator circuits, thereby inducing currents in the rotor.

When the rotor is wound with discrete coils as is the case for wound rotor induction machines, the calculation of parameters proceeds in the same fashion as has been developed for stator windings. This case need not be discussed further. However, when the machine is equipped with a squirrel cage, the problem is clearly quite different since the rotor currents do not flow in discrete coils but in shorted bars. Since the current in each of the S_2 rotor slots is independent, a completely general solution would require S_2 differential equations, one for each rotor mesh. Such a complicated solution is clearly impractical and it is useful to search for simpler equations which yet maintain the same effective behavior as the S_2 rotor currents. Clearly, it is not necessary to retain all S_2 currents since, by symmetry, the currents induced under

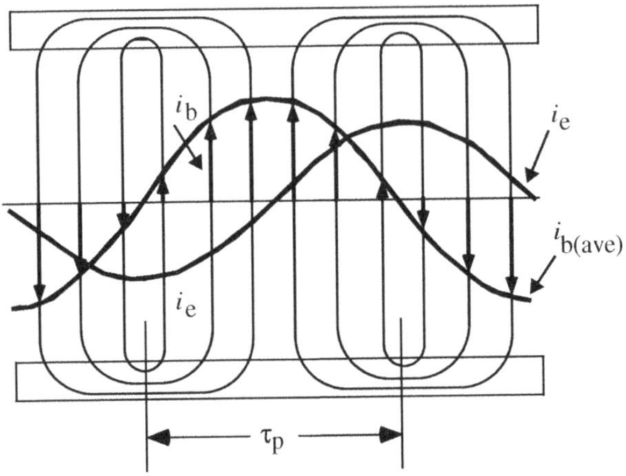

Figure 4.27 Current paths in squirrel-cage rotor having six bars per pole.

each stator pole will basically be the same. In addition, as the stator flux passes over the bars, the currents in adjacent bars will be nearly the same, differing only in the time phase related to the pitch of one rotor tooth.

For simplicity, assume that in response to the sinusoidally rotating B field, the current in each bar will vary sinusoidally. In general, the amplitude and speed of rotation of the stator B field are not fixed but can change in response to external influences such as line switching, load torque changes, etc. However, if the time constants of these changes are sufficiently small then the system can be assumed in the quasi-steady-state in which all of the rotor bar currents are of the same amplitude and that adjacent bars have currents displaced in phase by the pitch of one rotor slot expressed in electrical radians, that is, by the angle

$$\frac{\pi P}{S_2} = \frac{\pi \tau_r}{\tau_p} \tag{4.170}$$

where S_2 is the number of rotor slots and P is the number of stator (and rotor) poles. It is apparent that the squirrel cage effectively represents a polyphase winding with as many phases m_2 as there are slots per pole, that is, $m_2 = S_2/P$. In general, this number need not be an integer and indeed integer values are avoided in order to prevent the cogging torques previously described. An integer value will be assumed here only for convenience in depicting the bar current distribution, illustrated in Figure 4.27 for the simple case of six bars per pole.

The end-ring segments between bars are clearly made up of the difference in current between two adjacent bars and therefore are also sinusoidal and phase displaced by a fixed phase angle in the steady state. If one lets

R_b, L_b	=	resistance and leakage inductance of each bar
R_e, L_e	=	resistance and leakage inductance of each end-ring segment
i_b	=	current in the bars
i_e	=	current in end-ring segments
e_b	=	induced voltage per bar

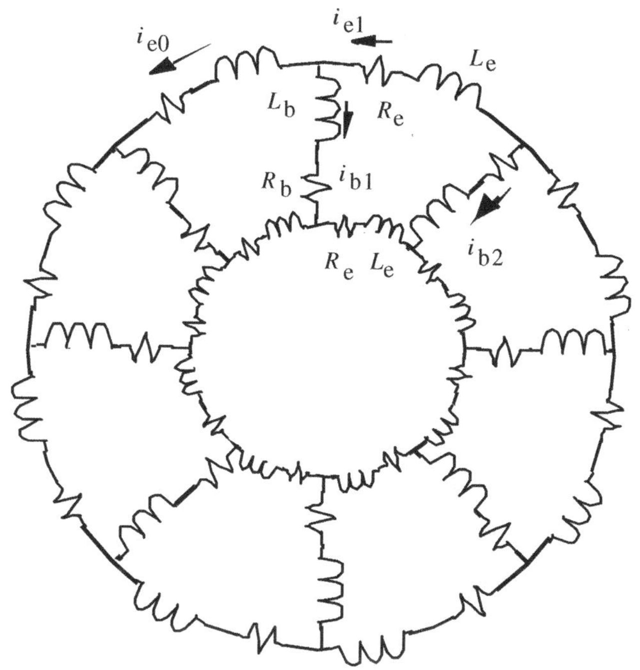

Figure 4.28 Meshes formed by squirrel cage with eight bars per pole pair.

then the squirrel cage of the machine can be represented by the planar circuit of Figure 4.28. Applying Kirchoff's voltage law to mesh 1

$$e_{b1} - e_{b2} = R_b(i_{b_1} - i_{b2}) + L_b\frac{d}{dt}(i_{b1} - i_{b2}) + 2R_e i_{e1} + 2L_e\frac{di_{e1}}{dt} \quad (4.171)$$

and

$$i_{b1} + i_{e0} = i_{e1} \quad (4.172)$$

If steady-state (or an approximation thereof) is assumed, then in phasor form

$$\widetilde{E}_{b2} = \widetilde{E}_{b1}e^{(j\pi P)/S_2} \quad (4.173)$$

$$\widetilde{I}_{b2} = \widetilde{I}_{b1}e^{(j\pi P)/S_2} \quad (4.174)$$

$$\widetilde{I}_{e1} = \widetilde{I}_{e0}e^{(j\pi P)/S_2} \quad (4.175)$$

where S_2 denotes the number rotor slots.

From equations (4.172) and (4.175),

$$\widetilde{I}_{b1} = \widetilde{I}_{e1} - \widetilde{I}_{e0} = \widetilde{I}_{e1}(1 - e^{(-j\pi P)/S_2}) \quad (4.176)$$

or

$$\widetilde{I}_{e1} = \frac{\widetilde{I}_{b1}}{1 - e^{(-j\pi P)/S_2}} \quad (4.177)$$

and from equations (4.173) and (4.174)

$$\widetilde{E}_{b1} - \widetilde{E}_{b2} = \widetilde{E}_{b1}(1 - e^{(j\pi P)/S_2}) \tag{4.178}$$

$$\widetilde{I}_{b1} - \widetilde{I}_{b2} = \widetilde{I}_{b1}(1 - e^{(j\pi P)/S_2}) \tag{4.179}$$

Substituting equations (4.173) to (4.178) in the equivalent phasor form of equation (4.171)

$$\widetilde{E}_{b1}(1 - e^{(j\pi P)/S_2}) = \widetilde{I}_{b1}(1 - e^{(j\pi P)/S_2})(R_b + j\omega L_b) + \frac{2\widetilde{I}_{b1}}{1 - e^{(-j\pi P)/S_2}}(R_e + j\omega L_e) \tag{4.180}$$

where ω denotes the angular frequency induced in the rotor meshes (i.e., slip frequency). That is,

$$\omega = \frac{\omega_e - (P\omega_{rm})/2}{\omega_e} = S\omega_e \tag{4.181}$$

where ω_e and ω_{rm} are the electrical angular frequency of the stator current and rotor mechanical speed in radians per second, respectively and S is the per unit slip

$$S = \frac{\omega_e - (P\omega_{rm})/2}{\omega_e} = \omega_e$$

Equation (4.180) can be written as

$$\widetilde{E}_{b1} = \widetilde{I}_{b1}(R_b + j\omega L_b) + \frac{2\widetilde{I}_{b1}}{(1 - e^{(j\pi P)/S_2})(1 - e^{(-j\pi P)/S_2})}(R_e + j\omega L_e) \tag{4.182}$$

But

$$(1 - e^{(j\pi P)/S_2})(1 - e^{(-j\pi P)/S_2}) = e^{\frac{j\pi P}{2S_2}}\left(e^{\frac{-j\pi P}{2S_2}} - e^{\frac{j\pi P}{2S_2}}\right)\left(e^{\frac{-j\pi P}{2S_2}}\right)\left(e^{\frac{j\pi P}{2S_2}} - e^{\frac{-j\pi P}{2S_2}}\right) \tag{4.183}$$

so that this term becomes

$$\begin{aligned} &= -\left(e^{\frac{(j\pi P)}{2S_2}} - e^{\frac{-j\pi P}{2S_2}}\right)^2 \\ &= -\left[2j\sin\left(\frac{(\pi P)}{2S_2}\right)\right]^2 \\ &= 4\sin^2\left(\frac{\pi P}{2S_2}\right) \end{aligned} \tag{4.184}$$

Hence, the mesh equation reduces to

$$\widetilde{E}_{b1} = \widetilde{I}_{b1}(R_b + j\omega L_b) + \widetilde{I}_{b1}\left[\frac{R_e + j\omega L_e}{2\sin^2[(\pi P)/(2S_2)]}\right] \tag{4.185}$$

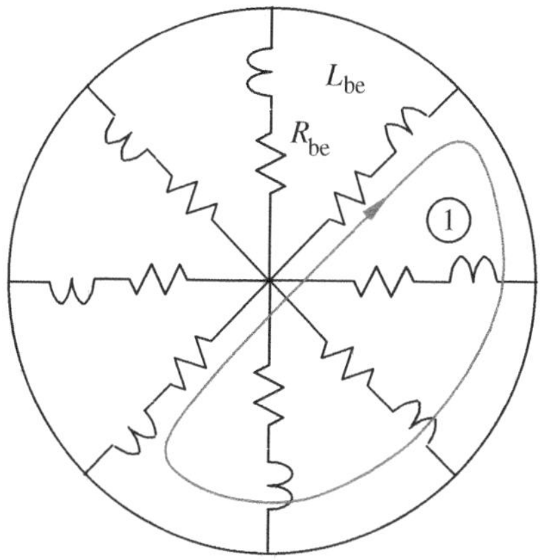

Figure 4.29 Equivalent eight-mesh circuit obtained by absorbing effect of end ring into rotor bar parameters.

Equation (4.185) shows that the effect of the end rings is to increase the bar resistance and inductance by an amount inversely proportional to the square of the sine of one-half the electrical angle between bars. It is useful to define quantities R_{be} and L_{be} as equivalent values of bar resistance and inductance. Then

$$R_{be} = R_b + \frac{R_e}{2\sin^2[(\pi P)/(2S_2)]} \tag{4.186}$$

$$L_{be} = L_b + \frac{L_e}{2\sin^2[(\pi P)/(2S_2)]} \tag{4.187}$$

This result implies that the mesh circuit of Figure 4.28 can be replaced by the equivalent circuit of Figure 4.29, where the resistance and inductance of the end rings have been absorbed into the rotor bar parameters. Because of the symmetry, the currents in this eight-mesh circuit can be solved by solving four meshes, one of which is shown.

4.13 FUNDAMENTAL COMPONENT OF ROTOR AIR GAP MMF

Consider now the fundamental component of the square wave air gap MMF set up by each one of these rotor meshes. If the machine has P poles rather than the simple two-pole example of Figure 4.29, meshes such as those shown exist for each pair of rotor poles. Hence, for mesh one and its companions in other pairs of rotor poles,

$$\mathcal{F}_{b1} = \frac{4}{\pi}\frac{i_{b1}}{2}\sin\frac{P\theta}{2} \qquad 0 \le \theta < 2\pi \tag{4.188}$$

where θ is now the mechanical angular measure along the rotor surface, $0 \leq \theta < 2\pi$, and

$$i_{b1} = I_{mr} \sin \omega t \tag{4.189}$$

For the next successive mesh

$$\mathcal{F}_{b2} = \frac{4}{\pi}\frac{i_{b2}}{2} \sin\left(\frac{P\theta}{2} - \frac{\pi P}{S_2}\right) \tag{4.190}$$

where

$$i_{b2} = I_{mr} \sin\left(\omega t - \frac{\pi P}{S_2}\right) \tag{4.191}$$

This process continues until the S_2/Pth mesh for which

$$\mathcal{F}_{bn} = \frac{4}{\pi}\frac{i_{bn}}{2} \sin\left[\frac{P\theta}{2} - \left(\frac{S_2}{P} - 1\right)\frac{\pi P}{S_2}\right] \tag{4.192}$$

$$i_{bn} = I_{mr} \sin\left[\omega t - \left(\frac{S_2}{P} - 1\right)\frac{\pi P}{S_2}\right] \tag{4.193}$$

The total fundamental component of MMF for all $S_2/2$ such rotor circuits is

$$\mathcal{F}_{pr} = \sum_{k=1}^{S_2/P} \mathcal{F}_{bn} \tag{4.194}$$

or,

$$\mathcal{F}_{pr} = \frac{2}{\pi} I_{mr} \sum_{k=1}^{S_2/P} \sin\left[\omega t - (k-1)\frac{\pi P}{S_2}\right] \sin\left[\frac{P\theta}{2} - (k-1)\frac{\pi P}{S_2}\right] \tag{4.195}$$

which can be written as

$$\mathcal{F}_{pr} = \frac{1}{\pi} I_{mr} \left\{ - \sum_{k=1}^{S_2/P} \cos\left[\omega t + \frac{P\theta}{2} - 2(k-1)\frac{\pi P}{S_2}\right] + \sum_{k=1}^{S_2/P} \cos\left(\omega t - \frac{P\theta}{2}\right) \right\} \tag{4.196}$$

The first summation adds to zero since it corresponds to a set of sine waves equally distributed in phase over P poles. Equation (4.196) therefore reduces to

$$\mathcal{F}_{pr} = \left(\frac{1}{\pi}\right)\left(\frac{S_2}{P}\right) I_{mr} \cos\left(\omega - \frac{P\theta}{2}\right) \tag{4.197}$$

Note that while S_2 is necessarily an integer, the number of bars per pole S_2/P (rotor phases per pole) need not be an integer. An integer number of bars per pole has been assumed in order to simplify the analysis. However, it can be shown that the result is the same for a fractional number of bars per pole.

A non-integer number of bars per pole is typically chosen to prevent subsynchronous cogging or locking torques which may prevent the machine from reaching rated speed but instead run continuously at a small fraction of rated speed. This phenomenon will be discussed in more detail in Chapter 6.

4.14 ROTOR HARMONIC LEAKAGE INDUCTANCE

Because of the presence of a finite number of rotor slots, the stator MMF impressed on the rotor induces additional MMF harmonics in addition to the fundamental component represented by equation (4.197). Since the fundamental component is of interest here, the temporal variation of the currents is again expressed by equation (4.193). However, the higher spatial harmonics are represented by multiplying the argument of equations (4.188), (4.190), etc., by h where the h's are the harmonic numbers which are to be determined. (As in Section 4.13, the values of h would equal 5, 7, 11, etc. if the number of bars per pole were an integer).

The MMF representing the additional spatial harmonic components produced by the rotor fundamental time harmonic component in the rotor bars can be written as

$$\mathcal{F}_{\text{prh}} = \frac{2}{\pi} \sum_{h} \sum_{k=1}^{S_2/P} I_{\text{rh}} \sin\left[\omega t - (k-1)\frac{\pi P}{S_2}\right] \sin\left[\frac{hP\theta}{2} - h(k-1)\frac{\pi P}{S_2}\right] \quad (4.198)$$

Using trigonometric identities, equation (4.198) can be expanded to form,

$$\mathcal{F}_{\text{prh}} = \frac{1}{\pi} \sum_{\text{h}} I_{\text{rh}} \left\{ \sum_{k=1}^{S_2/P} \cos\left[\omega t - \frac{hP\theta}{2} + (k-1)\gamma_1\right] - \cos\left[\omega t + \frac{hP\theta}{2} - (k-1)\gamma_1\right] \right\} \quad (4.199)$$

where

$$\gamma_1 = (h-1)\frac{\pi P}{S_2}$$

$$\gamma_2 = (h+1)\frac{\pi P}{S_2}$$

Expanding again the two cosine terms yields,

$$\mathcal{F}_{\text{prh}} = \frac{1}{\pi} \sum_{\text{h}} I_{\text{rh}} \left\{ \cos\left(\omega t - \frac{hP\theta}{2}\right) \sum_{k=1}^{S_2/P} \cos(k-1)\gamma_1 - \sin\left(\omega t - \frac{hP\theta}{2}\right) \sum_{k=1}^{S_2/P} \sin(k-1)\gamma_1 \right.$$
$$\left. - \cos\left(\omega t + \frac{hP\theta}{2}\right) \sum_{k=1}^{S_2/P} \cos(k-1)\gamma_2 - \sin\left(\omega t + \frac{hP\theta}{2}\right) \sum_{k=1}^{S_2/P} \sin(k-1)\gamma_2 \right\} \quad (4.200)$$

However, from Reference [6], Entries 420.2 and 420.1,

$$\sum_{k=1}^{q} \cos(k-1)\alpha = \frac{\sin\left(\frac{q\alpha}{2}\right)}{\sin\left(\frac{\alpha}{2}\right)} \cos\left[\frac{(q-1)}{2}\alpha\right]$$

$$\sum_{k=1}^{q} \sin(k-1)\alpha = \frac{\sin\left(\frac{q\alpha}{2}\right)}{\sin\left(\frac{\alpha}{2}\right)} \sin\left[\frac{(q-1)}{2}\alpha\right] \qquad \alpha \neq 0$$

Using these results in (4.200), after changing variables, yields

$$\mathcal{F}_{\text{prh}} = \frac{1}{\pi}\sum_{h=1}^{\infty} I_{\text{rh}} \left\{ \frac{\sin\left[(h-1)\frac{\pi}{2}\right]}{\sin(\gamma_1/2)} \left[\cos\left(\omega t - \frac{hP\theta}{2}\right)\cos\left(\frac{S_2 - P}{2P}\gamma_1\right) - \sin\left(\omega t - \frac{hP\theta}{2}\right)\sin\left(\frac{S_2 - P}{2P}\gamma_1\right)\right] - \frac{\sin\left[(h+1)\frac{\pi}{2}\right]}{\sin(\gamma_2/2)} \times\left[\cos\left(\omega t + \frac{hP\theta}{2}\right)\cos\left(\frac{S_2 - P}{2P}\gamma_2\right) + \sin\left(\omega t + \frac{hP\theta}{2}\right)\sin\left(\frac{S_2 - P}{2P}\gamma_2\right)\right]\right\} \tag{4.201}$$

where $\gamma_1, \gamma_2 \neq 2n\pi$ and where n is an integer or zero.

Examining this result more carefully, it should first be noted that the coefficients of the expression within the curly brackets are zero for all odd h, and therefore the sum is identically zero for all the harmonics generated by the spatial distribution of the bars which are all odd due to the regular spacing of the rotor slots. On the other hand if $\gamma_1 = 2n\pi$, then $\gamma_2 = 2n\pi + 2\pi(P/S_2)$ and equation (4.199) becomes

$$\mathcal{F}_{\text{prh}} = \frac{1}{\pi}\sum_{\text{h}} I_{\text{rh}} \left\{ \sum_{k=1}^{S_2/2} \cos\left[\omega t - \frac{hp\theta}{2} + (k-1)2n\pi\right] - \cos\left[\omega t + \frac{hP\theta}{2} - (k-1)\left(2n\pi + 2\pi\frac{P}{S_2}\right)\right]\right\} \tag{4.202}$$

which reduces to

$$\mathcal{F}_{\text{prh}} = \frac{S_2}{\pi P}\sum_{\text{h}} \left\{ I_{\text{rh}} \cos\left(\omega t - \frac{hP\theta}{2}\right)\right\}; \quad h = \frac{2nS_2}{P} + 1 \tag{4.203}$$

and $n = 1, 2, \ldots, \infty$. Similarly, if $\gamma_2 = 2n\pi$, equation (4.201) reduces to

$$\mathcal{F}_{\text{prh}} = \frac{S_2}{\pi P}\sum_{\text{h}} \left\{ I_{\text{rh}} \cos\left(\omega t - \frac{hP\theta}{2}\right)\right\}; \quad h = \frac{2nS_2}{P} - 1 \tag{4.204}$$

For each value of n, the index h takes on two values given by both equations (4.203) and (4.204). For example, if $S_2 = 36$ and $P = 2$ then when $n = 1$, $h = 35$ and 37. Note also that the fundamental component ($h = 1$) is obtained from equation (4.203) by setting $n = 0$.

In order to determine the inductance associated with this component of MMF, it is convenient to use the energy method employed in the study of zigzag leakage reactance. One can state that the energy stored in the field arising from the MMF is defined by

$$W_{\text{m}} = \frac{1}{2} L_{\text{r, har}} I_{\text{r,har}}^2 = \int_V \frac{1}{2}\mu_0 \sum_{\text{h}} H_{\text{h}}^2 dV$$

where $L_{r,har}$ and $I_{r,har}$ are an equivalent inductance and current which accounts for the energy stored in the field due to the harmonics.

The field energy associated with *one phase* of the S_2 bar rotor winding is thus,

$$W_{m/phase} = \frac{1}{2}\frac{L_{r,har}I_{r,har}^2}{(S_2/P)} = \int_V \frac{1}{2}\mu_0 \sum_h \frac{H_h^2}{(S_2/P)} dV$$

However, $H_h g_e = \mathcal{F}_{prh}$ where $\mathcal{F}_{prh}$ is the hth component of the MMF defined by equations (4.203) and (4.204). Twice the field energy per phase can be expanded to form

$$\frac{L_{r,har}I_{r,har}^2}{S_2/P} = \int_V \frac{\mu_0}{(S_2/P)} \sum_h \left(\frac{S_2 I_{rh}}{\pi P g_e}\right)^2 \left[\cos\left(\omega t \pm \frac{hP\theta}{2}\right)\right]^2 dV \qquad (4.205)$$

where either the plus or minus sign applies in the cosine term depending upon whether equation (4.203) or (4.204) has been satisfied. Upon integrating out the radial and longitudinal components,

$$\frac{L_{r,har}I_{r,har}^2}{(S_2/P)} = \int_0^{2\pi} \frac{\mu_0 S_2}{P}\left(\sum_h \left(\frac{I_{rh}}{\pi g_e}\right)^2 \left[\cos\left(\omega t \pm \frac{hP\theta}{2}\right)\right]^2\right)\left(\frac{D_{or}}{2}\right) l_e g_e d\theta \qquad (4.206)$$

Upon carrying out the final integration, it must be remembered that either a plus or a minus sign is attached to each value of h (not both). However, since the cosine term is squared, the net result of each term in the summation ultimately does not depend on the polarity of the phase angle $hp\theta/2$. Equation (4.206) reduces to

$$\frac{L_{r,har}I_{r,har}^2}{S_2/P} = \frac{\mu_0 S_2}{2P}\left[\sum_h \left(\frac{I_{rh}}{\pi g_e}\right)^2\right] D_{or} l_e g_e \pi$$

which can be written as

$$\frac{L_{r,har}}{(S_2/P)} = \mu_0 \left(\frac{D_{or} l_e}{\pi g_e}\right)\left(\frac{S_2}{2P}\right)\left[\sum_h \left(\frac{I_{rh}}{I_{r,har}}\right)^2\right] \qquad (4.207)$$

Recalling that

$$\pi D_{or} = P\tau_{p2}$$

equation (4.207) can be expressed as

$$\frac{L_{r,har}}{(S_2/P)} = \mu_0 \left(\frac{\tau_{p2} l_e}{2\pi^2 g_e}\right) S_2 \left[\sum_h \left(\frac{I_{rh}}{I_{r,har}}\right)^2\right] \qquad (4.208)$$

The rotor pole pitch is more accurately determined as the pitch as measured at the midpoint of the rotor slots.

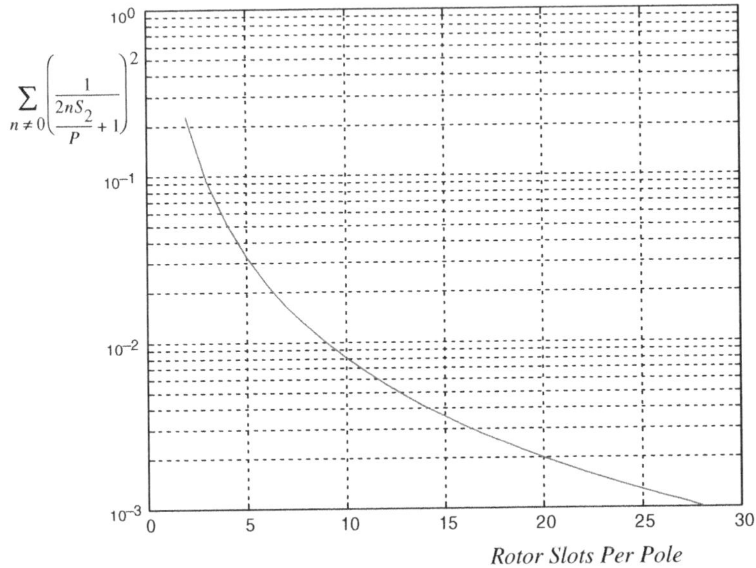

Figure 4.30 Harmonic leakage flux coefficient versus number of rotor slots per pole S_2/P (rotor phases).

Finally, if it can be assumed that the harmonic currents in the bars decrease inversely with the harmonic order h, then the belt harmonic leakage inductance per rotor phase can be written as

$$L_{\text{r,har/phase}} = \mu_0 \left(\frac{1}{2\pi^2}\right)\left(\frac{\tau_{\text{p2}} l_{\text{e}}}{g_{\text{e}}}\right) S_2 \left[\sum_{\text{h}} \left(\frac{1}{h^2}\right)\right] \tag{4.209}$$

where h is given by equations (4.203) and (4.204). The function $\sum_{\text{h}} (1/h^2)$ is plotted in Figure 4.30 as a function of the number of slots per pole S_2/P(rotor phases). Note that the function is continuous, that is, the number of slots per pole can take on non-integer values.

It is generally desirable to express equation (4.209) in terms of the inductance *per bar* rather than *per phase* since the calculation can be done at the same time as when calculating the slot and end leakage equation (4.187). Upon dividing equation (4.209) by the number of poles, results in

$$L_{\text{b(har)}} = \mu_0 \left(\frac{\tau_{\text{p2}} l_{\text{e}}}{g_{\text{e}}}\right)\left(\frac{1}{2\pi^2}\right)\left(\frac{S_2}{P}\right) \sum_{\substack{n=-\infty \\ (n\neq 0)}}^{\infty} \left(\frac{1}{\frac{2nS_2}{P}+1}\right)^2 \tag{4.210}$$

where the parameter h has been replaced with the number n by the use of equation (4.203). The index n takes on all positive and negative values but excludes $n = 0$. Note that the effective gap g_e is used here to account for the effect of slot openings, since this leakage effect contributes flux components in the air gap. The iron paths of the machine are not appreciably affected.

Clearly, the accuracy of this equation hinges upon how the actual harmonic current varies with frequency. While the higher harmonics in the stator MMF waveform also induce rotor currents which, in turn, result in leakage fluxes, this effect is usually relegated to those of a loss component. The subject of losses is treated in Chapter 5.

4.15 CALCULATION OF MUTUAL INDUCTANCES

The MMF that was given by equation (4.197) establishes a second flux density in the air gap which, in turn, links both itself as well as the stator windings. The mutual coupling resulting from rotor currents linking one phase of the stator winding must be calculated. Following the approach of Section 3.10, the air gap flux density corresponding to the rotor MMF, equation (4.197), is

$$B_{gr} = \mu_0 \frac{\mathcal{F}_{pr}}{g_e} \tag{4.211}$$

The flux linking one pole of one of the three stator phases due to all rotor phases is

$$\lambda_{psr} = \int_0^{(2\pi)/P} N_a(\theta) B_{gr}(\theta) l_e \left(\frac{D_{is}}{2}\right) d\theta \tag{4.212}$$

where, from previous work

$$N_a(\theta) = \frac{\mathcal{F}_a(\theta)}{(I_s/C)} = \frac{4}{\pi}\left(\frac{k_1 N_t}{P}\right)\cos\left(\frac{P\theta}{2}\right) \tag{4.213}$$

Equation (4.212) becomes, upon substitution for $B_{gr}(q)$ and $\mathcal{F}_a(q)$,

$$\lambda_{psr} = \frac{4}{\pi}\left(\frac{k_1 N_t}{P}\right)\left(\frac{S_2}{\pi P}\right)\int_0^{(2\pi)/P} I_r \cos\left(\frac{P\theta}{2}\right)\cos\left(\omega_e t - \frac{P\theta}{2}\right)\frac{\mu_0}{g_e} l_e \left(\frac{D_{is}}{2}\right) d\theta \tag{4.214}$$

Note that since the rotor is itself assumed to be rotating, the angular frequency of the rotating flux wave relative to the stator windings is assumed to be line frequency (ω_e).

The above expression reduces to

$$\lambda_{psr} = \frac{2}{\pi}\left(\frac{k_1 N_t}{P}\right)\left(\frac{S_2}{\pi P}\right)\left(\frac{\pi}{P}\right)\left(\frac{\mu_0 l_e D_{is}}{g_e}\right) I_r \cos\omega_e t \tag{4.215}$$

or, introducing the pole pitch

$$\tau_p = \frac{\pi D_{is}}{P}$$

equation (4.215) becomes

$$\lambda_{\mathrm{psr}} = \frac{2}{\pi^2}\left(\frac{k_1 N_{\mathrm{t}} S_2}{P^2}\right)\left(\frac{\mu_0 \tau_{\mathrm{p}} l_{\mathrm{e}}}{g_{\mathrm{e}}}\right) I_{\mathrm{r}} \cos \omega_{\mathrm{e}} t \tag{4.216}$$

The total flux linking all series-connected poles of one phase is

$$\begin{aligned}\lambda_{\mathrm{sr}} &= \frac{P}{C}\lambda_{\mathrm{psr}} \\ &= \frac{2}{\pi^2}\left(\frac{k_1 N_{\mathrm{t}} S_2}{PC}\right)\left(\mu_0\frac{\tau_{\mathrm{p}} l_{\mathrm{e}}}{g_{\mathrm{e}}}\right) I_{\mathrm{r}} \cos \omega_{\mathrm{e}} t\end{aligned} \tag{4.217}$$

The mutual inductance between stator and rotor phases is, by definition

$$L_{\mathrm{sr}} = \frac{\lambda_{\mathrm{sr}}}{I_{\mathrm{r}} \cos \omega_{\mathrm{e}} t} \tag{4.218}$$

or

$$L_{\mathrm{sr}} = \frac{2}{\pi^2}\left(\frac{k_1 N_{\mathrm{s}} S_2}{P}\right)\left(\frac{\mu_0 \tau_{\mathrm{p}} l_{\mathrm{e}}}{g_{\mathrm{e}}}\right) \tag{4.219}$$

where

$$N_{\mathrm{s}} = \frac{N_{\mathrm{t}}}{C}$$

This result can be compared with equation (3.80).

In a similar manner, the flux set up by the stator MMF which couples one of the rotor phases can be calculated. In this case, the total flux linking one of the S_2/P rotor meshes or phases is

$$\lambda_{\mathrm{sr}} = \int_0^{(2\pi)/P} N_{\mathrm{b1}}(\theta) B_{\mathrm{g1}}(\theta) l_{\mathrm{e}} \left(\frac{D_{\mathrm{or}}}{2}\right) d\theta \tag{4.220}$$

where

$$N_{\mathrm{b1}}(\theta) = \frac{2}{\pi} \sin \frac{P\theta}{2} \tag{4.221}$$

From equation (2.73), the MMF associated with the fundamental component of stator current of a three-phase, P-pole machine is

$$\mathcal{F}_{\mathrm{p1}} = \frac{6}{\pi}\frac{k_1 N_{\mathrm{s}}}{P} I_{\mathrm{s}} \sin\left(\frac{P\theta}{2} - \omega_{\mathrm{e}} t\right)$$

and, again

$$B_{\mathrm{g1}} = \mu_0 \frac{\mathcal{F}_{\mathrm{p1}}}{g_{\mathrm{e}}}$$

Setting $t = 0$ for convenience and upon solving, the flux linking one rotor circuit due to the stator MMF is

$$\lambda_{\mathrm{sr}} = \int_0^{(2\pi)/P} \frac{2}{\pi} \sin\left(\frac{P\theta}{2}\right)\left[\frac{\mu_0}{g_{\mathrm{e}}}\frac{6}{\pi}\frac{k_1 N_{\mathrm{s}}}{P} I_{\mathrm{s}} \sin\left(\frac{P\theta}{2}\right)\right] l_{\mathrm{e}} \left(\frac{D_{\mathrm{or}}}{2}\right) d\theta$$

so that

$$\lambda_{\mathrm{sr}} = \frac{6}{\pi^2}\left(\frac{k_1 N_{\mathrm{s}}}{P}\right) I_{\mathrm{s}} \left(\frac{\mu_0 l_{\mathrm{e}} \tau_{\mathrm{p}}}{g_{\mathrm{e}}}\right) \tag{4.222}$$

Hence,

$$L_{\mathrm{sr}} = \frac{6}{\pi^2}\left(\frac{k_1 N_{\mathrm{s}}}{P}\right) \frac{\mu_0 l_{\mathrm{e}} \tau_{\mathrm{p}}}{g_{\mathrm{e}}} \tag{4.223}$$

The quantity L_{sr} expresses the mutual coupling resulting from the flux produced by all stator phases with one of the S_2/P rotor meshes assuming balanced excitation of all three stator phases. Note that this expression is quite different than L_{sr} which expresses the coupling resulting from flux produced by all the rotor meshes linking one of the three stator phases.

The final inductance to be calculated is the magnetizing inductance of one of the rotor meshes (phases). This quantity is found by integrating the air gap flux density produced by all of the rotor meshes over one of the rotor meshes. In this manner, the "mutual coupling" between any two rotor meshes is automatically taken into account.

For example, choosing as reference the rotor phase defined by rotor bar "b1," one has, from equation (4.188)

$$\mathcal{F}_{\mathrm{b1}} = \frac{4}{\pi}\left(\frac{i_{\mathrm{b1}}}{2}\right) \sin\frac{P\theta}{2}$$

so that the *winding function* distribution is, from equation (1.147),

$$N_{\mathrm{b1}}(\theta) = \frac{2}{\pi}\sin\frac{P\theta}{2} \tag{4.224}$$

The flux linkages which couple this P-pole mesh arising from the air gap flux density B_{gr1} is

$$\lambda_{\mathrm{mr}} = \int_0^{2\pi} N_{\mathrm{b1}}(\theta) B_{\mathrm{gr1}}(\theta) l_{\mathrm{e}} \frac{D_{\mathrm{or}}}{2} d\theta \tag{4.225}$$

which reduces to

$$\begin{aligned}
\lambda_{\mathrm{mr}} &= \int_0^{2\pi} \left[\frac{2}{\pi}\sin\left(\frac{P\theta}{2}\right)\right]\left(\frac{\mu_0}{g_{\mathrm{e}}}\right)\left(\frac{S_2}{\pi P}\right) I_{\mathrm{r}} \cos\left(\omega_{\mathrm{e}} t - \frac{P\theta}{2}\right) l_{\mathrm{e}} \frac{D_{\mathrm{or}}}{2} d\theta \\
&= \frac{1}{\pi}\left(\frac{\mu_0 l_{\mathrm{e}} D_{\mathrm{or}}}{g_{\mathrm{e}}}\right)\left(\frac{S_2}{P}\right) I_{\mathrm{r}} \sin(\omega_{\mathrm{e}} t) \\
&= \left(\frac{1}{\pi^2}\right)\left(\frac{S_2}{P}\right)\left(\frac{\mu_0 \tau_{\mathrm{p}} l_{\mathrm{e}}}{g_{\mathrm{e}}}\right) I_{\mathrm{r}} \sin(\omega_{\mathrm{e}} t)
\end{aligned} \tag{4.226}$$

The magnetizing inductance associated with this flux linkage is

$$L_{\mathrm{mr}} = \frac{\lambda_{\mathrm{mr}}}{I_{\mathrm{r}} \sin(\omega_{\mathrm{e}} t)}$$

or

$$L_{mr} = \left(\frac{1}{\pi^2}\right)\left(\frac{S_2}{P}\right)\frac{\mu_0 \tau_p l_e}{g_e} \tag{4.227}$$

The point has now been reached where the total flux linking one of the stator phases as well as the flux which links one of the rotor meshes (phases) has been established. Since sinusoidal currents have been assumed in the rotor meshes, steady-state operation with sine wave excitation is implied. The form of these equations on a "per phase" basis is, in phasor form

$$\tilde{\lambda}_s = L_{ls}\tilde{I}_s + L_{ms}\tilde{I}_s + L_{sr}\tilde{I}_r \tag{4.228}$$

$$\tilde{\lambda}_r = L_{lr}\tilde{I}_r + L_{mr}\tilde{I}_r + L_{rs}\tilde{I}_s \tag{4.229}$$

where L_{ls} is the leakage inductance associated with one of the stator phases and is the sum of the slot, end winding, belt and zigzag leakage inductances (equations (4.89), (4.133), (4.143), and (4.166)) and L_{lr} is the leakage inductance of one rotor mesh (equations (4.36), (4.134), and (4.187) and (4.210)). That is

$$L_{lr} = 2(L_{be} + L_{b(har)}) \tag{4.230}$$

The inductances L_{ms}, L_{sr}, and L_{mr} are fixed by equations (3.80), (4.219), and (4.227), respectively. Substituting explicitly for L_{ms}, L_{sr}, and L_{mr},

$$\tilde{\lambda}_s = L_{ls}\tilde{I}_s + \frac{12}{\pi^2}\frac{(k_1 N_s)^2}{P}\mathcal{P}_p\tilde{I}_s + \frac{2}{\pi^2}\frac{k_1 N_s S_2}{P}\mathcal{P}_p\tilde{I}_r \tag{4.231}$$

$$\tilde{\lambda}_r = L_{lr}\tilde{I}_r + \frac{6}{\pi^2}\frac{(k_1 N_s)^2}{P}\mathcal{P}_p\tilde{I}_s + \frac{1}{\pi^2}\frac{S_2}{P}\mathcal{P}_p\tilde{I}_r \tag{4.232}$$

where the permeance per pole $\mathcal{P}_p$ is

$$\mathcal{P}_p = \mu_0\frac{\tau_p l_e}{g_e}$$

Equation (4.231) can be manipulated so that the magnetizing inductance coefficient is the same for both components of current if it is written as

$$\tilde{\lambda}_s = L_{ls}\tilde{I}_s + \frac{12}{\pi^2}\frac{k_1^2 N_s^2}{P}\mathcal{P}_p \cdot \left(\tilde{I}_s + \frac{S_2}{6k_1 N_s}\tilde{I}_r\right) \tag{4.233}$$

Recalling the turns ratio transformation of a transformer, it is clear that the quantity $S_2/(6k_1 N_s)$ has the appearance of being a turns ratio transformation. However, the issue is more involved in this case since the number of phases of the stator and rotor are different. The term is more easily interpreted if it is written as

$$\frac{S_2}{6k_1 N_s} = \frac{(S_2/P)}{(3)}\frac{(1/2)}{(k_1 N_s/P)} \quad \text{(Phases/pole)(turns/phase)} \tag{4.234}$$

One can again define a modified value of rotor current, referred to stator turns as

$$\tilde{I}_r' = \frac{S_2}{6k_1 N_s}\tilde{I}_r \tag{4.235}$$

Incorporating this modified value of current in the rotor flux linkage equation,

$$\tilde{\lambda}_r = \frac{6k_1N_s}{S_2}(2L_{be} + 2L_{b(har)})\tilde{I}'_r + \frac{6}{\pi^2}\frac{k_1N_s}{P}\mathcal{P}_p\tilde{I}_s + \frac{6k_1N_s}{\pi^2 P}\mathcal{P}_p\tilde{I}'_r \tag{4.236}$$

The mutual inductance coefficient can now be made the same as in equation (4.233) if the entire equation is multiplied by $2k_1 N_s$

$$2k_1N_s\tilde{\lambda}_r = \frac{12k_1^2N_s^2}{S_2}(2L_{be} + 2L_{b(har)})\tilde{I}'_r + \frac{12}{\pi^2}\frac{k_1^2N_s^2}{P}\mathcal{P}_P(\tilde{I}_s + \tilde{I}'_r) \tag{4.237}$$

One can now define

$$\tilde{\lambda}'_r = \frac{k_1N_s}{(1/2)}\tilde{\lambda}_r \tag{4.238}$$

$$L_m = \frac{12}{\pi^2}\frac{k_1^2N_s^2}{P}\mathcal{P}_p \tag{4.239}$$

$$L'_{lr} = \frac{24k_1^2N_s^2}{S_2}(L_{be} + L_{b(har)}) \tag{4.240}$$

The stator and rotor flux linkage equations can now be expressed in the form

$$\tilde{\lambda}_s = L_{ls}\tilde{I}_s + L_{ms}\left(\tilde{I}_s + \tilde{I}'_r\right) \tag{4.241}$$

$$\tilde{\lambda}'_r = L'_{lr}\tilde{I}'_r + L_{ms}\left(\tilde{I}_s + \tilde{I}'_r\right) \tag{4.242}$$

The primed variables are interpreted as equivalent quantities referred to the stator turns by an effective turns ratio which includes not only the number of stator and rotor turns but also the number of stator and rotor phases.

In an analogous manner the voltage equation which describes the current in a given mesh, that is, from equation (4.185)

$$\tilde{V}_{r1}(=0) = 2r_{be}\tilde{I}_r + j\omega\tilde{\lambda}_r \tag{4.243}$$

can be "transformed" to stator turns. The result is

$$\tilde{V}'_r(=0) = r'_r\tilde{I}'_r + j\omega\tilde{\lambda}'_r \tag{4.244}$$

where

$$\tilde{V}'_r = 2k_1N_s\tilde{V}_r \tag{4.245}$$

$$r'_r = \frac{12k_1^2N_s^2}{S_2}r_r \tag{4.246}$$

where $r_r = 2r_{be}$ and λ'_r are defined above.

4.16 EXAMPLE 5—CALCULATION OF ROTOR LEAKAGE INDUCTANCE PER PHASE

Using the results of Section 4.14, one can now attempt to establish the effect of the rotor leakage flux on a per phase rather than a per bar basis. Again the 250 HP machine will be used as the example. From equation (4.187), the effective leakage inductance per rotor mesh is

$$L_{\text{lr}} = 2(L_{\text{be}} + L_{\text{b(har)}})$$
$$= 2L_{\text{b}} + \frac{L_{\text{e}}}{\sin^2\left(\dfrac{\pi P}{2S_2}\right)} + 2L_{\text{b(har)}}$$

From Example 2, it was specified that the example machine has 97 rotor slots and eight poles. From Example 4, the slot leakage and the end-winding leakages per bar are computed as

$$L_{\text{b}} = 0.32\ \mu\text{H}$$
$$L_{\text{e}} = 0.01252\ \mu\text{H}$$

In addition, the phase belt rotor inductance is, from equation (4.210),

$$L_{\text{b(har)}} = \mu_0\left(\frac{\tau_{\text{p2}} l_{\text{e}}}{g_{\text{e}}}\right)\left(\frac{1}{2\pi^2}\right)\left(\frac{S_2}{P}\right)\sum_{n=-\infty,n\neq 0}^{\infty}\left(\frac{1}{\dfrac{2nS_2}{P}+1}\right)^2$$
$$= \mu_0\left[\left(\frac{9.425}{39.37}\right)\frac{8.68}{0.0677}\right]\left(\frac{1}{2\pi^2}\right)\left(\frac{97}{8}\right)0.006$$
$$= 0.142\ \mu\text{H}$$

The factor 0.006 is obtained from Figure 4.30. Hence the equivalent rotor phase inductance including the effect of the end-ring current and harmonic leakage is

$$L_{\text{lr}} = 2\left[0.32 + \frac{0.01252}{2\sin^2\left(\dfrac{8\pi}{2(97)}\right)} + 0.142\right] \times 10^{-6}$$
$$= 1.654\ \mu\text{H}$$

The rotor leakage per stator phase (referred to stator turns) is obtained from equation (4.240) as

$$L'_{\text{lr}} = \frac{12k_1^2 N_{\text{s}}^2}{S_2} L_{\text{lr}}$$
$$= \frac{(12)(0.91)^2(240)^2}{(97)}(1.654) \times 10^{-6}$$
$$= 9.7\ \text{mH}$$

4.17 SKEW LEAKAGE INDUCTANCE

Although the four types of leakage inductances previously discussed are unavoidable, the skew leakage inductance occurs from an intentional twisting or "skewing" of either stator or rotor windings in the circumferential direction. Skewing serves to prevent undesirable pulsations of flux which are, in general, caused by the presence of space harmonics of MMF. These pulsations, if not sufficiently suppressed, may result not only in objectionable noise but also in appreciable negative torques produced by the locking effect between stator and rotor harmonic fluxes which rotate opposite to the direction of the main field. These negative torques, also called parasitic torques, if not minimized, would have the effect of producing pronounced dips in the torque-speed curves and interfere with the smooth acceleration of the rotor during the starting period. These dips could be so severe that they may even result in subsynchronous operation or "crawling."

The beneficial effects of skewing are, on the other hand, offset by an inevitable increase in the total leakage inductance of the motor, thereby reducing both the starting torque and the breakdown torque. The reasons for this increased leakage inductance may be understood by considering the coupling that would occur between the stator and rotor without skewing, then introducing skewing and noting the effect on the rotating flux.

Without skewing, the equations which describe coupling between one of the stator phases, say the a phase and the corresponding rotor a phase are, from equations (4.241) and (4.242)

$$\lambda_{as} = L_{ls} i_{as} + L_{ms}(i_{as} + i'_{ar}) \tag{4.247}$$

$$\lambda'_{ar} = L'_{lr} i'_{ar} + L_{ms}(i_{as} + i'_{ar}) \tag{4.248}$$

Note that it has already been assumed that the rotor equation variables have been referred to the stator by the ratio of effective stator turns to rotor turns.

The introduction of skewing of the stator or rotor windings (or both) will affect only the mutual coupling between the stator and rotor. The flux linkages with skew can be written

$$\lambda_{as} = L_{ls} i_{as} + L_{ms} i_{as} + k_{s1} L_{ms} i'_{ar} \tag{4.249}$$

$$\lambda'_{ar} = L'_{lr} i'_{ar} + k_{s1} L_{ms} i_{as} + L_{ms} i'_{ar} \tag{4.250}$$

Equations (4.249) and (4.250) can be written in the form

$$\lambda_{as} = L_{ls} i_{as} + L_{ms} i_{as} + k_{s1}^2 L_{ms} \frac{i'_{ar}}{k_{s1}} \tag{4.251}$$

$$k_{s1} \lambda'_{ar} = k_{s1}^2 L'_{lr} \frac{i'_{ar}}{k_{s1}} + k_{s1}^2 L_{ms} \left(i_{as} + \frac{i'_{ar}}{k_{s1}} \right) \tag{4.252}$$

which in turn can be rearranged to the form

$$\lambda_{as} = L_{ls} i_{as} + (1 - k_{s1}^2) L_{ms} i_{as} + k_{s1}^2 L_{ms} \left(\frac{i'_{ar}}{k_{s1}} + i_{as} \right) \tag{4.253}$$

$$k_{s1}\lambda'_{ar} = k_{s1}^2 L'_{lr}\frac{i'_{ar}}{k_{s1}} + k_{s1}^2 L_{ms}\left(i_{as} + \frac{i'_{ar}}{k_{s1}}\right) \tag{4.254}$$

If one defines

$$\lambda''_{ar} \underline{\Delta}\, k_{s1}\lambda'_{ar}; \quad i''_{ar} \underline{\Delta}\, \frac{i'_{ar}}{k_{s1}}; \quad L''_{lr} \underline{\Delta}\, k_{s1}^2 L'_{lr}; \quad L''_{ms} \underline{\Delta}\, k_{s1}^2 L_{ms}$$

then

$$\lambda_{as} = L_{ls}i_{as} + (1 - k_{s1}^2)L_{ms}i_{as} + L''_{ms}(i''_{ar} + i_{as}) \tag{4.255}$$

$$\lambda''_{ar} = L''_{lr}i''_{ar} + L''_{ms}(i''_{ar} + i_{as}) \tag{4.256}$$

The double primed quantities are referred rotor variables which include the effects of rotor skew. Note the presence of the extra term in the stator flux linkage equation proportional to $(1 - k_{s1}^2)$. The additional leakage caused by skewing is clearly

$$L_{lsk} = (1 - k_{s1}^2)L_{ms} \tag{4.257}$$

$$= L_{ms}\left\{1 - \left[\frac{\sin(\alpha_s/2)}{\alpha_s/2}\right]^2\right\} \tag{4.258}$$

If α is sufficiently small then

$$\sin\alpha_s/2 = \frac{\alpha_s}{2} - \frac{1}{3!}\left(\frac{\alpha_s}{2}\right)^3$$

and

$$\left[\frac{\sin\alpha_s/2}{\alpha_s/2}\right]^2 = \left[1 - \frac{\alpha_s^2}{24}\right]^2$$

$$= 1 - \frac{\alpha_s^2}{12} + \frac{\alpha_s^4}{24^2} \tag{4.259}$$

$$= 1 - \frac{\alpha_s^2}{12}$$

so that approximately

$$L_{lsk} = \frac{\alpha_s^2}{12}L_{ms} \tag{4.260}$$

In like manner, the voltage equation for the rotor circuit can be written, for rotor phase a, as

$$k_{s1}v'_{ar} = k_{s1}^2 r'_r\left(\frac{i'_{ar}}{k_{s1}}\right) + \frac{d}{dt}(k_{s1}\lambda'_{ar}) \tag{4.261}$$

which becomes

$$v''_{ar} = 0 = r''_r i''_{ar} + \frac{d\lambda''_{ar}}{dt} \tag{4.262}$$

where

$$r_r'' = k_{s1}^2 r_r' \tag{4.263}$$

It is important to mention that the method used to incorporate skew leakage is not unique. In fact, it can be shown that if the effective number of turns in the mutual inductance is retained as $k_{s1}N_s$ then a skew leakage term appears in the rotor equation by

$$L_{lsk} = L_{ms}\frac{1-k_{s1}^2}{k_{s1}} \tag{4.264}$$

If $\sqrt{k_{s1}}N_s$ is used as the effective number of turns, then equal values of skew inductance appear in both the stator and rotor equations. This approach is sometimes preferred (e.g., by Alger [1]). Here, however, the skew leakage term will be lumped in the stator flux linkage equation.

4.18 EXAMPLE 6—CALCULATION OF SKEW LEAKAGE EFFECTS

Again the parameters of the 250 HP machine of the previous examples will be used. It is assumed now that the rotor slots are skewed one stator slot pitch. Since there are 15 stator slots per pole, the skew angle α in degrees is 1/15 of 180° or

$$\alpha_s = \frac{180}{15} = 12^\circ$$

The inductance due to skew is found from equation (4.258)

$$\begin{aligned}L_{lsk} &= L_{ms}\left\{1-\left[\frac{\sin(\alpha_s/2)}{\alpha/2}\right]^2\right\}\\ &= L_{ms}\left\{1-\left[\frac{\sin(\pi/30)}{\pi/30}\right]^2\right\}\\ &= 0.199(1-0.9963)\\ &= 0.736 \text{ mH}\end{aligned}$$

Because of skew it is important to remember that the magnetizing inductance and rotor leakage inductance must, in principle, be modified appropriately. When skew is considered

$$\begin{aligned}L_{ms}'' &= k_{s1}^2 L_{ms}\\ &= (0.9963)0.199\\ &= 0.198 \text{ mH}\end{aligned}$$

so that typically this modification can be neglected unless the skew is several slot pitches.

The total stator leakage inductance can be found as the sum of the slot, end winding, belt, zigzag, and skew leakage inductances or

$$\begin{aligned} L_{\mathrm{ls}} &= L_{\mathrm{lsl}} + L_{\mathrm{lew}} + L_{\mathrm{lbt}} + L_{\mathrm{lzz}} + L_{\mathrm{lsk}} \\ &= 3.4 + 0.796 + 0.0 + 1.094 + 0.736 \qquad (4.265) \\ &= 6.026 \text{ mH} \end{aligned}$$

For purposes of comparison, it is useful to express the machine inductances in per unit form. The base impedance of a 250 HP, 2400 V, 60 Hz machine is

$$\begin{aligned} Z_{\mathrm{b}} &= \frac{V_{\mathrm{llb}}^2}{P_{\mathrm{b}}} \\ &= \frac{2400^2}{(746)(250)} = 30.88\ \Omega \end{aligned}$$

Hence, the per unit values of stator leakage, rotor leakage, and magnetizing reactance are

$$\begin{aligned} X_{\mathrm{ls(pu)}} &= \frac{\omega_{\mathrm{e}} L_{\mathrm{ls}}}{Z_{\mathrm{b}}} \\ &= \frac{(377)0.00603}{30.88} = 0.074 \text{ p.u.} \\ X_{\mathrm{lr(pu)}'} &= \frac{(377)(0.0099)}{30.88} = 0.12 \text{ p.u.} \\ X_{\mathrm{m(pu)}} &= \frac{(377)(0.198)}{30.88} = 2.42 \text{ p.u.} \end{aligned}$$

Note that the total per unit leakage reactance in per unit is $0.074 + 0.12 = 0.194$. For most practical designs this value is typically in the range 0.15 and 0.3 p.u. For well-designed machines of this rating, the magnetizing inductance is generally between 2.0 and 3.0 p.u.

4.19 CONCLUSION

It is clear that the precise computation of the leakage inductances of a squirrel-cage induction machine is a challenging task. Much of the lore of machine design over the past century has been justly devoted to this topic. While modern computational tools such as finite-element methods can automate the calculation of the overall effects of leakage flux, little insight can be gained from an overall flux plot. This chapter is intended to provide the student with a good understanding of the complex issues associated with motor design which is, in turn, intimately related to the intelligent management of leakage inductances.

REFERENCES

[1] P. L. Alger, *The Nature of Induction Machines*, 2nd edition, Gordon and Breach Publishers, 1970.
[2] P. Hammond, "Electric and Magnetic Images," Monograph No. 379, Institution of Electrical Engineers, London, May 1960, pp. 306–313.

[3] D. Ban, D. Zarko, and I. Mandic, "Turbo generator end-winding leakage inductance calculation using a 3-D analytical approach based on the solution of Neumann integrals," *IEEE Transactions on Energy Conversion*, vol. 20, no. 1, March 2005, pp. 98–105.

[4] G. A. Campbell, "Mutual inductances of circuits composed of straight wires," *Physical Review*, vol. 5, 1915, pp. 452–458.

[5] M. Liwschitz-Garik and C. C. Whipple, *Electric Machinery, 2nd edition, vol. 2, AC Machines*, Van Nostrand Publishers, 1961.

[6] H. B. Dwight, *Tables of Integrals and Other Mathematical Data*, 4th edition, Macmillan, 1961.

[7] C. R. Paul, *Inductance Loop and Partial*, John Wiley & Sons, 2010.

[8] S. Williamson and M. A. Muller, "Calculation of the impedance of rotor cage end rings," IEE Proceedings-B, vol. 140, no. 1, January 1993.

[9] F. W. Grover, "Inductance Calculations," Dover Publications, New York, 1946.

CHAPTER 5

CALCULATION OF INDUCTION MACHINE LOSSES

THE FINAL STEP IN A COMPLETE description of the circuit parameters of an induction machine is the calculation of its resistances. At first, this computation may appear to be almost a trivial task since the resistance of a wire of known length and having uniform cross section can be calculated from elementary circuit theory. Alas, nothing could be farther from the truth. The computation of losses is probably the most difficult and challenging aspect of machine design. In fact, the subject is much too lengthy to treat in detail here and it will be necessary to discuss much of the subject in qualitative terms.

5.1 INTRODUCTION

In general, the power losses in an induction motor can be relegated to five elements

1. Stator conductor loss
2. Rotor conductor loss
3. Iron loss
4. Stray-load loss
5. Friction and windage loss

Aside from the friction and windage loss which accounts for mechanical losses, the losses of the machine arise from two basic phenomena: ohmic loss in the conductors (1 and 2 above) and magnetic loss in the laminations. The magnetic loss, in turn, consists of hysteresis loss which is a true magnetic loss and ohmic loss due to eddy currents (3 above). The fourth source of loss, namely stray load loss, lumps all of the electromagnetic losses which occur because the machine does not have a smooth air gap and sinusoidal winding distributions. The term is really a misnomer since this type of loss includes no-load as well as load losses. As can be imagined, this term is by far the most complex part of the loss picture.

Theoretically, it is possible to solve Maxwell's equations and thus find the magnetic and electric fields at every point in the iron and conductor. It would then be a short step to accurately calculate all the electromagnetic losses. Until recently, this

Introduction to AC Machine Design, First Edition. Thomas A. Lipo.

would be an impossible task. It remains, even in the age of high speed computers, a formidable assignment and has only recently been tackled. An alternate approach is to consider losses to be caused by separate, independent phenomena. Each phenomenon is treated in a relatively simple manner physically. The physical parameters can then be reduced to a set of elements such as resistances and reactances to be used in an equivalent circuit. Unfortunately, the loss effects are not, in fact, independent since saturation makes the phenomena nonlinear. Correlation of any newly developed theory with test results is very difficult since it is possible, in general, to measure only the total or some aggregation of several loss components. Although some experimental work has been done on separately measuring several of the more important loss components, this work has generally not progressed to the point of being useful to the machine designer.

5.2 EDDY CURRENT EFFECTS IN CONDUCTORS

Although ohmic losses in conductors are basically easy to calculate, the problem is complicated by the fact that the current distribution in the conductors is often nonuniform. The effect of slot leakage flux from current-carrying conductors placed in the slots has already been considered in Chapter 4. It was observed that more flux links with the conductor in the bottom of the slot than a conductor in the top portion of the slot. Thus, in effect, a higher voltage is induced in the bottom conductor resulting in a difference in potential between top and bottom conductors. If these two conductors are electrically insulated, the insulation medium will withstand this difference in potential. However, should these two conductors be connected together in parallel, this difference in potential drives a circulating current around the mesh formed by the two parallel conductors. These circulating currents are called eddy currents and they produce additional copper losses in the conductors which are known as eddy current losses. When this phenomenon acts to cause current flow near the surface of conductors, it is also known by the term *skin effect.*

To investigate this problem, consider initially a solid conductor having a height d and width b lying in an open slot of depth d_s and width b_s as shown in Figure 5.1. Within the conductor, Ampere's law states that

$$\nabla \times \boldsymbol{H} = \boldsymbol{J} \tag{5.1}$$

From the constituent equations one has, in SI units

$$\boldsymbol{H} = \frac{\boldsymbol{B}}{\mu_0} \tag{5.2}$$

and

$$\boldsymbol{J} = \boldsymbol{E}/\rho \tag{5.3}$$

where the resistivity $\rho = 1/\sigma$. Taking the curl of both sides of equation (5.1)

$$\nabla \times \nabla \times \left(\frac{\boldsymbol{B}}{\mu_0}\right) = \nabla \times \left(\frac{\boldsymbol{E}}{\rho}\right) \tag{5.4}$$

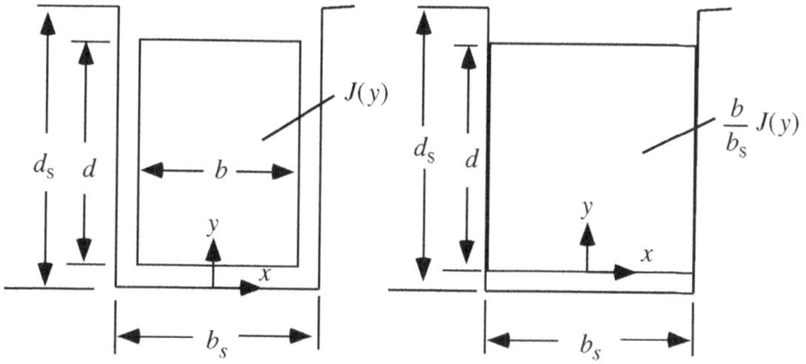

Figure 5.1 Rotor bar configuration and equivalent arrangement for purposes of analysis.

However, from Faraday's law in vector form

$$\nabla \times \boldsymbol{E} = -\frac{\partial \boldsymbol{B}}{\partial t}$$

In addition, an identity in vector algebra states that

$$\nabla \times \nabla \times \boldsymbol{V} = \nabla(\nabla \bullet V) - \nabla^2 \boldsymbol{V}$$

where $\boldsymbol{V}$ is any vector. Hence, equation (5.4) can be expressed as

$$\nabla \times (\nabla \bullet \boldsymbol{B}) - \nabla^2 \boldsymbol{B} = \frac{\mu_0}{\rho}\left(-\frac{\partial \boldsymbol{B}}{\partial t}\right) \tag{5.5}$$

Conservation of flux or Gauss' law for magnetic fields states that

$$\nabla \bullet \boldsymbol{B} = 0$$

Hence, equation (5.5) reduces to

$$\nabla^2 \boldsymbol{B} = \frac{\mu_0}{\rho}\frac{\partial \boldsymbol{B}}{\partial t} \tag{5.6}$$

If it is again assumed that all flux lines go straight across the slot as was done in Chapter 4, then the magnetic flux density has only an x component which is a function only of the depth y as shown on Figure 5.1. Hence, for the fields within a rectangular slot the partial differential equation to be solved is

$$\frac{\partial^2 B_x}{\partial y^2} = \frac{\mu_0}{\rho}\frac{\partial B_x}{\partial t} \tag{5.7}$$

For fields outside the current-carrying region, one can assume that equation (5.7) reduces to

$$\frac{\partial^2 B_x}{\partial y^2} = 0 \tag{5.8}$$

since the current density $\boldsymbol{J}$ in this region is zero and the field in the region above the slot can be found by other means.

Simultaneous solution of equations (5.7) and (5.8) is a difficult problem since the solutions in the two regions must be matched at the boundaries, that is, at

$x = b_s/2$ and $x = b/2$. In the application of interest, the space between the conductor and the walls of the slot are quite small. The actual problem will now be replaced by an equivalent one in which the current is uniformly distributed lengthwise over the slot width and is reduced in magnitude such that the total current is the same in the slot. That is,

$$J_e = \left(\frac{b}{b_s}\right) J_a \tag{5.9}$$

With this value of current density, it is now possible to solve equation (5.7) over the entire slot width $-b_s/2 < x < b_s/2$.

Since the excitation is sinusoidal, the response can also be assumed as sinusoidal in time, that is

$$B_x(x, y, t) = Re[\widetilde{B}_m(x, y)\, e^{j\omega t}]$$

and therefore

$$\frac{\partial B_x}{\partial t} = Re[j\omega \widetilde{B}_m(x, y)\, e^{j\omega t}]$$

Equation 5.7 can be written in the form

$$Re\left[\frac{\partial^2}{\partial y^2}\widetilde{B}_m(x, y)\, e^{j\omega t}\right] = \frac{\mu_0}{\rho} Re[j\omega \widetilde{B}_m(x, y)\, e^{j\omega t}] \tag{5.10}$$

If the real part of the bracketed quantity is assumed and the rotating complex time function $e^{j\omega t}$ canceled, equation (5.10) becomes

$$\frac{\partial^2}{\partial y^2}\widetilde{B}_m = \frac{j\omega\mu_0}{\rho}\widetilde{B}_m \tag{5.11}$$

where the flux density $\widetilde{B}_m$ is complex.

The general solution to equation (5.11) is

$$\widetilde{B}_m(x, y) = \widetilde{P}\cosh(\gamma_o y) + \widetilde{Q}\sinh(\gamma_o y) \tag{5.12}$$

where

$$\gamma_o = \sqrt{\frac{j\omega\mu_0}{\rho}} \tag{5.13}$$

Since $B = 0$, when $y = 0$,

$$\begin{aligned}\widetilde{B}_m(x, 0) &= \widetilde{P}\cosh(0) + \widetilde{Q}\sinh(0)\\ 0 &= \widetilde{P}\cdot 1 + \widetilde{Q}\cdot 0\end{aligned} \tag{5.14}$$

Hence $\widetilde{P} = 0$. Alternatively, when $y = d$, then all of the current is enclosed and

$$\frac{\widetilde{B}_m(x, d)}{\mu_0} b_s = I_m$$

so that

$$\frac{\mu_0 I_m}{b_s} = \widetilde{Q}\sinh(\gamma_o d)$$

and

$$\widetilde{Q} = \frac{\mu_0 I_m}{b_s} \frac{1}{\sinh(\gamma_o d)} \tag{5.15}$$

The solution for the amplitude of the flux density in complex form is therefore

$$\widetilde{B}_m = \frac{\mu_0 I_m}{b_s} \frac{\sinh(\gamma_o y)}{\sinh(\gamma_o d)} \tag{5.16}$$

From Ampere's law,

$$\nabla \times \overline{H} = \overline{J} \tag{5.17}$$

which for this case reduces to the complex form

$$\frac{\partial}{\partial y} \widetilde{H}_x = \widetilde{J}_e \tag{5.18}$$

or since

$$\widetilde{H}_x = \frac{\widetilde{B}_m}{\mu_0}$$

one obtains

$$\widetilde{J}_e = \frac{1}{\mu_0} \frac{\partial \widetilde{B}_m}{\partial y} \tag{5.19}$$

Substituting equation (5.16) into equation (5.19) the equivalent current density in the slots is

$$\widetilde{J}_e = \frac{\gamma_o I_m}{b_s} \left[\frac{\cosh(\gamma_o y)}{\sinh(\gamma_o d)} \right] \tag{5.20}$$

Recall that the actual current density is related to the equivalent current density by

$$\widetilde{J}_a = \frac{b_s}{b} \widetilde{J}_e$$

so that the actual current density in the conductor is

$$\widetilde{J}_a = \frac{\gamma_o I_m}{b} \left[\frac{\cosh(\gamma_o y)}{\sinh(\gamma_o d)} \right] \tag{5.21}$$

The total voltage drop along the bar can be found from Ohm's law and Faraday's law since these two effects act simultaneously on any current filament in the bar. The total IR drop across the length of the bar at any height y is

$$\widetilde{V}_R = \int_0^{l_i + n l_o} \widetilde{E} \bullet dl$$

Note that used here is the actual rotor length and therefore includes that portion of the bar (corresponding to the ducts and end extensions) which are not embedded in iron. Defining $l_b = l_i + nl_o$,

$$\widetilde{V}_R = \rho \widetilde{J}_a \int_0^{l_b} dl$$

or

$$\widetilde{V}_R = \frac{\gamma_o \rho I_m l_b}{b} \left[\frac{\cosh(\gamma_o y)}{\sinh(\gamma_o d)} \right] \tag{5.22}$$

Note that the resistive drop is a maximum at the top of the slot.

The total flux crossing the slot above the height y which therefore links the current below y is

$$\begin{aligned} \widetilde{\phi}_m(y) &= \int_0^{l_e} \int_y^d \widetilde{B}_m dy\, dz \\ &= \frac{\mu_0 I_m l_b}{\gamma_o b_s} \left[\frac{\cosh(\gamma_o d) - \cosh(\gamma_o y)}{\sinh(\gamma_o d)} \right] \end{aligned} \tag{5.23}$$

The total IX drop across the length of the bar at any height y is therefore

$$\widetilde{V}_L = \frac{\partial}{\partial t} \int_0^{l_e} \int_y^d \widetilde{B}_m dy\, dz$$

or in complex form

$$\widetilde{V}_L = j\omega \widetilde{\phi}_m$$

whereupon

$$\widetilde{V}_L = \frac{j\omega \mu_0}{\gamma_o} \frac{I_m l_b}{b_s} \left[\frac{\cosh(\gamma_o d) - \cosh(\gamma_o y)}{\sinh(\gamma_o d)} \right]$$

Note that the reactive drop is zero at the top of the slot and a maximum at the bottom of the slot.

The total voltage drop along the bar at an arbitrary height y is clearly the sum of the inductive and resistive drop or

$$\begin{aligned} \widetilde{V}_{bar} &= \widetilde{V}_R + \widetilde{V}_L \\ &= \frac{\gamma_o \rho I_m l_b}{b} \left[\frac{\cosh(\gamma_o y)}{\sinh(\gamma_o d)} \right] + \frac{j\omega \mu_0}{\gamma_o} \frac{I_m l_b}{b_s} \left[\frac{\cosh(\gamma_o d) - \cosh(\gamma_o y)}{\sinh(\gamma_o d)} \right] \end{aligned} \tag{5.24}$$

But

$$\frac{j\omega\mu_0}{\gamma_o} = \frac{j\omega\mu_0}{\sqrt{(j\omega\mu_0)/\rho}} = \rho\sqrt{\frac{j\omega\mu_0}{\rho}} = \rho\gamma_o \tag{5.25}$$

and if

$$\frac{l_b}{b} \approx \frac{l_e}{b_s} \tag{5.26}$$

then,

$$\tilde{V}_{bar} \approx \frac{\gamma_o \rho I_m l_b}{b} \frac{\cosh(\gamma_o d)}{\sinh(\gamma_o d)} \tag{5.27}$$

Note that the total voltage drop down the length of the bar is constant, which is to be expected since all of the current filaments are "connected" in parallel. While the assumption of equation (5.26) is rarely true, this result would have been an exact equality had simplifying assumptions not been made and the exact problem solved.

Assuming now an equality, equation (5.27) can be written in terms of the DC resistance of the bar if one recalls that

$$R_{dc} = \frac{\rho l_b}{bd} \tag{5.28}$$

Hence, the effective impedance of the bar is

$$\tilde{Z}_{bar} = \frac{\tilde{V}_{bar}}{I_m} = R_{dc}\left[\frac{(\gamma_o d)\cosh(\gamma_o d)}{\sinh(\gamma_o d)}\right] \tag{5.29}$$

The real portion of the impedance representing the AC resistance of the bar can be readily evaluated as

$$R_{ac} = \alpha_o d R_{dc}\left[\frac{\sinh(2\alpha_o d) + \sin(2\alpha_o d)}{\cosh(2\alpha_o d) - \cos(2\alpha_o d)}\right] \tag{5.30}$$

where

$$\alpha_o = \sqrt{(\omega\mu_0)/(2\rho)}$$

The imaginary component of the impedance represents the reactance of the bar and is

$$X_{ac} = \alpha_o d R_{dc}\left[\frac{\sinh(2\alpha_o d) - \sin(2\alpha_o d)}{\cosh(2\alpha_o d) - \cos(2\alpha_o d)}\right] \tag{5.31}$$

It is useful to consider the asymptotes of equations (5.30) and (5.31) for very small and very large values of αd (low and high values of frequency respectively). When $\alpha_o d$ is small, equations (5.30) and (5.31) become

$$R_{ac} \approx R_{dc}\left[1 + \frac{4}{45}(\alpha_o d)^4 - \frac{4}{4{,}725}(\alpha_o d)^8 + \dots\right] \quad (\alpha_o d) < 1.5 \qquad (5.32)$$

$$X_{ac} \approx \frac{2}{3}(\alpha_o d)^2 R_{dc}\left[1 - \frac{8(\alpha_o d)^4}{315} + \frac{32(\alpha_o d)^8}{31{,}185}\right] \quad (\alpha_o d) < 1.5 \qquad (5.33)$$

The coefficient in front of the bracketed term can be written as

$$\frac{2}{3}(\alpha_o d)^2 R_{dc} = \left(\frac{2}{3}\right)\left[\sqrt{\frac{\omega\mu_0}{2\rho}}d\right]^2 \frac{\rho l_b}{d \cdot b}$$
$$= \omega\left(\frac{\mu_0 d l_b}{3b}\right)$$

so that equation (5.33) becomes

$$X_{ac} \approx \omega L_{dc}\left[1 - \frac{8(\alpha_o d)^4}{315} + \frac{32(\alpha_o d)^8}{31{,}185} + \dots\right] \quad (\alpha_o d) < 1.5 \qquad (5.34)$$

Hence, as frequency is reduced the AC values of resistance and reactance tend toward their DC values.

If $\alpha_o d$ is large equations (5.30) and (5.31) can be approximated as

$$R_{ac} = \alpha_o d R_{dc} \qquad (5.35)$$

and

$$X_{ac} = \alpha_o d R_{dc} = \frac{3}{2}\frac{\omega L_{dc}}{\alpha_o d} \qquad (5.36)$$

Note that the resistance and reactance approach an equality as the frequency (or the bar depth) increases. The amplitude of the impedance increases as the square root of the frequency and the phase angle approaches 45°. Equation (5.35) can, by use of equation (5.28), be expressed in the form

$$R_{ac} = \frac{\rho l_b}{bd}(\alpha_o d)$$
$$= \rho\left(\frac{l_b}{b(1/\alpha_o)}\right) \qquad (5.37)$$

Hence, for sufficiently high frequencies, the AC resistance can be calculated in the same manner as for the DC resistance if one replaces the actual depth of the bar d by an equivalent depth $1/\alpha_o$. The quantity $1/\alpha_o$ is called the *skin depth* and the fact that the current distributes itself unevenly over a conductor due to sinusoidal excitation is called the *skin effect*.

At 75°C the resistivity of copper is 2.1 10^{-6} ohm-cm. Hence the critical thickness of a copper bar below which skin effect begins to dominate when excited at 60 Hz can be computed as

$$1/\alpha_o = \sqrt{\frac{(2)(2.1 \cdot 10^{-6})}{(4\pi \cdot 10^{-9})(377)}} = 0.9416\,\text{cm}$$
$$= 0.3707''$$

It is important to observe that since the reactance increases as the square root of frequency at high frequencies, then effectively the slot inductance actually decreases inversely with the square root of frequency. A plot of R_{AC}, X_{AC} and L_{AC} normalized with respect to their DC values is given in Figure 5.2. The plots have the same character as Bode plots of first-order lead or lag circuits but are instead a function of the square root of frequency.

When the skin effect phenomenon occurs in the rotor, it is usually referred to as the "deep bar effect." An improvement in the starting performance of a squirrel-cage induction motor can actually be achieved if proper use is made of this phenomenon. In general, two general categories of rotors which utilize skin effect are used, the *deep bar rotor* in which a single bar is shaped to have long bar depth (deep bar) and short width and the *double-cage rotor* in which a second bar is located below the shallow bar near the surface. A variety of rotor bar shapes have been used over the years and are illustrated in Figure 5.3. The area of the deep bar as well as the area of the two or three bars belonging to a unit in the double cage rotor is chosen so that the I^2r losses at small slip do not exceed the amount permissible with respect to the desired

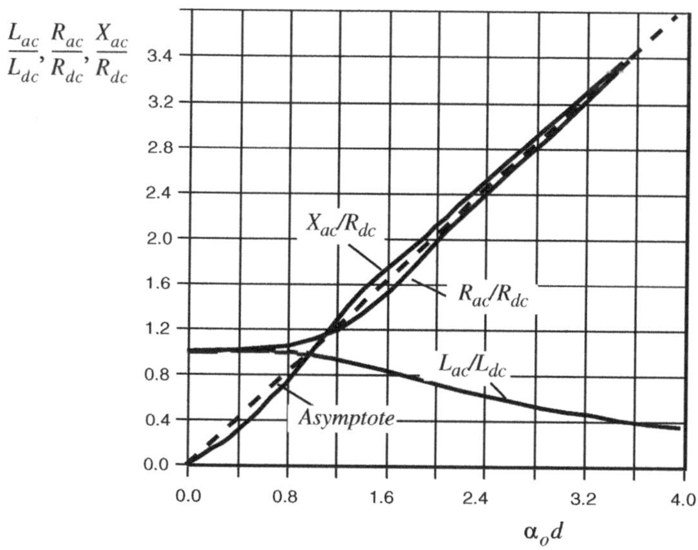

Figure 5.2 Sketch of normalized inductance, resistance and reactance of solid bar in a slot as a function of αd.

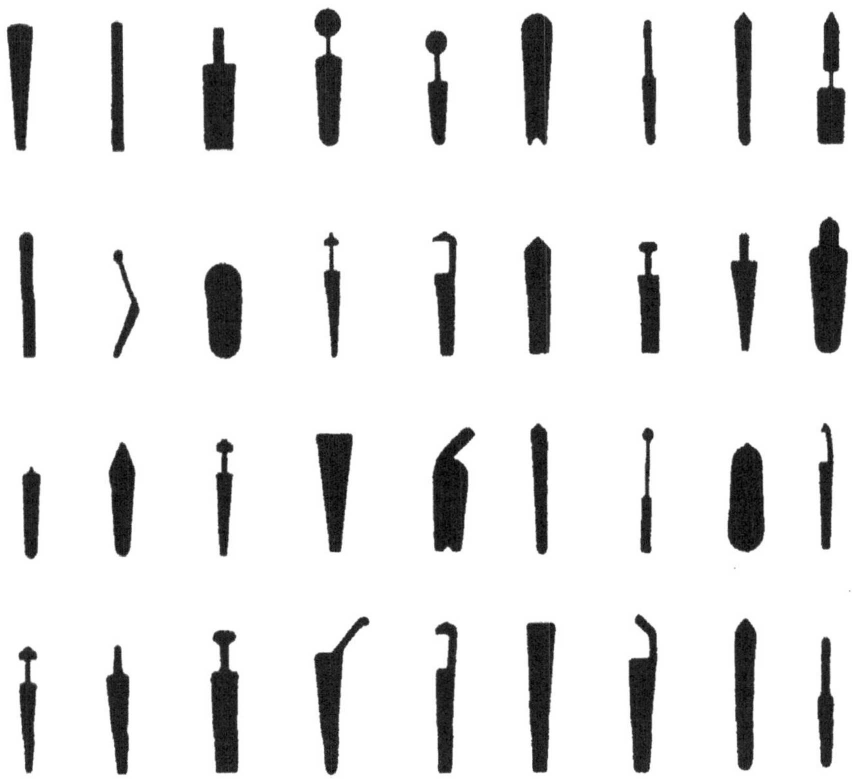

Figure 5.3 A variety of rotor shapes used to enhance skin effect.

efficiency and heat dissipation limit of the machine. Since the slip frequency is very low at normal speed where no skin effect occurs, the deep bar rotor and the double cage rotor behave at normal load just as the round bar or shallow bar rotor previously discussed. However, the behavior of both kinds of rotors is entirely different at high slips, for example, at standstill where the frequency of the rotor currents is the same as the line frequency. In this case, the current density given by equation (5.20) may be thought of as the superposition of a uniform (average) current and a circulating current flowing in an additive fashion at the top of the bar and negatively at the bottom, directed in such a way as to oppose the time rate of change of slot leakage flux. The skin effect forces current to flow in the part of the conductor area which lies at the top of the slot and the effective resistance is many times (usually three to four times) as large as the round bar or shallow bar rotor. In the case of the double bar rotor, the resistance of the top bar usually has a higher resistance than the bottom bar making the effective resistance at starting even larger if desired.

As the motor comes up to speed, the rotor frequency decreases and therefore the leakage reactance as well as the resistance of the rotor bar arrangement decreases and the influence of the skin effect becomes smaller. At low slips and at normal slip, the frequency of the rotor current is very small and the leakage reactance and resistance take on their DC values. The current is now uniformly distributed over the deep bar or divided in the ratio of the DC resistances of both cages in the case of the double cage rotor. The starting performance of a machine with considerable skin effect is therefore similar to that of the slip ring motor connected to an external resistance for starting.

5.3 CALCULATION OF STATOR RESISTANCE

Although skin effect in the rotor bars is sometimes useful in the design of machines with high starting torque, the presence of skin effect in the stator conductors is always detrimental since it produces nothing but extra losses. Since the skin effect increases with the height of the conductor, it is possible to reduce this effect by subdividing the conductor into subunits called "strands." Clearly, stranding will only be effective when the stands are insulated from each other.

In very large machines, the size of the smallest feasible strand results in a strand thickness which still yields an appreciable amount of skin effect. The skin effect can always be reduced to a very small amount if the stands of which the conductor is constructed are carried through the slot in such a manner that each strand takes on all possible positions in the depth of the slot. A sketch of such a transposition of strands is given in Figure 5.4. This kind of transposed conductor, called a *Roebel bar*, is very expensive to manufacture and is used only in very large machines in which good efficiency is paramount.

With proper use of stranding and transposition, the eddy currents in a stator coil can usually be reduced to a small value. The resistance can be calculated by first estimating an average length for each coil, then adding up the coils in series and parallel as appropriate. A good estimate for the length of one turn in a coil is simply the sum of the six straight portions of the diamond-shaped configuration. If l_c denotes the mean length of the coil, then approximately

$$l_c = 2l_s + 4l_{e2} + 4l_{e3} \tag{5.38}$$

where l_s is the length of the stator stack including ducts, l_{e2} is the straight extension of the coils beyond the stack, and l_{e3} is the diagonal portion of the end winding. By

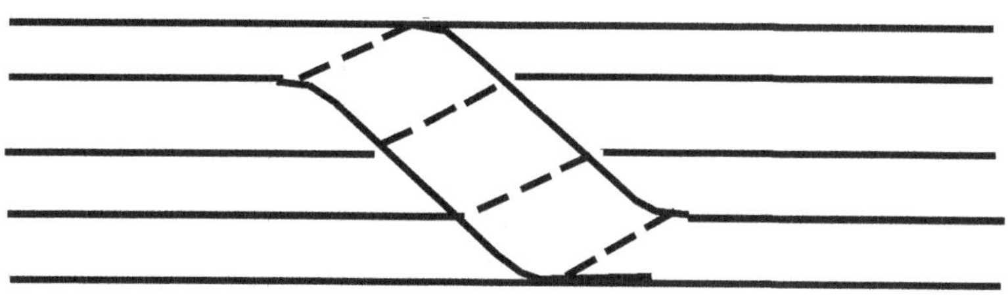

Figure 5.4 Roebel strand transposition.

reference to Figure 4.21, it is readily verified that l_{e3} can be obtained from parameters already defined as

$$l_{e3} = \frac{p\tau_{p(ave)}}{2} \frac{\tau_{s(ave)}}{\sqrt{\tau^2_{s(ave)} - (b_c + t_e)^2}} \tag{5.39}$$

The resistance of one coil is clearly

$$r_c = \frac{\rho n_c l_c}{A_c}$$

where ρ is the resistivity of the conductor and A_c is the cross-sectional area of one conductor in the coil. At 75°C, the resistivity of copper is

$$\rho_{cu} = 0.825 \times 10^{-6}\,\Omega\,\text{in.}$$
$$= 2.1 \times 10^{-6}\,\Omega\,\text{cm}$$

The resistivity at 75°C is usually used for the calculation of resistance since U.S.A. standards are based on parameter measurements at this temperature. The resistance of the coils connected in series over one pole pitch is, for a three phase machine

$$r_{cp} = \frac{S_1}{3P} r_c$$

If P/C poles are connected in series, then the resistance of one parallel branch of a C branch circuit is

$$r_b = \frac{P}{C} r_{cp} = \frac{S_1}{3C} r_c$$

The net resistance per phase is the resistance of these C circuits connected in parallel or

$$r_s = \frac{r_b}{C} = \frac{S_1}{3C^2} r_c \tag{5.40}$$

$$= \left(\frac{S_1}{3C^2}\right) \frac{\rho n_c l_c}{A_c} \tag{5.41}$$

Equation (5.41) can be written in a simpler form if it is recalled that for a two layer winding the number of turns per slot is twice the number of turns per coil. The number of turns per slot is related to the total number of turns for a three phase machine by equation (4.41),

$$n_s = \frac{6CN_s}{S_1}$$

Substituting these results into equation (5.41), one has the following alternative forms for the resistance per phase

$$r_s = \frac{S_1}{3C^2}\left(\frac{n_s}{2}\right)\frac{\rho l_c}{A_c}$$

$$= \frac{S_1}{3C^2}\left(\frac{3CN_s}{S_1}\right)\frac{\rho l_c}{A_c}$$

$$= \left(\frac{N_s}{C}\right)\rho\frac{l_c}{A_c} \tag{5.42}$$

5.4 EXAMPLE 7—CALCULATION OF STATOR AND ROTOR RESISTANCE

Again the 250 HP machine of the previous examples will be used. From information already provided, the following dimensions and constants can be determined which are necessary for the calculation of the stator resistance.

Total number of turns = $N_t = 240$

Number of circuits = $C = 1$

Pitch = $p = 0.8$

Pole pitch measured from the slot radial midpoint = $\tau_{p(ave)} = 10.324''$

Slot Pitch Measured from the Radial Midpoint of the

Slots = $\tau_{s(ave)} = 0.688''$

Gross core length including duct = $l_i + 2l_o = 9.25''$

Coil extension (Section 4.11) = $l_{e2} = 1.25''$

Spacing between coils (Figure 4.21) = $t_e = 0.0625''$

Coil thickness (Figure 4.21) = $b_c = 0.220''$

Area of coil (Section 4.11) = $A_c = (0.129)(0.204)(0.97) = 0.02553$ in.^{2}

The factor 0.97 in the area of the conductor takes rounding of the corners into account. The length of the diagonal portion of the end winding is calculated from equation (5.39) as

$$l_{e3} = \frac{(0.8)(10.32)(0.688)}{2\sqrt{(0.688)^2 - (0.0625 + 0.22)^2}} = 4.527''$$

The mean length of a coil is, therefore, from equation (5.38)

$$l_c = (2)(9.25) + (4)(1.25) + 4(4.527) = 41.61''$$

The stator resistance is now obtained from equation (5.42) as

$$r_s = \frac{(0.825)(10^{-6})(240)(41.61)}{(1)(0.02553)} = 0.323\,\Omega \quad @75°\text{C}$$

The influence of skin effect may now be considered. From the discussion in Section 5.2, the skin depth in a copper bar at 75° with 60 Hz excitation is

$$1/\alpha_o = 0.3707''$$

so that

$$\alpha_o = 2.6976\,\text{in.}^{-1}$$

Hence, in this example, since the thickness of each stator bar is 0.129″,

$$\alpha_o d = (2.6976)(0.129)$$
$$= 0.348$$

Since αd is much less than 1.5 it is possible to use the approximation for the AC resistance, equation (5.32), so that

$$r_{s(ac)} \approx r_{s(dc)} \left[1 + \left(\frac{4}{45}\right)(\alpha_o d)^4\right] = 0.323\,[1 + 0.0013]$$

or

$$r_{s(ac)} = 0.323\,\Omega \ (\textit{corrected for skin effect})$$

As is usual, the impact of skin effect on the stator resistance is essentially negligible. Comparing the coefficient of equation (5.33) (i.e., 8/315) with the coefficient of equation (5.32) (i.e., 4/45) it can be concluded that the influence of skin effect on slot leakage inductance is much less than for resistance and can safely be neglected as well.

The following parameters are taken from Examples 3 and 4 and can be used to calculate the rotor resistance:

Actual length including ducts	$l_{b1} = 9.25''$
Effective stator and rotor length	$l_{es} = l_{er} = 8.67''$
Bar width	$b = 0.375''$
Bar depth	$d = 0.5625''$
End ring area	$t_{be} = 1.0''$
	$d_{be} = 0.75''$
Rotor bar extension	$l_{be} = 0.375''$
Avg. rotor pole pitch	$\tau_{p2} = 8.604''$
Rotor slots	$S_2 = 97$

From this information, one can determine that the length of one rotor bar not including the portion over the end ring is

$$9.25 + 2\,(0.375) = 10.50''$$

In addition to this length, it is necessary to include some portion of the part over the end ring since the current curves from the bar into the end ring over the distance t_{be}. If one takes this value as $(1/3)t_{be}$, then the total active length of the rotor bars including the slight additional length due to skew is

$$l_b = \frac{\left[10.5 + 2\left(\frac{1}{3}\right)(1)\right]}{\cos[(2\pi)/120]} = 11.18''$$

The resistance of one rotor bar is, therefore

$$r_b = \frac{(11.18)(0.825) \times 10^{-6}}{(0.375)(0.5625)} = 43.7\ \mu\Omega$$

The tooth pitch at the middle of the end ring can be obtained from τ_{p2} as

$$\begin{aligned}\tau_{r2} &= \frac{P}{S_2}\tau_{p2}\\ &= \frac{8}{97}(8.604)\\ &= 0.71''\end{aligned}$$

The resistance of the end ring portion over one rotor slot pitch is

$$\begin{aligned}r_e &= \frac{\tau_{r2}\rho}{t_{be}d_{be}}\\ &= \frac{0.71(0.825\times10^{-6})}{1\,(0.75)} = 0.781\times10^{-6}\end{aligned}$$

The effective bar resistance is, therefore, from equation (4.186),

$$\begin{aligned}r_{be} &= r_b + \frac{r_e}{2\sin^2\left(\frac{\pi}{2}\frac{P}{S_2}\right)}\\ &= \left[43.7 + \frac{0.781}{2\sin^2\left(\frac{\pi\cdot 8}{2\cdot 97}\right)}\right]\times10^{-6} = (47.7+23.40)\times10^{-6}\\ &= 67.10\ \mu\Omega\end{aligned}$$

The resistance per rotor mesh is

$$r_r = 2r_{be} = 134.2\ \mu\Omega$$

The rotor resistance referred to the stator is obtained from equation (4.246)

$$\begin{aligned}r_r' &= \frac{12\,k_1^2\,N_s^2}{S_2}r_r\\ &= \frac{(12)\,(0.91)^2(240)^2}{97}134.2\times10^{-6}\\ &= 0.792\ \Omega\end{aligned}$$

Finally, it is necessary to remember to include the effects of skew. The coefficient needed to compensate for skew is the same as used for rotor leakage inductance. From Section 4.17, equation (4.263), one has the result that

$$\begin{aligned}r_r'' &= k_{s1}^2 r_r' = (0.9963)^2\,(0.792)\\ &= 0.786\ \Omega\end{aligned}$$

This value of resistance is valid over the full range of load conditions since near synchronous speed the frequency of the currents induced in the rotor bars is low. However, during starting, the frequency of the currents which flow in the rotor bars is 60 Hz and skin effect is a definite possibility. The possibility of skin effect can be evaluated by recalling that the value of $\alpha_o d$ of a rotor bar is 1.52. Reference to Figure 5.2 indicates that skin effect will indeed be important for the starting condition.

Although some skin effect exists in the air portion of the current path, the effect is less prominent than for the slot portion. Hence, it will be necessary to approximate the actual situation by assuming that skin effect only takes place in the slot region and is dictated by the effective length corresponding to rotor slot leakage flux.

The DC resistance of the portion of the bar including the iron and the duct portions is

$$r_{\mathrm{b(dc)}} = \frac{l_e \rho}{bd}$$

From previous work, the effective length for rotor slots (not including ducts) but including skew is

$$l_{\mathrm{er}} = 8.68''$$

so that, taking the ratio, the iron portion of the bar is

$$\begin{aligned} r_{\mathrm{bi(dc)}} &= \left(\frac{8.68}{11.18}\right)(43.7) \times 10^{-6} \\ &= 33.93\ \mu\Omega \end{aligned}$$

Since $\alpha_o d > 1.5$ the approximate formula, equation (5.32) will be inaccurate. The AC resistance computed from equation (5.30) is

$$\begin{aligned} r_{\mathrm{bi(ac)}} &= (\alpha_o d) r_{\mathrm{bi(dc)}} \left[\frac{\sinh(2\alpha_o d) + \sin(2\alpha_o d)}{\cosh(2\alpha_o d) - \cos(2\alpha_o d)}\right] \\ &= (1.517)(33.93)\left[\frac{10.366 + 0.1074}{10.414 + 0.9942}\right] \times 10^{-6} \\ &= 47.25\ \mu\Omega \end{aligned}$$

The effective resistance per rotor mesh will be equal to twice the sum of the resistance of the portion of the bar in the slots plus the resistance of the bar traversing the slot ducts and overhang plus the resistance due to the end rings.

$$r_r = 2\left(r_{\mathrm{bi(ac)}} + \frac{l_b - l_{\mathrm{er}}}{l_b} r_{\mathrm{b(dc)}} + \frac{r_e}{2\sin^2\left(\frac{\pi}{2}\frac{P}{S_2}\right)}\right)$$

$$\begin{aligned} r_r &= 2\left[47.25 + \frac{11.18 - 8.68}{11.18}(43.7) + 23.40\right] \times 10^{-6} \\ &= 160.8\ \mu\Omega \end{aligned}$$

The corresponding value referred to the stator by the turns ratio and skew is

$$\begin{aligned} r_r'' &= \frac{(12)(0.91)^2(0.9963)^2(240)^2}{(97)(1)^2}(160.8) \times 10^{-6} \\ &= 0.942\ \text{ohms}\ (\textit{starting value}) \end{aligned}$$

Note that skin effect has increased the resistance at starting by approximately 20%. The skin effect, of course, varies continuously as the rotor comes up to speed so that the rotor resistance varies smoothly. For low values of $\alpha_o d$, the AC resistance of the bar is related to the DC resistance by

$$r_{ac} \approx r_{dc}\left[1+\frac{4}{45}(\alpha_o d)^4\right] \tag{5.43}$$

If one defines

$$\alpha_o = \sqrt{\frac{2\pi \cdot f_e \mu_0}{2\rho}} \tag{5.44}$$

then for any slip S, the above equation for AC resistance can be written as

$$r_{ac} = r_{dc}\left[1+\frac{4}{45}(\alpha_o d)^4 S^2\right]$$

This equation can be solved in the form

$$\frac{r_{ac}-r_{dc}}{r_{dc}} = \frac{4}{45}(\alpha_o d)^4 S^2 \tag{5.45}$$

Hence, for low values of $\alpha_o d$, the resistance can be assumed to vary with the square of the slip. The curve passes through the value 0.942 for $S = 1$ and the point 0.786 for $S = 0$. A sketch of the rotor resistance as a function of slip is given in Figure 5.5.

When the skin effect is not large and follows essentially equation (5.45), it can be conveniently modeled in the per phase equivalent circuit as follows. The increase in rotor resistance above the DC value can, from equation (5.45), be expressed as

$$\Delta r_r = r_{ac} - r_{dc} = \frac{4}{45}(\alpha_o d)^4 S^2 r_{dc} \tag{5.46}$$

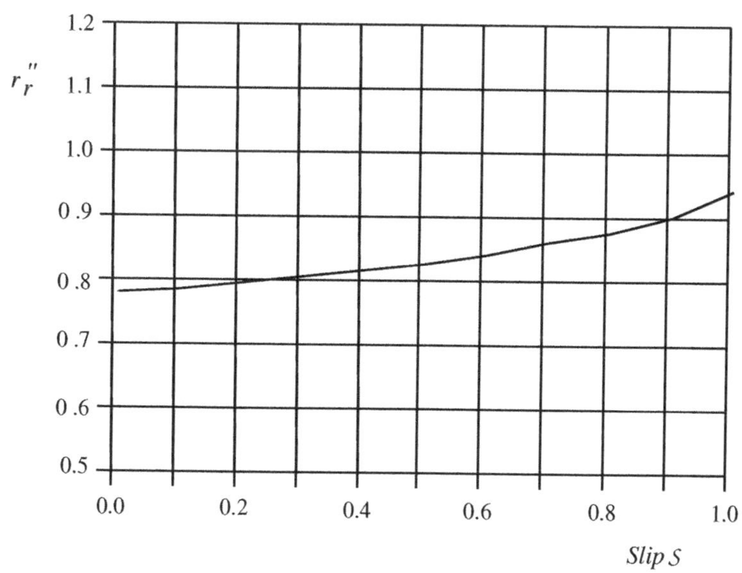

Figure 5.5 Variation of rotor resistance of 250 HP machine with slip due to deep bar effect.

so that

$$r_{ac} = r_{dc} + \Delta r_r \tag{5.47}$$

In the per phase equivalent circuit, the equivalent rotor resistance is,

$$\frac{r_{ac}}{S} = \frac{r_{dc}}{S} + \Delta r_{r0} S \tag{5.48}$$

where Δr_{r0} is the change in rotor resistance between $S = 1$. That is, Δr_{r0} is obtained by simply setting S in equation (5.46). In the case of the example 250 HP machine, $\Delta r_{r0} = (0.942 - 0.786) = 0.156$.

Having completed the study of how stator and rotor resistance vary with frequency, it must be remembered that the rotor leakage inductance is also influenced by skin effect. The corresponding AC value of leakage inductance can be computed from equation (5.31) as

$$\begin{aligned}
X_{ac} &= (\alpha_o d) r_{dc} \left[\frac{\sinh(2\alpha_o d) - \sin(2\alpha_o d)}{\cosh(2\alpha_o d) - \cos(2\alpha_o d)}\right] \\
&= (1.517)(33.93)\left[\frac{10.366 - 0.1074}{10.414 + 0.9942}\right] \times 10^{-6} \\
&= 46.29\ \mu\Omega
\end{aligned}$$

Hence

$$\begin{aligned}
L_{ac} &= X_{ac}/\omega_e = (46.29/377) \times 10^{-6} \\
&= 0.1228\ \mu\text{H}
\end{aligned}$$

It is important to realize that the value of AC inductance just computed is a correction only for the leakage flux which passes through the bar so that the leakage inductances associated with fluxes which cross the slot and link the total bar are unaffected. By reference to Example 4, the leakage inductance of one bar was obtained by first calculating the permeance per unit length from

$$p_{sl} = \mu_0 \left[\frac{d_{3r}}{3b_{sr}} + \frac{d_{1r}}{(b_{sr} - b_{or})} \ln\left(\frac{b_{sr}}{b_{or}}\right) + \frac{d_{or}}{b_{or}}\right]$$

Note that only the first term in this equation is associated with leakage fluxes which partially link the bar. The permeance for the remainder is

$$\begin{aligned}
p_{sl(air)} &= (\mu_0)\left[\frac{0.047}{0.385 - 0.09} \ln\left(\frac{0.385}{0.09}\right) + \frac{0.040}{0.09}\right] \\
&= 0.849 \times 10^{-6}\ \text{H/m}
\end{aligned}$$

The effective slot leakage per bar is therefore

$$\begin{aligned}
L_b &= p_{ls(air)} l_e + L_{ac} \\
&= \left[(0.849)\left(\frac{8.68}{39.37}\right) + 0.1228\right] \times 10^{-6} \\
&= 0.310\ \mu\text{H}
\end{aligned}$$

The effective leakage inductance per bar including the effect of the end ring and harmonic leakage is

$$L_{\text{lr}} = 2\left[L_{\text{b}} + \frac{L_{\text{e}}}{2\sin^2\left(\frac{\pi}{2}\frac{P}{S_2}\right)} + L_{\text{b(har)}}\right]$$

From which

$$L_{\text{lr}} = 2\left[0.310 + \frac{0.01251}{2\sin^2\left[\frac{\pi}{2}\left(\frac{8}{97}\right)\right]} + 0.142\right] \times 10^{-6}$$

So that for an entire mesh or rotor phase

$$L_{\text{lr}} = 1.654\ \mu\text{H}$$

The rotor leakage inductance referred to the stator turns is

$$\begin{aligned} L''_{\text{lr}} &= \frac{12\,k_1^2\,k_{\text{S1}}^2\,N_{\text{s}}^2}{S_2} L_{\text{lr}} \\ &= \frac{(12)\,(0.91)^2(0.9963)^2(240)^2}{(97)} 1.654 \times 10^{-6} \\ &= 9.7\ \text{mH (starting value)} \end{aligned}$$

This result can be compared with the previous value of 9.9 mH obtained in Example 5, corresponding to only a 2.2% reduction in the rotor leakage inductance.

For low values of αd, the AC reactance is related to the DC reactance by

$$X_{\text{ac}} = \omega L_{\text{ac}} = \omega L_{\text{dc}}\left[1 - \frac{8}{315}(\alpha d)^4\right]$$

or

$$L_{\text{ac}} = L_{\text{dc}}\left[1 - \frac{8}{315}(\alpha d)^4\right]$$

If one expresses $\alpha_o d$ in terms of the αd which occurs at starting, then

$$\alpha_o d = (\alpha_0 d)\sqrt{S}$$

or

$$L_{\text{ac}} = L_{\text{dc}}\left[1 - \frac{8}{315}(\alpha_0 d)^4 S^2\right]$$

$$\frac{L_{\text{ac}} - L_{\text{dc}}}{L_{\text{dc}}} = -\frac{8}{315}(\alpha_0 d)^4 S^2$$

Hence the deviation of the AC inductance from the DC inductance decreases as the square of the slip. A sketch of the equivalent circuit for the 250 HP motor without an iron loss term (to be discussed) is given in Figure 5.6. Note the presence of the slip-dependent rotor parameters. The variation of rotor leakage inductance is clearly

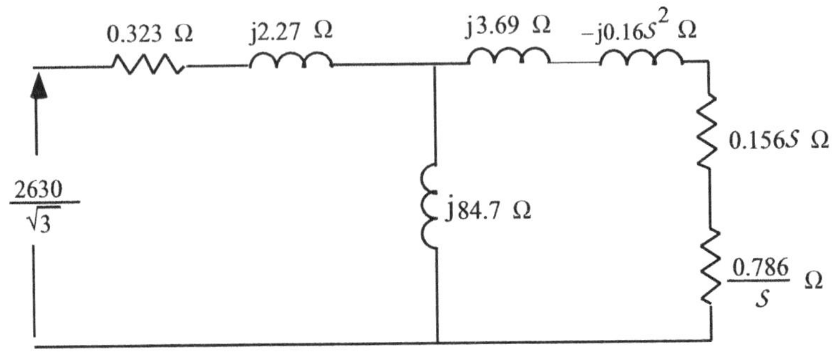

Figure 5.6 Per phase equivalent circuit for 250 HP machine example with 2630 $V_{l\text{-}l}$, 60 Hz excitation, deep bar effect included.

small and could have been neglected in this example but is included for the sake of completeness.

5.5 ROTOR PARAMETERS OF IRREGULARLY SHAPED BARS

By examining the behavior of a rectangular bar placed in iron, it has been shown that skin effect in rotor bars has the effect of increasing equivalent resistance and decreasing equivalent inductance as the frequency of current increases. The increase of rotor resistance has a beneficial effect in aiding starting capability of squirrel-cage machine and nearly all conventional machine designs (NEMA Designs B, C and D) enhance motor starting capability by specially selected rotor bar shapes. Figure 5.3 shows a typical variety of such shapes.

Although closed form solutions are usually impossible except for carefully selected exceptions, the following approach can be readily applied to rotor bars of arbitrary cross-sectional shape. In this algorithm, the rotor bar is divided into n layers illustrated in Figure 5.7. All layers of the bar are assumed to have the same axial length. If sufficient number of layers are chosen the layers can be assumed to be essentially rectangular in which case each layer p has a unique resistance r_p determined by

$$r_{\mathrm{p}} = \frac{\rho l_{\mathrm{e}}}{b_{\mathrm{p}} h_{\mathrm{p}}} \tag{5.49}$$

Likewise, each layer has a unique reactance determined by the layer geometry, μ_0, and the slip frequency.

$$X_{\mathrm{p}} = S\omega_{\mathrm{e}}\mu_0 \frac{l_{\mathrm{e}} h_{\mathrm{p}}}{b_{\mathrm{p}}} \tag{5.50}$$

If I_k is the current in layer k, the flux linking the pth layer can be expressed,

$$\tilde{\Phi}_{\mathrm{p}} = \mu_0 \frac{l_{\mathrm{e}} h_{\mathrm{p}}}{b_{\mathrm{p}}} \sum_{\mathrm{k=p+1}}^{n} \tilde{I}_{\mathrm{k}} \tag{5.51}$$

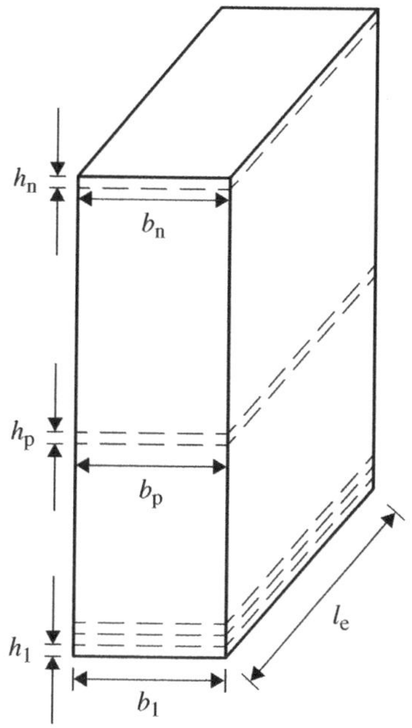

Figure 5.7 Rectangular bar broken into *n* sections. Rectangular bar is shown but methodology applies to any bar shape.

where, in the steady state

$$r_{p+1}\widetilde{I}_{p+1} - r_p\widetilde{I}_p = jS\omega\Phi_p \tag{5.52}$$

Considering the above equations, an expression for the current in the *p* rotor bar layer can be determined. This expression utilized as information concerning the previous rotor bar layers,

$$\widetilde{I}_p = \frac{r_{p+1}}{r_p}\widetilde{I}_{p+1} - j\frac{X_p}{r_p}\sum_{k=1}^{p}\widetilde{I}_k \tag{5.53}$$

This process is equivalent to forming loop equations of an equivalent circuit as illustrated in Figure 5.8. The computation of all the reactance values requires an initial assumption about the current in the first (outer) rotor bar layer. It is assumed that an arbitrary value of current flows through this layer. A value of one ampere is

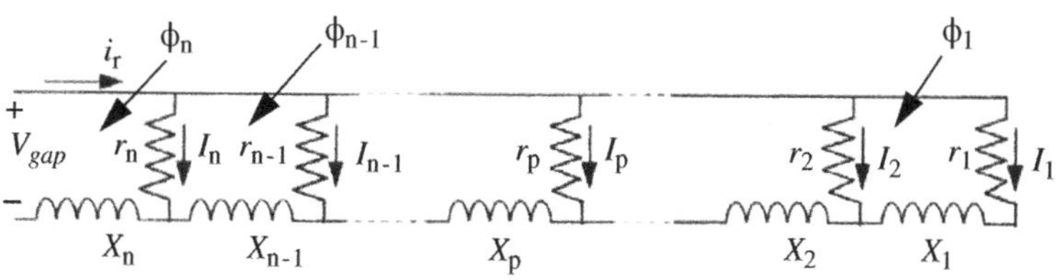

Figure 5.8 Rotor equivalent circuit with multilayer representation of a rotor bar.

convenient for calculation purposes. The values for all the reactances and resistances for the layers can then be determined. At the same time, the currents flowing through all the components is determined.

The voltage (V_{gap}) seen at the "outer" terminals of the equivalent circuit can be determined by the following,

$$\widetilde{V}_{gap} = r_n \widetilde{I}_n + jX_n \widetilde{I}_r \tag{5.54}$$

where

$$\widetilde{I}_r = \sum_{k=1}^{n} \widetilde{I}_k \tag{5.55}$$

The equivalent impedance $\widetilde{Z}_{in}$ of the rotor bar equivalent circuit can now be determined,

$$\widetilde{Z}_{b,in} = \frac{\widetilde{V}_{gap}}{\widetilde{I}_r} = r_{b,in} + jS\omega_e L_{b,in} \tag{5.56}$$

The real component of $\widetilde{Z}_{b,in}$ is the equivalent resistance of the rotor bar while the imaginary component of $\widetilde{Z}_{b,in}$ is the equivalent reactance of the rotor bar. The equivalent inductance of the rotor bar can be determined utilizing the slip frequency.

$$L_{b,in} = \frac{Imag(\widetilde{Z}_{b,in})}{S\omega_e} \tag{5.57}$$

The equivalent resistance and inductance of a single rotor bar can now be computed for every value of slip. The values for the leakage inductance and resistance of an individual rotor bar permits the computation of the equivalent circuit parameters X'_{lr} and r'_r by the approach of Section 4.15. Items such as rotor end ring resistance, rotor end-ring leakage inductance, harmonic leakage and effective rotor to stator turns ratio also need to be considered.

The skin effect algorithm can be programmed in MATLAB® and the code is shown in Appendix A. The rotor bar of a typical 250 HP machine was used with this process. The stator resistance and leakage inductance, as well as the magnetizing inductances are assumed to be the same as Figure 5.6. In general, the number of layers subdividing the rotor bar must be increased until convergence is guaranteed. In this example at least 10 layers were found to be necessary.

The range of the calculated equivalent rotor resistance over the operating range of the motor was 0.82 to 0.946 ohms and highest at high slip frequencies as expected. The calculated rotor leakage inductance ranges from 0.0197 H under high slip conditions to 0.0198 under small slip conditions.

Steady state torque–speed curves were produced for three cases to determine the effectiveness of the skin effect algorithm. The rotor resistance and rotor leakage inductance is plotted versus slip for each case in Figure 5.9 and Figure 5.10 respectively. The three cases are

A. Base Case: Same as Figure 5.6.

B. Wide Case: Only the dimensions of the rotor bars are altered. The bars are twice as wide and half as deep compared with the base case. (This is an impractical case since no room is left for the rotor teeth but was chosen simply for purpose of illustration).

C. Deep Case: Only the dimensions of the rotor bars are changed. The bars are twice as deep and half as wide compared with the base case.

The cases were designed to provide the identical rotor resistance under very low slip conditions. The effects of skin effect dramatically increased as the relative depth of the rotor bars increased. The following two figures illustrate how the rotor resistance and rotor leakage inductance change with variations in the bar shape.

In general, the algorithm to incorporate the deep bar effect can be readily implemented in a machine design code since the computations involved for a specific bar shape are relatively straightforward. Alternatively, the computations can be done "off line" and the data inputted to the design program either as a table lookup or a functional polynomial to fit the computed curves. The software MATHEMATICA® is particularly convenient for computing a polynomial fit. A plot of two frequently used bar shapes is given in Figures 5.11 and 5.12. They can also be used to approximate the frequency dependence for various other bar geometries. Note that skin effect is much more dominant for the coffin-shaped bar of Figure 5.12 since the narrow portion of the bar is near the rotor surface. The solution for rotor

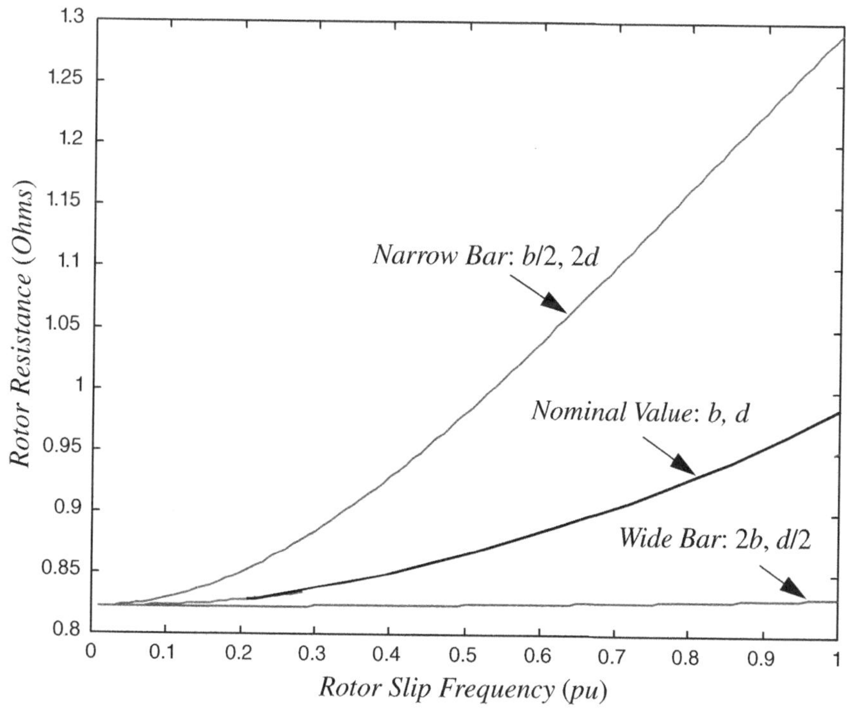

Figure 5.9 Rotor bar resistance vs. slip frequency for three different bar shapes.

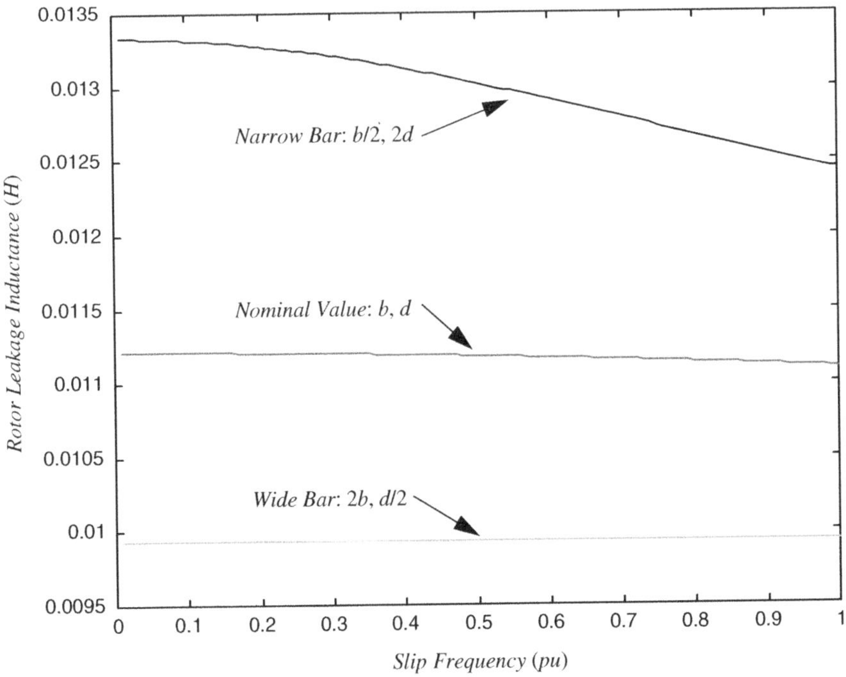

Figure 5.10 Rotor bar leakage inductance vs. slip frequency for three different bar shapes.

leakage inductance is somewhat lower than the value predicted from the analytical solution since the effect of the slot leakage inductance in the portion above the bar was neglected in the computer solution.

5.6 CATEGORIES OF ELECTRICAL STEELS

Electrical steels are generally divided into three categories: (a) grain-oriented (GO) silicon, (b) non-oriented (NO) silicon, and (c) motor lamination quality (MLQ) steels. GO silicon steels have a strongly preferred texture in the rolling direction. The preferred crystal orientation in the rolling direction gives the steel superior magnetic properties in that direction when compared with the transverse direction. This stems from the superior magnetic properties of iron along the crystal lattice cube edge. GO sheets are used in large high efficiency power and distribution transformers and large high speed generator segments.

NO silicon steels have essentially a random texture with nearly isotropic magnetic properties. These steels are used in rotating electric machines such as motors and generators. NO steels are available in semi-processed (SP) and fully processed (FP) form. End users must anneal SP steels in order to develop the desired magnetic properties. FP steels are shipped from the steel mill fully annealed with fully developed magnetic properties. NO steels generally contain silicon to increase the resistivity of the steel thereby reducing eddy current losses. They are also usually provided with an insulating coating to reduce eddy currents between laminations.

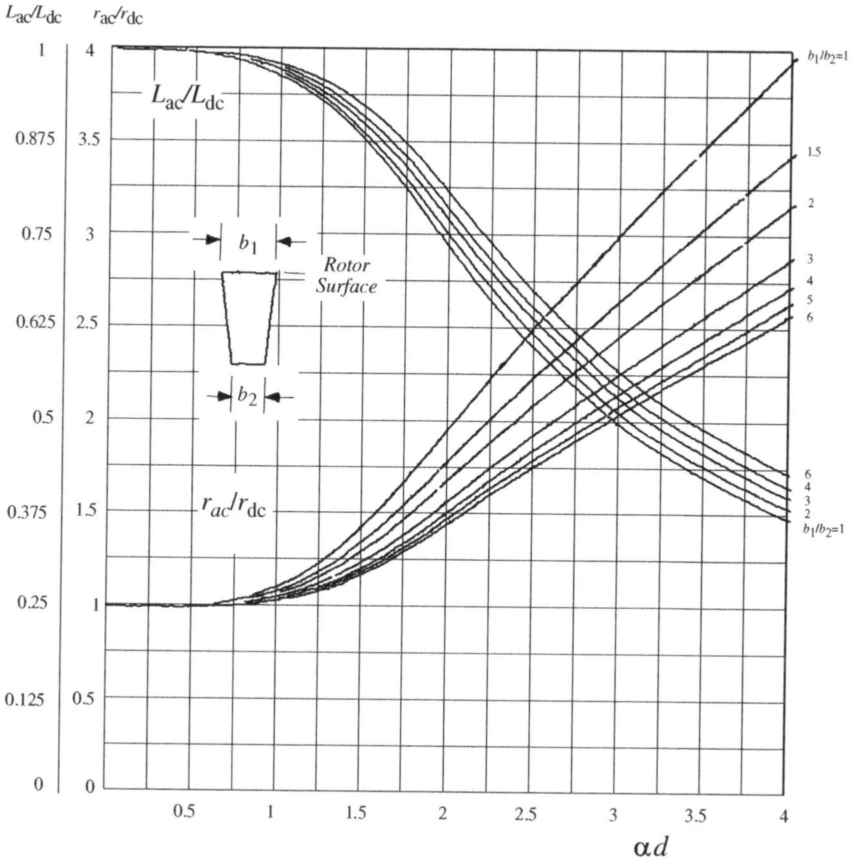

Figure 5.11 Variation of bar resistance and inductance for inverted coffin shaped bar with αd where d is the bar depth and $\alpha_o = \sqrt{(\omega\mu_0)/(2\rho)}$.

MLQ steels are a subgroup of NO steels. In the 1950s, cold-rolled carbon (CRC) steels were used in the cores of fractional horsepower and intermittent duty motors. Use of expensive silicon steels was not justified, since low core losses and high electrical efficiencies were not deemed important in the small motor industry at that time. Carbon steels had a relatively low cost in comparison to silicon steels; however, their magnetic properties were relatively poor. MLQ steels evolved from the need for an electrical steel that would cost less than silicon steels, but offered magnetic properties superior to those of CRC steels.

5.7 CORE LOSSES DUE TO FUNDAMENTAL FLUX COMPONENT

If the magnetic flux density in the core of an AC machine is sinusoidal, two types of losses are produced; eddy current and hysteresis losses. The eddy current loss arises

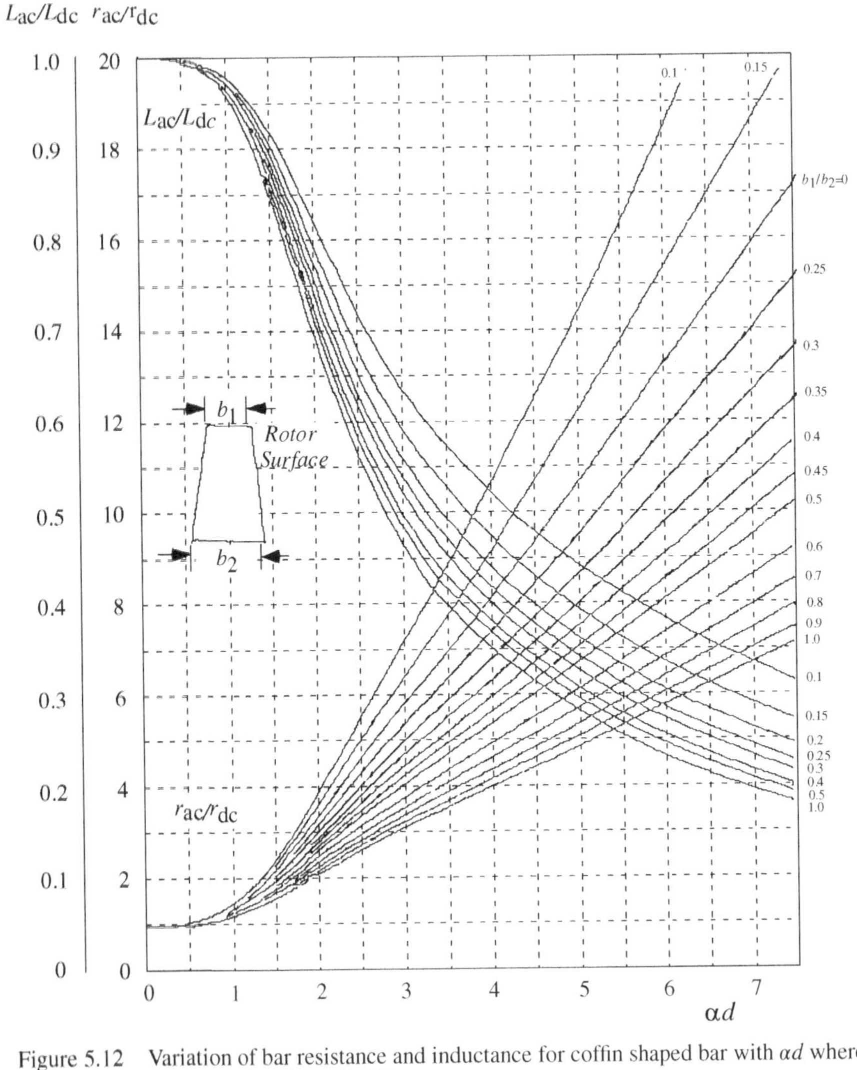

Figure 5.12 Variation of bar resistance and inductance for coffin shaped bar with αd where d is the bar depth and $\alpha_o = \sqrt{(\omega\mu_0)/(2\rho)}$.

from precisely the same phenomena that result in eddy current losses in conductors. These losses account for the currents which circulate within the laminations of the stator and rotor. These currents are induced by the time-varying magnetic field. These currents are again non-uniform and tend to be largest at the surface of the laminations.

The eddy current effect in iron laminations can be understood by reference to Figure 5.13 which shows a thin rectangular conductor impressed with a sinusoidal flux density $B_m \sin \omega t$. The flux enclosed within the current path shown at a distance x from the center is

$$\Phi(x) = 2x[L - (t - 2x)]B_m \tag{5.58}$$

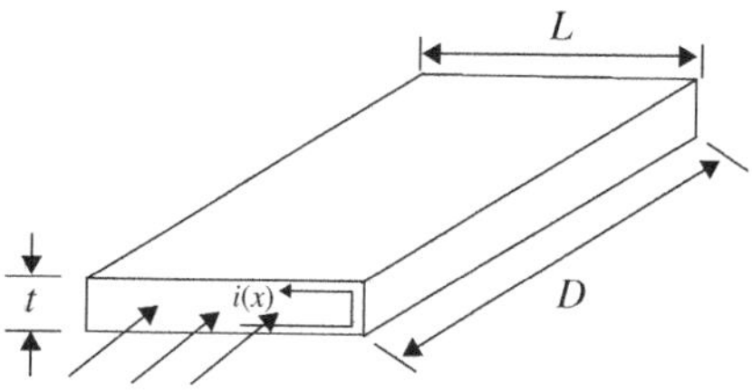

Figure 5.13 Eddy current path in a thin rectangular iron sheet.

The amplitude of the voltage induced within this current path is

$$E_m = \omega B_m 2x[L - (t - 2x)] \tag{5.59}$$

The resistance to current flow around this path is

$$dR\,(x) = \rho_{\text{iron}} \left(\frac{2[L - (t - 2x)]}{Ddx} + \frac{4x}{Ddx} \right) \tag{5.60}$$

or,

$$dR\,(x) = \rho_{\text{iron}} \left(\frac{2L - 2t + 8x}{Ddx} \right) \tag{5.61}$$

The corresponding loss is therefore

$$dP_{\text{e}} = \frac{E_m(x)^2}{2dR\,(x)} = \frac{\omega^2 B_m^2 \{2x\,[L - (t - 2x)]\}^2 Ddx}{2\rho_{\text{iron}}\,(2L - 2t + 8x)} \tag{5.62}$$

If t and x are small compared to L then

$$dP_{\text{e}} = \frac{\omega^2 B_m^2 x^2 DL}{\rho_{\text{iron}}} dx \tag{5.63}$$

The total eddy current loss in the lamination is

$$P_{\text{e}} = \int_0^{\frac{t}{2}} \frac{\omega^2 B_m^2 x^2 DL}{\rho_{\text{iron}}} dx \tag{5.64}$$

$$= \omega^2 B_m^2 \frac{t^3}{24\rho_{\text{iron}}} DL \tag{5.65}$$

The eddy current loss per unit volume is therefore given by

$$p_{\text{e}} = \frac{\pi^2 f_{\text{e}}^2 B_m^2 t^3}{6\rho_{\text{iron}}} \ \text{w/m}^3 \tag{5.66}$$

where ρ_{iron} is the resistivity of the magnetic material, B_m is the peak flux density, f_{e} is the frequency and t is the thickness of the lamination. The thickness is clearly of considerable importance in limiting the eddy current losses since it appears in equation (5.66) as a squared term. The eddy current can be controlled by making the laminations very thin. However, the increased cost in assembly together with an increasing penalty in space factor k_i prevents widespread use of laminations thinner than 0.014″ with the thinner laminations generally being preferred for the higher frequency machines such as 400 Hz alternators.

The second type of iron loss, hysteresis loss, accounts for the energy involved in continually reversing the molecular dipoles of the magnetic material. This loss is proportional to the area of the hysteresis loop of the material times the number of times this area is traversed per second, that is, the frequency. These losses are usually expressed in the form

$$p_{\text{hys}} = K_{\text{hys}} f_e B_m^{k_{\text{hys}}} \ \text{w/m}^3 \tag{5.67}$$

The exponent k_{hys} is called the Steinmetz coefficient and, during Steinmetz's era, was equal to 1.6. This coefficient of 2 is chosen for convenience for most modern magnetic materials.

The total iron loss is the sum of the hysteresis and eddy current loss and is generally expressed in the form

$$p_i = \left(K_{\text{hys}} f_e + K_e f_e^2\right) B_m^2 \ \text{w/m}^3 \tag{5.68}$$

where the value K_{hys} has been taken as 2. A plot of the eddy current and hysteresis losses in low carbon, cold-rolled steel is given in Figures 5.14 and 5.15. While given in English units, recall that 1.0 tesla = 64.5 kilolines/in.^{2} Also 1 in.^{3} = 16.387 cm^3. Note that when the lamination thickness is 0.025″, the eddy current loss is about 50% greater than the hysteresis loss. When the lamination thickness decreased to 0.0185″, the hysteresis losses begin to exceed the eddy current losses. A set of curves for 3% silicon steel is given in Figure 5.16.

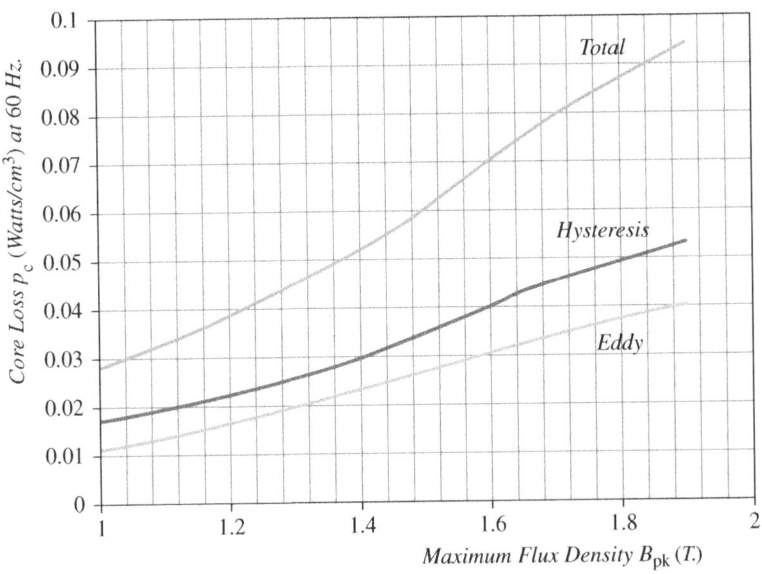

Figure 5.14 Core loss curves for 0.0185″ cold rolled, low carbon steel.

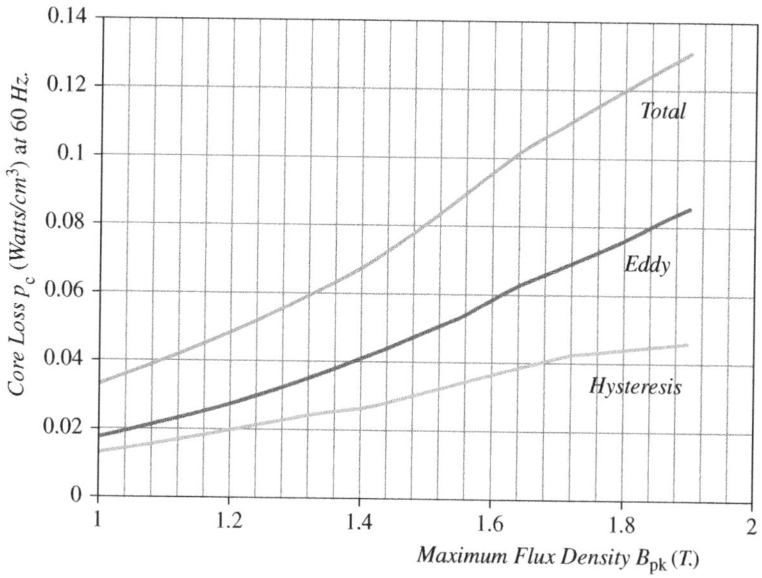

Figure 5.15 Core loss curves for 0.025″ cold rolled, low carbon steel.

Calculations of iron losses with the aid of equation (5.68) unfortunately give good values only in the case of transformers and the like and not in the case of electric machinery. The reason for this problem is that space harmonics produce flux densities whose rate of rise, *dB*/*dt*, is much greater than that of a sine wave of the same flux density. This causes the eddy current losses to rise much faster than simply the square of the fundamental component of flux density and the actual exponent in highly saturated

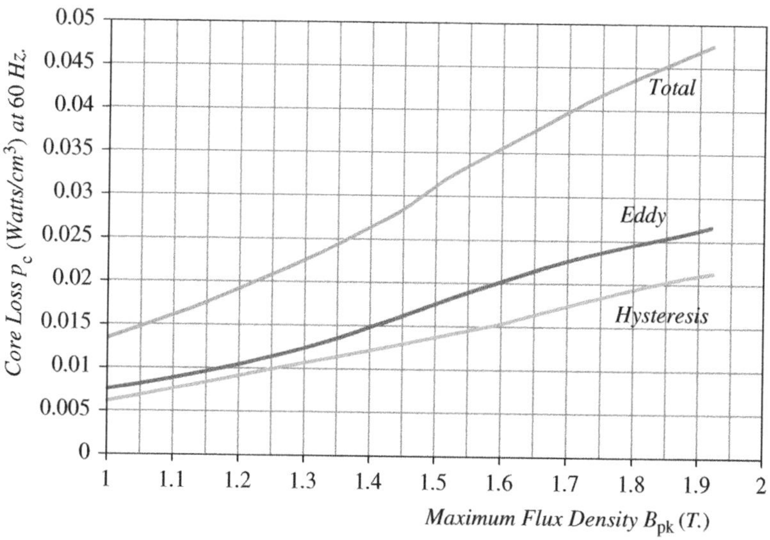

Figure 5.16 Core loss curves for 0.025″ 3% silicon steel.

portions of the machine may be 6 or more rather than 2. In addition, it is not possible to confine flux to the plane of the laminations, for example, in the end-winding region, thereby producing additional eddy current losses. Finally, the constants K_{hys} and K_e are typically obtained in the laboratory by testing prepared specimens and cannot be relied upon as the losses build up in cores of actual machines. Additional losses are encountered which include burrs on edges of stamping causing contact between lamination and attendant eddy current losses. Localized stresses in the lamination sheets, for example, near bolt holes and at the surface of the teeth due to the punching process, cause increased hysteresis losses in these regions. These effects are generally accounted for by adding on an extra 30–40% to the losses measured from laboratory specimens. Additional losses can influence the efficiency of a small induction machine by as much as 2.5%. If accurate measurements are made at one level of flux density and frequency, the following equation is a reasonably accurate estimate of fundamental core loss at frequencies and flux densities not differing widely from the reference value,

$$\text{Fund. Core Loss}_2 = \text{Fund. Core Loss}_1 \left(\frac{\text{volts/Hz}_2}{\text{volts/Hz}_1} \right)^{2.5} \left(\frac{f_2}{f_1} \right)^{1.4} \tag{5.69}$$

5.8 STRAY LOAD AND NO-LOAD LOSSES

The calculation of the stray load and stray no-load losses of an induction machine is a very complex problem which deals precisely with the issues that have been so far avoided in this development, namely the fact that the rotor and stator surfaces are slotted and not smooth and also the fact that the windings are located in these slots and therefore result in MMF harmonics higher than the first harmonic. Calculation of such loss phenomena is typically done with the aid of a more detailed equivalent circuit than the accustomed two-loop circuit, shown in Figure 5.17.

In order to gain an appreciation for the complexity involved, it is useful to discuss briefly the significance of the various terms in this equivalent circuit.

a. Permeance Variation Loss. When two magnetic surfaces, one slotted and one smooth, with an MMF across the gap between them, move relative to each other, the smooth surface sees a locally varying magnetic field due to the periodic permeance variations caused by the slots in the opposite surface. The permeance amplitude variation is related to the geometry of the slots and the frequency from the speed of rotation. In particular, the stator slot openings causes a current to flow in the rotor bars which results in an ohmic loss. The current which flows in the rotor is limited primarily by the reactance of the rotor bars which is much larger than its resistance. In addition to the copper losses, the permeance variation of the stator slots also causes eddy currents to flow in the rotor iron laminations causing additional losses. When the rotor as well as the stator is slotted, these additional losses occur in both the stator and rotor teeth. The problem is represented in the circuit by two meshes obtained

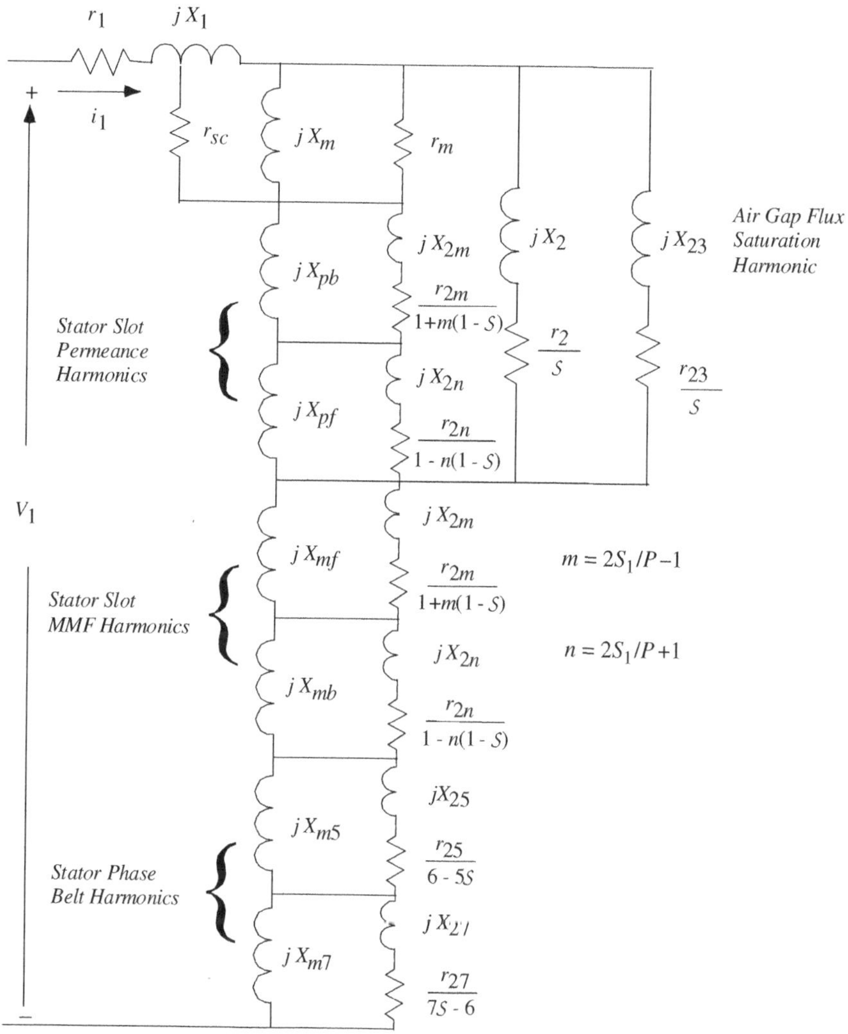

Figure 5.17 General equivalent circuit of a polyphase induction motor.

by breaking up the permeance wave into a positive rotating and a negative rotating component. In general, the permeance wave can be represented by the function

$$\mathcal{P}_s = \mathcal{P}_0 + \mathcal{P}_1 \cos(S_1\theta) \tag{5.70}$$

Restricting oneself to only the positively rotating MMF wave, then, if balanced sinusoidal currents flow, the fundamental component of stator MMF can, from equation (2.73), be represented as a traveling wave of the form,

$$\mathcal{F}_a + \mathcal{F}_b + \mathcal{F}_c = \frac{3}{2}\left(\frac{4}{\pi}\right)\left(\frac{k_1 N_t I_m}{CP}\right)\sin\left(\frac{P}{2}\theta - \omega_e t\right) \tag{5.71}$$

Multiplying these two expressions together yields an expression for the flux rotating in the air gap due to stator currents. The term containing the permeance $\mathcal{P}_0$ is the normal constant amplitude, sinusoidally distributed and synchronously rotating wave. The second term is of the form,

$$\Phi_p = \Phi_m \sin\left(\frac{P}{2}\theta - \omega_e t\right)\cos(S_1\theta) \tag{5.72}$$

where

$$\Phi_m = \frac{3}{2}\left(\frac{4}{\pi}\right)\left(\frac{k_1 N_t I_m}{CP}\right)\mathcal{P}_1 \tag{5.73}$$

This expression can be expanded to form

$$\Phi_p = \frac{\Phi_m}{2}\left[\sin\left(\frac{P}{2}\theta - \omega_e t + S_1\theta\right) + \sin\left(\frac{P}{2}\theta - \omega_e t - S_1\theta\right)\right] \tag{5.74}$$

The first term represents a wave which is traveling backward with respect to the synchronously rotating main flux wave but yet in the forward direction at an electrical angular speed of $\omega_e/(1 + (2S_1)/P)$. The second term corresponds to a wave which is traveling backward at a speed of $\omega_e/((2S_1)/P - 1)$. The slip corresponding to the forward traveling wave is

$$\begin{aligned} S_f &= \frac{\omega_e/(1 + (2S_1)/P) - \omega_r}{\omega_e/(1 + (2S_1)/P)} \\ &= 1 - \left(1 + \frac{2S_1}{P}\right)(1 - S) \end{aligned} \tag{5.75}$$

where S is the slip corresponding to the fundamental component. In a similar manner, it can be readily determined that the slip for the backward travelling wave is

$$S_b = 1 + \left(\frac{2S_1}{P} - 1\right)(1 - S) \tag{5.76}$$

The harmonic components of the pulsating gap flux density over each tooth can be found from the same conformal mapping solution used to obtain the Carter Factor discussed in Chapter 3. This theory assumes that each tooth is excited by an *MMF* which can be treated as essentially a constant over one tooth pitch. The appropriate constant can be taken as the average value of the flux density estimated at the center line of the slot in question after the fundamental component has been removed as shown in Figure 5.18. The presence of the slot clearly causes the flux to dip over the center line of the slot opening causing harmonics in the flux waveform as well as the decrease in the average flux previously discussed. The first three slot harmonic components of flux in the gap are given in Figure 5.19 as a function of the slot opening to tooth pitch ratio. The quantity β_o in this figure is one-half the peak to peak pulsation

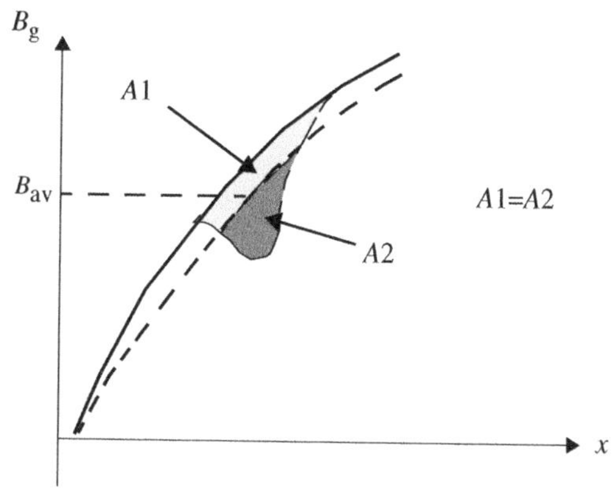

Figure 5.18 Illustrating method for choosing average flux density over each tooth span.

in the flux over one tooth divided by the maximum flux and is plotted in Figure 5.20 and approximated in equation form as

$$\beta_o = \frac{1}{2}\left[1 - \frac{1}{\sqrt{1 + \left(\frac{b_o}{2g}\right)^2}}\right] \tag{5.77}$$

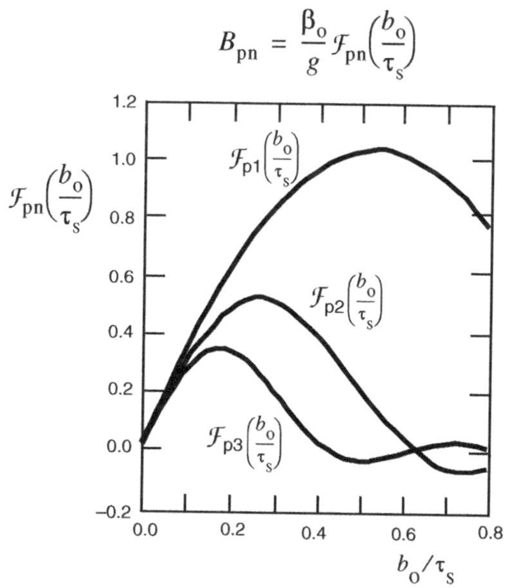

Figure 5.19 First three harmonics of slot ripple flux density, b_o = slot opening, τ_s = slot pitch.

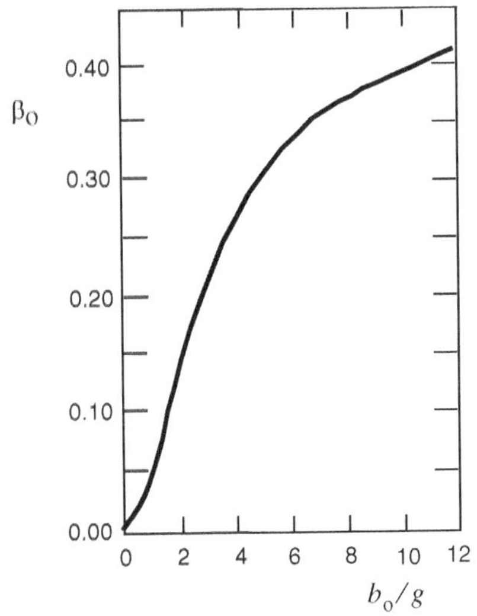

Figure 5.20 Parameter β_o as a function of the ratio of slot opening to the air gap length.

These harmonics relate to the Fourier coefficients of the slot pulsation and should not be confused with the fundamental component of the useful flux B_{g1}.

The permeance $\mathcal{P}_1$ in equation (5.70) can be obtained from Figure 5.19. Since the amplitude of the flux Φ_m in equation (5.73) is, in general

$$\Phi_m = B_{p1} A_{s+t} = B_{p1} \tau_s l_e \tag{5.78}$$

and, from Figure 5.19,

$$\Phi_m = \frac{\beta_o}{g} \mathcal{F}_{p1} \left(\frac{b_o}{\tau_s} \right) \tau_s l_e \tag{5.79}$$

whereupon,

$$\mathcal{P}_1 = \frac{\Phi_m}{\mathcal{F}_{p1}} = \beta_o \frac{\tau_s l_e}{g} \tag{5.80}$$

The effect of the slot permeance harmonics can be incorporated into the per phase equivalent circuit as shown in Figure 5.17. The quantity X_{pb} in Figure 5.17 represents the reactance corresponding to the voltage generated at the air gap by the backward traveling permeance harmonic ($m = 2S_1/P - 1$). This reactance is obtained by converting the permeance of one slot to the phase inductance by the same procedure used to obtain the slot leakage inductance, equation (4.42). The elements acted upon by that voltage have values equal to the reactance X_{2m} and the resistance r_{2m}. The quantity X_{2m} is the reactance which limits the flow of current in the squirrel-cage rotor. The resistance r_{2m} is the resistance in series with X_{2m} which is used to calculate the permeance harmonic loss. These two quantities are obtained in the same manner as used to obtain L_{lr} and r_r in Section 4.15. However, the frequency dependence due to deep bar effect as in Section 5.5 is clearly necessary. The quantities X_{pf}, X_{2n} and r_{2n} have the same meaning for the forward traveling permeance harmonic. Since the slot induced ripple is a function of air gap flux rather than current, the circuit elements are placed in series with the magnetizing branch of Figure 5.17.

b. Slot MMF Loss. The higher harmonics which are present in the MMF waveform due to the concentration of the current in stator slots also cause losses in the rotor copper and steel. These harmonics are of the same order as the permeance harmonics as can be seen by reference to equation (2.35). The slot MMF harmonic losses are proportional to the load current instead of the magnetizing current, since the harmonics involved arise out of the Fourier analysis of the MMF wave as opposed to the air gap flux density wave. Losses again occur in both the rotor conductors and in the rotor laminations. The problem is again analyzed by breaking up the slot MMF harmonic into positively and negatively rotating components. The result is very similar since the slot harmonic MMFs are multiples of the number of slots per pole-pair as were the permeance harmonics.

The reactance X_{mb} represents the voltage which is generated by the backward traveling slot MMF harmonic ($m = 2S_1/P - 1$). The quantity X_{2m} is again the zigzag reactance which limits the flow of current acted upon by the MMF harmonic voltage. The term r_{2m} is the series resistance which accounts for the losses. The combination X_{2m} and r_{2m} are present in both cases since physically, the squirrel cage is acted upon by harmonics of the same order from two different sources, namely the permeance variations and the steps in the MMF wave. In this case, the circuit elements must be placed in series with the stator resistance and inductance since these losses are clearly dependent on the amplitude of the stator MMF.

c. Phase Belt Harmonic Losses. It has been mentioned in Chapter 4 that a leakage reactance for higher harmonics of the stator MMF do not exist for squirrel-cage machines since the cage acts to short out these harmonic fluxes. The flow of this current does, however, produce losses and account must be taken of its effect in the form of phase belt harmonic losses. In Figure 5.17, X_{m5} represents the fifth harmonic voltage generated by the stator phase belt. The quantity X_{m7} represents the seventh harmonic voltage generated. Because the phase belt harmonics fall off rapidly, MMF harmonics between the fifth and seventh harmonic are typically neglected. The reactances X_{25} and X_{27} limit the current that flows due to the phase belt harmonics and r_{25} and r_{27} are used to calculate the phase belt harmonic losses. Since the wave travels backward and has five times the number of poles as the fundamental, the harmonic slip for the fifth harmonic is

$$\begin{aligned} S_5 &= \frac{\dfrac{\omega_e}{5} + \omega_r}{\dfrac{\omega_e}{5}} \\ &= 1 + 5\frac{\omega_r}{\omega_e} \\ &= 6 - 5S \end{aligned}$$

The corresponding result is easily obtained for the positively rotating seventh harmonic wave. The relevant 5th and 7th harmonic inductances can be calculated from equation (4.141).

d. Saturation Induced Loss. This loss occurs as a result of the fact that the air gap flux density is typically non-sinusoidal as a result of saturation, typically in the teeth. As a result, the flux density has a significant third harmonic component (and other odd harmonics) which, in turn, causes currents to flow in the rotor cage. Since these harmonics necessarily rotate synchronously with the fundamental component, they are different from the slot harmonic which rotates at sub-synchronous speeds. A treatment of saturation-induced losses and the equivalent circuit parameters for this case are discussed in Reference [1].

5.9 CALCULATION OF SURFACE IRON LOSSES DUE TO STATOR SLOTTING

The problem concerned with loss in the iron due to the effect of slotting can be visualized by referring to Figure 5.21a in which the air gap flux density is plotted as a function of the gap periphery. The average value for each slot pitch associated with the DC offset in Figure 5.21b is accounted for by Carter's coefficient.

An approximation to the air gap slot ripple flux density using only the first harmonic of the ripple obtained from Figure 5.18 is sketched in Figure 5.22. The air gap flux is assumed to be essentially sinusoidal but perturbed by the effect of stator slotting. The rotor surface is assumed smooth initially. Note that the MMF harmonics have been neglected here since they will be accounted for separately. If one subtracts the fundamental component from this waveform slot ripple field of Figure 5.21b

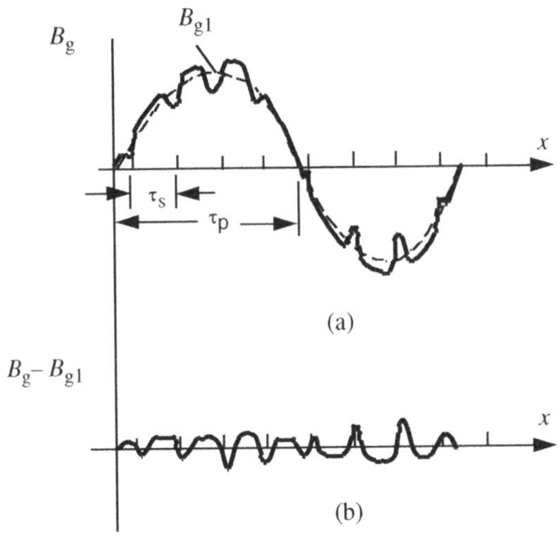

Figure 5.21 (a) Air gap flux density distribution along the air gap periphery neglecting the higher *MMF* harmonics, (b) flux density distribution with the fundamental component (useful component) removed.

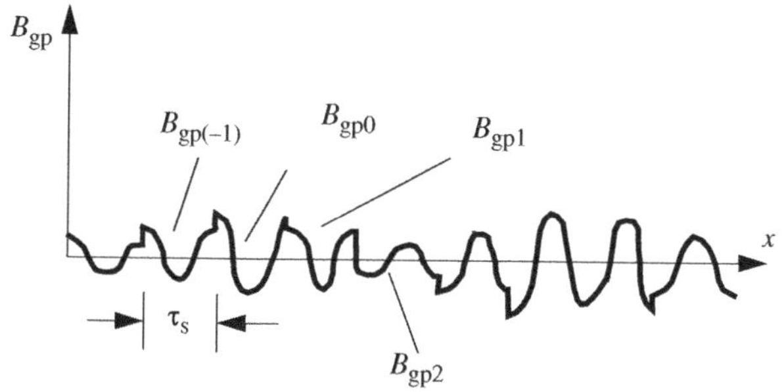

Figure 5.22 Approximation of slot ripple flux using the first harmonic components calculated for each individual slot.

results. It can be observed that the slot ripple field varies somewhat from slot to slot since the ripple component in a particular tooth is related to the average flux density over the tooth span encompassing that tooth. The waveform shown in Figure 5.21b is given for only one time instant and each tooth eventually takes on the slot ripple waveform of all the individual teeth shown in Figure 5.21 as the fundamental flux rotates around the air gap.

It is now time to attempt to solve for the flux distribution on the rotor surface resulting from the first slot harmonic. Recall that since the flux wave is rotating, over the course of one cycle each individual stator tooth will pulsate at each of the "half cycle" waveforms of Figure 5.21b. In fact, this amplitude variation of the flux pulsation varies smoothly in time. The pulsating flux which occurs over one stator tooth pitch due to the first slot harmonic can therefore be written in the form

$$B_{gp0} = B_{p1} \cos\left(\frac{2\pi x}{\tau_s}\right) \cos(\omega_e t) \qquad -\tau_s/2 < x < \tau_s/2 \tag{5.81}$$

where B_{p1} corresponds to the ripple flux density obtained over the slot having the peak gap fundamental flux density. Equation (5.81) states that the flux density in the gap is sinusoidally distributed in space but pulsates in time at line frequency. The pulsating component of flux density in the gap over any of the other stator teeth is the same as equation (5.81) but phase shifted in time by the appropriate number of slot pitches. For example, the flux density over the adjacent tooth in the direction of flux rotation is

$$B_{gp(1)} = B_{p1} \cos\left(\frac{2\pi x}{\tau_s}\right) \cos\left(\omega_e t - \frac{\pi\tau_s}{\tau_p}\right) \qquad \tau_s/2 < x < 3\tau_s/2 \tag{5.82}$$

In general, then, over any pole pitch n

$$B_{gp(n)} = B_{p1} \cos\left(\frac{2\pi x}{\tau_s}\right) \cos\left(\omega_e t - \frac{m\pi\tau_s}{\tau_p}\right) \qquad m = 1,2,\ldots,2\tau_p/\tau_s$$

$$(2n-1)\frac{\tau_s}{2} < x < (2n+1)\frac{\tau_s}{2} \tag{5.83}$$

As a particular point on the rotor surface rotates past the stator teeth at rotor speed, it sequentially encounters each of the one cycle waveforms representing one

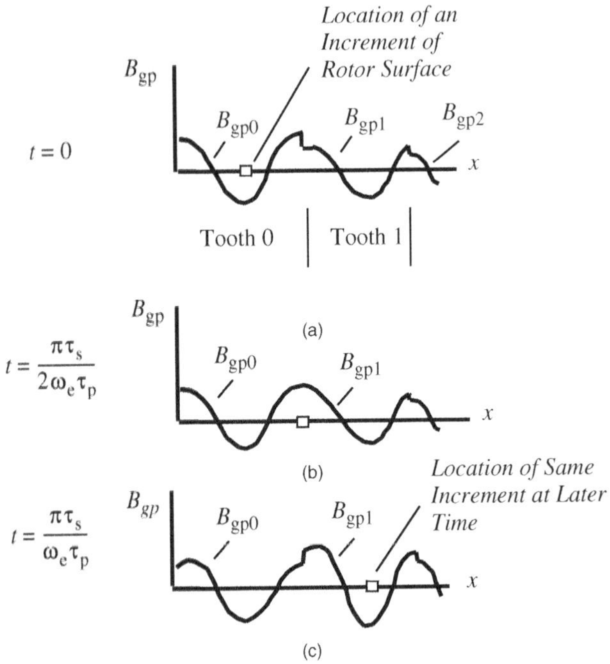

Figure 5.23 Increment of rotor surface moving between two stator slots. (a) alignment at $t = 0$ and corresponding flux density distribution, (b) alignment at an intermediate point, (c) alignment at $t = (P\pi)/(\omega_e S_1)$.

stator slot pitch at a rate corresponding to slip frequency. However, should the rotor rotate synchronously, the pattern of Figure 5.22 is essentially "locked" to the rotor surface since this pattern too rotates at synchronous speed. An exact solution for the flux over the rotor surface appears to remain a formidable task. To gain insight, it is useful to consider more carefully the instantaneous rotor flux distribution considering only one stator tooth span. Choosing for convenience the case in which the rotor is rotating synchronously, the region under the tooth can be selected as having maximum ripple flux density (tooth 0 corresponding to equation (5.81)) as the region of interest.

At $t = 0$ the location of a small rotor surface area is assumed to be centered under tooth 0 as shown in Figure 5.23a. The harmonic ripple flux density is instantaneously a minimum at this point. At time $t = \pi\tau_s/2\omega_e\,\tau_p$, this increment of the rotor surface has moved midway between teeth 0 and 1 as shown in Figure 5.23b. The harmonic ripple flux density is clearly a maximum at this instant. When $t = \pi\tau_s/\omega_e\,\tau_p$, the increment of surface has moved under tooth #1 as in Figure 5.23c. It appears as if the increment has moved onto another (somewhat lower) "cycle" of Figure 5.22. However, it can be recalled that the rotor is moving synchronously with the stator excitation. Therefore, at this instant the windings in slot #1 are loaded with precisely the same value of current as existed in slot #0 at time $t = 0$. The same argument holds for each progressive slot pair. For example, the current in slot #2 is now the same

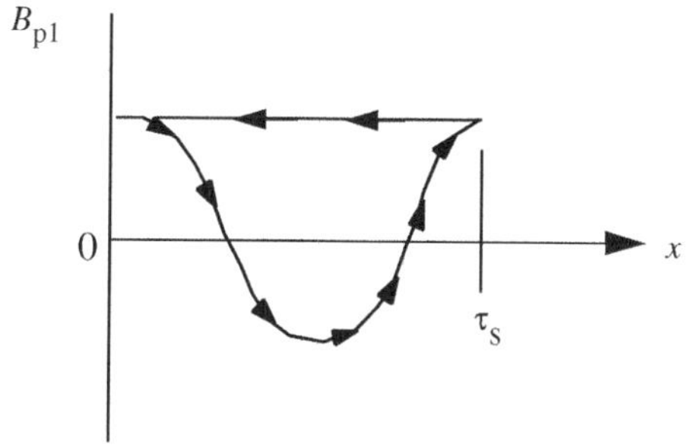

Figure 5.24 Illustrating cyclic motion of flux density over an increment of rotor surface.

as what appeared in slot #1 at $t = 0$. Hence, the flux density penetrating the incremental surface area under consideration has therefore returned to the same value as what existed previously at $t = 0$. The flux density in effect continuously cycles over the same cosinusoidal waveform as illustrated in Figure 5.24. Since the rotor surface immediately to the left of the sample increment experiences the same variation only slightly delayed in time, it follows that this effect can be interpreted as a traveling wave impinging on the rotor surface of the form

$$B_{gp1} = B_{p1} \cos\left[\frac{2S_1}{P}\left(\omega_e t - \frac{\pi x}{\tau_p}\right)\right] \tag{5.84}$$

Let x, y and z denote the tangential, radial and axial directions of an equivalent cartesian coordinate system. The solution for the flux density in the rotor iron is described by the Maxwell equation

$$\nabla \times \boldsymbol{B} = \boldsymbol{J}/\mu_i \tag{5.85}$$

Since B is directed radially, this would imply that a current density is induced in the axial direction. However, since the iron is laminated, current cannot flow in this direction (except for eddy currents which are assumed to sufficiently small that they do not materially affect the field). Hence, the z (axial) component of equation (5.85) becomes

$$\frac{\partial}{\partial y}B_{px} - \frac{\partial}{\partial x}B_{py} = 0 \tag{5.86}$$

Due to symmetry, one can deduce that the z component of B is zero and, since the flux density is everywhere continuous

$$\nabla \bullet \boldsymbol{B} = 0$$

or

$$\frac{\partial}{\partial x}B_{px} + \frac{\partial}{\partial y}B_{py} = 0 \tag{5.87}$$

Differentiating equations (5.86) and (5.87) with respect to x and subtracting the result yields

$$\frac{\partial^2}{\partial x^2}B_{px} + \frac{\partial^2}{\partial y^2}B_{px} = 0 \tag{5.88}$$

If equations (5.86) and (5.87) are differentiated with respect to y and added, one obtains

$$\frac{\partial^2}{\partial x^2}B_{py} + \frac{\partial^2}{\partial y^2}B_{py} = 0 \tag{5.89}$$

The form of these equations clearly suggests a harmonic solution. Since the flux density must be continuous across the rotor surface, then if increasing y denotes penetration into the rotor

$$B_{py}(y)\Big|_{y=0} = B_{py}(0) = B_{p1}\cos\left[\frac{2S_1}{P}\left(\omega_e t - \frac{\pi x}{\tau_p}\right)\right] \tag{5.90}$$

Therefore, it is assumed that for any point x, y within the rotor body

$$B_{py}(x,y) = B_{p1}b_y(y)\cos\left[\frac{2S_1}{P}\left(\omega_e t - \frac{\pi x}{\tau_p}\right)\right] \tag{5.91}$$

Substituting this expression in equation (5.89), performing the indicated differentiation and simplifying,

$$\frac{\partial^2}{\partial y^2}b_y(y) - \left[\left(\frac{2S_1}{P}\right)\frac{\pi}{\tau_p}\right]^2 b_y(y) = 0 \tag{5.92}$$

Since $B_{py}(x,y) \to 0$ as $y \to \infty$, the solution of equation (5.92) is of the form

$$b_y(y) = e^{-\left(\frac{2S_1}{P}\frac{\pi}{\tau_p}y\right)} \tag{5.93}$$

Hence,

$$B_{py}(x,y) = B_{p1}e^{-\left(\frac{2S_1}{P}\frac{\pi}{\tau_p}y\right)}\cos\left[\frac{2S_1}{P}\omega_e t - \frac{\pi x}{\tau_p}\right] \tag{5.94}$$

From equation (5.86), it is readily established that

$$B_{py}(x,y) = -B_{p1}e^{-\left(\frac{2S_1}{P}\frac{\pi}{\tau_p}y\right)}\sin\left[\frac{2S_1}{P}\omega_e t - \frac{\pi x}{\tau_p}\right] \tag{5.95}$$

It is useful to define the *penetration constant* or *equivalent depth* as

$$d_p = \frac{\tau_p}{\pi}\frac{P}{2S_1} = \frac{\tau_s}{2\pi} \tag{5.96}$$

It can be noted that the fields B_{px} and B_{py} are everywhere 90° out of phase. Therefore, the amplitude of the field at a point x,y inside the rotor is

$$B_{ro}(x,y) = \sqrt{2}B_{p1}e^{-\frac{y}{d_e}} \tag{5.97}$$

which indicates that the flux density is everywhere uniform with x (tangential direction) but decreases exponentially with y (radial direction into the rotor). It can be recalled that when the rotor is rotating synchronously, equation (5.97) applies only over the stator slot pitch corresponding to maximum (and negative maximum) flux density. The subscript 0 in equation (5.97) is used as a reminder that the solution applies to a stator slot pitch of the rotor material spanned by the pitch of stator tooth 0 in Figure 5.24.

If d_p is used as an equivalent skin depth, the losses in the rotor can be written as the loss density times the rotor volume in question. Recall from Section 5.5 that the iron losses vary as the flux density squared. Hence, the loss for this particular portion of the rotor surface is

$$P_{r0} = \left(\frac{2\pi D_{or}}{S_1}\right) k_{ir} l_i d_p \left(\frac{B_{p1}}{B_{g1}}\right)^2 C_{ir} \tag{5.98}$$

where l_i is the length of rotor iron, k_{ir} is the stacking factor for rotor iron, B_{g1} is the peak value of fundamental flux density in the gap and C_{ir} is the loss per unit volume of the rotor magnetic material measured at the value B_{g1} and at frequency $(2S_1/P)(\omega_e/2\pi)$.

The losses can be obtained over other stator pole pitches by the same manner. Since the average flux density over adjacent teeth varies sinusoidally, so also does the ripple flux density. The solution for the other teeth can be written by inspection. For example, for tooth #1,

$$B_{r1} = \sqrt{2}B_{p1}\cos\left(\frac{\pi P}{S_1}\right)e^{\frac{-y}{d_e}} \tag{5.99}$$

The losses for this portion of the rotor are

$$P_{r1} = \left(\frac{2\pi D_{or}}{S_1}\right) k_{ir} l_i d_p \left(\frac{B_{p1}}{B_{g1}}\right)^2 \cos^2\left(\frac{\pi P}{S_1}\right) C_{ir} \tag{5.100}$$

The total rotor surface loss can be written as

$$P_{r(surf)} = \sum_{n=0}^{S_1/P-1} P\left(\frac{2\pi D_{or}}{S_1}\right) k_{ir} l_i d_p \left(\frac{B_{p1}}{B_{g1}}\right)^2 \cos^2\left(\frac{n\pi P}{S_1}\right) C_{ir} \tag{5.101}$$

which, after some factoring becomes

$$P_{r(surf)} = 2[\pi D_{or} k_{ir} l_i d_p]\left(\frac{B_{p1}}{B_{g1}}\right)^2 C_{ir} \sum_{n=0}^{S_1/P-1} \frac{\cos^2\left(\frac{n\pi P}{S_1}\right)}{(S_1/P)} \tag{5.102}$$

The coefficient inside the square brackets is recognized as the total rotor surface volume. It can be shown that the summation of equation (5.102) is identically equal to 0.5 regardless of the number of slots/pole whereupon

$$P_{\mathrm{r(surf)}} = \pi D_{\mathrm{or}} k_{\mathrm{ir}} l_{\mathrm{i}} d_{\mathrm{p}} \left(\frac{B_{\mathrm{p1}}}{B_{\mathrm{g1}}}\right)^2 C_{\mathrm{ir}} \tag{5.103}$$

It is convenient to replace the equivalent depth d_{p} in equation (5.103) by its equivalent in terms of the stator slot pitch. Also, the presence of possible rotor slot openings has so far been neglected in the solution. Clearly, one must reduce the volume corresponding to surface losses by the ratio of the thickness of the rotor tooth to the rotor slot pitch to correct for this missing material. Hence, finally,

$$P_{\mathrm{r(surf)}} = 0.5 D_{\mathrm{or}} k_{\mathrm{ir}} l_{\mathrm{i}} \tau_{\mathrm{e}} \left(\frac{t_{\mathrm{or}}}{\tau_{\mathrm{r}}}\right) C_{\mathrm{ir}} \left(\frac{B_{\mathrm{p1}}}{B_{\mathrm{g1}}}\right)^2 \tag{5.104}$$

where t_{or} denotes the equivalent width of a rotor tooth top at the air gap surface.

Thus far only the first slot harmonic has been considered. Analogous solutions can be obtained readily for all higher harmonics since the form of the result, equation (5.104), is identical. An expression which considers all harmonics is

$$P_{\mathrm{r(surf)}} = 0.5 D_{\mathrm{or}} k_{\mathrm{ir}} l_{\mathrm{i}} \tau_{\mathrm{s}} \left(\frac{t_{\mathrm{or}}}{\tau_{\mathrm{r}}}\right) \sum_{n=1}^{\infty} C_{\mathrm{ir},n} \left(\frac{B_{pn}}{B_{\mathrm{g1}}}\right)^2 \tag{5.105}$$

Note that the loss coefficient C_{ir} is included in the summation since it is a function of frequency. The solution as written is inconvenient since the loss coefficient C_{ir} is now a function of two variables, the gap flux density B_{g1} as well as the frequency. From equation (5.68), it is evident that the losses vary with frequency (hysteresis loss) as well as frequency squared (eddy current loss). The relative weighting of the hysteresis and eddy current losses depends upon the thickness of the material but clearly the exponent associated with frequency-dependent losses is between 1 and 2. If C_{ir} is assumed to vary with frequency to the νth power, and since the slot harmonic frequencies are integer ratios of the fundamental, equation (5.105) can be written as

$$P_{\mathrm{r(surf)}} = 0.5 D_{\mathrm{or}} k_{\mathrm{ir}} l_{\mathrm{i}} \tau_{\mathrm{s}} \left(\frac{t_{\mathrm{or}}}{\tau_{\mathrm{r}}}\right) C_{\mathrm{ir}} \left(\frac{B_{\mathrm{p1}}}{B_{\mathrm{g1}}}\right)^2 \left[1 + 2^{\nu}\left(\frac{B_{\mathrm{p2}}}{B_{\mathrm{p1}}}\right)^2 + 3^{\nu}\left(\frac{B_{\mathrm{p3}}}{B_{\mathrm{p1}}}\right)^2 + \ldots\right]$$

$$n = 1, \ldots, \infty \tag{5.106}$$

It is conventional to define a pole face loss factor for stator slot openings as

$$K_{\mathrm{pfs}} = \left(\frac{B_{\mathrm{p1}}}{B_{\mathrm{g1}}}\right)^2 \sum_{n=1}^{\infty} n^{\nu} \left(\frac{B_{\mathrm{pn}}}{B_{\mathrm{p1}}}\right)^2 \tag{5.107}$$

so that equation (5.106) becomes

$$P_{\mathrm{r(surf)}} = 0.5 D_{\mathrm{or}} k_{\mathrm{ir}} l_{\mathrm{i}} \tau_{\mathrm{s}} \left(\frac{t_{\mathrm{or}}}{\tau_{\mathrm{r}}}\right) C_{\mathrm{ir}} K_{\mathrm{pfs}} \tag{5.108}$$

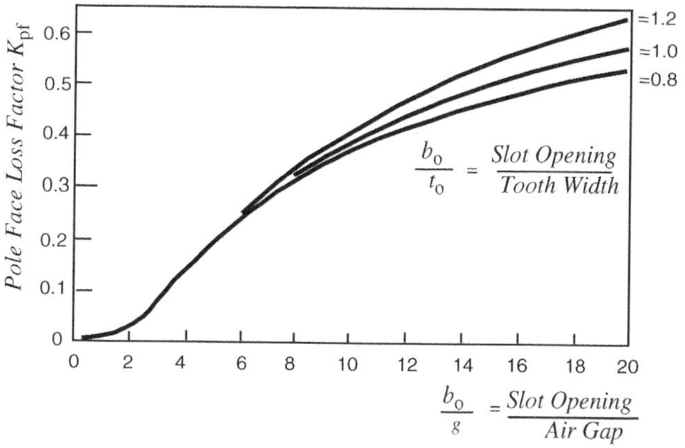

Figure 5.25 Pole face loss factor K_{pf} as a function of the ratio of slot opening to air gap length.

The quantity K_{pf} is plotted in Figure 5.25 for three different values of slot opening. The subscript "s" has been dropped from Figure 5.26 since the curves are equally valid for the losses in the rotor as well as the stator teeth.

It can be recalled that the loss coefficient C_{ir} represents the rotor loss per unit volume at the frequency $2S_1 f_e/P$ and at the flux density B_{g1}. Since the losses vary as

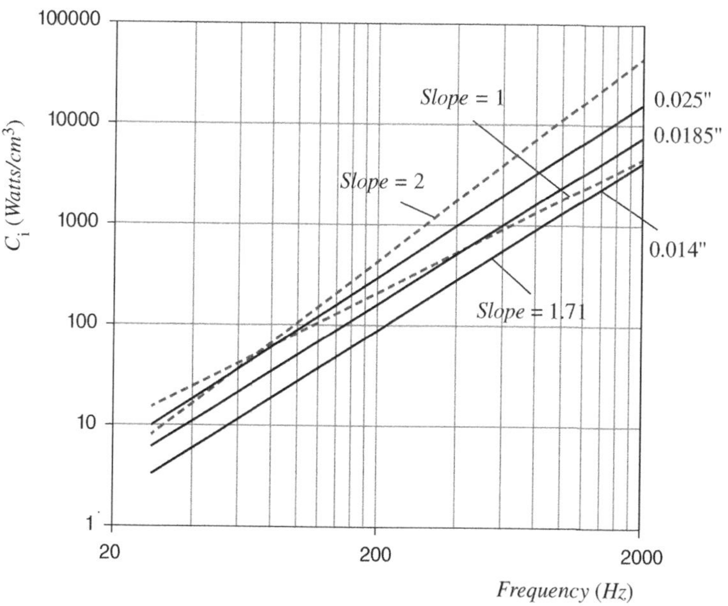

Figure 5.26 Loss coefficient C_i for silicon steel as a function of frequency, peak flux density maintained at 1.55 T (100 $Klines/in^2$).

B_2 it is not necessary to measure C_{ir} at each value of gap flux density. Rather, equation (5.108) can be put in the form

$$P_{r(surf)} = 0.5D_{or}k_{ir}l_i\tau_s\left(\frac{t_{or}}{\tau_r}\right)C_{ir}K_{pfs}\left(\frac{B_{p1}}{B_{ref}}\right)^2 \qquad (5.109)$$

where C_{ir} is now measured at a reference value of flux density B_{ref}. B_{ref} is generally taken as 100 kilolines/in.^{2}(1.55 tesla). A plot of C_{ir} as a function of frequency is shown in Figure 5.26 for three different lamination thicknesses. Here again, the subscript "r" has been dropped since the curves can be used for stator as well as rotor tooth losses. Figure (5.26) can be used to estimate the value of "v" contained in equation (5.106) as $v = 1.71$.

When the rotor has slotting as well as the stator, losses also appear on the stator surface. The proper equation is simply obtained from equation (5.109) by inspection. The relevant equation is

$$P_{s(surf)} = 0.5D_{is}k_{is}l_i\tau_r\left(\frac{t_{os}}{\tau_s}\right)C_{is}K_{pfr}\left(\frac{B_{p1}}{B_{ref}}\right)^2 \text{ watts} \qquad (5.110)$$

Whereas equations (5.109) and (5.110) can be considered as the final result, a short amount of additional discussion is useful. Recall that it was decided to calculate the losses for the specific case of synchronous speed operation. It could be questioned whether the results change when the motor is loaded since synchronous speed operation implies an unloaded condition. Consider again the rotor material under one stator tooth pitch used for the previous derivation. As the motor begins to slip, the material begins to encounter the one cycle cosinusoidal waveforms shown in Figure 5.23. A proper expression for the losses in this rotor material would now be a weighted average of the losses associated with each of the waveforms. Examination of equation (5.106), however, indicates that just such an average is already included in the expression by taking the summation shown. Hence, if the slip frequency is small equations (5.109) and (5.110) remain accurate. It is clear, however, that a frequency modulation effect actually occurs so that these equations become less accurate for large slips. Clearly, when the rotor is stationary the "slot frequency" becomes equal to the line frequency (see equation (5.81)). If desired, this frequency-dependent effect could be corrected by using a value for C_{ir} and C_{is} corresponding to the frequency

$$\left[\frac{2S_1}{P} + \left(1 - \frac{2S_1}{P}\right)S\right]f_e \qquad (5.111)$$

where S is the per unit slip and f_e is the line frequency. Alternatively, equations (5.109) and (5.110) could be made speed-dependent by including a term of the form

$$\left[1 + \left(\frac{P}{2S_1} - 1\right)S\right]^v$$

In this case, C_{ir} and C_{is} remain values of loss density corresponding to the slot frequency at synchronous speed, namely $2S_1 f_e/P$.

Since many machine designers choose to use the average rather than the peak value of gap flux density as a key parameter, equation (5.110) can be written alternatively as

$$P_{r(surf)} = (\pi^2/8)D_{is}k_{is}l_i\tau_s\left(\frac{t_{or}}{\tau_r}\right)C_{ir}K_{pfs}\left(\frac{B_{g1(ave)}}{B_{ref}}\right)^2 \tag{5.112}$$

The coefficient $\pi^2/8 = 1.234$. This coefficient has variously appeared as 1.65 and 2.0 in other authors' works using different assumptions [2]. Because the gap of an induction machine is small, the surface losses tend to be the major contributor to stray losses, typically 20–30%.

5.10 CALCULATION OF TOOTH PULSATION IRON LOSSES

In addition to the surface losses, other types of losses also occur which are caused by the slotting. If the rotor and the stator had equal numbers of slots then the only high-frequency loss in the rotor would be the surface iron loss just calculated. However, if for example, the rotor slots were twice the number of stator slots and if the rotor slots were open, nearly all the stator harmonic flux would tend to encircle the rotor slots. Hence, the harmonic flux which is interrupted by the open slot can, in general, also find a path which encircles the slot and the amount of flux encircling the slot depends on the difference between the number of stator and rotor slots per pole. Clearly, additional losses are encountered in the rotor conductors and in the iron surrounding the slot. These losses are typically calculated separately.

The problem of tooth pulsation losses can be explained by reference to Figure 5.27 which illustrates the alignment of the stator and rotor teeth for two positions of the rotor. It is assumed for purposes of the example that the stator tooth width is larger than the rotor tooth width. At some time $t = 0$ a particular rotor tooth is aligned with a stator tooth as shown in Figure 5.27a. A short instant later at $t = t_1$ the center line of the rotor tooth becomes aligned with a stator slot. At time $2t_1$ the situation becomes the same as at $t = 0$. It is apparent by comparison of the first two cases that the flux entering the rotor slot at $t = 0$ will be larger than $t = t_1$. The difference between these two cases constitutes a flux pulsation which passes down the tooth body and ultimately links one of the rotor conductors. Hence, additional rotor copper as well as iron losses in the stator and rotor are created.

In order to investigate this problem, it is again assumed that the rotor is rotating synchronously. Again the rotor surface is assumed initially smooth. The value of pulsating flux that occurs for the rotor tooth in question is again proportional to the average flux linking the tooth in much the same manner as the previous problem. The rotor tooth which follows the peak of the flux density wave B_{g1} is chosen. When the rotor tooth is lined up with a stator tooth, the flux density over the tooth is equal to B_{g1}. The flux remains at this value until the edge of the stator slot begins to sweep over the rotor tooth. The flux linking the rotor tooth then abruptly drops to a smaller value which again remains fairly constant until the stator slot sweeps off of the other

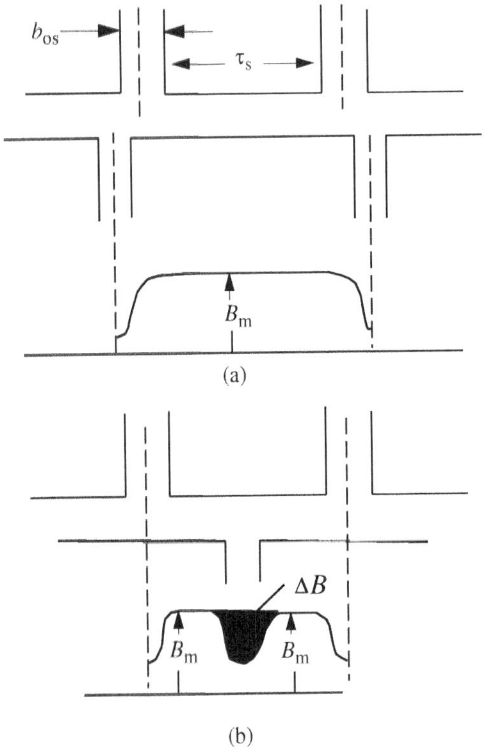

Figure 5.27 Illustration of slot flux pulsation showing two limiting alignments of stator and rotor teeth.

side of the rotor tooth. The flux penetrating the rotor tooth as a function of time is approximately a rectangular function as shown in Figure 5.28. The average (negative) component of this waveform is again accounted for by Carter's Coefficient. Since the harmonic components of Figure 5.28 contribute to the losses, it is apparent that the problem is most severe when the rotor slot pitch is exactly half the stator slot pitch (twice as many rotor slots as stator slots).

The flux change $\Delta\Phi$ which occurs whenever a stator slot passes by the rotor slot is found by solving for the area ΔB shown in Figure 5.27. Since the flux density

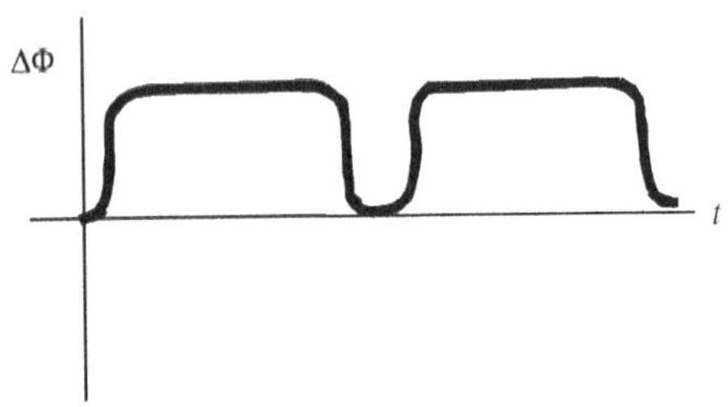

Figure 5.28 Waveform of slot flux pulsation as a function of time.

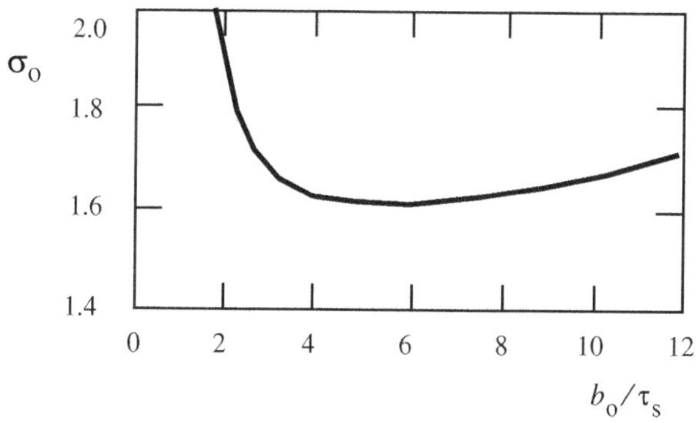

Figure 5.29 Parameter σ_o as a function of the ratio of slot opening to the air gap length.

distribution is complicated the relationship between the flux pulsation and the maximum flux across the slot must be determined numerically. It can be shown that the flux pulsation over a span of one rotor tooth can be written in the form [3].

$$\Delta\Phi = b_{os} l_{es} \beta_o \sigma_o B_{g1} \tag{5.113}$$

where B_{g1} is taken as the maximum flux density over the slot under consideration. The functions β_o and σ_o are plotted in Figure 5.20 and Figure 5.29 respectively.

The fundamental component of the flux change can be found by taking the first Fourier harmonic of Figure 5.28. The result is

$$\Delta\Phi_{1p} = \left(\frac{2}{\pi}\right) b_{os} l_e \beta_o \sigma_o B_{g1} \sin\left(\frac{S_1}{S_2}\pi\right) \tag{5.114}$$

The frequency associated with the first harmonic is again equal to the stator slot angular frequency $[(2S_1)/P]\omega_e$.

The flux density in the slot itself is

$$B_{\rm lt} = \frac{\Delta\Phi_{1p}}{l_i t_{r(ave)}}$$

where $t_{r(ave)}$ is a representative average value of rotor tooth thickness. Or,

$$B_{lt} = \left(\frac{2}{\pi}\right) \frac{b_{os}}{t_{r(ave)}} \frac{l_e}{l_i} \beta_o \sigma_o B_{g1} \sin\left(\frac{S_1}{S_2}\pi\right) \tag{5.115}$$

The iron losses associated with the rotor teeth are again found by adding up the losses in each rotor tooth, properly weighting the terms by the sine of the location of the tooth with respect to a pole pitch. The solution becomes

$$P_{tr} = \pi D l_i d_{tr} C_{ir} \left(\frac{B_{1t}}{B_{g1}}\right)^2 \sum_{n=0}^{S_2/P-1} \frac{\cos^2\left(\frac{n\pi P}{S_2}\right)}{\frac{S_2}{P}} \tag{5.116}$$

where d_{tr} is the depth of the rotor tooth. The summation is again equal to 1/2. The final expression can be written as

$$P_{tr} = \left(\frac{2}{\pi}\right) D l_e d_{tr} C_{ir} \left(\frac{l_e}{l_i}\right) \left(\frac{b_{os}}{t_{r(ave)}}\right)^2 \beta_o^2 \sigma_o^2 \sin^2\left(\frac{S_1}{S_2}\pi\right) \tag{5.117}$$

The loss density C_{ir} again pertains to the slot frequency $2S_1f/P$ and the gap flux density B_{g1}. The loss density can again be referred to a reference value by assuming that the losses vary as the square of flux density. An alternative expression is

$$P_{tr} = \left(\frac{2}{\pi}\right) D l_e d_{tr} C_{ir} \left(\frac{l_e}{l_i}\right) \left(\frac{b_{os}}{t_{r(ave)}}\right)^2 \beta_0^2 \sigma_o^2 \left(\frac{B_{g1}}{B_{ref}}\right)^2 \sin^2\left(\frac{S_1}{S_2}\pi\right) \tag{5.118}$$

Again B_{ref} is typically taken as 1.55 tesla (100 kilolines/in.^{2}). An analogous expression can be written for the stator tooth pulsation loss should the rotor be slotted. Also, the tooth flux density can also be corrected for higher harmonics in the waveform. Since the harmonics of a square wave are inversely proportional to n, a more accurate version of B_t is

$$B_t^2 = B_{1t}^2 \left[1 + \left(\frac{1}{3}\right)^{3-v} + \left(\frac{1}{5}\right)^{5-v} + \ldots\right] \tag{5.119}$$

The bracketed term in equation (5.119) can be included as an extra coefficient in equation (5.116). Because of the inaccuracies in approximating the flux pulsation as a square wave, only the first several terms in equation (5.119) have much significance. Again the result obtained is valid at small slip frequencies as well as at zero slip (synchronous speed). If desired, the loss term can be corrected for frequency changes by the same factor as used in Section 5.9, that is equation (5.110).

The tooth pulsation iron loss is sometimes neglected by other authors since it is claimed that the currents which flow in the short-circuited bar (to be computed) tend to cancel the flux pulsation. However, this occurs only for a purely inductive bar which is clearly not the case at slot ripple frequencies due to skin effect. In fact, it can be shown that the when the pulsation frequency is sufficiently large such that the high frequency approximations are valid (equations (5.36) and (5.37)), the tooth flux produced by the slot currents could, in fact, increase the net flux in the tooth.

As already mentioned, the pulsating flux in the rotor teeth also results in harmonic losses in the short-circuited rotor bars of a squirrel-cage machine. The current which flows in a given rotor bar is a result of the time rate of change of the flux linking the bar. It is useful to refer to Figure 5.30 which now shows an array of rotor teeth sweeping along the stator. Again the rotor is assumed essentially smooth with no significant slotting. Two adjacent rotor teeth embracing one specific rotor slot can be identified as shown on Figure 5.30a. Again it is assumed that the rotor rotates synchronously. The left-hand tooth (tooth #1) slot is taken to correspond to the slot rotating synchronously with the peak of the air gap flux wave. Flux changes induced in these two rotor slots are shown in Figure 5.30b. Note that the flux pulsation induced in the right-hand tooth (tooth #2) lags in time by a time corresponding to the rotation of one rotor slot pitch. In addition, the amplitude of the flux pulsation in tooth #2 is reduced since the average gap flux density varies sinusoidally in space.

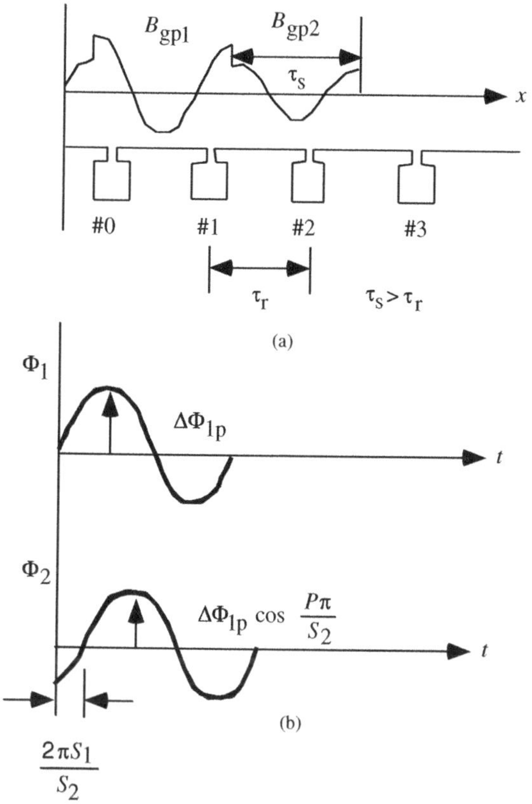

Figure 5.30 Flux pulsations of two adjacent teeth vs. time.

The first harmonic of the flux pulsation associated with tooth #1 is

$$\Delta\Phi_1(t) = \Delta\Phi_{1p}\cos\left(\frac{2S_1}{P}\omega_e t\right) \tag{5.120}$$

where $\Delta\Phi_{1p}$ is given by equation (5.114). The flux penetrating tooth #2 can be written as

$$\Delta\Phi_2(t) = \Delta\Phi_{1p}\cos\left(\frac{2S_1}{P}\omega_e t - 2\pi\frac{S_1}{S_2}\right)\cos\left(\frac{P\pi}{S_2}\right) \tag{5.121}$$

The pulsating flux linking the bar can be found by subtracting equation (5.121) from equation (5.120). After a certain amount of algebra, the result can be written as

$$\begin{aligned}\Delta\Phi_1(t) - \Delta\Phi_2(t) = -\Delta\Phi_{1p}&\left[\sin\left(\frac{2S_1}{P}\omega_e t - \frac{S_1}{S_2}\pi + \frac{P\pi}{2S_2}\right)\sin\left(\frac{S_1}{S_2}\pi - \frac{P\pi}{2S_2}\right)\right.\\ &\left. + \sin\left(\frac{2S_1}{P}\omega_e t - \frac{S_1}{S_2}\pi - \frac{P\pi}{2S_2}\right)\sin\left(\frac{S_1}{S_2}\pi + \frac{P\pi}{2S_2}\right)\right]\end{aligned} \tag{5.122}$$

Two terms appear because of the modulation of the slot frequency with the fundamental frequency.

The EMF induced in the bar between teeth #1 and #2, say bar #1, can be represented in phasor form as

$$\tilde{E}_{\mathrm{bp(bar1)}} = j\left(\frac{2S_1}{P}\right)\omega_{\mathrm{e}}(\Delta\tilde{\Phi}_1 - \Delta\tilde{\Phi}_2) \tag{5.123}$$

which can be written in terms of two voltage components

$$\tilde{E}_{\mathrm{bp(bar1)}} = \tilde{E}_{\mathrm{bp1}} + \tilde{E}_{\mathrm{bp2}} \tag{5.124}$$

where

$$\tilde{E}_{\mathrm{bp1}} = \left(\frac{2S_1}{P}\right)\omega_{\mathrm{e}}\Delta\tilde{\Phi}_{\mathrm{1p}}\sin\left(\frac{S_1}{S_2}\pi - \frac{P\pi}{2S_2}\right)e^{j\left[-\frac{S_1}{S_2}\pi+\frac{P\pi}{2S_2}\right]} \tag{5.125}$$

and

$$\tilde{E}_{\mathrm{bp2}} = \left(\frac{2S_1}{P}\right)\omega_{\mathrm{e}}\Delta\tilde{\Phi}_{\mathrm{1p}}\sin\left(\frac{S_1}{S_2}\pi - \frac{P\pi}{2S_2}\right)e^{j\left[-\frac{S_1}{S_2}\pi-\frac{P\pi}{2S_2}\right]} \tag{5.126}$$

The solution for the pulsating currents flowing in the cage of the machine follows in much the same manner as the analysis previously carried out for the fundamental current component, Section 4.12. However, the solution must be carried out twice, once for each component of equation (5.124).

For example, the flux linking bar #2 next to bar #1 can be written

$$\begin{aligned}\Delta\Phi_2(t) - \Delta\Phi_3(t) = -\Delta\Phi_{\mathrm{1p}}\Bigg[&\sin\left(\frac{2S_1}{P}\omega_{\mathrm{e}}t - \frac{3S_1}{S_2}\pi+\frac{3P}{2S_2}\pi\right)\sin\left(\frac{S_1}{S_2}\pi - \frac{P}{2S_2}\pi\right)\Bigg]\\ &+\sin\left(\frac{2S_1}{P}t - \frac{S_1}{S_2}\pi - \frac{3P}{2S_2}\pi\right)\sin\left(\frac{S_1}{S_2}\pi+\frac{P}{2S_2}\pi\right)\end{aligned} \tag{5.127}$$

The EMF induced in bar #2 can be expressed in phasor form as

$$\tilde{E}_{\mathrm{bp(bar2)}} = \tilde{E}_{\mathrm{bp1}}e^{j\pi\left[-\frac{3S_1}{S_2}-\frac{3P}{S_2}\right]} + \tilde{E}_{\mathrm{bp2}}e^{j\pi\left[-\frac{3S_1}{S_2}-\frac{3P}{S_2}\right]} \tag{5.128}$$

The solution proceeds in much the same manner as for the fundamental currents except that a different expression is used for the phase shift. The effects of the end ring can be included in the same manner as before. After including the effects of the end ring, the equations which define the solution for the current that flows in the bar reduces to

$$\tilde{E}_{\mathrm{bp1}} = \tilde{I}_{\mathrm{bp1}}\left[R_{\mathrm{b}} + j\left(\frac{2S_1}{P}\right)\omega_{\mathrm{e}}L_{\mathrm{b}}\right] + \tilde{I}_{\mathrm{bp1}}\left[\frac{R_{\mathrm{e}} + j\left(\frac{(2S_2)}{P}\right)\omega_{\mathrm{e}}L_{\mathrm{e}}}{2\sin^2\left(\frac{S_1}{S_2}\pi-\frac{P\pi}{2S_2}\right)}\right] \tag{5.129}$$

$$\tilde{E}_{\mathrm{bp2}} = \tilde{I}_{\mathrm{bp2}}\left[R_{\mathrm{b}} + j\left(\frac{2S_1}{P}\right)\omega_{\mathrm{e}}L_{\mathrm{b}}\right] + \tilde{I}_{\mathrm{bp2}}\left[\frac{R_{\mathrm{e}} + j\left(\frac{(2S_2)}{P}\right)\omega_{\mathrm{e}}L_{\mathrm{e}}}{2\sin^2\left(\frac{S_1}{S_2}\pi+\frac{P\pi}{2S_2}\right)}\right] \tag{5.130}$$

where the total current in the bar is the vector sum of $\widetilde{I}_{\text{bp1}}$ and $\widetilde{I}_{\text{bp2}}$. The "(#1)" portion of the subscripts have been removed from these equations since they apply equally to any bar. The power loss in the S_2 rotor bars is, therefore,

$$P_{\text{prc}} = 0.5Re\left[\widetilde{E}_{\text{bp1}}\widetilde{I}^*_{\text{bp1}} + \widetilde{E}_{\text{bp2}}\widetilde{I}^*_{\text{bp2}}\right] \tag{5.131}$$

where the asterisk denotes the complex conjugate of the quantity. The 0.5 coefficient appears since peak values have been assumed rather than RMS quantities. It is important to emphasize that skin effect is to be included in these calculations so that R_{b} and L_{b} must be evaluated at the slot frequency $2S_1(\omega_e/P)$. This loss exists only for the squirrel-cage rotor configuration and can be omitted for the stator conductors or for the rotor conductors of wound rotor machines since the induced EMFs cancel out when the conductors in the slots are connected in series.

The issue of stray losses is clearly complex indeed and analytical work is continuing on this problem. It should be emphasized here that only a small portion of the iron loss picture has been investigated here, namely rotor iron and copper losses due to stator slot openings (only one of 4 identified loss components mentioned at the onset of Section 5.8. Fortunately, these losses in different classes of machines remain roughly constant so that they can be measured and used for a class of machines. It is general practice to make corrections to the usual (fundamental component) iron and copper losses to account for these losses. If measured at one value of current and frequency, the stray losses can be converted to another pair of values by

$$\text{Stary Losses}_2 = \text{Stary Losses}_1\left(\frac{I_2}{I_1}\right)^2\left(\frac{f_2}{f_1}\right)^{1.4} \tag{5.132}$$

A tabulation of the stray load and no-load losses for a variety of machines is given in Table 5.1.

TABLE 5.1 Iron loss distribution in AC machines

	Fundamental losses (%)	Additional losses in % due to non-sinusoidal waveform (punching stresses, burrs, etc.)	Stray no-load losses (%)	Stray load loss in % of output power
Induction motor with semi-open stator and rotor slots	100	30–40	50–70	0.3–0.6
Induction motor with semi-open stator slots	100	30–40	120–160	0.3– 0.6
Salient pole synchronous machines	100	40–60	50–70	0.1–0.2
Round rotor turbogenerators (four-pole)	100	30–40	40–50	0.05– 0.15
(Two-pole)	100	15–25	25–35	0.05–0.15

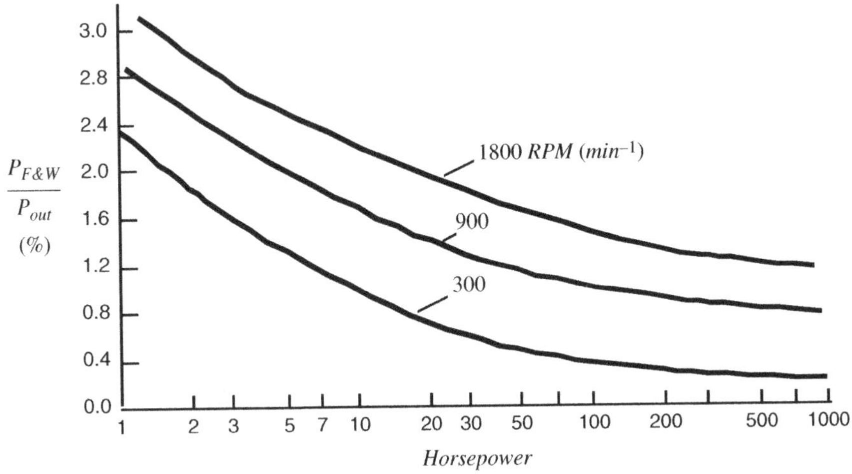

Figure 5.31 Friction and windage losses in synchronous machines as a function of output power and speed [4].

5.11 FRICTION AND WINDAGE LOSSES

Bearing friction losses are an unavoidable loss in a rotating machine. The amount of the bearing friction losses depends upon the pressure on the bearing, the peripheral speed of the shaft at the bearing and the coefficient of friction between bearing and shaft. The windage losses are also produced by rotation and depend upon the peripheral speed of the rotor, the rotor diameter, the core length and generally upon the construction of the machine. While friction losses can be determined more or less accurately, the windage losses have to be determined on the basis of an actual test. Figure 5.31 and Figure 5.32 show typical windage and friction losses for different classes of machines and were derived by tests on a large number of machines. They can be used to obtain an approximate estimate of these losses. As a rule of thumb, the friction and windage losses can be converted from a known measured speed to another speed by the expression

$$F\&W_2 = F\&W_1 \left(\frac{speed_2}{speed_1} \right)^2 \tag{5.133}$$

5.12 EXAMPLE 8—CALCULATION OF IRON LOSS RESISTANCES

The iron loss characteristic curves for 0.025″, 3% Si steel was given in Figure 5.16. Note that these curves combine the effects of eddy current and hysteresis and are given in units of W/lb rather than W/in.^{3}. In order to convert to useful units the density of the laminations must be specified. This value can be taken as 0.276 lb/in.^{3}. Consider

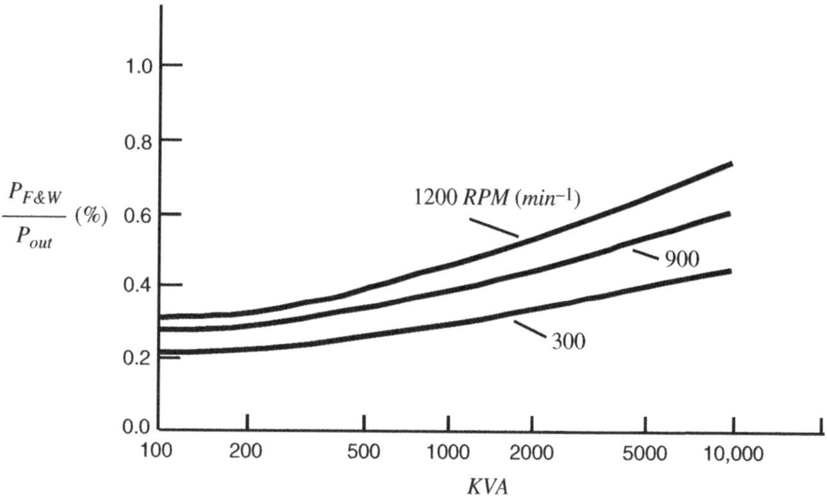

Figure 5.32 Friction and windage losses in synchronous machines as a function of output power and speed [4].

first the calculation of the equivalent iron loss resistance which accounts for the loss in the stator teeth of the example 250 HP machine.

The volume of iron used for the S_1 stator teeth can be expressed as the following integral

$$V_{\text{teeth}} = S_1 k_{\text{is}} l_{\text{is}} \int_0^{d_s} \left[t_t + \left(\frac{t_t + t_r}{d_s} \right) x \right] d_s$$

The numerical values of the various quantities can be determined by reference to Figure 3.11. This integral can be evaluated as

$$V_{\text{teeth}} = S_1 k_{\text{is}} l_{\text{is}} \left(\frac{t_t + t_r}{2} \right) d_s$$

which becomes

$$\begin{aligned} V_{\text{teeth}} &= (120)(0.97)(8.5)\left(\frac{0.256 + 0.369}{2}\right)(2.2) \\ &= 680.2\ \text{in}^3 \end{aligned}$$

From Example 2 in Chapter 3 it was determined that for an air gap fundamental component of flux density of 0.775 tesla, the density in the top, the middle, and the bottom of the tooth are

$$\begin{aligned} B_t &= 1.86\,\text{T} = 119{,}790\ \text{lines/in.}^2 \\ B_m &= 1.52\,\text{T} = 98{,}040\ \text{lines/in.}^2 \\ B_b &= 1.29\,\text{T} = 83{,}205\ \text{lines/in.}^2 \end{aligned}$$

Note that these values relate to the flux density at a point 30° away from the maximum value of B_g. However, since the machine is heavily saturated for the condition used to evaluate the parameters, these values will also be taken as the maximum value of B in the teeth.

From Figure 5.16 the corresponding values of power density are

$$p_t = 0.72\,\text{W/in.}^3$$
$$p_m = 0.52\,\text{W/in.}^3$$
$$p_b = 0.38\,\text{W/in.}^3$$

Using Simpson's approximation the loss in the stator teeth is therefore approximately

$$\begin{aligned} P_{\text{teeth}} &= \left(\frac{1}{6}p_e + \frac{2}{3}p_m + \frac{1}{6}p_r\right)V_{\text{teeth}} \\ &= \left(\frac{0.72}{2} + \frac{2}{3}(0.52) + \frac{0.38}{6}\right)680.2 \\ &= 360.5\ \text{watts} \end{aligned}$$

This figure is an uncorrected value which does not take into account manufacturing imperfections and non-sinusoidal waveform.

Reference to Table 5.1 indicates that for an induction machine with open stator slots and semi-closed rotor slots this value can be increased by 30–40% to account for these additional losses. The exact value clearly depends on a number of unknowns such as manufacturing processes, and would usually be determined for a particular line of machines by test. If a figure of 35% is estimated then a corrected value of power lost in the teeth is

$$\begin{aligned} P_{\text{teeth}} &= (1.35)(360.5) \\ &= 487\ \text{watts} \end{aligned}$$

From Example 2 it was established that the voltage across the machine which results in a gap flux density of 0.775 tesla is 2550 volts, line-to-line. At no load, this voltage appears across the sum of the stator leakage inductance plus the magnetizing inductance if resistance is neglected. The stator tooth loss is roughly dependent on air gap flux since the slot leakage flux takes a short cut across the slot and the end-winding leakage is not involved. Hence, the equivalent value of resistance which results in a power loss in the teeth of 487 watts is, from Example 3,

$$\begin{aligned} r_t &= \frac{V_{\text{gap}}^2}{P_{\text{teeth}}} = \frac{(2550)^2}{487} \\ &= 13{,}350\,\Omega \end{aligned}$$

Consider now the loss which occurs in the stator core. The volume of iron which occupies the stator core is

$$\begin{aligned} V_{\text{core}} &= k_{is} l_i \frac{\pi}{4}\left[D_{os}^2 - (D_{os} - 2h_{cs})^2\right] \\ &= \pi k_{is} l_i h_{cs}(D_{os} - h_{cs}) \end{aligned}$$

which can be evaluated as

$$V_{\text{core}} = (\pi)(0.97)(8.5)(1.76)(32 - 1.76)$$
$$= 1378.6\,\text{in}^3$$

By reference to Example 2, the maximum value of flux density in the stator core is 1.69 tesla or, changing units,

$$B_{\text{cs}} = 109{,}000\,\text{lines/in.}^3$$

which, from Figure 5.16 results in a loss density of

$$p_{\text{c}} = 0.65\,\text{watts/in.}^3$$

If again a factor of 35% is used to correct for additional fundamental losses then

$$P_{\text{core}} = (1.35)\,p_{\text{c}}V_{\text{core}}$$
$$= (1.35)(0.65)(1378.6) = 1210\,\text{watts}$$

Because the stator slot leakage fluxes take a return path through the stator core, then, neglecting resistance, the total voltage also appears across the flux producing core loss which corresponds to both the core flux plus the slot leakage flux. Hence, an equivalent value of resistance which results in a power loss in the core of 930 watts is

$$r_{\text{c}} = \left(\frac{L_m + L_{\text{lsl}}}{L_m + L_{\text{ls}}}\right)^2 \frac{V_{1-1}^2}{P_{\text{core}}}$$
$$= \frac{(0.223 + 0.003355)^2(2550)^2}{(0.223 + 0.00651)^2 1210}$$
$$= 5226\,\Omega$$

It should be noted that iron loss also exists for the rotor teeth and rotor core. However, during normal operation the frequency associated with this loss is much lower than 60 Hz and is therefore much smaller than the stator iron loss. This loss is effectively included in the 35% "imponderables" factor.

Calculations must now be made to account for the stray losses. First a computation can be carried out for a value of resistance which accounts for the stray *load* losses. Reference to Table 5.1 indicates that for a machine with open stator slots and semi-open rotor slots the power dissipated due to stray load losses is between 0.3 and 0.6% of rated output. For the 250 HP machine under study a figure of 0.4% will be used. The rated current which flows under rated load conditions can be approximated by taking rated output power and dividing by rated voltage and an estimated power factor and efficiency, or

$$I_{\text{b}} = \frac{P_{\text{s}}}{\sqrt{3}V_{(1-1)}\,(\text{powerfactor})\,(\text{efficiency})}$$
$$= \frac{(746)(250)}{\sqrt{3}(2400)(0.93)(0.9)}$$
$$= 53.6\text{ A}$$

The resistance which accounts for load losses is a function of current and is therefore in series with the load current rather than in parallel across the terminal voltage. The equivalent resistance which produces a loss of 0.4% of rated power is

$$\begin{aligned} r_{\rm sll} &= \frac{(0.004)\,(P_{\rm b})}{3I_{\rm b}^2} \\ &= \frac{0.004\,(746)\,(250)}{(3)\,(53.6^2)} \\ &= 0.086\,\Omega \end{aligned}$$

The final resistance to be calculated is to account for no-load losses. In this case a resistor is to be calculated which appears across the magnetizing inductance since this loss component results from fluxes in the air gap. It should be noted that the presence of the resistor to account for load losses is in series with the stator current and will already produce a loss at no load since magnetizing current must flow through this resistor as well as load current. Since the series resistances are small the magnetizing current of the machine is readily approximated by the expression

$$I_m = \frac{V_{\rm l-l}}{\sqrt{3}(X_{\rm ls} + X_m)}$$

where $X_{\rm ls}$ and X_m are the 60 Hz reactance corresponding to $L_{\rm ls}$ and L_m. This quantity can be evaluated as

$$\begin{aligned} I_m &= \frac{2550}{(\sqrt{3})\,(377)\,(0.006654 + 0.223)} \\ &= 17.0\,\text{A} \end{aligned}$$

The power consumed by the load loss resistor $r_{\rm ll}$ at no load is therefore

$$\begin{aligned} P_{\rm l} &= 3\,I_m^2\,r_{\rm ll} \\ &= 3(17)^2\,(0.086) \\ &= 74.6\ \text{watts} \end{aligned}$$

Reference to Table 5.1 shows that for a machine with open stator and semi-closed rotor slots the no-load losses as a percent of fundamental losses is 120 to 160%. A value of 140% will be used here in this calculation. The stray no- load loss is therefore

$$\begin{aligned} P_{\rm nl} &= (1.4)\,(487 + 1210) \\ &= 2376\ \text{watts} \end{aligned}$$

The losses which are thus far unaccounted for are

$$\begin{aligned} P_{\rm nl} - P_{\rm l} &= 2376 - 74.6 \\ &= 2301\ \text{watts} \end{aligned}$$

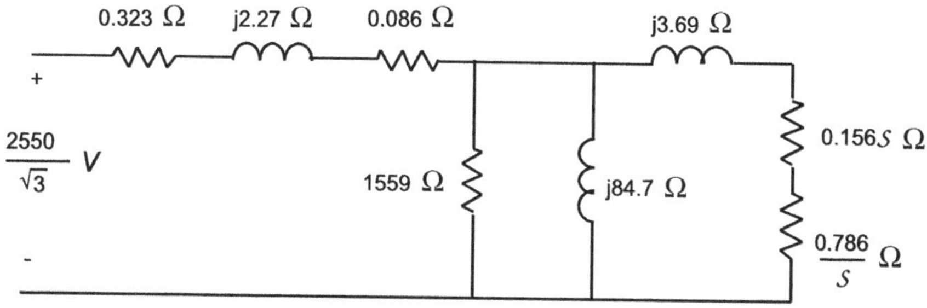

Figure 5.33 Per phase equivalent circuit of the 250 HP induction machine at 60 Hz including fundamental component iron loss and stray loss terms.

Hence, the resistance which will produce these losses when connected across the magnetizing branch is

$$\begin{aligned} r_{nl} &= \left(\frac{V_{l-l}^2}{P_{nl} - P_l}\right)\left(\frac{L_{ms}}{L_{ms} + L_{ls}}\right)^2 \\ &= \frac{(2550)^2}{2301}\left(\frac{0.223}{0.223 + 0.006654}\right)^2 \\ &= 2665\,\Omega \end{aligned}$$

Since the resistance r_c depends upon the total stator core flux, it can logically be placed across the stator circuit so as to include the slot leakage portion of stator leakage reactance. However, this refinement is rarely used since the slot leakage inductance is a very small fraction of the magnetizing inductance. The resistances r_{nl}, r_c, and r_t can then be combined in parallel to form one equivalent iron loss resistor r_i which accounts for the air gap flux-dependent iron loss. This resistance is

$$\begin{aligned} r_i &= \frac{1}{1/r_{nl} + 1/r_c + 1/r_t} \\ &= 1/(1/2665 + 1/5226 + 1/13{,}350) \\ &= 1559\,\Omega \end{aligned}$$

A final version of the equivalent circuit per phase for the 250 HP example induction machine is shown in Figure 5.33. As a practical matter, the numerical values have been rounded off to three significant digits. The small effect of the slip frequency on the rotor leakage inductance has been neglected. It is important to mention that the calculation has been based on an assumption of a gap flux density of 0.775 tesla (Example 2). In Example 3, this value was determined to produce an air gap voltage of 9% above rated voltage (2400 V_{l-l}). In practice, the process must be repeated several times to account for a range of voltage ±10% around rated voltage. The variation of magnetizing inductance with voltage is typically presented in terms of a plot of no-load (zero-slip) voltage versus magnetizing current.

5.13 CONCLUSION

The losses in an electrical machine are an inevitable outcome of the electromechanical energy conversion process. Since thermal issues are the ultimate limiting aspect of most motor designs, accurate prediction of the losses becomes a critical factor in determining whether the machine will satisfy its intended purpose. Even today, "safety factors" are often applied after losses are computed by the best means available to ensure that the machine meets its specifications. This chapter has served merely as a brief introduction to this complex issue.

REFERENCES

[1] Y. Liao and T.A. Lipo, "Effect of saturation third harmonic on the performance of squirrel-cage induction machines," *Electric Machines and Power Systems*, vol. 22, no. 2, March/April 1993, pp. 155–172.

[2] P. L. Alger, G. Angst, and E. J. Davies, "Stray-load losses in polyphase induction machines," *Trans AIEE*, June 1959, pp. 349–355.

[3] B. Heller and V. Hamata, *Harmonic Field Effects in Induction Machines*, Elsevier Scientific Publishing Co., Amsterdam, 1977.

[4] M. Liwschitz-Garik and C.C. Whipple, *Electric Machinery, Vol. 2 AC Machines*, 2nd edition, Van Nostrand Publishers, 1961.

CHAPTER 6

PRINCIPLES OF DESIGN

THE DESIGN OF ELECTRICAL MACHINERY is both an art and a science. There are so many factors involved in the design process that it is not possible to work along rigid lines. Any design must be a compromise between a large number of conflicting requirements. Thus there is no unique solution to a design problem, and designs for the same specification will differ because of different emphasis being placed on each requirement by the particular designer.

To a great extent, design is an iterative process; that is, parts of the design are repeated in order to obtain the desired solution. For instance, in the design of a motor of a given horsepower rating, an initial estimate has to be made of the efficiency. When the design has been completed, only then can the efficiency be checked. The initial and final values must agree to some desired tolerance. If they do not, the initial values and portions of the design must be adjusted as necessary. Similar iterations may be involved in optimizing other parts of the design. As a designer gains experience, he or she will be able to deal with the iterative process more quickly, generally because one makes better initial assumptions. This dependence on an iterative process has, to a large extent, led to the use of high speed digital computers for the design of electric machinery.

6.1 DESIGN FACTORS

Those factors which influence the design of a machine may be categorized as follows.

1. Economic Factors. In most cases this is the overriding consideration, since all other design factors being equal, this factor will decide who sells the machine and who does not. Clearly, in order to be competitive, there must be an absolute minimum of material in the machine and the machine must be designed such that manufacturing cost is minimized. Design of the machine should always be compatible with the equipment available for machining and assembly already in place in the plant and with materials which are in stock without time consuming and costly special purchase. Better performance is nearly always possible at more cost but the best machine for the job is that in which the first cost plus the cost of losses and maintenance over the life expectancy is a minimum. The rising costs of electrical energy have recently prompted a resurgence of interest in the trade-off between first cost and running cost resulting in machines of much higher efficiency that customarily used in the past.

Introduction to AC Machine Design, First Edition. Thomas A. Lipo.

2. *Material Limitations*. The technical and economic limits of materials generally determine the performance and dimensions of the machine. Progress in magnetic and insulating materials has been continuous. These new materials have had a dramatic effect on machine design in the past and will continue to do so in the future.

3. *Specifications*. The design, performance, and materials used are often subject to specifications issued by IEEE or similar bodies. In this country, the motor manufacturers have banded together to form the National Electrical Manufacturers Association (NEMA) to set standards for size, performance, and testing of AC and DC machines from 1.0 up to 450 HP. Also, standards of the individual motor manufacturer also play an important role since there will often be standard wire sizes, insulation thicknesses, etc., dictated by manufacturing economy. NEMA also sets standard diameters which sometimes may necessitate ingenuity in determining the best size for a given application.

4. *Special Factors*. In some applications special considerations may exist which have an overriding concern. For example, design of aircraft generators requires a design of minimum weight with maximum reliability. For design of traction motors the emphasis is on reliability and ease of servicing. For typewriter motors the most important factor may be minimum noise. Motors for starting large compressors may have to deal with a large inertia in which the heating during starting is severe. Hence, the starting torque per ampere may be a major consideration for this application.

5. *Theoretical Factors*. In this category are all the quantifiable details of electrical and mechanical design. These considerations are discussed in more detail in the subsequent sections.

6.2 STANDARDS FOR MACHINE CONSTRUCTION

It might be said that rarely is the machine designer presented with a problem in which he is not confronted with geometric constraints. Frequently these constraints are application dependent, for example, it is essential that a traction motor on a locomotive fit in the space available under the rail car. The need for standard shapes and sizes of motor and generators has resulted in NEMA standards. It is necessary for general purpose electrical machines to satisfy certain criterion in order to bear their stamp of approval. The most recent ANSI/NEMA standards is Publication MG1-2009 (R2010), which includes a number of revisions. NEMA standards specify the power, speed, voltage, and frequency rating of machines for a number of applications. Standard designs are classified according to performance. Typical torque-speed curves for various NEMA Design Classes for squirrel-cage induction machines are shown in Figure 6.1.

Four classes of motors are identified. *Class A* motors are the "vanilla" flavor motor having normal starting torque and normal starting current. The motor is designed to have 200–250% pull-out torque. Larger sizes usually require special starting equipment. This type of motor is frequently replaced by a *Class B* motor. *Class B* motors have a normal starting torque but a lower starting current. They can be started online up to 300 HP. These motors have slightly poorer efficiency and power factor

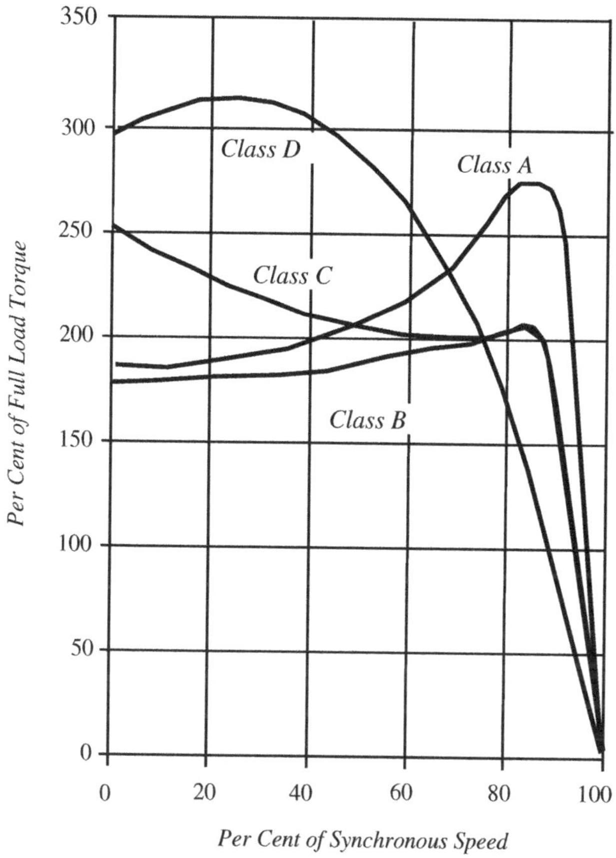

Figure 6.1 Torque-speed curves for NEMA squirrel-cage induction machines.

than *Class A* motors as well as alightly lower pull-out torque. They usually contain deep bars in the rotor. These motors are probably the most popular type of induction motor. *Class C* motors have a high starting torque and a low starting current. The machines usually employ a double squirrel cage. Starting torque is on the order of 175–200% while pull-out torque is greater or equal to 185% of rated torque. The motor has low slip, high efficiency and is suitable for high starting torque applications. *Class D* motors are machines which have very high starting torque and are designed to operate with high slip. These motors are used for accelerating large inertia loads or which operate with a flywheel (e.g., a punch press). The high resistance rotor cage is frequently accomplished by use of brass bars. Typical starting current and starting torque values for the most common type of machines, NEMA *Class A* and *B* are summarized below in Tables 6.1 and 6.2. Figure 6.2 shows torque versus speed curves for a family of 1800 RPM, NEMA *Class B* machines.

Motors are also designated according to their type of enclosure. Twenty types of enclosures are defined by NEMA in two major categories: Open Machines, which have ventilating openings which permits passage of external cooling air to the

TABLE 6.1 Typical per unit starting current values (NEMA Class B)

HP = 1	2	3	5	10	25	200
I_{start} = 10.3	9	7.6	6.9	5.8	5.5	5.5

TABLE 6.2 Typical starting torque values in per unit of rated torque (NEMA Class A and B)

HP	P = 2	4	6	8	20	12
5	1.35	1.35	1.35	1.25	1.2	1.15
20	1.35	1.35	1.35	1.25	1.2	1.15
40	1.25	1.35	1.35	1.25	1.2	1.15
100	1.0	1.35	1.35	1.25	1.2	1.15
200	1.0	1.15	1.35	1.25	1.2	1.15

machine, and Totally Enclosed Machines in which the internal air is enclosed within the housing. The most popular type of Open Machine is the drip-proof machine which is protected from water or liquid falling essentially vertically on the machine (ODP or open drip proof). Other types of open machines are Weather Protected Type 1 and Type II machines which add extra screens or louvers to the design and Splash-Proof which offers greater protection than drip proof but is still regarded as an indoor-type enclosure. The most frequently used Totally Enclosed Machine is the Totally Enclosed Fan Cooled Machine (TEFC) which is a machine equipped with exterior cooling by means of a fan integral within the machine but external to the enclosing

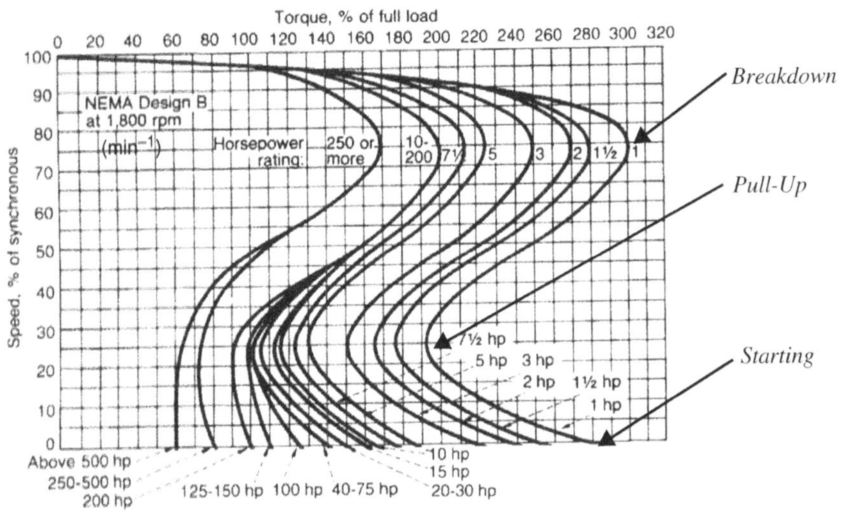

Figure 6.2 Speed versus torque curves showing breakdown, pull-up, and starting torques of 1800 RPM, NEMA Class B machines for various ratings.

TABLE 6.3 Recommended voltages for electric motors

Rated voltage (V)	Recommended power range (HP)
230 or 460	Up to 100
460 or 575	100–600
2300	200–4000
4000	400–7000
6600	1000–12,000
13,200	3,500—25,000

parts. Special strengthening of the frames and special fittings are utilized to make some totally enclosed motors "explosion-proof" and "dust-explosion proof." Other types of enclosures include TEAAC (totally enclosed air to air cooled), TEFV (totally enclosed force ventilated), and TEWAC (totally enclosed water to air cooled), and TEPC (totally enclosed pipe cooled).

NEMA specifies the following horsepowers as standard: 0.5, 0.75, 1, 1.5, 2, 3, 5, 7.5, 10, 20, 25, 30, 40, 50, 60, 75, 100, 125, 150, 200, 250, 300, 350, 400, 450, and 500 horsepower. It can be noticed that there exists a rather logarithmic increase in horsepower sizes. From the point of view of a machine designer of NEMA machines, the major concern is to utilize the same laminations or the least number of laminations having the same outside diameter D_{os} to produce a family of induction machines. Standard voltages for NEMA machines as a function of size are summarized in Table 6.3.

One primary purpose of NEMA standards is to standardize the external dimensions of electrical machines according to "frame sizes" so that original equipment manufacturers (OEMs) as well as end point users can more easily select between various manufacturers without drastic changes in shaft couplings, mounting brackets, etc. A typical page from the NEMA standards showing standard letter designations of a drip-proof-type enclosure is shown in Figure 6.3. Most of the dimensions of Figure 6.3 are not specified by NEMA and most of those that do concern practical matters such as spacing of bolt holes. The most important parameter from a designer's point of view is the shaft height since this quantity essentially fixes the diameter of the machine for a given frame size. A table showing the specific value of a number of this key dimension is given in Table 6.4. Note that the first two numbers of the frame size are exactly equal to four times the dimension D corresponding to the shaft height in Figure 6.4. Therefore the stator outer diameter in inches of a machine of a given frame size is roughly equal to one-half the first two digits of the frame size.

6.3 MAIN DESIGN FEATURES

The major theoretical factors in design of an electric machine can be broken into five areas: electrical, magnetic, dielectric, thermal, and mechanical.

1. Electrical. In order to make the machine compatible with the electrical supply the voltage, frequency, and number of phases of the machine are specified. In

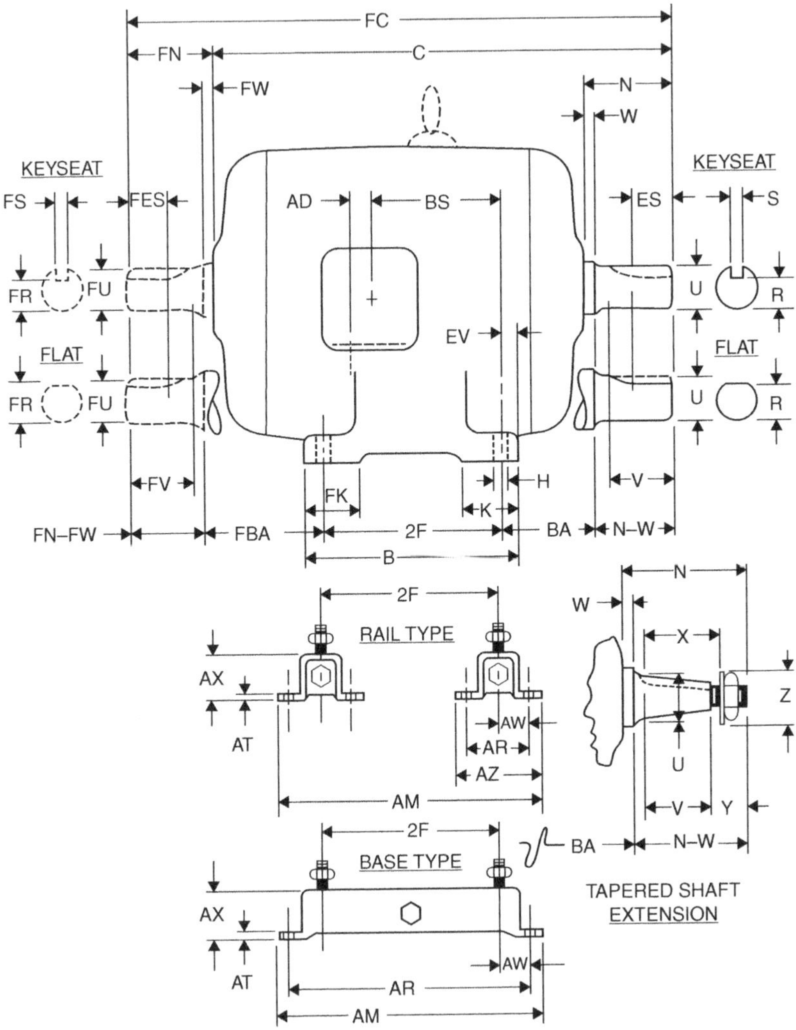

Figure 6.3 Standard designations for an open drip proof squirrel-cage machine—side view.

TABLE 6.4 NEMA frame sizes and corresponding shaft height D and diameter of the shaft extension U (both in inches)

Frame	D	U	Frame	D	U
140	3.50	7/8	360	9.00	2 3/8
160	4.00	1	400	10.00	2 7/8
180	4.50	1 1/8	440	11.00	3 3/8
210	5.00	1 3/8	500	12.50	
250	6.25	1 5/8	580	14.50	
280	7.00	1 7/8	680	17.00	
320	8.00	2 1/8			

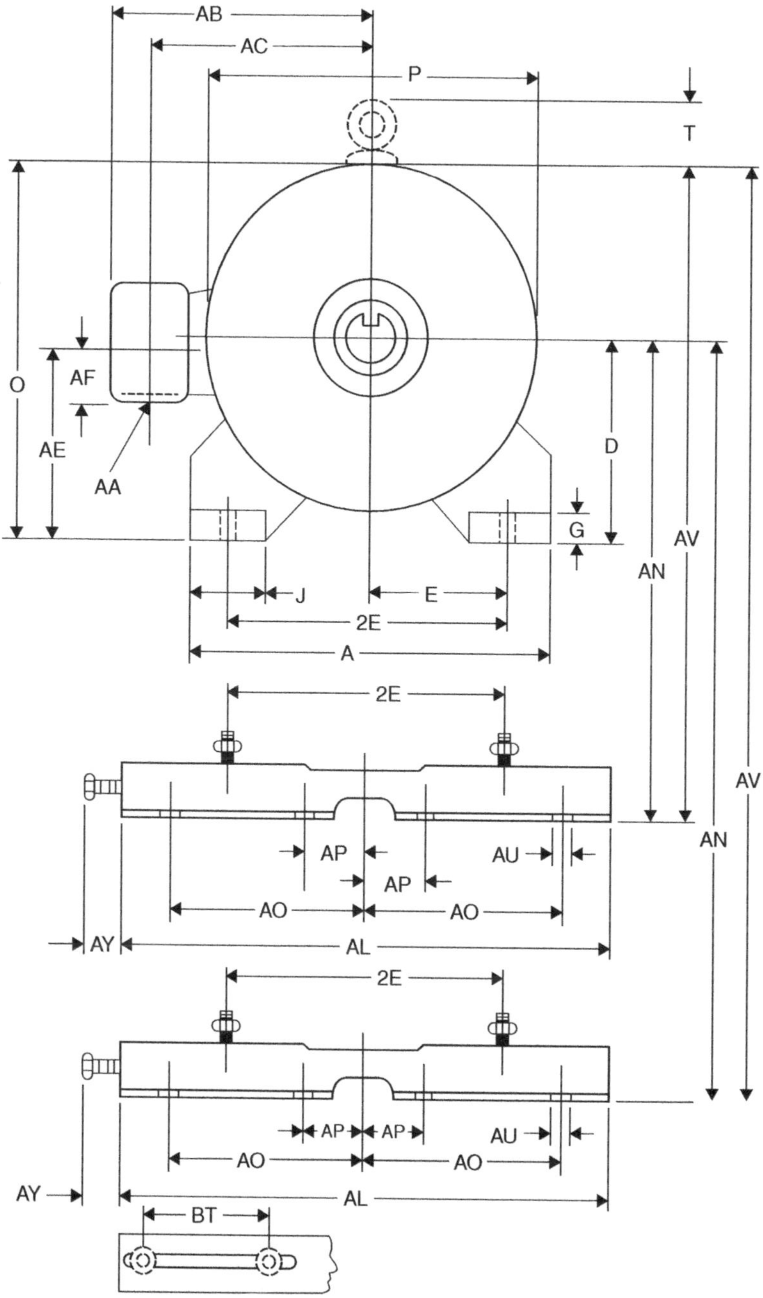

Figure 6.4 Standard designations for an open drip proof squirrel-cage machine—end view.

addition, the minimum power factor at rated load may also be specified. From this data the designer must determine the type of connection (wye or delta), the type of winding (wave or lap, one or two layer, random or form wound, etc.), and winding factors. Other important electrical features are the current density in the windings, the copper losses, and the short-circuit current.

2. *Magnetic.* In the category of magnetic considerations are such factors as fixing maximum flux densities in the teeth and core, the resulting iron losses, the effect of saturation on prescribed over-voltages, calculation of magnetizing and leakage inductances, determination of slot and tooth shapes, and calculation of harmonic effects (stray-load and no-load losses and harmonic torques).

3. *Dielectric.* The influences of the electric fields also have an important effect on the design of the machine. Important considerations involve selection of proper insulation thickness for strand to strand, coil to coil, and coil to ground insulation to survive continuous as well as surge voltages (i.e., lightning strokes). Other related considerations concern proper routing of the windings to the outside world and the selection of bushings to avoid flashovers.

4. *Thermal.* The heat produced within a machine will, without proper venting, cause the machine to self-destruct. Although the considerations here intrigue mechanical rather than electrical engineers, the problems are no less important. Important considerations involve selection of the coolant (air, water, hydrogen), selection and spacing of ducts, design of centrifugal fan, calculation of temperature rise, and design of cooling tank and radiators (where applicable).

5. *Mechanical.* Major mechanical considerations involve calculation of critical rotational speed, modes of acoustical vibration, mechanical stresses on rotating shaft during normal and overspeed conditions, determination of moment of inertia, and calculation of forces on windings particularly in the end-winding portion during short circuits.

One of Murphy's laws, of course, dictates that all of the above considerations interact. A good understanding of all of these factors is, indeed, a lifetime undertaking. It is apparent that many of the subtleties can never be relegated to a computer algorithm so that machine design will always remain a stimulating and challenging engineering profession.

6.4 THE D^2L OUTPUT COEFFICIENT

It is well known that the torque produced in any electrical machine is proportional to the product of the MMF per pole and the flux per pole. Since, for a given amount of copper and iron, these two quantities cannot be increased without limit, it follows that the "densities" of these two quantities have definite upper values and that the torque produced can also be expressed in terms of these densities. An important interrelation between the output power and the densities of the current and flux can be developed if one assumes that stator resistance and stator leakage inductance drops to be negligibly small.

6.4.1 Essen's Rule

Consider a three-phase machine having C circuits per phase. For sinusoidal excitation, the power input to the machine is then given in phasor form by

$$P_S = \frac{3}{2} Re(\widetilde{V}_s \widetilde{I}_s^\dagger) \tag{6.1}$$

where V_s and I_s are the rated peak per-phase values of stator voltage and current and "†" denotes the complex conjugate. If stator resistance is neglected then the power as measured at the air gap is the same, that is,

$$P_{gap} = P_s = \frac{3}{2} Re\left(\widetilde{V}_{gap} \widetilde{I}_s^\dagger\right) \tag{6.2}$$

The amplitude of the voltage is related to the amplitude of the air gap flux linkages by

$$\begin{aligned} V_{gap} &= \omega_e \lambda_m \\ &= \omega_e k_1 N_s \Phi_g \end{aligned} \tag{6.3}$$

where Φ_g is the peak value of flux crossing the gap over each pole. Hence, neglecting stator resistance, the power input to the machine measured at the air gap can be expressed as

$$P_s = \frac{3}{2} \omega_e k_1 N_s \Phi_g I_s \cos \phi_{gap} \tag{6.4}$$

The value of the gap flux per pole is related to the peak fundamental air gap flux density in the gap by

$$\begin{aligned} \Phi_g &= \frac{2}{\pi} B_{g1} (\tau_p l_e) \\ &= \frac{2}{\pi} B_{g1} \left(\frac{\pi D_{is} l_e}{P} \right) = \frac{2 D_{is} l_e}{P} B_{g1} \end{aligned} \tag{6.5}$$

where, in equation (6.5), l_e is the effective length of the stator iron stack including the interlaminar air space and the effect of fringing but excluding ventilation ducts. The density of the conductors in the stator slots can be depicted by forming an expression for the average amperes per unit length of the air gap surface. Since the three phases are assumed balanced, the average peak current is the same in each coil of the machine. Incorporating the winding factor and assuming that each slot is equally filled, the peak component of AC current per unit length of the stator circumference is,

$$K_{s1} = \frac{S_1 n_s k_1 (I_s/C)}{\pi D_{is}} = \frac{S_1}{\pi D_{is}} \left(\frac{6 C N_s}{S_1} \right) k_1 \left(\frac{I_s}{C} \right) \tag{6.6}$$

$$= \frac{6 k_1 N_s I_s}{\pi D_{is}} \tag{6.7}$$

Inserting equations (6.5) and (6.7) into equation (6.4), the following expression for volt-amperes is obtained

$$VA_{\text{gap}} = \frac{3}{2}\omega_e k_1 N_S \left(\frac{2D_{is}l_e B_{g1}}{P}\right)\left(\frac{\pi D_{is} K_{s1}}{k_1 6N_s}\right)$$

or

$$VA_{\text{gap}} = \left(\frac{\pi}{2}\right)\frac{\omega_e}{P}\left(D_{is}^2 l_e\right) B_{g1} K_{s1} \tag{6.8}$$

The frequency of the machine can be related to the rotational synchronous speed by

$$f_e = \frac{\Omega_s P}{120} \tag{6.9}$$

where Ω_s is the synchronous speed of the machine in revolutions per minute or RPM with units (min^{-1}), so that equation (6.8) can also be written as

$$VA_{\text{gap}} = \left(\frac{\pi^2}{120}\right)\Omega_s\left(D_{is}^2 l_e\right) B_{g1} K_{s1} \tag{6.10}$$

The power at the gap of the machine is defined by

$$P_{\text{gap}} = VA_{\text{gap}} \cos\phi_{\text{gap}} \tag{6.11}$$

where $\cos\phi_{\text{gap}}$ is the power factor as measured at the gap. At the output shaft,

$$P_{\text{mech}} = VA_{\text{gap}}\eta_{\text{gap}} \cos\phi_{\text{gap}} \tag{6.12}$$

where η_{gap} is the efficiency as seen from the air gap, that is, including rotor iron and copper loss and stray losses. So that, finally,

$$P_{\text{mech}} = \left(\frac{\pi^2}{120}\right)\Omega_s\left(D_{is}^2 l_e\right) B_{g1} K_{s1}\eta_{\text{gap}} \cos\phi_{\text{gap}} \tag{6.13}$$

Equations (6.10) and (6.13) express the volt-ampere input and power output of any AC or DC machine and is sometimes called *Essen's Rule* [1]. It is useful to examine these results carefully in order to note the effect of each term on the machine design. Clearly, for a given volt-ampere input, the output power is maximized by making $\cos\phi_{\text{gap}}$ unity which can only be approximated by having a very small rotor leakage inductance. This implies that the peak gap flux density B_{g1} and MMF per pole $\mathcal{F}_{p1}$ then act at right angles internally at the air gap of the machine. The input and output power can be also raised by increasing one of four quantities.

1. The peak fundamental component of the AC surface current density K_{s1}.
   The surface current density is limited by the I^2R losses in the conductor, the effectiveness of the cooling media and the permissible temperature rise in the insulating materials.
2. The peak fundamental component of magnetic flux density B_{g1}.
   The flux density is limited by the saturation point of the material used, the hysteresis and eddy current losses, the stray-load and no-load losses, and the effectiveness of the coolant.

3. The rotational synchronous speed in $\min^{-1}$ Ω_s.
   The rotational speed is limited by rotational stresses in the rotating member, by brush problems in the cases of DC, synchronous or wound rotor induction machines, and by performance requirements of the machine for a given application.

4. The quantity $D_{is}^2 l_e$
   Without ducts, the physical length l_s is essentially equal to l_e (length of actual iron plus effects of fringing) and this term differs from the volume of the portion of the machine occupied by the rotor only by the constant $\pi/4$. This quantity is also roughly proportional to the total active volume of the machine since the volume of the stator copper and iron is roughly equal to the volume used for rotor copper and iron.

6.4.2 Magnetic Shear Stress

Since K_{s1}, B_{g1}, and Ω_s have definite limits, it is apparent that increased size is inevitably achieved by an increased volume of iron and copper. Equation (6.13) can be written in the form [2]

$$\frac{P_{mech}}{(D_{is}^2 l_e)\Omega_s \eta_{gap} \cos\phi_{gap}} = \left(\frac{\pi^2}{120}\right) K_{s1} B_{g1} \tag{6.14}$$

Since the right-hand side varies over relatively small limits, the quantity expressed by the right-hand side of equation (6.14) is sometimes called the *output constant* of a machine [2]. In practice, the output constant is not truly a constant but increases with the rating. This fact is due to many factors including the use of better magnetic and insulating materials as size increases, the fact that large machines have a larger pole pitch for the same speed resulting in better cooling than a machine with a smaller rating, and due to the larger extension of the end winding and larger air velocity in machines of higher rating. Also, the voltage is important since the higher the voltage the more space is necessary for the insulation and less space is left for the copper.

The torque as a function of $D_{is}^2 l_e$ is readily obtained from equation (6.13) by dividing by rotor speed. The result is,

$$T_e = \frac{P_{mech}}{2\omega_e/P} \tag{6.15}$$

$$T_e = \left(\frac{\pi}{4}\right)\left(D_{is}^2 l_e\right) B_{g1} K_s \eta_{gap} \cos\phi_{gap} \tag{6.16}$$

It is useful to note here that the torque of conventional induction and synchronous machines are fundamentally independent of the number of poles of the machine.

The quantity $(K_{s1}B_{g1})/2$ of equation (6.14) represents the product of the useful surface current density and air gap flux density and is also often used as a figure of merit to compare machines. The magnetic shear stress has nominal units (amperes/meter) × (teslas). However, reference to equation (1.6) in Chapter 1 shows that in basic units, teslas = newtons/ampere-meter so that the product has basic units of newtons/meter^2 which is alternatively termed a Pascal (Pa). Hence, this quantity is equivalent to a pressure and appropriately termed as the *magnetic shear stress* σ_m.

Formally defining

$$\sigma_m = \frac{K_{s1}B_{g1}}{2} = K_{s(rms)}B_{g1(rms)} \tag{6.17}$$

the output torque in newton meters is then related to the shear stress and key remaining motor parameters by

$$T_e = \left(\frac{\pi}{2}\right)\left(D_{is}^2 l_e\right)\sigma_m \eta_{gap} \cos\phi_{gap} \tag{6.18}$$

and the output power by

$$P_{mech} = \left(\frac{\pi^2}{60}\right)\sigma_m \left(D_{is}^2 l_e\right)\Omega_s \eta_{gap} \cos\phi_{gap} \tag{6.19}$$

or since the rotor volume is

$$V_{rotor} = \frac{\pi D_{or}^2 l_e}{4} \approx \frac{\pi D_{is}^2 l_e}{4} \tag{6.20}$$

the torque can also be expressed as

$$T_e = 2V_{rotor}\sigma_m \eta_{gap} \cos\phi_{gap} \tag{6.21}$$

where the small effect of the air gap length is neglected.

The dependence of the magnetic shear stress of an induction machine on pole pitch is shown in Figure 6.5 [2]. Compared to other machines the values of shear stress are relatively low. This disadvantage is due to the fact that the corresponding rotor surface current density is nearly equal to that of the stator under-rated load. Thermal constraints limit the total current loading. However, high performance servo motors often have a shear stress reaching 20 kN/m^2 (20 kPa). Permanent magnet machines which do not have a rotor current disadvantage typically fall in the range from 40 to 60 kN/m^2. As an upper limit, liquid-cooled machines having shear stresses of up to 100 kN/m^2 have been reported. The magnetic shear stress is frequently expressed in imperial units. For this purpose it can be mentioned that 1 kN/m^2 = 0.145 lbf/in.^{2} or, inversely, 1 lbf/in.^{2} = 6.895 kN/m^2.

Having selected the magnet shear stress σ_m, it is possible to determine the required volume of the machine when the volt-ampere input and speed are given or the VA input when D_{is}, l_e, and Ω_s are given. When the power output rather than VA input is given, its input must be determined by estimating the power factor and efficiency of the machine. These figures generally come from experience with similar machines or from published curves. Such typical curves are shown in Figures 6.6 and 6.7 for a NEMA Class B machines, corresponding to machines having starting torques between 115% and 150% rated and starting currents between 500% and 1000% rated. Similar curves exist for other NEMA classes.

Using a power regression technique, it is shown in [3] that the power factor curves of Figure 6.6 can be approximated by the expression

$$\cos\phi = 1.131 P_{mech}^{0.015} P^{-0.08} f^{-0.07} \tag{6.22}$$

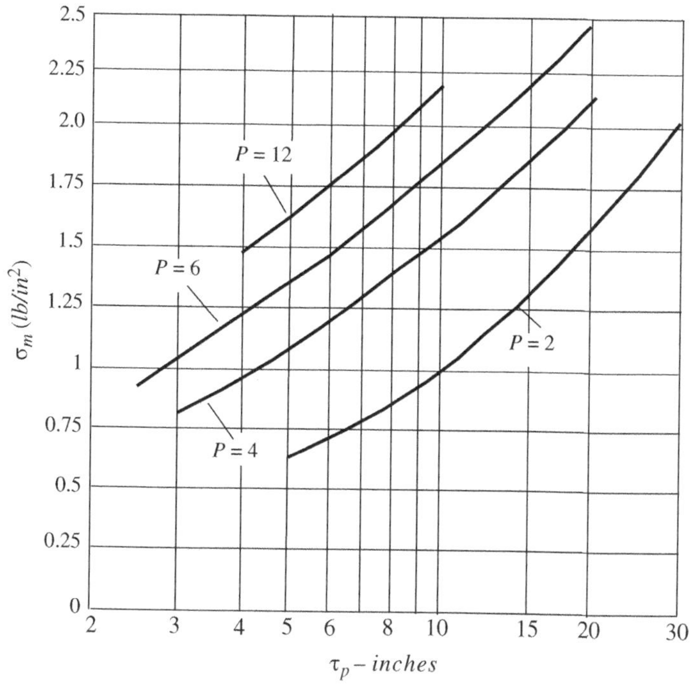

Figure 6.5 Magnetic shear stress σ_m of typical induction motors in lbf/ in.^{2} as a function of pole pitch τ_p (Adapted from [2]).

where P_{mech} is the rated mechanical output power in watts, P is the number of poles, and f is the frequency (60 has been assumed).

The terminal power factor can be referred to the air gap power factor noting

$$V_{gap}I_s \cos \phi_{gap} = V_s I_s \cos \phi - r_s I_s^2 \tag{6.23}$$

so that

$$\cos \phi_{gap} = \frac{V_s}{V_{gap}} \cos \phi - \frac{r_s I_s}{V_{gap}} \tag{6.24}$$

Assuming that the stator side impedance drop is essentially inductive

$$\cos \phi_{gap} = \frac{1}{1 - \frac{I_s X_{ls}}{V_s}} \cos \phi - \frac{(r_s I_s)/V_s}{1 - \frac{I_s X_{ls}}{V_s}} \tag{6.25}$$

Further assuming that the rated current and voltage correspond to base or reference values

$$\cos \phi_{gap} = \frac{1}{1 - \frac{X_{ls}}{Z_b}} \cos \phi - \frac{\frac{r_s}{Z_b}}{1 - \frac{X_{ls}}{Z_b}} \tag{6.26}$$

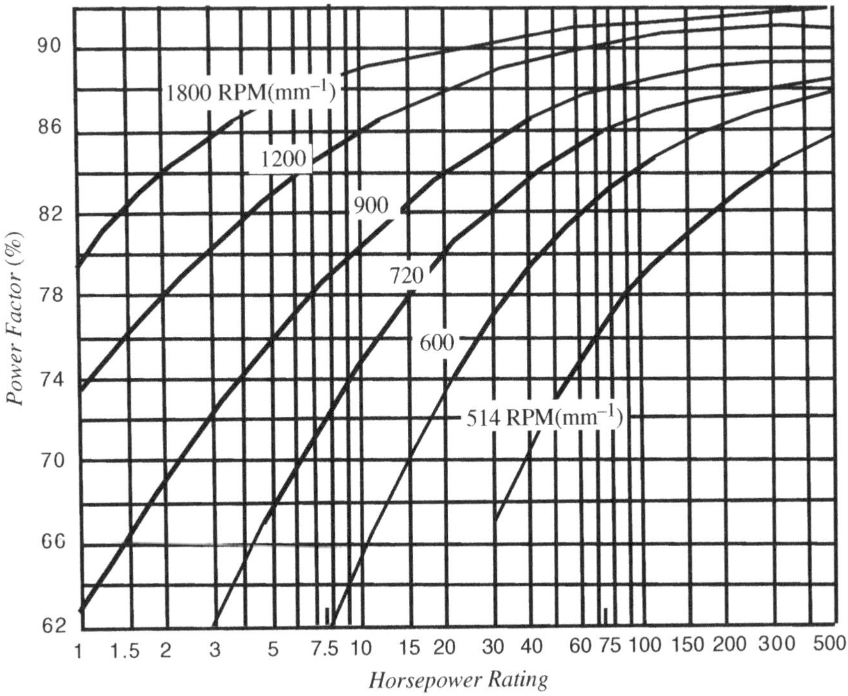

Figure 6.6 Approximate power factors of NEMA Class B three-phase 60-Hz induction motors.

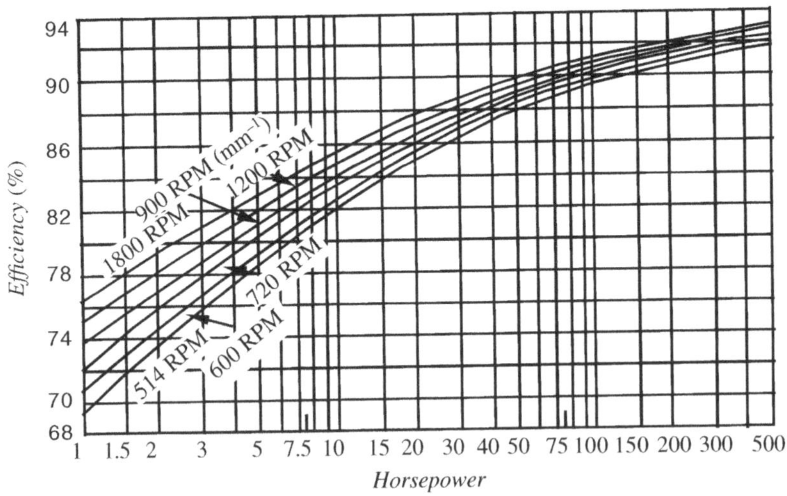

Figure 6.7 Approximate efficiencies of NEMA Class B three-phase, 60-Hz induction motors.

Finally,

$$\cos\phi_{gap} = \frac{1}{1 - X_{ls,pu}}\cos\phi - \frac{r_{s,pu}}{1 - X_{ls,pu}} \tag{6.27}$$

Equation (6.27) can be used to correct the terminal power factor to that measured at the gap. The values of per unit reactance typically vary from 0.08 to 0.15 while the per unit resistance varies from 0.02 to 0.05. It is generally acceptable to neglect the second term in (6.27) for simplicity.

The NEMA efficiency data of Figure 6.7 can also be estimated analytically as follows [3],

$$\eta = 0.712 P_{mech}^{0.03} p^{-0.025} f^{-0.01} \tag{6.28}$$

The expression can be corrected to an estimated air gap value by adding the term $r_{s,pu}/(1 - X_{ls,pu})$ to the efficiency measured at the terminals.

Comparison of electrical machine performance is frequently done on the basis of its magnetic shear stress. However, it is important to note that for the case of an induction machine the figure of merit for current density K_s is concerned only with the stator. A more realistic comparison should include the sheet current density for both stator and rotor so that copper losses are the same for all machines. In this case, one can write that the total current density

$$\begin{aligned} K_{sr} &= K_s + K_r \\ &= K_s\left(1 + \frac{K_r}{K_s}\right) = K_s(1 + k_{sr}) \end{aligned} \tag{6.29}$$

An estimated expression for k_{sr} has also been established as [3]

$$k_{sr} = 1.107 P_{mech}^{0.0116} p^{-0.062} f^{-0.054} \tag{6.30}$$

Measured efficiency and power factors for a range of standard and high efficiency machines for a single manufacturer have been reported in Reference [4]. Figures 6.8–6.11 show these nominal results and can be used as indicators of expected power factor and efficiency of actual implementations.

6.4.3 The Aspect Ratio

In Figure 6.12 is an auxiliary curve which gives the ratio between the effective core length and the pole pitch, the so-called *aspect ratio*. This ratio cannot be chosen arbitrarily because it is of importance for the cooling of the machine. In order to demonstrate this fact it is first observed that thermal considerations specify that for a given amount of losses and cooling air, the bore surface area cannot be made smaller than

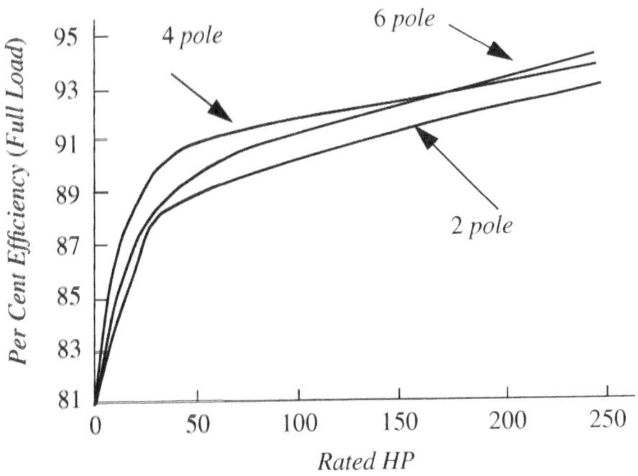

Figure 6.8 Efficiencies of standard class B squirrel-cage induction machines of one manufacturer [4].

a definite minimum value. Hence, one can state that

$$P\tau_p l_e = A_{bore} \quad \text{(a constant)} \tag{6.31}$$

or

$$\pi^2 \frac{D_{is}^2}{P}\left(\frac{l_e}{\tau_p}\right) = A_{bore}$$

It follows that the bore diameter

$$D_{is} = \frac{1}{\pi}\sqrt{PA_{bore}\frac{1}{l_e/\tau_p}} \tag{6.32}$$

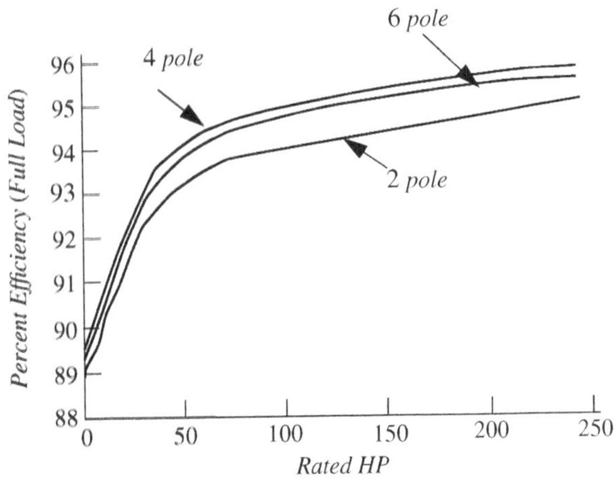

Figure 6.9 Efficiencies of high-efficiency squirrel-cage induction motors from one manufacturer [4].

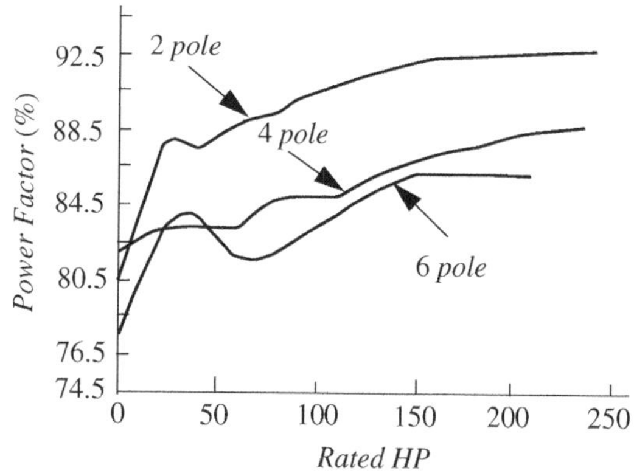

Figure 6.10 Power factors of standard class B induction machines corresponding to Figure 6.8 [4].

Hence the inner bore diameter varies inversely with the square root of the aspect ratio. A large l_e/τ_p is desirable when weight is a major consideration, since the rotor volume is

$$\begin{aligned} V_r &= \frac{\pi}{4} D_{is}^2 l_e = A_{bore} \frac{D_{is}}{4} \\ &= \frac{\sqrt{P}}{4\pi} A_{bore}^{3/2} \frac{1}{\sqrt{l_e/\tau_p}} \end{aligned} \tag{6.33}$$

Therefore, for a given bore surface S_0 determined by thermal considerations, the weight (i.e., volume) and hence cost of the machine decreases with increasing l_e/τ_p. Increasing the aspect ratio also clearly decreases the inertia of the rotor. This

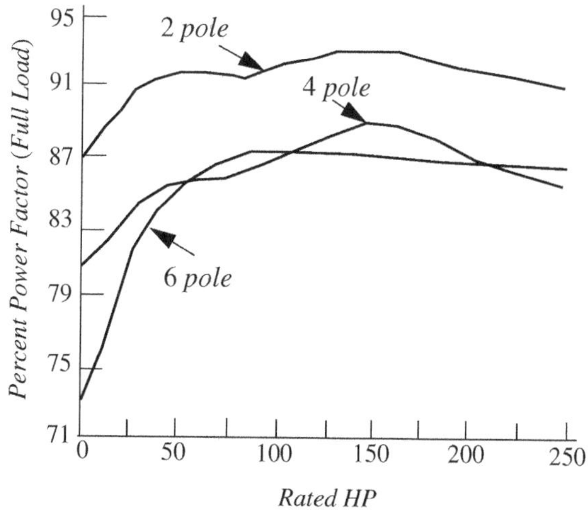

Figure 6.11 Power factor of high-efficiency machines of Figure 6.9 [4].

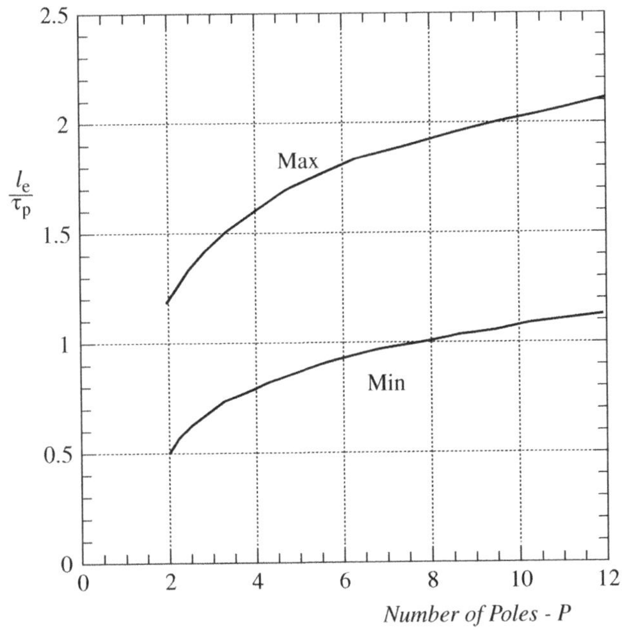

Figure 6.12 Maximum and minimum suggested ratio of equivalent core length to pole pitch (aspect ratio) of an induction machine as a function of the number of poles (from [2]).

is especially useful for control and servo applications. The penalty paid for a large l_e/τ_p is an increase in the stator surface current density K_s. If the aspect ratio is too large the machine is long and tubular and is difficult to cool whereas if the ratio is too small the machine is pillbox shaped. In this case, leakage inductance severely affects the available breakdown torque due to the large end-winding leakage inductance. The aspect ratio l_e/τ_p is a compromise between these two extremes and is typically in the range of 1.0–2.0 as shown in Figure 6.12.

6.4.4 Base Impedance

From equation (6.3), the peak voltage at the air gap is

$$V_{gap} = \omega_e k_1 N_s \Phi_p \tag{6.34}$$

where the flux per pole is

$$\Phi_p = \frac{2D_{is}l_e}{P}B_{g1} \tag{6.35}$$

whereupon

$$V_{gap} = \frac{2}{P}N_s k_1 B_{g1} D_{is} l_e \omega_e \quad \text{V} \cdot \text{pk} \tag{6.36}$$

For convenience, this quantity can be adopted as the per unit base for voltage V_b. The stator phase current is related to the surface current density by equation (6.7),

$$K_{s1} = \frac{6k_1 N_s I_s}{\pi D_{is}} \tag{6.37}$$

whereupon

$$I_s = \frac{\pi D_{is} K_{s1}}{6k_1 N_s} \quad \text{A} \cdot \text{pk} \tag{6.38}$$

Taking the ratio of equations (6.36) and (6.38) and using rated values of B_{g1} and K_{s1} the base impedance for the motor is

$$Z_b = \frac{V_b}{I_b} = \frac{12(k_1 N_s)^2 B_{g1} l_e \omega_e}{\pi P K_{s1}} \quad \text{ohms} \tag{6.39}$$

This quantity can be used to normalize or per unitize the motor parameters in terms of the key design quantities.

6.5 THE D^3L OUTPUT COEFFICIENT

While the D^2L equation has been used extensively to design electrical machines, it does not consider several key factors. For example, machine size is affected by the complete stator geometry which includes the relative proportions of the stator inner and outer diameters, slot and tooth dimensions, flux densities in the iron parts as well as the gap and also the actual current densities in the conductor themselves. The magnetic shear stress contains only the air gap quantity B_{g1} and the stator surface current density K_s. There are additional relationships connecting these air gap quantities with the flux and current densities existing within the machine's interior. Also, the value D^2L obtained from equation (6.14) only gives an estimate of the air gap diameter, rather than the machine's outer diameter, which is more often the constrained variable than the rotor diameter. This problem can be resolved by introducing a modified form of the output coefficient based instead on the true current density within the conductors themselves, namely $J_{s(rms)}$ in A/mm^2 [5].

If D_{is}, D_{os}, t_s, and d_{cs} denote the stator inner and outer diameters, tooth thickness, and core depth, respectively as illustrated in Figure 6.13, then the flux densities in the stator tooth and core can be expressed as

$$B_{ts} = \left(\frac{\pi}{2}\right) \frac{P\Phi_p}{S_1 t_s k_{is} l_i} \tag{6.40}$$

$$B_{cs} = \frac{\Phi_p}{2} \left(\frac{1}{d_{cs} k_{is} l_i}\right) \tag{6.41}$$

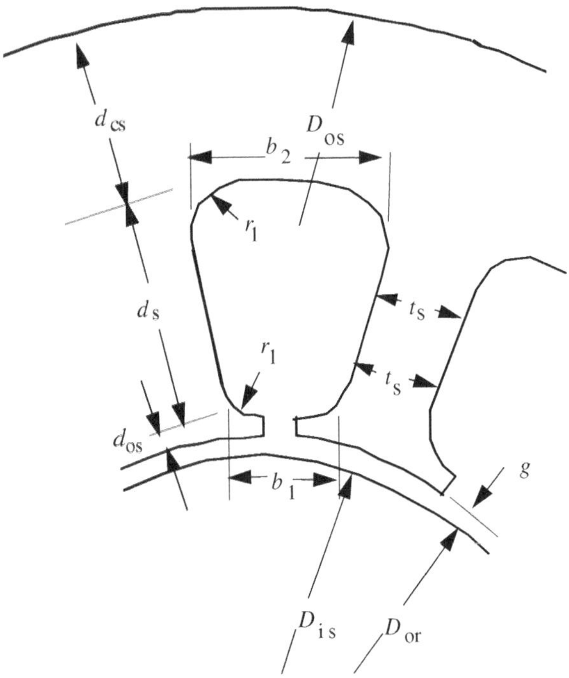

Figure 6.13 Dimensional data for stator slot shape.

where k_{is} represents the ratio between the actual length of stator iron (excluding ducts but including air space between laminations) and the physical length of iron l_i, Φ_p denotes the flux per pole, (equation 6.4). In addition, from equation (6.5),

$$B_{g1} = \frac{P\Phi_p}{2D_{is}l_e} \tag{6.42}$$

Upon taking ratios, it can be determined that

$$t_s = \frac{\pi D_{is}}{S_1 k_{is}}\left(\frac{B_{g1}}{B_{ts}}\right)\left(\frac{l_e}{l_i}\right) \tag{6.43}$$

$$d_{cs} = \frac{D_{is}}{Pk_{is}}\left(\frac{B_{g1}}{B_{cs}}\right)\left(\frac{l_e}{l_i}\right) \tag{6.44}$$

The stator outer diameter and slot depth can be introduced by means of the equation

$$D_{os} = D_{is} + 2(d_{os} + d_s + d_{cs}) \tag{6.45}$$

and the slot widths and core depth by

$$\pi(D_{is} + 2d_{os}) = (t_s + b_1)S_1 \tag{6.46}$$

$$\pi(D_{os} - 2d_{cs}) = (t_s + b_2)S_1 \tag{6.47}$$

Upon solving these equations for b_1, b_2, and d_s one obtains

$$b_1 = \frac{\pi}{S_1}\left[D_{is}\left(1 - \frac{B_{g1}}{k_{is}B_{ts}}\frac{l_e}{l_i}\right) + 2d_{os}\right] \tag{6.48}$$

$$b_2 = \frac{\pi}{S_1}\left[D_{os} - D_{is}\left(\frac{B_{g1}}{k_{is}B_{ts}}\frac{l_e}{l_i} + \frac{2}{P}\frac{B_{g1}}{k_{is}B_{cs}}\frac{l_e}{l_i}\right)\right] \tag{6.49}$$

and

$$d_s = \frac{S_1}{2\pi}(b_2 - b_1) \tag{6.50}$$

The area of the stator slot is, approximately,

$$A_s = \frac{d_s}{2}(b_1 + b_2) \tag{6.51}$$

Inserting equations (6.48)–(6.50) into equation (6.51) results in a quadratic in terms of the ratio of inner to outer stator diameters sometimes called the *split ratio* as,

$$a\left(\frac{D_{is}}{D_{os}}\right)^2 - 2b\left(\frac{D_{is}}{D_{os}}\right) + 1 = \frac{S_1A_s}{\frac{\pi D_{os}^2}{4}} + \frac{\delta_1}{D_{os}^2} \tag{6.52}$$

where

$$a = \left(\frac{B_{g1}}{k_{is}B_{ts}}\frac{l_e}{l_i} + \frac{2}{P}\frac{B_{g1}}{k_{is}B_{cs}}\frac{l_e}{l_i}\right)^2 - \left(1 - \frac{B_{g1}}{k_{is}B_{ts}}\frac{l_e}{l_i}\right)^2$$

$$b = \left(\frac{B_{g1}}{k_{is}B_{ts}} + \frac{2}{P}\frac{B_{g1}}{k_{is}B_{cs}}\right)\frac{l_e}{l_i}$$

$$\delta_1 = 4\left[d_{os}^2 + D_{is}d_{os}\left(1 - \frac{B_{g1}}{k_{is}B_{ts}}\frac{l_e}{l_i}\right)\right] \tag{6.53}$$

Note that once the flux density ratios between gap and tooth and between gap and core have been selected, a and b become constants. If both numerator and denominator of the first term on the right-hand side of equation (6.52) are multiplied by the iron length l_i then the numerator of this term essentially corresponds to the volume of the stator core reserved for copper (volume of the stator slots), while the denominator corresponds to the volume of the purchased steel. The second term on the right-hand side is a correction factor to account for the material used for the slot opening and is relatively small.

From equation (6.4), the volt-ampere rating of the machine can be expressed as

$$VA_{gap} = 3\pi f_e k_1 N_s \Phi_p I_s \tag{6.54}$$

where

$$\Phi_p = \frac{2D_{is}l_e}{P}B_{g1} \tag{6.55}$$

The peak phase current I_s can be expressed in terms of the current density as

$$J_{s(rms)} = \frac{3\sqrt{2}N_s I_s}{S_1 A_{cu}} \tag{6.56}$$

where A_{cu} is the portion of the slot area occupied by copper. That is,

$$A_{cu} = k_{cu} A_s$$

In terms of the air gap flux density, current density, and slot area, equation (6.54) can, therefore, be written as

$$VA_{gap} = \sqrt{2}\pi f_e k_1 \frac{D_{is} l_e}{P}(S_1 k_{cu} A_s) B_{g1} J_{s(rms)} \tag{6.57}$$

Neglecting the second term on the right-hand side of equation (6.52), solving for $S_1 A_s$, and then inserting into equation (6.57) results in the following expression in terms of the split ratio

$$VA_{gap} = \frac{\sqrt{2}\pi^2}{4} f_e \frac{k_1 k_{cu}}{P} \left[a\left(\frac{D_{is}}{D_{os}}\right)^3 - 2b\left(\frac{D_{is}}{D_{os}}\right)^2 + \frac{D_{is}}{D_{os}} \right] \left(D_{os}^3 l_e \cdot B_{g1} J_{s(rms)}\right) \tag{6.58}$$

which represents a sizing equation in terms of $D_{os}^3 l_e$. In terms of mechanical speed, from equation (6.9),

$$\frac{VA_{gap}}{\Omega_{mech}} = \frac{\sqrt{2}\pi^2}{480}(k_1 k_{cu}) \left[a\left(\frac{D_{is}}{D_{os}}\right)^3 - 2b\left(\frac{D_{is}}{D_{os}}\right)^2 + \frac{D_{is}}{D_{os}} \right] (D_{os}^3 l_e) \left[B_{g1} J_{s(rms)}\right] \tag{6.59}$$

The polynomial in the square brackets is clearly a function which should be maximized in order to produce a minimum value of $D_{os}^3 l_e$ for a given volt-ampere and speed requirement. Since the input volt-amperes are related to the output power in watts by

$$P_{out} = VA_{gap} \eta_{gap} \cos\theta_{gap} \tag{6.60}$$

where η_{gap} is the efficiency measured at the air gap, that is. neglecting stator copper and stator iron loss. In a similar manner, $\cos\theta_{gap}$ is the power factor when viewed at the air gap (i.e., θ_{gap} is the phase angle of the stator current with respect to the air gap voltage). Equation (6.59) can also be written in terms of the output power as

$$\frac{P_{out}}{\Omega_{mech}} = \frac{VA_{gap}}{\Omega_{mech}} \eta_{gap} \cos\theta_{gap} \tag{6.61}$$

or, from equation (6.59)

$$\frac{P_{out}}{\Omega_{mech}} = \frac{\sqrt{2}\pi^2}{480}(k_1 k_{cu})(\eta_{gap}\cos\theta_{gap}) \left[a\left(\frac{D_{is}}{D_{os}}\right)^3 - 2b\left(\frac{D_{is}}{D_{os}}\right)^2 + \frac{D_{is}}{D_{os}} \right] \times (D_{os}^3 l_e)(B_{g1} J_{s(rms)}) \tag{6.62}$$

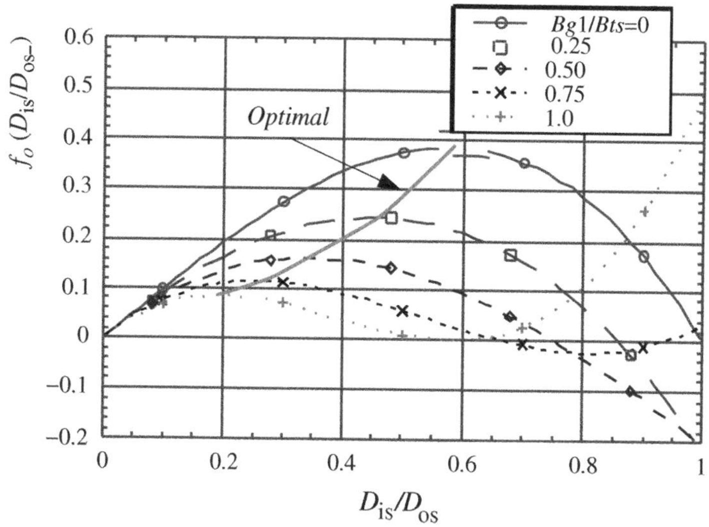

Figure 6.14 The output function $f(D_{is}/D_{os})$ versus D_{is}/D_{os} for the case where $B_{cs} = 0.8B_{ts}$ and $P = 4$.

or simply,

$$\frac{P_{out}}{\Omega_{mech}} = \xi_o D_{os}^3 l_e \tag{6.63}$$

The quantity ξ_o is called the D_0^3L output coefficient. The quantity in square brackets in equation (6.62) is clearly important in the optimization of the machine design. Defining the function

$$f_o(D_{is}/D_{os}) = \left[a\left(\frac{D_{is}}{D_{os}}\right)^3 - (2b)\left(\frac{D_{is}}{D_{os}}\right)^2 + \frac{D_{is}}{D_{os}} \right] \tag{6.64}$$

The function $f_o(D_{is}/D_{os})$ is plotted in Figure 6.14 as a function of D_{is}/D_{os}. The peak of one of the curves represents an optimum (minimum) value of $D_{os}{}^3 l_e$ for a given ratio of $B_{cs}/B_{ts} = 0.8$ and a particular value of B_{g1}/B_{ts} and D_{os} [5]. The optimal value of D_{is} can be determined by differentiating equation (6.64) with respect to D_{is} while holding D_{os} constant, and setting the result to zero. Hence,

$$\frac{\partial f_o(D_{is}/D_{os})}{\partial D_{is}} = 0 = 3a\frac{D_{is}^2}{D_{os}^3} - 4b\frac{D_{is}}{D_{os}^2} + \frac{1}{D_{os}} \tag{6.65}$$

so that

$$\left.\frac{D_{is}}{D_{os}}\right|_{optimal} = \frac{2b \pm \sqrt{4b^2 - 3a}}{3a} \tag{6.66}$$

Figure 6.15 shows how the optimal value of $f_o(D_{is}/D_{os})$ varies with pole number. Note that the optimum value of D_{is}/D_{os} increases as the number of poles increase indicating a tendency toward a "ring-shaped" stator. The maximum value

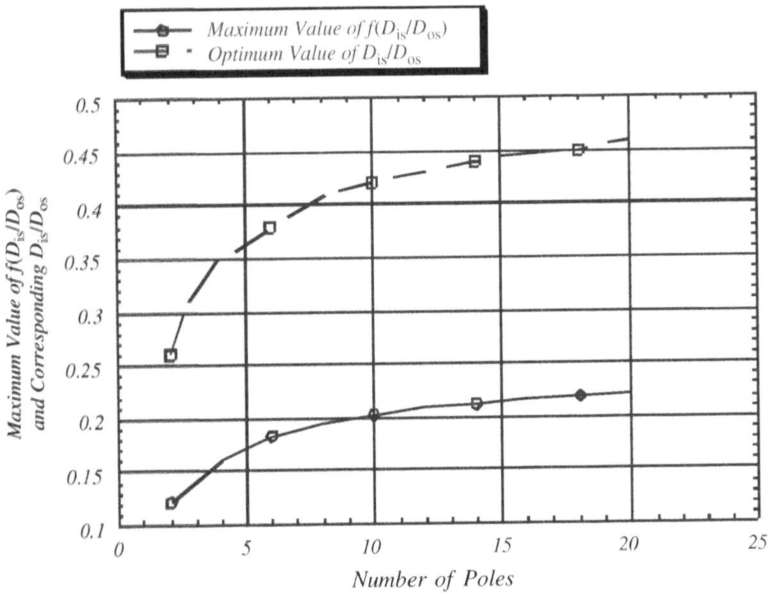

Figure 6.15 Effect of pole number on the output function $f_o(D_{is}/D_{os})$. Flux densities $B_{cs} = 0.8\,B_{ts}$, $B_{g1}/B_{ts} = 0.5$.

of $f_o(D_{is}/D_{os})$ also increases slowly beyond a pole number of six, indicating that the amount of stator iron slowly decreases with pole number beyond six. This is primarily due to the fact that the flux per pole decreases with pole number thereby necessitating a progressively smaller core cross-sectional area as the pole number increases.

While the solution plotted in Figure 6.15 is optimal, it is optimal only when the current density $J_{s(rms)}$ is held constant and not the surface current density $K_{s(rms)}$. In the case of machines of low pole number, the optimal values obtained from equation (6.66) are good designs. However, when the pole number increases, the values obtained from this result yield progressively poorer designs since the usual limitation on surface current density (heating) becomes exceeded. This problem can be overcome by introducing an inequality constraint in addition to equation (6.64). The surface current density is related to the current density by

$$K_{s(rms)} = \frac{k_{cu} A_s S_1}{\pi D_{is}} J_{s(rms)} \tag{6.67}$$

and, from equation (6.52), neglecting δ_1,

$$a\left(\frac{D_{is}}{D_{os}}\right)^2 - 2b\left(\frac{D_{is}}{D_{os}}\right) + 1 = \frac{4 S_1 A_s}{\pi D_{os}^2} \tag{6.68}$$

Solving for A_s and substituting the result into equation (6.67),

$$K_{s(rms)} = \frac{k_{cu} J_{s(rms)}}{4}\left(a D_{is} - 2b D_{os} + \frac{D_{os}^2}{D_{is}}\right) \tag{6.69}$$

The problem now is to find $D_{is(opt)}$ which maximizes

$$f(D_{is}) = a\frac{D_{is}^3}{D_{os}^2} - 2b\frac{D_{is}^2}{D_{os}} + D_{is} \tag{6.70}$$

subject to

$$g(D_{is}) = K_{s(max)}^* - \frac{k_{cu}J_{s(rms)}}{4}\left(aD_{is} - 2bD_{os} + \frac{D_{os}^2}{D_{is}}\right) \geq 0 \tag{6.71}$$

where $K_{s(max)}^*$ is a prescribed maximum value of surface current density.

The problem involving the inequality can be converted into one with an equality constraint by the method of Lagrange multipliers [6]. In terms of this formulation, the problem is to solve,

$$\frac{\partial f(D_{is})}{\partial D_{is}} = \zeta\frac{\partial g(D_{is})}{\partial D_{is}} \tag{6.72}$$

satisfying

$$g(D_{is}) = 0 \tag{6.73}$$

and where ζ is the Lagrange multiplier. Equations (6.72) and (6.73) can be written explicitly as

$$3a\left(\frac{D_{is}}{D_{os}}\right)^2 - 4b\left(\frac{D_{is}}{D_{os}}\right) + 1 = -\zeta\frac{k_{cu}J_{s(rms)}}{4}\left(a - \frac{D_{os}^2}{D_{is}^2}\right) \tag{6.74}$$

satisfying

$$K_{s(max)}^* - \frac{k_{cu}J_{s(rms)}}{4}\left(aD_{is} - 2bD_{os} + \frac{D_{os}^2}{D_{is}}\right) = 04 \tag{6.75}$$

Multiplying equation (6.75) by D_{is} forms the quadratic equation,

$$\frac{k_{cu}J_{s(rms)}}{4}\left[a\frac{D_{is}^2}{D_{os}^2} - 2b\frac{D_{is}}{D_{os}} + 1\right] - \frac{K_{s(max)}^*}{D_{os}}\left(\frac{D_{is}}{D_{os}}\right) = 0 \tag{6.76}$$

which can be solved as

$$\frac{D_{is}}{D_{os}} = \frac{b}{a} + \frac{2K_s^*}{ak_{cu}J_{s(rms)}D_{os}} \pm \sqrt{\left(\frac{b}{a} + \frac{2K_s^*}{ak_{cu}J_{s(rms)}D_{os}}\right)^2 - \frac{1}{a}} \tag{6.77}$$

Solving equation (6.74) for ζ,

$$\zeta = \frac{3a\left(\frac{D_{is}}{D_{os}}\right)^2 - 4b\left(\frac{D_{is}}{D_{os}}\right) + 1}{\frac{k_{cu}J_{s(rms)}}{4}\left[\left(\frac{D_{os}}{D_{is}}\right)^2 - a\right]} \tag{6.78}$$

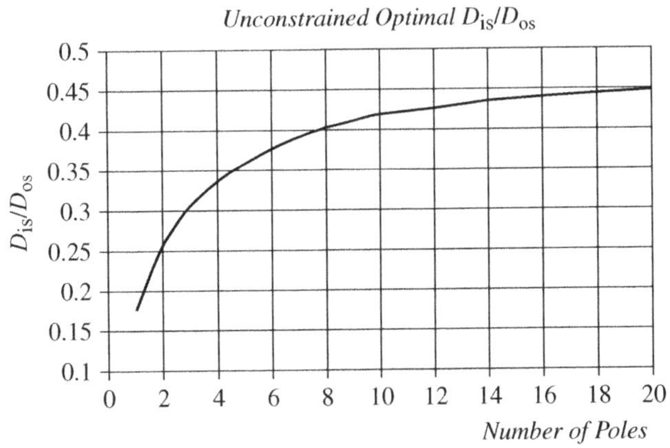

(a)

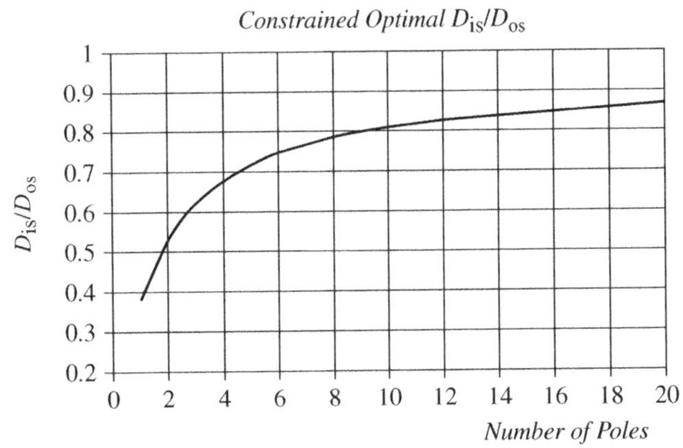

(b)

Figure 6.16 Comparison of diameter ratio D_{is}/D_{os} (a) without and (b) with constraints on the RMS surface current density $K_{s(rms)}$.

The procedure for finding the optimal value of D_{is}/D_{os} is to first assume a tentative value of D_{os}, then to solve equation (6.76) for D_{is}/D_{os} and to insert this solution into equation (6.78). Since the plus sign in equation (6.77) yields a value of $D_{is}/D_{os} > 1$, the negative sign should be selected. If ξ is negative, then the solution for D_{is}/D_{os} is optimal. If ξ is positive, then $J_{s(rms)}$ must be reduced or D_{os} increased to obtain a feasible solution. Figure 6.16 compares the optimal values obtained with K_s as unconstrained and constrained, wherein $B_t/B_{g1} = 0.5$, $B_c/B_{g1} = 0.8$, $K_{s(rms)} = 50$ A/mm (1270 A/in.), and $J_{s(rms)} = 5$ A/mm^2 (3225 A/in.^{2}). Note that with the constraint, D_{is}/D_{os} approaches 1.0 asymptotically rather than 0.5 indicating a ring-shaped stator of small radial thickness when the surface current constraint is taken into account. Finally, knowing D_{os} and D_{is}/D_{os} the corresponding value of D_{is} can be calculated. If this value of D_{is} is not the same as the required value determined, for example by equation (6.13), iteration can now be used to converge on the desired value.

6.6 POWER LOSS DENSITY

It has already been mentioned that limits must be imposed on the surface current density to prevent overheating. Since the cooling capability of a machine varies with the area, a convenient expression describing loss from the point of view of optimization is the power loss per unit area. From equation (6.6), the current per unit length of stator circumference (surface current density) is

$$K_s = \frac{S_1 n_s (I_s/C)}{\pi D_{is}} = \frac{I_s}{C}\frac{n_s}{\pi(D_{is}/S_1)} \tag{6.79}$$

$$= \frac{I_s}{C}\frac{n_s}{\tau_s} \tag{6.80}$$

The resistance of the slot portion of each conductor is

$$r_{sl} = \frac{\rho l_s}{A_{cu}} \tag{6.81}$$

where l_s is the length of the stack including ducts. Thus the copper loss in each slot is

$$P_{sl} = n_s^2 \frac{I_s^2}{2C^2} r_{sl} = n_s^2 \frac{\rho l_s}{A_{cu}} \frac{I_s^2}{2C^2} \tag{6.82}$$

The heat produced in a given slot is dissipated over the surface of one slot pitch. The power loss per unit area of air gap surface is, therefore,

$$P_{diss} = \frac{P_{sl}}{\tau_s l_s}$$

$$= \left(\frac{I_s}{\sqrt{2}C}\frac{n_s}{\tau_s}\right)\left(\frac{n_s I_s}{\sqrt{2}CA_{cu}}\right)\rho \tag{6.83}$$

$$= K_{s(rms)} J_{s(rms)} \rho = \frac{K_{s(rms)} J_{s(rms)}}{\sigma} \tag{6.84}$$

where $K_{s(rms)}$ is the RMS of the peak value given by equation (6.7) and $J_{s(rms)}$ is the current density per unit area of the conductor cross section, defined by equation (6.56).

6.7 THE $D^{2.5}L$ SIZING EQUATION

Multiplication of the two sizing equations for $D_{os}^3 l_e$ and $D_{is}^2 l_e$ and taking the square root of the result yields another useful expression for the output torque of the machine

$$\frac{P_{out}}{\Omega_{mech}} = \sqrt{\xi_o \xi_r} \cdot \sqrt{D_{os}^3 D_{is}^2} \cdot l_e \tag{6.85}$$

where

$$\xi_r = \frac{\sqrt{2}\pi^2}{120} k_1 (K_{s(rms)} B_{g1}) \eta_{gap} \cos\phi \tag{6.86}$$

and, from equation (6.62)

$$\xi_0 = \frac{\sqrt{2}\pi^2}{480}(k_1 k_{cu}) \left[a\left(\frac{D_{is}}{D_{os}}\right)^3 - 2b\left(\frac{D_{is}}{D_{os}}\right)^2 + \frac{D_{is}}{D_{os}} \right] J_{s(rms)} B_{g1} (\eta_{ag} \cos\phi_{ag}) \tag{6.87}$$

Equation (6.85) can be written as

$$\frac{P_{out}}{\Omega_{mech}} = \sqrt{\xi_o \xi_r \left(\frac{D_{is}}{D_{os}}\right)^2} \cdot D_{os}^{2.5} l_e \tag{6.88}$$

or simply

$$\frac{P_{out}}{\Omega_{mech}} = \xi_{o(2.5)} \cdot D_{os}^{2.5} l_e \tag{6.89}$$

where

$$\xi_{o(2.5)} = \frac{\sqrt{2}\pi^2}{240} k_1 B_{g1} \eta_{gap} \cos\phi_{gap} \sqrt{k_{cu} K_{s(rms)} J_{s(rms)}} \left[\frac{D_{is}}{D_{os}} \sqrt{f_o\left(\frac{D_{is}}{D_{os}}\right)} \right] \tag{6.90}$$

The function $f_o(D_{is}/D_{os})$ is defined by equation (6.64). It has been shown in Section 6.6 that the quantity $K_{s(rms)}J_{s(rms)}$ is proportional to the stator I^2R loss per unit surface area (neglecting the portion of the loss due to the end windings) and hence is closely related to the temperature rise of the machine. Thus, if this product is kept constant the cooling capability is preserved as the machine dimensions are changed to satisfy other criteria. A plot of the function

$$\left(\frac{D_{is}}{D_{os}} \sqrt{f_o\left(\frac{D_{is}}{D_{os}}\right)} \right)$$

is shown in Figure 6.17.

6.8 CHOICE OF MAGNETIC LOADING

The choice of magnetic loading is influenced by a number of factors which apply to all types of machines.

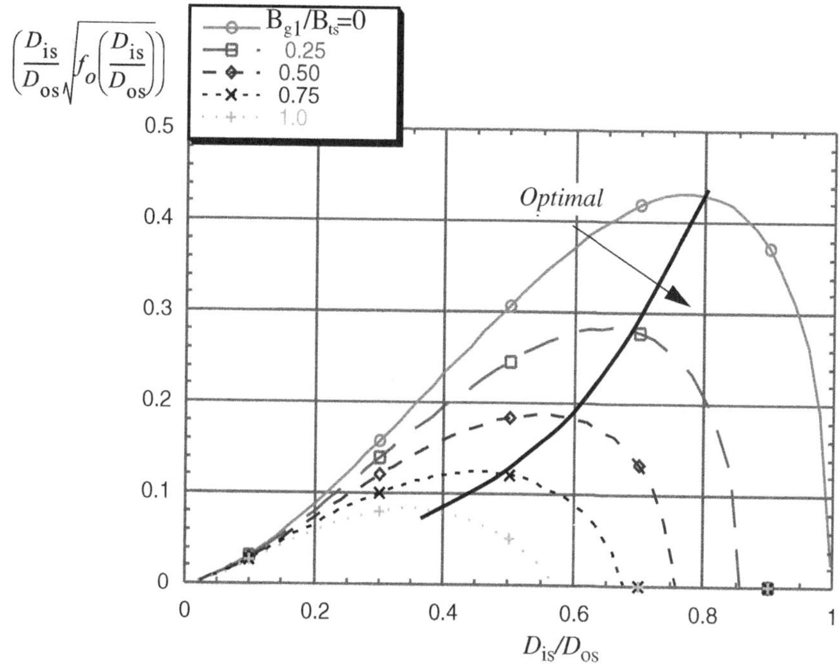

Figure 6.17 The output function $\left(\frac{D_{is}}{D_{os}}\sqrt{f_o\left(\frac{D_{is}}{D_{os}}\right)}\right)$ versus D_{is}/D_o for the case where B_{cs} = 0.8, B_{ts} and $P = 4$.

6.8.1 Maximum Flux Density in Iron

The maximum flux density in any iron component must clearly be kept below a well-defined maximum to avoid saturation. In a well-designed machine, the highest flux density usually occurs in the teeth. The flux density entering a tooth is related approximately to the flux density in the gap by the ratio of tooth width to slot pitch, that is,

$$B_o \approx \frac{\tau_s}{t_o} B_{g1} \tag{6.91}$$

Clearly, if the flux density at the narrowest point in the teeth is to remain below a certain limit, the maximum value of B_{g1} is also fixed. The maximum value for tooth flux density is typically in the range 1.55–1.9 teslas (100,000 to 123,000 lines/in.^{2}) for common steel while the flux density in the core normally ranges over values from 1.4 to 1.7 teslas (90,000 to 110,000 lines/in.^{2}). These values are for 60 Hz operation and must, of course, be adjusted for operation at higher frequencies. The effect of frequency on permissible tooth flux density can be estimated roughly from [3]

$$B_{ts} = 5.47 f^{-0.32} \tag{6.92}$$

The corresponding value of air gap flux density is

$$B_{g1} = \frac{B_{ts}}{k_{is}}\left(1 - \frac{b_o}{\tau_s}\right) \tag{6.93}$$

where b_o is the slot opening, k_{is} is the stator stacking factor, and τ_s is the slot pitch. As a starting value one can assume simply that $b_o/\tau_s \cong 0.5$, $k_{is} \cong 1$ in which case $B_{g1} = 0.5B_{ts}$.

In large machines which have large diameters the taper of teeth is not significant. However, in small machines which have smaller diameters the taper of the teeth is very pronounced and consequently saturates first at the tip of the stator teeth and the root of the rotor teeth. In addition, higher slot fill factors are possible for large machines. Therefore, less space is lost to insulation resulting in a consequent increase in space available for the teeth. Hence, small machines typically have lower values of gap flux density than large machines. With ordinary electric steels, the maximum value of B for induction machines should typically not exceed the following levels at 60 Hz:

Stator core	$B_{cs} = 1.6$ teslas $\approx$ 105,000 lines/in.^{2}
Rotor core	$B_{cr} = 1.7$ teslas $\approx$ 110,000 lines/in.^{2}
Stator tooth	$B_{ts} = 1.8$ teslas $\approx$ 115,000 lines/in.^{2}
Rotor tooth	$B_{tr} = 1.9$ teslas $\approx$ 120,000 lines/in.^{2}

6.8.2 Magnetizing Current

The magnetizing current of a machine is directly proportional to the MMF required to force the flux through the air gap and iron portions of the machine. The MMF required for the air gap is simply proportional to the air gap flux density. However, the flux density in the iron depends upon the value of specific magnetic loading. If a small value of loading is chosen, the flux density in the iron is low and therefore these parts are worked on the linear portion up to the "knee" of the B–H curve. Operation in this region requires a small or even negligible value of MMF. However, such operation represents inefficient utilization of the available iron. If larger values of magnetic loading (flux density) are assumed the MMF required rises rapidly. Thus a large value of magnetic loading results in increased values of magnetizing MMF and hence magnetizing current. A good compromise involves design of a mildly saturated machine with sufficient reserve to accommodate an overvoltage of 10% (NEMA standard).

The iron loss is also frequency dependent since the hysteresis and eddy current losses vary with frequency and frequency squared, respectively. It follows that for high speed machines (high frequency machines), the magnetic loading must be reduced in order to obtain reduced iron losses so that a reasonable value of efficiency may be achieved. Whereas a typical value of peak air gap flux density for a 60-Hz machine may be 0.8 tesla (53,000 lines/in.^{2}), the corresponding value in a 400-Hz servo motor may be only half this value. The magnetic loading can be increased somewhat since the power output increases faster than the magnetic losses with increased dimension. This topic will again be discussed in Section 6.12. Figure 6.18 shows typical air gap flux densities for both standard and high efficiency induction machines for a range of ratings from 3 to 150 HP [7]

Without better available information the following equation may be used as a first guess for air gap flux density, assuming operation at 50 or 60 Hz.

$$B_{g1} = 0.464\tau_p^{1/6} \tag{6.94}$$

where τ_p is given in centimeters and B_{g1} in teslas.

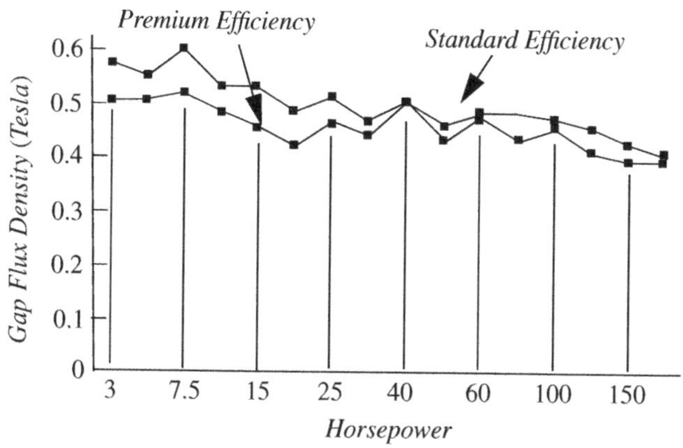

Figure 6.18 Air gap flux density for a range of motors from one manufacturer (64.5 klines/in.^{2} = 1 tesla).

6.9 CHOICE OF ELECTRIC LOADING

Although the saturation phenomena of magnetic circuits do not exist in the case of electric circuits, the losses associated with the conductors again play a major role in fixing the electric loading. Major factors in limiting the electric loading of the machine are the following.

6.9.1 Voltage Rating

Contrary to what one might expect, more of the slot is typically occupied by air than by the copper conductor. In addition to assembly problems which dictate that some "play" remains in order to slip the coils in the slots, the conductors must be insulated between strands, between separate coils, and with respect to ground, that is, the core. The insulation required with respect to ground is, in itself, a major portion of the available slot area since this insulation must withstand surges due to lightning and line switching in addition to normal overvoltage conditions. Standard IEEE tests require that the machine withstand, for 1 min, a voltage of two times the rated phase voltage plus 1000 V at its normal working temperature. The amount of copper cross-sectional area expressed as a per unit of the available cross-sectional area of a stator slot is typically in the range 0.2–0.4. The amount of available conductor area decreases as the voltage rating of the machine increases, since ground wall and coil separator insulation as well as insulation buildup on the conductor wires increase as shown in Table 6.5.

Modern pulse width modulated inverters to vary the speed of AC motors has presented the motor designer with new challenges in the area of insulation. In such cases, the motor insulation system is subjected to voltage step changes at a rate of up to 20,000 times per second. These changes are, in reality, very oscillatory indicating resonance with the capacitance of the motor insulation system. Overshoots of double the inverter DC bus voltage can occur. Hence, when a nominal 460 V AC supply

TABLE 6.5 Table of round copper wire for random-wound machines

AWG	Bare wire Diameter (in.)	Heavy (in.)	Triple (in.)	Quadruple (in.)	Glass (in.)	Weight /1000 ft. at @ 25°C (lb)	Ohms/ 1000 ft. (Ω)
	0.0605	0.0634	0.0652	0.0662	0.0662	11.10	2.80
15	0.0517	0.0600	0.0617	0.0628	0.0628	9.87	3.24
	0.0538	0.0557	0.0585	0.0594	0.0594	8.76	3.65
16	0.0508	0.0537	0.0552	0.0563	0.0563	7.81	4.10
	0.0480	0.0508	0.0524	0.0534	0.0534	6.97	4.59
17	0.0453	0.0481	0.0498	0.0506	0.0506	6.21	5.15
	0.0427	0.0454	0.0470	0.0480	0.0480	5.52	5.80
18	0.0403	0.0430	0.0445	0.0455	0.0455	4.91	6.51
	0.0380	0.0406	0.0422	0.0432	0.0432	4.37	7.32
19	0.0359	0.0385	0.0399	0.0410	0.0410	3.90	8.21
	0.0339	0.0364	0.0379	0.0390	0.0390	3.48	9.20
20	0.0320	0.0345	0.0358	0.0370	0.0370	3.10	—
	0.0302	0.0326	0.0340	0.0352	0.0352	2.76	—
21	0.0285	0.0309	0.0322	0.0334	0.0334	2.46	—
22	0.0253	0.0276	0.0288	0.0301	0.0301	1.94	—
23	0.0226	0.0248	0.0260	0.0273	0.0273	1.55	—
24	0.0201	0.0222	0.0230	0.0243	—	1.22	—
25	0.0179	0.0199	0.0212	0.0219	—	0.97	—

is replaced by a similarly rated inverter, the rectified 650 V DC can result in over 1200 V across the motor winding from line to line. The magnitude of this transient combined with the high repetition rate can reduce the life of the insulation system by as much as 90% statistically increasing rapidly when the inverter switching rate exceeds 5000 Hz. NEMA standard MG1 defines the acceptable wave fronts for a general purpose motor as $V_{peak} \leq 1000$ V with rise time ≥ 2 μs and for an "inverter grade" motor as $V_{peak} \leq 1600$ V with rise time ≥ 0.1 μs.

A new magnet wire has been developed that adds a shield coat between the original base coat and the top coat on the wire. This style of coating of three different materials achieves a greatly improved protection from rapid rise time, high frequencies, and elevated voltages while maintaining good flexibility and processability. Accelerated laboratory testing has indicated that this new wire is up to 200 times more resistant to inverter-transient damage than normal magnet wire [7, 8].

6.9.2 Current Density Constraints

There are several possible criteria for determining the appropriate value of the surface current density K_s in the stator winding. The steady-state value could be constrained by the ability of the cooling system to conduct away the heat produced in the stator (winding plus core loss) within the temperature limits of the winding insulation. Another criterion might be the attainment of a specified efficiency, permitting only a specified per unit loss in the winding. In either case, the task is to choose motor and winding dimensions to achieve a specified torque to within some limit on the stator winding loss P_{sw}.

Consider a stator winding with N_s series-connected turns. The peak fundamental current density for all three phases is obtained from equation (6.7)

$$K_{s1} = \frac{6}{\pi}\frac{k_1 N_s}{D_{is}} I_s \quad \text{A/m} \tag{6.95}$$

One can now allocate space for this winding to achieve a specified winding loss [9]. The slot width is typically $b_s \approx 0.5\,\tau_s$ to optimize the force per unit of gap area. The slot depth is defined as d_s. The ratio of conductor area to slot area or space factor k_{cu} is usually in the range 0.3–0.5 for random-wound machines. Thus, the winding has an equivalent continuous depth d_e of solid conductor of

$$d_e \approx k_{cu} d_s \frac{b_s}{\tau_s} \tag{6.96}$$

Figure 6.19 shows one turn of a typical stator coil spanning approximately 1/P of the inner stator periphery. The ratio υ_e of the length of the end-winding portion of this coil to the pole pitch τ_p ranges from 1.4 to 1.8. The stator winding can be considered as equivalent to a solid conductor of axial length l_c where

$$l_c = l_s + \upsilon_e \frac{(\pi D_{is})}{P} \tag{6.97}$$

The power loss in the stator copper can now be expressed as

$$P_{s,cu} = \rho_{cu} \pi D_{is} d_e l_c J^2_{s(rms)} \quad \text{W} \tag{6.98}$$

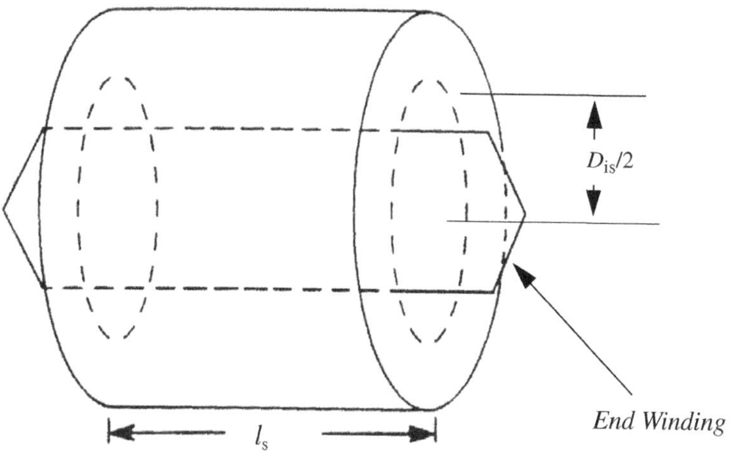

Figure 6.19 Typical stator coil.

where ρ_{cu} is the resistivity of the copper conductor. In terms of the continuous equivalent depth, the area current density $J_{s(rms)}$ *is* related to the peak surface current density by

$$J_{s(rms)} = \frac{K_{s1}}{\sqrt{2}d_e} \quad \text{A/m}^2 \tag{6.99}$$

Combining equations (6.95), (6.97), (6.98), and (6.99), the stator copper loss $P_{s,cu}$ is directly related to the surface current density K_s by

$$P_{s,cu} = \left(\frac{\pi}{2}\right)\frac{D_{is}\rho_{cu}}{k_{cu}d_s}\left(\frac{\tau_s}{b_s}\right)\left[l_s + \upsilon_e\frac{(\pi D_{is})}{P}\right]K_{s1}^2 \quad \text{W} \tag{6.100}$$

This result can be expressed in per unit of the rated mechanical power of the motor. From equation (6.13), the rated power is

$$P_{mech} = \left(\frac{\pi}{2}\right)\frac{\omega_e}{P}k_1\left(D_{is}^2 l_e\right)B_{g1}K_{s1}\eta_{gap}\cos\phi_{gap} \tag{6.101}$$

and the values of B_{g1} and $K_{s(rms)}$ are assumed to be at their rated value. Thus, dividing equation (6.100) by (6.101)

$$P_{s,cu(pu)} = \frac{(P\rho_{cu})\left(\frac{\tau_s}{b_s}\right)\left[l_s + \upsilon_e\frac{(\pi D_{is})}{P}\right]K_{s1}}{k_{cu}k_1\omega_e(D_{is}l_e)d_sB_{g1}\eta_{gap}\cos\phi_{gap}} \quad \text{per unit} \tag{6.102}$$

Several additional factors deserve consideration before motor dimensions are chosen, but it may be useful at this stage to consider a sample design of a 20 kW, 4-pole, 1800 r/min motor with a stator-winding loss of 0.02 per unit.

Thus, one is given

$$\begin{aligned} P_{s,cu(pu)} &= 0.02 \\ P &= 4 \\ \rho_{cu} &\approx 1.68\cdot 10^{-8}\ \Omega - \text{m} \\ \omega_e &= 377\ \text{rad/s} \end{aligned}$$

The following values are chosen:

$$\begin{aligned} B_{g1} &= 0.8 \\ l_s &= D_{is} \quad (\text{chosen shape } l_s/\tau_p = P/\pi) \\ l_e &= 1.05\, l_s \\ \upsilon_e &= 1.6 \\ \frac{\tau_s}{b_s} &= 2 \\ k_{cu} &= 0.7 \\ d_s &= 0.11\, D_{is} \quad (\text{chosen ratio}) \\ k_1 &= 0.95 \\ \eta_{gap} &= 0.93 \\ \cos\phi_{gap} &= 0.9 \end{aligned}$$

The efficiency and power factor at the gap are determined by first designing the rotor assuming an impressed flux density of B_{g1}.

Insertion of these values into equations (6.100) and (6.101) gives

$$D_{is}K_{s1}^2 = 2.586 \cdot 10^8 \text{ A}^2/\text{m} \tag{6.103}$$

$$D_{is}^3 K_{s1} = 192.1 \text{ A-m}^2 \tag{6.104}$$

from which one obtains

$$K_{s1} = \sqrt[5]{\frac{(2.586 \cdot 10^8)^3}{192.1}} = 39.0 \times 10^3 \text{ A/m} \tag{6.105}$$

$$D_{is} = \sqrt[5]{\frac{192.1^2}{2.586 \cdot 10^8}} = 0.170 \text{ m} \tag{6.106}$$

$$J_{s(rms)} = \frac{K_{s1}}{\sqrt{2}d_e} = \frac{K_{s1}}{\sqrt{2}k_{cu}d_s\left(\dfrac{b_s}{\tau_s}\right)}$$

$$J_{s(rms)} = \frac{39000}{\sqrt{2} \times 0.7 \times (0.11 \times 0.170) \times 0.5} = 4.21 \times 10^6 \text{ A/m}^2 \tag{6.107}$$

6.9.3 Representative Values of Current Density

Representative values of stator surface current density $K_{s(rms)}$ and stator current density $J_{s(rms)}$ for open drip proof (ODP) and TEFCs of one manufacturer are given in Tables 6.6 and 6.7, respectively. Without information available from similar designs, the following equation may be used as a starting point to select $K_{s(rms)}$

$$K_{s(rms)} = 100\,\tau_p^{2/3} \tag{6.108}$$

where τ_p is given in meters and $K_{s(rms)}$ expressed in RMS A/mm.

In squirrel-cage rotors, the current in the bars and the end-ring segments are normally limited to 8 kA(rms)/m (5000 A(rms)/in.^{2}). For a slip-ring machine the current density in the rotor windings is typically allowed to be 20–30% higher than in the

TABLE 6.6 Current densities $K_{s(rms)}$ and $J_{s(rms)}$ in RMS A/mm and RMS A/mm^2, respectively for open drip proof machines of NEMA design

	$P = 2$		$P = 4$		$P = 6$	
HP	$K_{s(rms)}$	$J_{s(rms)}$	$K_{s(rms)}$	$J_{s(rms)}$	$K_{s(rms)}$	$J_{s(rms)}$
1	14	4.8	14	4.35	14	4.0
5	21	4.35	21	4.0	21	3.7
10	24	4.35	24	4.0	24	3.7
50	32	4.35	28	4.0	27	3.7
100	34	4.35	31	4.0	28	3.7
500	36	4.35	32	4.0	29	3.7

TABLE 6.7 Current densities $K_{s(rms)}$ and $J_{s(rms)}$ in RMS A/mm and RMS A/mm^2, respectively for totally enclosed, fan-cooled machines for typical NEMA design

		$J_{s(rms)}$			
HP	$K_{s(rms)}$	$P = 2$	$P = 4$	$P = 6$	$P = 8$
1	11.5	5.4	4.5	3.7	3.4
5	21	5.1	4.3	3.6	3.25
10	24	5.0	4.2	3.6	3.1
50	26	3.1	3.7	3.4	3.0
100	26	2.3	2.3	2.5	2.6
500	26	2.3	2.3	2.0	2.0

stator winding. Totally enclosed machines are typically in the range 1–5 kA(rms)/m whereas fan-cooled machines can reach roughly 8–10 kA(rms)/m. Liquid-cooled machines can attain much higher values depending upon the cooling scheme reaching up to 10–20 kA(rms)/m for frame-cooled and 30–40 kA(rms)/m for conductor-cooled machines. The maximum tolerable value is, of course, set by thermal considerations which will be dealt with in Chapter 7.

It should be mentioned that the values of current density given here are only representative values. The amount in which the current density can be pushed is entirely dependent on the type of cooling used. Generally speaking, in totally enclosed induction machines with no external cooling, aggressive designs can be made with current densities up to 5 A(rms)/mm^2. Machines with forced air cooling over the stator surface can tolerate a current density of 7.5–9 A(rms)/mm^2. When the air cooling is done through the use of ducts or vents embedded either axially or radially in the stator (and, perhaps rotor) laminations, the current density can be increased to about 14–15 A(rms)/mm^2. Finally, if liquid cooling is employed (using water, oil, WEG (water, ethylene–glycol), either in ducts or by spraying on the end windings) the current density can be increased to values reaching 20 A(rms)/mm^2 or greater. The value of surface current density is correspondingly increased in each case.

If equations (6.94) and (6.108) are used as representative values for air gap flux density and surface current, respectively, the pole pitch can be found as a function of output horsepower directly from equation (6.13) if the efficiency, power factor, and aspect ratio are assumed. Figure 6.20 shows the result if the efficiencies and power factors of Figures 6.6 and 6.7 are used and if the aspect ratio l_e/τ_p is taken as the dashed line of Figure 6.12.

It is interesting to observe how the rating of the machine increases if the rules set down for flux density and current density (equations 6.94 and 6.108) are observed. From equation (6.10), it is apparent that the volume of the machine increases with pole pitch squared so that if the pole pitch (i.e., diameter) is increased while the length is held constant then

$$T_e = \frac{P_{mech}}{\Omega_s} \propto \tau_p^{(2+1/6+2/3)} = \tau_p^{17/6} \tag{6.109}$$

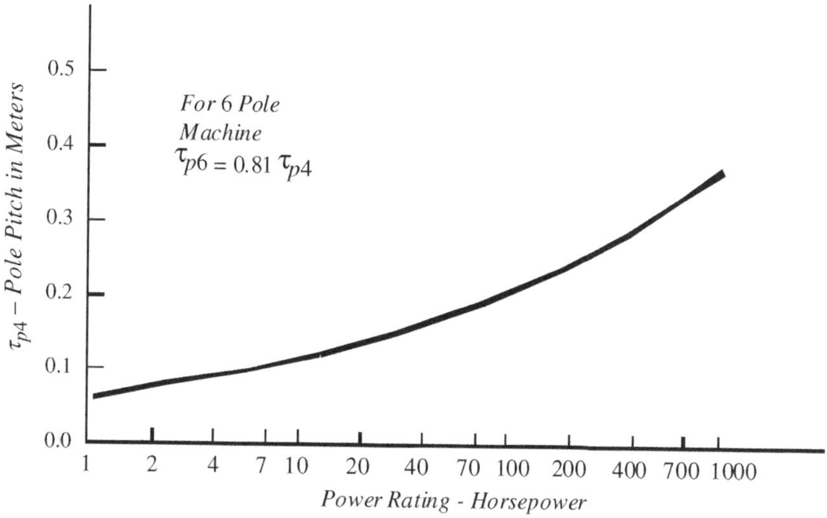

Figure 6.20 Estimate for pole pitch as a function of horsepower output.

or nearly as the cube of pole pitch or diameter. On the other hand, if the machine length is increased while the pole pitch (i.e., diameter) is held constant then since B_{g1} and $K_{s(rms)}$ remain constant,

$$T_e = \frac{P_{mech}}{\Omega_s} \propto l_i \tag{6.110}$$

so that a linear variation results. Equations (6.109) and (6.110) can be used as scaling rules in order to arrive at a satisfactory design.

6.10 PRACTICAL CONSIDERATIONS CONCERNING STATOR CONSTRUCTION

In general, two types of stator slots are in widespread use: open slot and semi-closed slot. An open slot, shown in Figure 6.21, is designed to be used with form-wound coils whereas the semi-closed slot is used with random-wound coil construction. The open slot construction is generally used on larger machines, approximately 100 HP or larger, particularly when the number of turns per coil is low or where the voltage is high, that is, 2300 V and up. Considerations for high voltages necessitate the application of ground insulation prior to insertion of the coil in the slot. Hence, the need for open slot construction. The open slot arrangement significantly increases the effective air gap, resulting in a smaller magnetizing reactance and a relatively lower power factor compared with the closed slot. The open slot construction may also accentuate the importance of slot and permeance harmonics and their influence on torque pulsations and stray losses. A further advantage of the open slot is that it permits use

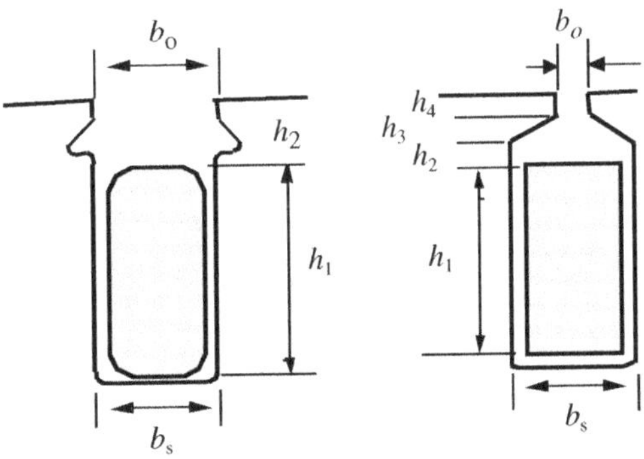

Figure 6.21 Open and semi-closed slot construction.

of a recessed wedge away from the air gap, thereby allowing an additional path for ventilating air.

The semi-closed slot construction is quite limited as to creepage or insulation thickness since it is approximately the thickness of the slot wedge to ground and also the thickness of the phase insulation between phases in the slot. Accordingly, semi-closed slot construction is seldom used for applications above 550 V. By virtue of its limited creepage in the slot, random-wound coils are usually not built with large creepages in the end turn region also. For example, slot tubes usually extend only 1/4″ beyond the end of the core and even this nominal value may be reduced if the tubes are not centered. Sometimes a cuff is used on the end of the tube which can be effective in providing some additional insulation. However, if moisture exposure is to be expected, random-wound motors of open construction are a risky application and it is better to use the TEFC design. The slot opening of the open slot bears an important relation to the insulated diameter of the wire to be used. In general, the opening should be over twice this diameter to avoid crossed wires from being damaged during insertion and to materially reduce the insertion time.

6.10.1 Random Wound vs. Formed Coil Windings

In addition to advantages of higher power factor and less stray losses, random windings are lower in cost. These coils are preformed to their approximate final shape and then inserted turn by turn into a slot ground wall insulator which has been previously inserted into the slot. In regard to material utilization, the amount of copper in the gross area of the slot is typically 30% for random-wound construction as compared to about 50% for form-wound machines of the same voltage rating and these numbers should be used for design purposes unless more accurate information concerning slot fill is available. The space advantage of form-wound coils is somewhat offset by poorer iron utilization in the teeth since parallel-sided slots result in tapered teeth with a resulting non-uniform flux density.

An additional reason for limiting the use of random winding to the smaller machine sizes arises from the difficulty in bracing the coil ends. The forces due to inrush currents or short-circuit currents may be quite high in larger machines. Tying alone is insufficient to support random windings, particularly in two-pole machine, since the motion of even one wire may result in loosening of the coil bundle. As a result, varnish dip must be relied upon to anchor the end turns rigidly. This, in turn, places a limit upon the time for which a random-wound machine may remain stalled before the overload protection removes power since the varnish softens as it becomes heated. Even with form-wound coils in large machines, bracing of the end windings is often a serious problem, particularly for machines with a small number of poles. In this case, the end-winding length is large and the ampere-turns per pole and consequent forces between coils are likely to be large. In addition to the usual interleaved tying which is done on all machines, additional bracing in the form of one or more supporting rings is often used to provide additional reinforcement for tying cords and to enhance structural rigidity. The cords themselves are glass fiber tape treated with permafil epoxy which is subsequently cured by baking.

Still another difference between random-wound and form-wound stators is in the method of connection. In the former, particularly in sizes up to 25 HP, an entire phase of a number of coils may be wound without cutting the wire. Such an assembly of coils, together with other assemblies for the other phases, may then be inserted coil by coil without the necessity of any wire cutting. The resulting wound stator is then free from many of the connections which must otherwise be made resulting in greater yield. On the other hand, wound coils require a separate connection operation for all the coils in the machine.

6.10.2 Delta vs. Wye Connection

Machines of both delta and wye or star connection are in widespread use. In general, the choice usually depends on which connection results in the optimum number of turns per coil. This consideration becomes more important as machines become larger since the turns per coil become smaller so that the options available from series or parallel connection of phase belts become restricted. Textbooks often make a point of the fact that third harmonic voltages are induced in a delta connection which can result in the corresponding currents circulating in the phases of a delta connection. While this is the case, careful choice of pitch factor and control of saturation can keep these losses to a negligibly small value. One major advantage of wye-connected machines, however, involves applications where the overload protection may be marginal. If a motor is stalled due to an open circuit of one line, the line current of a wye-connected machine will be the same as the equivalent delta-connected machine. However, in the delta-connected machine the phase across the line will have twice the current of the other two phases which are connected in series across the line. The net result is that the delta-connected machine will have the same losses in the three phases connected across the line than the equivalent wye connection has in its two active phases. However, the maximum current in the delta phase connected across the line will be 15% greater than the maximum current in either of the two remaining wye-connected phases. This will result in a rate of temperature rise in one of the delta phases which is 33% greater when compared to the wye-connected machine. Overload protection must be adjusted accordingly if this aspect is at all critical.

6.10.3 Lamination Insulation

Insulation between laminations is of considerable importance in induction machines as it is with any type of motor. In general, flux densities and frequencies are such that the voltages induced across the laminations are in the order of 1 1/2 volts per inch for two-pole, 60-Hz motors. Short core length motors, six inches length or less, can usually rely upon the oxide coating of the steel laminations which occur during the annealing process. However, when core lengths reach 12 inches or more, a more reliably controlled insulation system is warranted and core plate enamel should be considered. As to the rotor, the interlamination voltage is very small due to the small slip frequencies. However, insulation may still be necessary or harmonic fluxes will produce significant increases in load loss. For example, the effect of skewing the rotor slots will be essentially nullified if the insulation between laminations and that from the rotor bars to the laminations is not effective.

6.10.4 Selection of Stator Slot Number

The number of stator slots S_1 for a given size of punching is usually determined by the requirements of the motor including breakdown torque, starting current, allowable temperature rise, and so forth. With small motors and few poles, an integral number of slots per pole per phase is invariably chosen in order to maintain a balance between phases. The slot numbers 24, 36, 48, 54, 60, and 72 are very commonly used.

In general, it might be imagined that it is desirable to keep the number of slots as small as possible, since manufacturing costs clearly rise with the number of slots and therefore coils to be inserted. However, many factors tend to favor a large number of slots. First, and perhaps most important, is the fact that the breakdown torque varies inversely with the leakage reactance of the machine. Since the slot leakage, end winding leakage, and zigzag leakage reactances vary inversely with the number of stator slots, it is desirable to keep the number of stator slots large. In addition, a large number of slots permits the design of a winding distribution which has a small MMF harmonic content, thereby reducing the belt leakage as well as the stray-load losses. Finally, concentration of coils into a small number of slots results in a concentration of heat in these slots leading to cooling problems.

On the other hand, the number of slots cannot increase arbitrarily due to structural problems arising from narrow stator teeth. Also, it is more difficult to load the slots to the same current loading $K_{s(rms)}$ since the space lost to insulation, slot opening, packing factor, etc., increases as the slots become smaller. The thickness of the stator teeth also depends upon the breadth of the slot selected relative to the available slot pitch. Clearly, narrow teeth relative to the slot pitch tends to result in teeth which are highly saturated. In addition, problems with end turns arise since the angle α which defines the angle of departure of the end turns from the stack is directly related to the ratio of slot breadth to slot pitch (see Figure 4.21). As a result, the number of stator slots is usually chosen such that the thickness of the stator teeth lies between 1/4″ and 1″. The ratio of slot breadth to slot pitch generally lies in the range 0.4–0.6 with 0.5 taken as a recommended starting point.

6.10.5 Choice of Dimensions of Active Material for NEMA Designs

Although special designs may permit the choice of motor diameter and length purely from application considerations, the motor designer is more often restricted to designs which conform to the NEMA standard frame sizes. However, the NEMA standards only specify the overall length and width of the machine and shed little light on the working dimensions important to a machine designer, namely air gap diameter and active iron length. Tables 6.8 and 6.9 below can be used as a guide for selecting important motor-active dimensions which will allow for a frame, cooling fan, standard shafts, etc., and usually result in a machine of acceptable design. The duct lengths, if any, are not considered.

TABLE 6.8 Recommended diameters, maximum lengths, and air gaps for typical NEMA standard machines, dimensions in inches

Frame size	Poles	Maximum stator OD (D_{os})	Minimum rotor ID (D_{ir})	Maximum iron length (l_i)	Air gap
250	2	10 - 1/2	1 - 7/8	4 - 1/2	0.025
	4	10 - 1/2	1 - 7/8	5 - 1/4	0.017
	6 / 8	10 - 1/2	1 - 7/8	4	0.018
280	2	11 - 3/4	2 - 1/4	5 - 1/2	0.027
	4	11 - 3/4	2 - 1/4	6 - 1/2	0.018
	6 / 8	11 - 3/4	2 - 1/4	7	0.020
320	2	13 - 1/2	2 - 5/8	6	0.030
	4	13 - 1/2	2 - 5/8	8	0.020
	6 / 8	13 - 1/2	2 - 5/8	8	0.022
360	2	15 - 1/4	3 - 1/4	6 - 3/4	0.033
	4	15 - 1/4	3 - 1/4	7 - 1/4	0.028
	6 / 8	15 - 1/4	3 - 1/4	7 - 1/2	0.024

TABLE 6.9 Recommended diameters, maximum lengths, and air gaps for NEMA standard machines, dimensions in inches

Frame size	Poles	Maximum stator OD(D_{os})	Minimum rotor ID(D_{ir})	Maximum iron length (l_i)	Air gap
400	2	17	3 - 7/8	7 - 1/2	0.045
	4	17	3 - 7/8	8 - 1/4	0.032
	6 / 8	17	3 - 7/8	8 - 1/4	0.025
440	2	18 - 3/4	4 - 3/8	10 - 1/2	0.050
	4	18 - 3/4	4 - 3/8	14	0.035
	6 / 8	18 - 3/4	4 - 3/8	14	0.028

6.10.6 Selection of Wire Size

Even the largest motor manufacturing plant is confronted with an inventory problem so that the motor designer must frequently make imaginative use of the available stock in order to realize a good design. Frequently when the desired wire size is not available, the required cross-sectional area is achieved by using two or more conductors of the same or dissimilar diameters. Conductors used in parallel in this manner are usually called strands and the process of using such parallel conductors termed "wires in hand." While not all of the following wire diameters and insulation thicknesses may be available in a given situation, the following table, Table 6.5, can be considered as representative. The symbol AWG denotes a standard size (American Wire Gauge) but as can be observed, copper wire is not necessarily made only in standard AWG sizes. Heavy insulated wire is typically used with stators having B and F insulation up to approximately 100 HP, while quadruple-coated wire is used for stators of machines larger than 100 HP and in most wound rotors. Triple-coated wire is used in stators with class H insulation and in wound rotors with class H insulation. Glass fiber insulated wire is frequently used for high voltage (2300 V) machines. Classes A and B consist mainly of organic materials such as silk, cellulose, asbestos, and mica whereas classes F and H consist of inorganic materials such as fiberglass, Teflon, and silicon compounds.

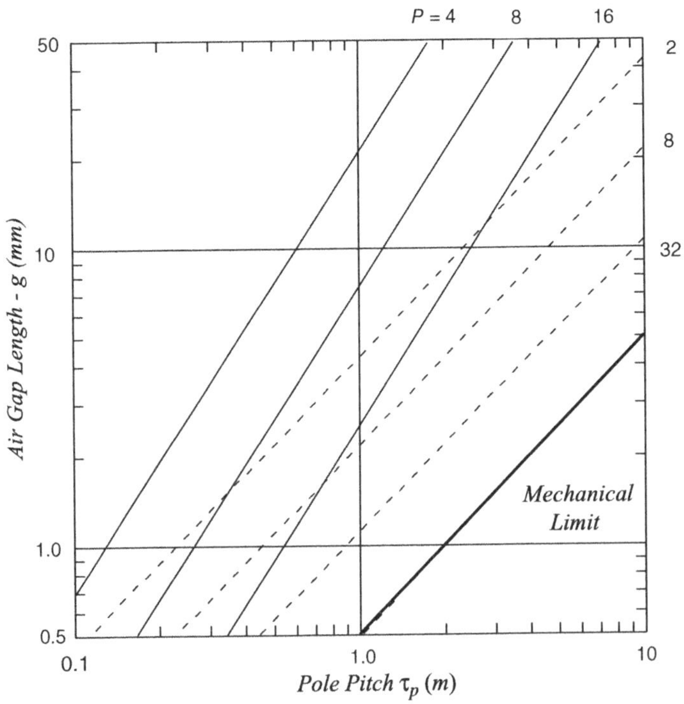

Figure 6.22 Suggested air gap length of squirrel-cage induction machines (dashed lines) and synchronous machine (solid lines) versus pole pitch τ_p and number of poles P [10].

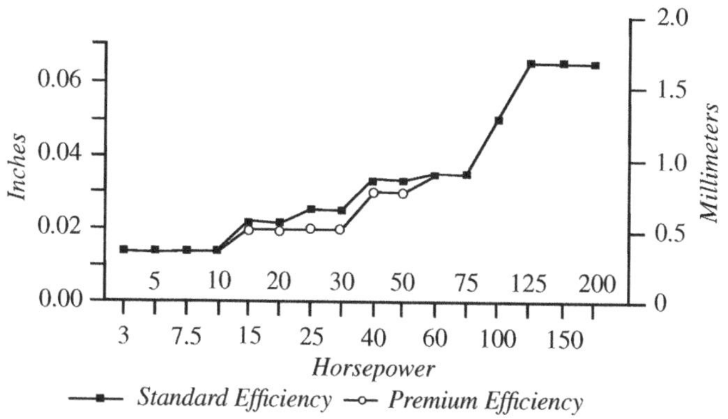

Figure 6.23 Air gap in inches and millimeters for standard and high-efficiency induction machines.

6.10.7 Selection of Air Gap

Figure 6.22 shows a plot of recommended air gaps for induction machines having poles between 2 and 16 [9]. Also shown is the mechanical limit below which it is not possible to maintain the required spacing between stator and rotor due to mechanical tolerances. Air gaps for high efficiency motors are often somewhat greater than for standard motors as shown, in order to reduce the stray loss. This often results in a slightly poorer power factor since the magnetizing inductance is necessarily reduced. The variation in the air gap as a function of pole pitch and number of poles can be approximated by [3]

$$g = 3 \times 10^{-3} \left(\sqrt{\frac{P}{2}} \right) \tau_{\text{pmm}}. \tag{6.111}$$

A plot showing air gaps of machines designed over a range from 3 to 200 HP is given in Figure 6.23.

6.11 ROTOR CONSTRUCTION

The major problem concerned with rotor construction centers around the selection of the number of rotor slots S_2. Literature on this subject abounds. In many cases, the rules are based on qualitative appraisals of the results of harmonic analysis, but when they are put to test the assumptions inherent in many of these analyses lose significance. It is unfortunate but true that it is extremely difficult to make quantitative calculations. Rules for the choice of rotor slot number can only be classified as good or bad when specifically related to some attribute of interest. Frequently, a choice categorized as good for one purpose is bad for another. The following information has been culled from the writings of Alger, Kuhlman, and Veinott [11–13].

6.11.1 Slot Combinations to Avoid

The following slot combinations have been shown to be noisy or to produce vibrations:

$$\begin{aligned} S_1 - S_2 &= \pm 2 \\ &= \pm(P \pm 1), \text{ and} \\ &= \pm(P \pm 2). \end{aligned}$$

The following slot combinations may have cusps in the torque versus speed curve:

$$\begin{aligned} S_1 - S_2 &= \pm P \\ &= -2P, \text{ or} \\ &= -5P \end{aligned}$$

The following may have a cogging problem which tends to make the motor "hang up" at zero speed:

$$S_1 - S_2 = 0 \quad \text{or} \quad = \pm mP$$

where m is an integer. Problems concerned with cogging indicate a problem concerned with a torque component, a function of angular position whereas cusps result from torques produced by a function of speed. Cusps, therefore, are essentially induction motor torques arising from a magnetic pole number different than the fundamental component (MMF harmonics) while cogging is the result of synchronous motor type torques arising from reluctance variation.

In general, if S_2 meets the criterion above and is divisible by the number of poles, motor noise is a minimum but some cogging difficulties may result. Quietness is also helped if S_2 differs from S_1 by 20% or more. If S_2 is larger than S_1 the rotor leakage inductance and resistance when referred to the stator are reduced and vice versa resulting in relatively high breakdown torque and starting current. On the other hand, the manufacturing cost increases with the number of rotor slots particularly for cast rotors. For cast rotors, or when rotor bar insulation may be a problem, stray-load losses can be reduced if S_2 is smaller than S_1. The amount of reduction, however, is only on the order of 15%. The most usual combinations found today in smaller sizes have $S_1 - S_2 = \pm 2P$ and the rotor slots are skewed one rotor slot in order to reduce the torque cusps inherent in this combination. A table showing desirable stator/rotor slot combinations for various pole numbers obtained from various sources is given in Table 6.10.

6.11.2 Rotor Heating During Starting or Under Stalled Conditions

It may be just as likely that limiting thermal stresses will be reached first in the rotor of the machine than in the stator. Foam-like holes in cast aluminum rotors may cause localized high current densities which may melt like a fuse under abnormal stalled

TABLE 6.10 Recommended stator/rotor slot combinations

Pole number	Stator/rotor slot number
2	36/26, 28, or 44 48/38, 40, or 56 54/46 60/52, 68, or 78
4	36/26, 28, or 44 48/34, 38, 40, or 56 60/34, 44, 46, or 76 72/58
6	36/46 or 48 48/58, 60, 64, or 68 54/42 or 66 72/54, 58, 84, or 88
8	36/48 or 52 48/58 or 64 54/70 72/58 or 88
10	72/58, 88, or 92
12	72/58 or 92

conditions. Double-cage rotors in particular, by the nature of their design, have most of the current flowing in the top bar of the cage which is typically a much smaller area than the bottom bar. Such windings are quite vulnerable to failure. Modern designs provide a thermal path for heat to easily flow from the top cage to the lower one by means of an integrally cast connection between the two windings as shown in Figure 6.24. Note that the connection between the top and bottom bars is "off center." By reversing rotor punchings every half inch or so, a poor electrical connection is made to exist axially in order to produce the beneficial effects of a double-cage machine. However, a good radial connection is maintained thermally in order to carry the heat to the bottom bar.

6.12 THE DESIGN PROCESS

This chapter has outlined a variety of approaches to arriving at an acceptable design of an induction machine. Because of the many constraints that may be imposed on

Figure 6.24 Asymmetric rotor slot construction for double-cage machine with good thermal heat capacity.

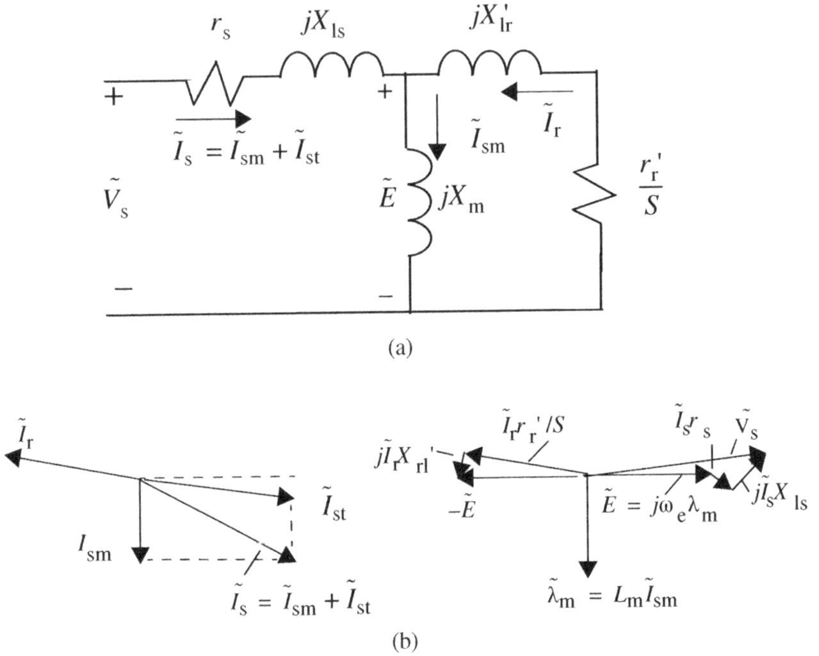

Figure 6.25 (a) Squirrel-cage induction motor equivalent circuit and (b) phasor diagrams for current and voltage.

a design, a generic approach which can cover all situations is clearly impossible. However, outlined below is one approach that has been used successfully if a limit on the outer diameter is imposed as is frequently the case.

Before commencing, consider first the classical equivalent circuit and corresponding phasor diagram of Figure 6.25. It is seen that the stator current is composed of two components, one of which magnetizes the iron in the machine while the other corresponds to the component producing torque. The current $\tilde{I}_{s,mag}$ is the circuit equivalent of the MMF $\mathcal{F}_{p1}$ of Chapter 2. The resulting voltage at the air gap $\tilde{E}$ produces the rotor current $\tilde{I}'_r$. The defined polarities of rotor current and air gap voltage result in the relationship

$$-\tilde{E} = \frac{\tilde{I}'_r r'_r}{S} + j\tilde{I}'_r X'_{lr} \tag{6.112}$$

so that the air gap flux produced by the rotor current $\tilde{I}'_r$ is completely cancelled by the equal and opposite components of stator current $\tilde{I}_{s,load}$. Thus the allowable surface current density K_s is effectively divided between a component $K_{s,mag}$ and a torque producing component $K_{s,load}$. These two components are essentially (but not quite) orthogonal. The torque-producing stator surface current is cancelled out by an equal (but opposite in phase) rotor surface current K_r.

Consider now that you are given the task of designing an induction motor, given the following characteristics: line voltage $V_{ll,(rms)}$, power output P_{out}, angular frequency ω_e, poles P and outer diameter D_{os}

1. Calculate the required electromagnetic torque from equation (6.15).

2. Assume initially that the slot width and tooth width are equal and that the stator tooth flux density B_t is limited to a prescribed value using the *B–H* curve for the stator lamination. A value of 1.6–1.8 is typical for low carbon steel. See Figure 1.16.

3. Calculate the peak fundamental air gap flux density. A simple but reasonably accurate value is to assume that the air gap flux density is the average of the stator slot and stator tooth flux density, that is, $B_{g1} = B_t(b_s/\tau_s) = 0.5\,B_t$. See Figure 6.18.

4. Determine the required value of linear current density K_s. See Figure 6.5 and Tables 6.6 and 6.7 Alternatively, select a value of shear stress σ_m and determine the corresponding value of K_s given the value of B_{g1} obtained in step 4 using equation (6.17). Typical values of shear stress range from 5–20 kPa with increasing values requiring increasingly demanding cooling.

5. Set K_s of step 5 equal to K_s^*, select a desired value of $J_{s(rms)}$ and solve for the optimal value of D_{is}/D_{os} using equation (6.77).

6. Determine if the value of D_{is}/D_{os} is feasible using equation (6.78). If not feasible, $J_{s(rms)}$ must be changed.

7. Knowing D_{os} and D_{is}/D_{os}, solve equation (6.62) for the effective length l_e. Check the aspect ratio $l_e/\tau_p = (Pl_e)/(\pi D_{is})$ to determine if the machine is excessively tubular. If so, the machine may be difficult to cool and the constraint on D_{os} relaxed.

8. Determine the MMF per pole $\mathcal{F}_{p1}$ needed to obtain the required peak flux density B_{g1} using the procedure of Chapter 3. A quick estimate of $\mathcal{F}_{p1}$ is to use equation (3.77) letting g_e equal 2–3 times the physical gap g.

9. Determine the magnetizing component of the stator surface current density K_{sm} corresponding to $\mathcal{F}_{p1}$. The correspondence is readily determined from equations (3.63) and (6.7) to be

$$K_{sm} = \frac{P}{D_{is}}\mathcal{F}_{p1}$$

10. Calculate the torque component of K_s, assuming that the two components are orthogonal

$$K_{st} = \sqrt{K_s^2 - K_{sm}^2}$$

11. Determine values of the permeances corresponding to the magnetizing inductance (equation 3.76) and the stator leakage inductance (equation (4.265) and sections of Chapter 4).

$$\mathcal{P}_m = \frac{L_{ms}}{N_s^2}$$

$$\mathcal{P}_{ls} = \frac{L_{ls}}{N_s^2}$$

12. Calculate the required number of turns to achieve a terminal voltage of V_{ll}. One method to accomplish this step is to first find the phase voltage $V_s = V_{ll}/(\sqrt{3})$ and then solve for the ratio

$$\frac{V_s}{V_m} = \frac{L_{ls}(I_{st} + I_{sm}) + L_{ms}I_{sm}}{L_{ms}I_{sm}}$$

or

$$= \frac{L_{ls}}{L_{ms}} + \frac{L_{ls}}{L_{ms}}\left(\frac{I_{st}}{I_{sm}}\right) + 1$$

to determine V_m. However, I_{st} and I_s are proportional to K_{st} and K_{sm} respectively and the ratio L_{ls}/L_{ms} is independent of the number of turns in which case

$$\frac{V_s}{V_m} = \frac{\mathcal{P}_{ls}}{\mathcal{P}_{ms}} + \frac{\mathcal{P}_{ls}}{\mathcal{P}_{ms}}\left(\frac{K_{st}}{K_{sm}}\right) + 1$$

The ratio of permeances can be calculated or simply estimated. A typical value is 0.15.

13. The air gap flux linkage λ_m corresponding to V_m is

$$\lambda_m = \frac{V_m}{\omega_e}$$

14. Now since

$$\lambda_m = L_{ms}I_{sm}$$

one has, from equations (3.76) and (3.87)

$$\lambda_m = \left[\left(\frac{3}{2}\right)\left(\frac{8}{\pi^2}\right)\frac{k_1^2 N_s^2}{P}(\tau_p l_e)\left(\frac{B_{g1}}{\mathcal{F}_{p1}}\right)\right]\left[\frac{\mathcal{F}_{p1}}{\left(\frac{3}{2}\right)\left(\frac{4}{\pi}\right)\left(\frac{k_1 N_s}{P}\right)}\right]$$

which becomes ultimately

$$\lambda_m = \left[(2)(k_1 N_s)\left(\frac{D_{is} l_e}{P}\right)(B_{g1})\right]$$

This result is then used to solve for the necessary number of series-connected turns N_s. The entire stator design can now be completed with attention given to winding layout, number of circuits, number of parallel strands per turn, etc., to achieve a good design.

15. To begin the design of the rotor, it must be recalled that the load component of stator ampere-turns are cancelled by an equivalent but oppositely directed component of rotor ampere-turns, thus

$$K_r = K_{st}$$

16. The rotor inductance and resistance referred to the stator are given by equations (4.240) and (4.246) in terms of the bar shape and slot number which must now be decided. Since

$$\frac{\lambda_{\mathrm{lr}}}{\lambda_{\mathrm{m}}} = \frac{L_{\mathrm{lr}} I_{\mathrm{r}}}{L_{\mathrm{ms}} I_{\mathrm{sm}}} = \left(\frac{L'_{\mathrm{lr}}}{L_{\mathrm{ms}}}\right)\left(\frac{K_{\mathrm{st}}}{K_{\mathrm{sm}}}\right)$$

this equation can be used to solve for the rotor leakage flux linkage using the results of step 14 and consequently the voltage drop across the rotor leakage inductance

$$V_{\mathrm{lr}} = \omega_{\mathrm{e}} \lambda_{\mathrm{lr}}$$

The ratio $L'_{\mathrm{lr}}/L_{\mathrm{ms}}$ typically lies between 0.1 and 0.2.

17. The portion of the air gap voltage devoted to producing torque is then

$$V_{\mathrm{r}} = \sqrt{V_{\mathrm{m}}^2 - V_{\mathrm{lr}}^2}$$

18. The torque is consequently

$$T_{\mathrm{e}} = \frac{3P}{4} \frac{V_{\mathrm{r}}^2}{r'_{\mathrm{r}}} S$$

This expression can be used to calculate the value of slip needed to produce the required torque given an estimated value of r'_{r} given by equation (4.246). The rotor I^2R loss is, of course

$$P_{\mathrm{r,loss}} = \frac{3V_{\mathrm{r}}^2}{2r'_{\mathrm{r}}}$$

and if this value is too large, the rotor bar shape must be adjusted to reduce the loss and the slip to the values desired. This can be done by repeating steps 17–18.

6.13 EFFECT OF MACHINE PERFORMANCE BY A CHANGE IN DIMENSION

Because of the considerable effort involved, design of a line of machines is often accomplished by a detailed design of one machine, then a design of similar machines but different rating by simple scaling. Consider two machines of the same design with all dimensions in the ratio k:1 and having the same speed, flux density, and current density.

1. Weight. Clearly since the linear dimension of both the iron and copper parts has increased by k the weight of the second machine will increase by k^3.

2. Terminal Voltage. The core area, the flux per pole increases by k^2, and therefore if the number of turns is not changed, the voltage will increase by a factor of k^2. In most cases, the number of turns must be changed to satisfy a given terminal voltage constraint. In this case the number of stator turns must decrease inversely by k^2.

3. Load Current. The current-carrying capacity of the machine, assuming adequate ventilation increases in proportion to the area of the conductors, or by a factor of k^2. When the stator turns are reduced by k^2 and the amount of copper is kept the same, the current-carrying capacity of the machine increases by a factor of k^4.

4. Input KVA. Since the current and voltage increase as k^2 the power input will increase by a factor of k^4. This factor remains the same when the number of turns are adjusted to keep the voltage rating the same.

5. Resistance. The copper cross section will increase by a factor of k^2 when the turns are not changed. However, the conductor length will increase by k, so that the resistance will decrease by a factor of k. When the number of turns are reduced by k^2 the cross section will increase by an overall factor of $k^2 * k^2 = k^4$ while the conductor length will change by an overall $k/k^2 = 1/k$. Hence, when the voltage rating of the new machine is held constant the resistance will decrease by a factor of k^5.

6. Copper Loss. Since the current squared goes up by k^4 and the resistance goes down by k, the copper loss will increase by a factor of k^3. When the voltage is maintained constant, the current squared goes up by k^8 while the resistance goes down by k^5, making the copper loss increase by the same factor of k^3.

7. Iron Loss. Since the flux density in the iron remains constant and the volume of iron increases by k^3, the iron loss will also increase by a factor of k^3. The iron loss will not change with changes in the number of turns if the resulting flux density remains unchanged.

8. Power Output. Since power input increases by k^4 and the losses increase by k^3 the power output increases as $k^4 - ak^3$ where a is the per unit losses at the design $k = 1$. Hence, the power output increases at a rate slightly less than k^4. Note that this is essentially the same result as would be indicated by equation (6.109) if both τ_p and l_i are changed by the same amount.

9. Efficiency. Since the efficiency is the ratio of power output to power input, then efficiency varies as $(k^4 - ak^3)/k^4 = 1 - a/k$. Hence, as k becomes increasingly large the efficiency approaches unity. This result clearly shows that, with electric and magnetic loading held constant, an increase in size will always improve the efficiency of the machine. This also explains in part why fractional horsepower machines have an efficiency on the order of 60% while the efficiency of a large turbo-alternator is about 98%. Note that efficiency and power output remains nominally unchanged with a change in the number of stator turns.

10. Inductance. Since the area of the magnetic circuit will increase by the factor k^2 while its length increases by k, the inductances of the machine increase by k whereas the resistances have been shown to decrease by k when the number of turns are held constant. When the stator turns are changed by $1/k^2$ to keep the rated voltage constant, the inductances of the machine will decrease by k^3 rather than increase by a factor of k.

11. Magnetizing Current. If the gap flux density is kept constant the MMF required to drive the flux across the air gap is proportional to the gap g. Thus the magnetizing MMF varies with k whereas the load current varies with k^2 so that the magnetizing current as a percentage of load current decreases with $1/k$. This helps explain why induction machines get better as they get larger. It also helps explain why permanent magnet motors become less desirable as the size of the machine becomes larger since the magnetizing current penalty which affects the no-load losses of the induction machine become relatively unimportant as the machine becomes large.

12. Stored Magnetic Energy. Since magnetizing current increases linearly with the dimension and the inductance also increases to the first power, the stored energy increases with k^3. The stored energy remains unchanged with a modification in the number of turns. On the other hand, the load current increases with the k^2 so that the stored energy associated with the leakage field increases as k^5. This points to the problem involved with the protection of large machines, particularly during faults.

13. Power Factor. Neglecting the effect of leakage inductance, the power factor can be estimated from the ratio $I_{\text{load}}/(\sqrt{I_{\text{load}}^2 + I_{\text{mag}}^2})$. Since I_{load} varies as k^2 and I_{mag} as k the power factor varies as

$$pf = \frac{k^2}{\sqrt{k^4 + k^2}} = \frac{k}{\sqrt{1 + k^2}}$$

which slowly increases as the size increases.

14. Time Constants. Since the resistance decreases and the inductance increases, the electrical time constants of the machine (transient, subtransient, etc.) will increase by a factor k^2. When the stator turns ratio is changed and the amount of copper is kept constant, the electrical time constant remains changed by the same factor, k^2. This fact has an important bearing on the transient behavior of a given machine and explains why a large machine is more difficult to control than a small machine.

15. Breakdown Torque. Since the voltage increases as k^2 and the inductance as k, the breakdown torque increases as k^3. However, since power output increases nearly as k^4, this points to a reduction in the overload capability of the machine. This conclusion indicates that simple scaling will ultimately lead to an unsatisfactory design if k is too large. The problem remains the same if the stator turns are changed.

Control of leakage reactance is an important aspect of any type of machine design and, in general, means must be found to maintain the leakage inductance constant as the size of the machine increases. This is typically accomplished by increasing the number of slots in a manner roughly proportional to the air gap diameter.

16. Power Density. Since the volume increases by k^3 and the power input increases by k^4 the power density increases by k. This result indicates that large machines are more effective in utilizing the available magnetic material and explains the trend toward ever larger turbo-generators.

17. Temperature. Since the losses increase as k^3 whereas the available surface area for cooling increases only as k^2, the temperature will increase by a factor k. Hence, a simple scale of all dimensions by k will not necessarily result in a satisfactory thermal design since the temperature limits of the machine may be exceeded. This result points to the need for improved cooling techniques as the ratings of the machines increase.

TABLE 6.11 The trend of motor quantities as a function of l and D with constant B and J

Parameter	Symbol and/or proportionality	Constant B, J, and N	Constant B, J, and V
Voltage	$V \propto N \cdot D \cdot l$	$l \cdot D$	1
Turns	N	1	$1/(l \cdot D)$
Current	$I \propto D^2/N$	D^2	$l \cdot D^3$
Power	$V \cdot I$	$l \cdot D^3$	$l \cdot D^3$
Resistance	$R \propto N^2 \cdot l/D^2$	$l \cdot D^2$	$1/(l \cdot D^4)$
Losses	I^2R	$l \cdot D^2$	$l \cdot D^2$
Power out	$V \cdot I - I^2R$	$l \cdot (D^3 - aD^2)$	$l \cdot (D^3 - aD^2)$
Magnetizing inductance	$L_m \propto N^2 \cdot A_m/l_{mpath}$	l	$1/(l \cdot D^2)$
Leakage inductance	$L_l \propto N^2 \cdot A_l/l_{lpath}$	l	$1/(l \cdot D^2)$
Magnetizing current	$I_m \propto V/L_m$	D	$l \cdot D^2$
Magnetizing field energy	$W_m \propto L_m \cdot I_m^2$	$l \cdot D^2$	$l \cdot D^2$
Leakage field energy	$W_l \propto L_l \cdot I^2$	$l \cdot D^4$	$l \cdot D^4$
Electrical time constant	$\tau_e \propto L_m/R$	D^2	D^2
Breakdown torque	$T_{pk} \propto V^2/L_l$	$l \cdot D^2$	$l \cdot D^2$
Power density	$(VA)/(Vol)$	D	D
Temperature rise	$\Theta \propto (I^2R)/A_{surf}$	D	D
Power factor	I/I_m	D	D

Frequently design modifications involve changing only the length or the diameter of a previous design but not both. Table 6.11 shows how the same parameters, previously discussed, vary explicitly with length l and diameter D when N is held constant and when N is changed to keep the terminal voltage constant.

6.14 CONCLUSION

This chapter has served as an introduction to the complexity of producing a machine design which meets all of its performance requirements. The reader is cautioned that the barest minimum of issues has been dealt with here. Many other factors are of importance including audible noise, mechanical resonances, rotor rotational stresses, critical mechanical speeds, manufacturing tolerances etc., not to mention thermal issues which have been delayed until the next chapter. The task of a machine designer is obviously a challenging and exacting profession but clearly an intellectually rewarding one as well. Hopefully, this text will contribute to sending at least a few of its readers down its rewarding path.

REFERENCES

[1] W. B. Essen, "Notes on the design of multipolar dynamos," *Journal of the Institution of Electrical Engineers*, vol. 20, no. 93, 1891, pp. 265–293.

[2] M. Liwschitz-Garik and C. C. Whipple, *Electric Machinery*, 2nd edition, vol. 2, Van Nostrand Publishers, 1961.

[3] S. Huang, J. Luo, F. Leonardi, and T. A. Lipo, "A general approach to sizing and power density equations for comparison of electrical machines," *IEEE Transactions on Industry Applications*, vol. 34, no. 1, January/February 1998, pp. 92–97.

[4] P. Pillay, "Applying energy-efficient motors in the petrochemical industry," *IEEE Industry Applications Magazine*, vol. 3, no. 1, January/February 1997, pp. 32–40.

[5] V. B. Honsinger, "Sizing equations for electrical machinery," *IEEE Transactions on Energy Conversion*, vol. EC-2, no. 1, March 1987, pp. 116–121.

[6] J. C. Hsu and A. U. Meyer, *Modern Control Principles and Applications*, McGraw-Hill Book Co., New York, 1968.

[7] A. H. Bonnett, "Quality and reliability of energy-efficient motors," *IEEE Industry Applications Magazine*, vol. 3, no. 1, January/February 1997, pp. 22–31.

[8] D. Van Son, and J. C. Kauffman, "Inverter-fed motors improved with transient resistant windings," *Advanced Motor Drive News*, vol. 3, no. 3, Summer 1996, pp. 4–5.

[9] T. Sebastian, G. R. Slemon, and M. A. Rahman, "Modelling of permanent magnet synchronous motors," *IEEE Transactions on Magnetics*, vol. MAG-22, no. 5, September 1986, pp. 1069–1071.

[10] H. De Jong, *AC Motor Design: Rotating Magnetic Fields in a Changing Environment*, Hemisphere Publishing Co., 1989.

[11] P. L. Alger, *Induction Machines: Their Behavior and Uses*, 2nd edition, Gordon and Breach Science Publishers, New York, 1965.

[12] J. H. Kuhlman, *Design of Electrical Apparatus*, 2nd edition, John Wiley and Sons, New York, 1940.

[13] C. G. Veinott, *Theory and Design of Small Induction Motors*, McGraw-Hill Book Co., New York, 1959.

CHAPTER 7

THERMAL DESIGN

AN ELECTRIC MACHINE IS not only a complex three-dimensional electromagnetic structure but also a complex spatial fluid dynamic structure with cooling gas (typically air) streaming through its rotating and stationary parts taking out heat from spatially distributed heat sources. The three-dimensional heat flow in a machine is determined not only by the thermal conductivity of electric conductors, iron, and insulating materials, but also by spatially distributed contact surfaces in the machine, upon which the heat exchange takes place.

Knowledge of the temperature distribution in an electric machine is extremely important for its design. This is especially the case when the machine is designed to operate at or near its thermal limits. An optimized thermal design can help increase machine-rated power substantially, almost without any increase of its manufacturing costs. Lately, finite element programs have been successfully used for three-dimensional gas flow and thermal computations. However, classical methods of thermal design are still used in everyday practice and remain a good starting point.

7.1 THE THERMAL PROBLEM

The process of electromechanical energy conversion in an electric machine is inevitably accompanied by a unidirectional transformation of electrical and mechanical energy into thermal energy. Typical current densities vary from 2.5 to 15 A/mm^2 and these current densities result in volume loss densities from 0.15 up to 5 MW/m^3. Losses in the machine inevitably increase the temperature of all portions of the machine. The purpose of cooling is to limit this temperature increase by removing this heat. The temperature rise in the machine, especially in its windings, must be limited in order to realize the expected lifetime of the machine. An elevated temperature deteriorates the mechanical and electrical strength of the winding insulation, and will lead to its premature failure. Also, extreme time and/or spatial temperature cycling can introduce unacceptable motion between conductors in the slots and iron, resulting in separation of the insulation from the conductors and its consequent destruction.

To determine the temperature distribution in an electric machine, one must have a good estimate of the machine electrical (losses) properties of the cooling fluid and thermal characteristics of the magnetic, conductive, and insulating portions of

Introduction to AC Machine Design, First Edition. Thomas A. Lipo.

the machines. First, the distribution of heat sources has to be found based upon the machine geometry and dissipated losses. The cooling gas flow (particularly the gas velocities at various portions of the machine) then must be determined as a function of the fan capability, gas thermal properties, and machine geometry. Gas velocities obtained in this manner are used to determine the heat convection coefficients on gas–iron and gas–conductor contact surfaces in the machine. The heat convection and conduction coefficients along with the heat source distribution define finally the temperature distribution in the machine.

7.2 TEMPERATURE LIMITS AND MAXIMUM TEMPERATURE RISE

The life of an insulation system is strongly temperature dependent and even the best insulation will fail quickly if operated sufficiently far from its rated value. Figure 7.1 shows how thermal aging affects the expected life for the four insulation classes. These values were derived from the fact that winding insulation life is doubled for every 10 °C reduction in average operating temperature. The nominal life of 20,000 hours is generally used to define the expected operating temperature of the insulation classes, making the temperature hot-spot "ratings" of insulation classes *A*, *B*, *F*, and *H* to be 105 °C, 130 °C, 155 °C, and 180 °C, respectively as given in Table 7.1

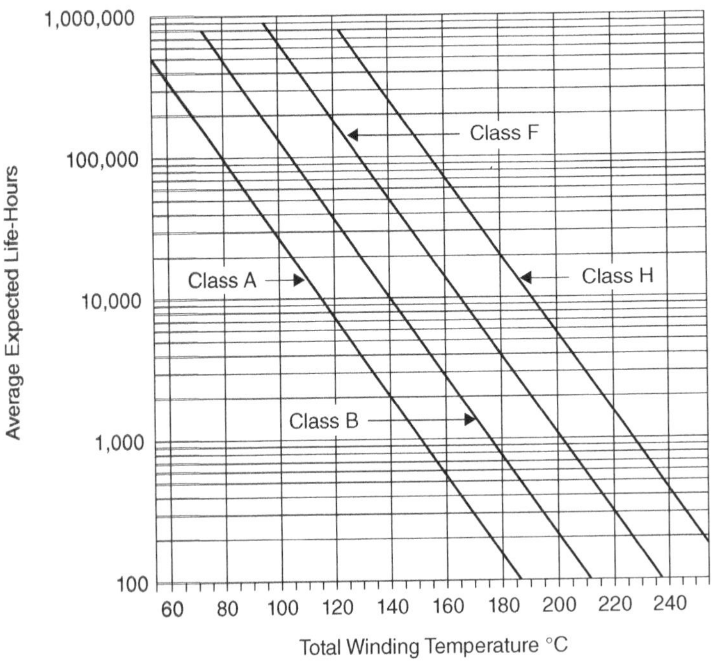

Figure 7.1 Temperature versus. life curves for insulation systems (per IEEE 177 and 101 standards).

TABLE 7.1 Maximum temperature at rated load for drip proof and totally enclosed induction motors

Class A insulation	105 °C
Class B insulation	130 °C
Class F insulation	155 °C
Class H insulation	180 °C

The winding insulation has direct contact with hottest parts of the machine, the current-carrying conductors. The temperature of a winding can be determined based upon measurements of its electrical resistance. The resistance measurement gives, however, only an average winding temperature. In order to estimate the hot-spot temperature (the highest possible insulation temperature), the measured winding temperature has to be increased. Recommended values are shown in Table 7.2. (Temperature differences in °C and °K are identical.)

7.3 HEAT CONDUCTION

Heat conduction is characterized by heat flow from a region with higher temperature to a region with lower temperature. The steady-state temperature distribution $\Theta(x, y, z)$ is, in general, defined by the heat diffusion partial differential equation or simply the *heat equation* [1]

$$\nabla \bullet (\nabla \Lambda \Theta) + q_{\text{h}} = \rho_{\text{m}} c_{\text{p}} \frac{\partial \Theta}{\partial t} \tag{7.1}$$

where Θ is the temperature in °C or °K, and q_{h} represents the heat generation rate in watts/m^3. Also, Λ is the thermal conductivity in watts/(meter-°K). ρ_{m} is the density in kg/m^3 and c_{p} is the specific heat in joules/(kg-°K). In the steady state, assuming Cartesian coordinates and anisotropic media, equation (7.1) can be written

$$\Lambda_x \frac{\partial^2 \Theta}{\partial x^2} + \Lambda_y \frac{\partial^2 \Theta}{\partial y^2} + \Lambda_z \frac{\partial^2 \Theta}{\partial z^2} = -q_{\text{h}}(x, y, z) \tag{7.2}$$

with Λx, Λy, and Λz denoting thermal conductivities in x, y, and z directions, respectively. For isotropic media ($\Lambda x = \Lambda y = \Lambda z$) and for problems without heat sources ($q_{\text{h}} = 0$), equation (7.2) can be written as

$$\frac{\partial^2 \Theta}{\partial x^2} + \frac{\partial^2 \Theta}{\partial x^2} + \frac{\partial^2 \Theta}{\partial x^2} = 0 \tag{7.3}$$

Equation (7.3) is the Laplace differential equation, in which $\nabla \bullet \nabla \Theta = 0$.

TABLE 7.2 Temperature increase for hot-spot temperature estimation

Insulation class	A	B	F	H
$\Delta\Theta$ [°C or °K]	5	10	15	15

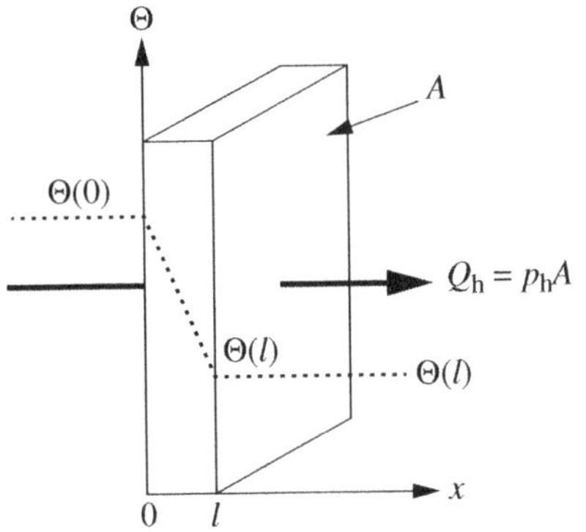

Figure 7.2 One-dimensional heat flow through a plate with thickness l.

If no heat is internally generated in the body then Laplace's Equation can be solved by setting the gradient $\nabla\Theta$ to a constant. This constant can be expressed in terms of a heat flow gradient p_{h}, which corresponds to the rate at which heat is transferred across a surface per unit area. One can write, for isotropic media,

$$p_{\mathrm{h}} = -\Lambda\nabla\Theta \tag{7.4}$$

For a simple one-dimensional heat distribution, one can further write

$$p_{\mathrm{hx}} = -\Lambda_x \frac{\partial\Theta}{\partial x} \tag{7.5}$$

which can be illustrated by example of a sourceless heat flow through a plate with thickness l and cross-sectional surface area A (Figure 7.2).

7.3.1 Simple Heat Conduction Through a Rectangular Plate

Discarding the redundant subscript x, for the simple plate in Figure 7.2 one can write equation (7.5) as

$$p_{\mathrm{h}} = \frac{Q_{\mathrm{h}}}{A} = -\Lambda\frac{\partial\Theta}{\partial x} \tag{7.6}$$

or,

$$\frac{\partial\Theta}{\partial x} = -\frac{Q_{\mathrm{h}}}{A\Lambda} \tag{7.7}$$

The solution of this first-order differential equation is

$$\Theta(x) = -\frac{Q_{\mathrm{h}}}{A\Lambda}x + C \tag{7.8}$$

where constant C is determined from the boundary condition $x = 0$ resulting in $\Theta(x) = \Theta(0)$, so that $C = \Theta(0)$. The temperature in a plate decreases in proportion to the distance from the side of the plate at which the heat enters.

TABLE 7.3 Analogies between heat flow and current flow

Electric circuit	Heat conduction
Electric current in a resistive load $$I = \frac{VA\sigma}{l} = \frac{V}{R}$$	Heat flow $$Q_h = \frac{\Theta A\Lambda}{l} = \frac{\Theta}{R_h}$$
Current I (A)	Heat flow Q_h(W)
Voltage V (V)	Temperature Θ(°K)
Area of conductor A (m^2)	Area of heat path A (m^2)
Length of conductor l (m)	Length of heat path l (m)
Electrical conductivity $\sigma[A/(V-m)]$	Thermal conductivity $\Lambda[W/(°K-m)]$
Electrical resistance $R = \frac{l}{\sigma A}(V/A)$	Thermal resistance $R_h = \frac{l}{\Lambda A}(°K/W)$
Electrical capacitance $C = \varepsilon A/d(A-s/V)$	Thermal capacitance $C_h = \rho_m c_p(W-s/°K)$
Current density $$J = \frac{I}{A}$$	Heat flow density $$p_h = \frac{Q_h}{A}$$
Current density J (A/m^2)	Heat flow density p_h(W/m^2)

The temperature drop over the plate $\Delta\Theta = \Theta(0) - \Theta(l)$ can now be expressed as

$$\Delta\Theta = \frac{l}{A\Lambda}Q_h \tag{7.9}$$

Defining the heat resistance R_h of the plate as

$$R_h = \frac{l}{A\Lambda} \tag{7.10}$$

one obtains Ohm's law for heat conduction as

$$\Delta\Theta = R_h Q_h \tag{7.11}$$

The heat conduction law in Cartesian coordinates is fully analogous to Ohm's law for electric currents as illustrated in Table 7.3.

7.3.2 Heat Conduction Through a Cylinder

Being essentially a cylindrical body, heat flow in most electrical machines takes place both axially and radially. Typical radial heat flow occurs through the supporting structure, for example, the stator frame. When no heat is generated in the component itself, then the heat flow can again be considered simply as a gradient. In cylindrical coordinates, the heat flow equation in the radial direction is

$$\frac{1}{r}\frac{\partial}{\partial r}\left(\Lambda_r r\frac{\partial\Theta}{\partial r}\right) = 0 \tag{7.12}$$

The general solution is of the form

$$\Theta(r) = C_1 \ln(r) + C_2 \tag{7.13}$$

Applying the boundary conditions to the solution

$$\Theta_i = C_1 \ln\left(\frac{D_i}{2}\right) + C_2 \tag{7.14}$$

and

$$\Theta_o = C_1 \ln\left(\frac{D_o}{2}\right) + C_2 \tag{7.15}$$

where Θ_i and Θ_o are known temperatures of the inner and outer surfaces, whereupon, solving for C_1 and C_2, the resultant distribution can be written as

$$\Theta(r) = \frac{\Theta_o - \Theta_i}{\ln\left(\frac{D_o}{D_i}\right)} \ln\left(\frac{2r}{D_o}\right) + \Theta_o \tag{7.16}$$

The heat flow per unit area for radial heat flow in cylindrical coordinates is

$$\Lambda\nabla\Theta = \Lambda_r\left(\frac{\partial\Theta}{\partial r}\right) = p_h \tag{7.17}$$

Performing the differentiation and inserting into equation (7.17) results in

$$\Lambda_r\left(\frac{\Theta_o - \Theta_i}{\ln\left(\frac{D_o}{D_i}\right)}\right)\frac{1}{r} = p_h \tag{7.18}$$

The total heat flow through any cylindrical area between $r = D_i/2$ and $r = D_o/2$ must be the total heat flow Q_h. Hence,

$$Q_h = p_h 2\pi r l = \Lambda_r\left(\frac{\Theta_o - \Theta_i}{\ln\left(\frac{D_o}{D_i}\right)}\right)2\pi l \tag{7.19}$$

If one defines the thermal resistance for radial heat flow in cylindrical geometry then

$$R_{h,rad} = \frac{\ln\left(\frac{D_o}{D_i}\right)}{\Lambda_r(2\pi l)} \tag{7.20}$$

Equation (7.19) again takes on the form

$$Q_h = \frac{\Theta_o - \Theta_i}{R_{h,rad}} \tag{7.21}$$

When heat flows in the axial direction, the cross-sectional area through which the heat flows is

$$A = \pi\left(\frac{D_o^2}{4} - \frac{D_i^2}{4}\right) \tag{7.22}$$

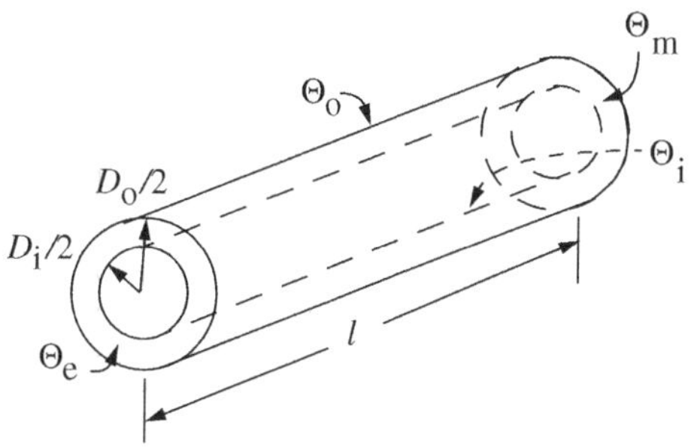

Figure 7.3 General cylindrical component.

and the temperature change in axial direction is simply

$$\Delta\Theta = \Theta_{\mathrm{m}} - \Theta_{\mathrm{e}} = \frac{4Q_{\mathrm{h}}}{\Lambda_{\mathrm{ax}}} \frac{l}{\pi\left(D_{\mathrm{o}}^2 - D_{\mathrm{i}}^2\right)} \tag{7.23}$$

where Θ_{m} and Θ_{e} denote the temperatures at the two ends of the cylinder. In this case, the thermal resistance to heat flow in the axial direction is

$$R_{\mathrm{h,ax}} = \frac{4l}{\Lambda_{\mathrm{ax}}\pi\left(D_{\mathrm{o}}^2 - D_{\mathrm{i}}^2\right)} \tag{7.24}$$

7.3.3 Heat Conduction with Simple Internal Heat Generation

In most parts of the machine heat is generated within the material itself. For example consider an iron core with eddy current and hysteresis losses, cooled on both sides as shown in Figure 7.4. With only one-dimensional variation of the temperature, the

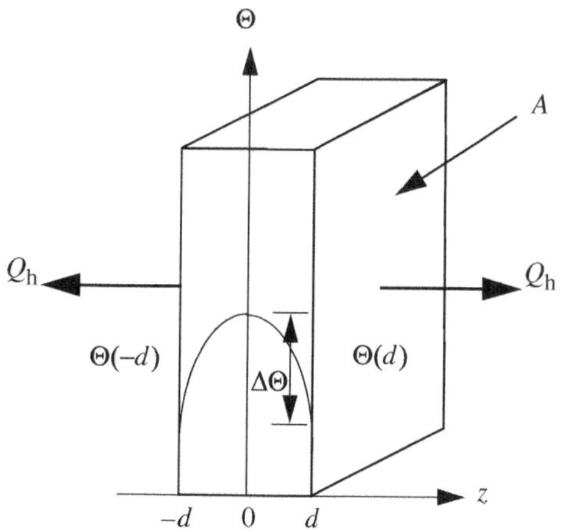

Figure 7.4 Dissipative body cooled on both sides.

Poisson partial differential equation can be written as

$$\Lambda_z \frac{\partial^2 \Theta}{\partial z^2} = -q_\mathrm{h} \tag{7.25}$$

where the heat density q_h has the unit watts per cubic meter. The direction "z" typically represents heat flow in the axial direction of an electrical machine.

Upon integrating twice results in

$$\Theta(z) = -\frac{q_\mathrm{h}}{2\Lambda_z} z^2 + C_1 z + C_2 \tag{7.26}$$

at $z = 0$ it is assumed that $\Theta = \Theta_\mathrm{i}$, so that

$$C_2 = \Theta_\mathrm{i}$$

Also it is assumed that there is no heat flow through the surface at $z = 0$ so that, from equation (7.5)

$$-\Lambda_z \frac{\partial}{\partial z}\Theta(z) = p_{\mathrm{h},z}(z) \tag{7.27}$$

or, from equation (7.26),

$$\Lambda_z \left(\frac{q_\mathrm{h}}{\Lambda_z} z - C_1 \right) = p_{\mathrm{h},z}(z) \tag{7.28}$$

At $z = 0$, $p_{\mathrm{h},z}(0) = 0$, so that clearly,

$$C_1 = 0$$

The solution becomes

$$\Theta(z) = -\frac{q_\mathrm{h}}{2\Lambda_z} z^2 + \Theta_\mathrm{i} \tag{7.29}$$

Since the temperature on the left side of the plate is equal to the temperature on its right side, that is, $\Theta(d) = \Theta(-d)$, one obtains for temperature distribution in the plate

$$\Theta(z) = \Theta(d) + \frac{q_\mathrm{h}}{2\Lambda_z}(d^2 - z^2) \tag{7.30}$$

The amount of rise for the highest temperature (at $z = 0$) is

$$\Delta\Theta = \frac{q_\mathrm{h}}{2\Lambda_z} d^2 \tag{7.31}$$

Since the result is the same for axial heat flow in a cylinder, this result can be used to determine the heat rise in the center of the machine with respect to the end-winding portion.

The power loss per unit volume associated with the plate can be written as

$$q_\mathrm{h} = \frac{Q_\mathrm{p}}{V_\mathrm{p}} = \frac{Q_\mathrm{p}}{A_\mathrm{p} 2d} \tag{7.32}$$

where V_p is the volume of the plate and Q_p is the total heat loss in the plate. Thus the temperature drop from the middle of the plate to the end can be expressed as

$$\Delta\Theta_s = Q_p\left(\frac{d}{4\Lambda_z A_p}\right) \tag{7.33}$$

where Λ_z denotes the thermal conductivity in the z-direction. Since $2d = l$, the total length of the plate is

$$\Delta\Theta_s = Q_p\left(\frac{l}{8\Lambda_z A_p}\right) \tag{7.34}$$

or, simply

$$\Delta\Theta_s = Q_p R_p \tag{7.35}$$

where R_p is a thermal resistance for the plate given by

$$R_{s,cu} = \frac{l}{8\Lambda_z A_p} \tag{7.36}$$

A similar procedure can be used to determine a temperature drop and thermal resistance for any element of the machine with heating in which the geometry does not change in the axial (z) direction and where no external heat inputs into the body exist. For example, the conductor losses in both the stator winding and the rotor cage can be computed by setting A_p equal to the cross-sectional areas of a stator coil or rotor bar. Also, this equation can be used to determine the temperature drop along the stator and rotor teeth and core in the axial direction when no heat enters externally. In the case of the stator core, the area A_p corresponds to the annular cross-sectional area of the core, $(\pi/4)(D_{os}^2 - D_{bs}^2)$, where D_{os} and D_{bs} denotes the outer diameter of the cores and an inner diameter at the bottom of the stator slot, respectively.

7.3.4 Example 9—Stator Winding Heating

A small electric machine is indirectly cooled in such a manner that the generated heat flows axially through the stator from the middle of the machine toward its ends, where the heat exchange takes place. The winding temperature rise between the middle of the machine and its end winding can be calculated knowing the following data:

- current density in the winding: $J = 4$ A/mm^2;
- machine axial length of iron: $l_i = 0.4$ m;
- machine-winding temperature: 100 °C.
- thermal conductivity of copper: $\Lambda_{cu} = 360$ W/°K-m
- electrical resistivity of copper at 100 °C: $\rho = 1/\sigma = 2.27 \cdot 10^{-8}\ \Omega\text{-}m$

Solution:

The copper losses per winding volume is

$$q_h = \frac{J^2}{\sigma} = (4 \cdot 10^6)^2 \cdot 2.27 \cdot 10^{-8} = 363\ \text{kw/m}^3 \tag{7.37}$$

The temperature rise in the middle of the machine from ambient is, from equation (7.31),

$$\Delta\Theta = \frac{q_h}{2\Lambda_z} d^2 = \frac{363 \cdot 10^3}{2 \cdot 360} \cdot 0.2^2 = 20\ ^\circ\text{K} \tag{7.38}$$

7.3.5 One-Dimensional Conductive Heat Flow with Distributed Internal Heat Generation

In most practical cases, external heat flow enters into one portion of the body and the results of Section 7.3.3 are not valid. Again the rectangular plate of Figure 7.2 is considered with even heat generation in the plate but this time with additional heat entering one side of the body. It is necessary to again deal with Poisson's equation

$$\Lambda_x \frac{\partial^2\Theta}{\partial x^2} + \Lambda_y \frac{\partial^2\Theta}{\partial y^2} + \Lambda_z \frac{\partial^2\Theta}{\partial z^2} = -q_h(x, y, z) \tag{7.39}$$

If the temperature distribution is again one-dimensional and if the heat q_h is generated evenly throughout the body,

$$\Lambda_z \frac{\partial^2\Theta}{\partial z^2} = -q_h \tag{7.40}$$

Upon integrating,

$$\Lambda_z \frac{\partial\Theta}{\partial z} = -q_h z + C_1 \tag{7.41}$$

Integrating a second time

$$\Lambda_z \Theta(z) = -q_h \frac{z^2}{2} + C_1 z + C_2 \tag{7.42}$$

The constants C_1 and C_2 can be solved by considering the boundary conditions. At $z = 0$, it is assumed that $\Theta(z) = \Theta_1$ so that $C_2 = \Lambda_z \Theta_1$, whereupon

$$\Lambda_z \Theta(z) = -q_h \frac{z^2}{2} + C_1 z + \Lambda_z \Theta_1 \tag{7.43}$$

At $z = d$, assume $\Theta(z) = \Theta_2$ in which case

$$\Lambda_z \Theta_2 = -q_h \frac{d^2}{2} + C_1 d + \Lambda_z \Theta_1 \tag{7.44}$$

or

$$C_1 = \frac{\Lambda_z}{d}(\Theta_2 - \Theta_1) + \frac{q_h d}{2} \tag{7.45}$$

The solution is therefore

$$\Lambda_z \Theta(z) = -\frac{q_h z^2}{2} + \frac{\Lambda_z}{d}(\Theta_2 - \Theta_1) z + \frac{q_h d}{2} z + \Lambda_z \Theta_1 \tag{7.46}$$

The average value of temperature between surfaces 1 and 2 is

$$\Lambda_z \Theta_{12\text{ave}} = \frac{1}{d}\int_0^d \Lambda_z \Theta(z)(dz) \tag{7.47}$$

$$= -\frac{q_\text{h}}{2d}\int_0^d z^2 dz + \frac{1}{d}\left[\frac{\Lambda_z}{d}(\Theta_2 - \Theta_1) + \frac{q_\text{h} d}{2}\right]\int_0^d z dz + \Lambda_z \Theta_1 \tag{7.48}$$

$$= -\frac{q_\text{h} d^2}{6} + \Lambda_z\left(\frac{\Theta_2 - \Theta_1}{2}\right) + \frac{q_\text{h} d^2}{4} + \Lambda_x \Theta_1 \tag{7.49}$$

$$= -\frac{q_\text{h} d^2}{12} + \Lambda_z\left(\frac{\Theta_2 - \Theta_1}{2}\right) \tag{7.50}$$

Letting $Q_\text{h} = A2dq_\text{h}$ denote the total heat source generated in the body, where A is the cross-sectional area then

$$\Theta_{12,\text{ave}} = \frac{Q_\text{h} d}{24\Lambda_z A} + \left(\frac{\Theta_2 - \Theta_1}{2}\right) \tag{7.51}$$

Defining the z component of thermal resistance as

$$R_{\text{h},z} = \frac{d}{24\Lambda_z A} \tag{7.52}$$

$$\Theta_{12,\text{ave}} = \frac{\Theta_1 - \Theta_2}{2} + Q_\text{h} R_{\text{h},z} \tag{7.53}$$

It is again useful to develop an equivalent circuit which produces equation (7.53). Consider the circuit of Figure 7.5 for which the thermal resistance R is as yet unknown. From the equivalent circuit, one can write

$$\Theta_1 = R_{\text{h},z} Q_1 + R_{\text{h},z} Q_2 + \Theta_2 \tag{7.54}$$

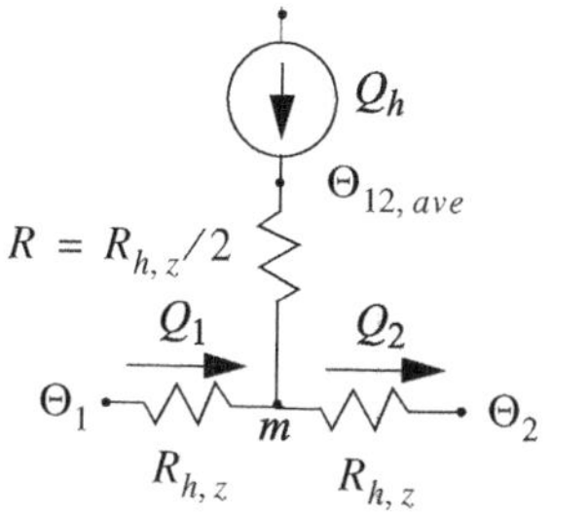

Figure 7.5 Equivalent circuit for one-dimensional heat conduction with internal heat generation.

and

$$\Theta_1 = R_{h,z}Q_1 - RQ_h + \Theta_{12,ave} \tag{7.55}$$

However,

$$Q_2 = Q_1 + Q_h \tag{7.56}$$

Substituting this constraint into equation (7.54) yields

$$\Theta_1 = 2R_{h,z}Q_1 + R_{h,z}Q_h + \Theta_2 \tag{7.57}$$

Using equation (7.57) to eliminate Q_1 from equation (7.55) results in

$$\Theta_1 = [(\Theta_1 - \Theta_2) - R_{h,z}Q_h]\frac{1}{2} - RQ_h + \Theta_{12,ave} \tag{7.58}$$

which can be written as

$$\frac{\Theta_1 + \Theta_2}{2} + \left(\frac{R_{h,z}}{2} + R\right) Q_h = \Theta_{12,ave} \tag{7.59}$$

Choosing $R = \frac{1}{2}R_{h,z}$, then

$$\frac{\Theta_1 + \Theta_2}{2} + R_{h,z}Q_h = \Theta_{12,ave} \tag{7.60}$$

which is the same as equation (7.53).

It should be noted that since the branch containing the thermal resistance $R_{h,z}/2$ is fed by a heat source (current source equivalent), it is useful in the solution of the circuit only if it is desired to estimate the average temperature within the plate. If this is not necessary, the heat source can be connected directly to the center point "m" on Figure 7.5.

7.3.6 Two- and Three-Dimensional Conductive Heat Flow with Internal Distributed Heat Generation

Consider now the case of two-dimensional heat flow. In general, the exact analytical temperature distribution is complicated [1–8]. However, a reasonable approximation can be obtained by assuming that the heat flow in the two directions are independent. The circuits in the two directions can then be connected as shown in Figure 7.6. In some cases, the heat flows in the two directions over nearly equal cross-sectional areas in which case $h \cong w$. The equivalent circuit of Figure 7.7 then becomes a reasonable approximation to the circuit of Figure 7.6.

It is clear that this concept can also be extended to three-dimensional heat conduction. The means that by which this case can be extended from the two-dimensional case should be clear. However, the possibility of significant errors increases with this complexity. Fortunately, a two-dimensional equivalent circuit is generally sufficient to account for all of the major heat paths in a typical machine.

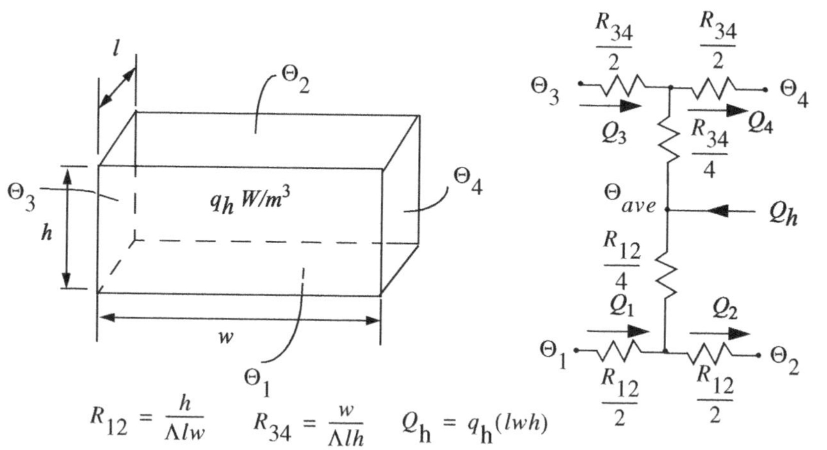

Figure 7.6 Two-dimensional heat conduction with internal heat generation and resulting equivalent circuit.

7.3.7 Application of Two-Dimensional Heat Flow to Stator Teeth

The heat produced in a stator tooth is a typical example of two-dimensional heat flow. Figure 7.8 shows a sketch of a typical stator tooth geometry. For rough estimates of heat distribution, the actual tooth geometry can be replaced by an equivalent tooth having the same cross-sectional area and the same length in the radial direction shown as the dotted line in Figure 7.8a. More accurate models of tooth geometries are available but without much improvement in accuracy [4, 5].

The thermal circuit for one stator tooth is shown in Figure 7.8. In most practical cases, the heat generated in all of the S_1 stator teeth are identical. Thus the temperatures on the sides of the tooth will be identical. In effect, it can be stated that heat flow in the stator and rotor teeth are normally not circumferential. As a result, heat flow in the plane of the laminations of each of the stator teeth can be simplified. The

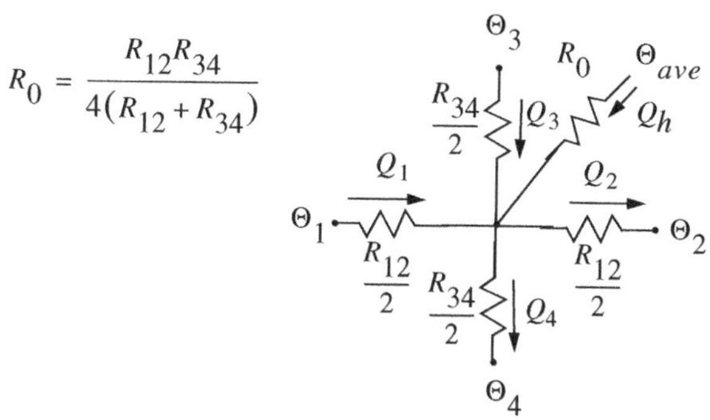

Figure 7.7 Simplified two-dimensional equivalent circuit when $w \cong h$.

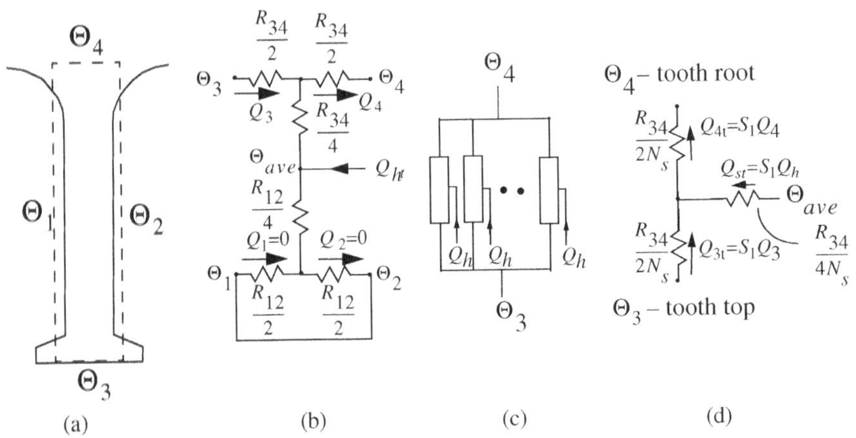

Figure 7.8 Stator tooth geometry (solid line) and equivalent rectangular tooth (dashed), (b) thermal circuit for the tooth, (c) resulting S_1 branch circuit for all stator teeth, (d) simplified thermal circuit for all stator teeth accounting for heat flow in the radial direction.

equivalent circuit describing all of the stator teeth becomes a circuit with S_1 identical branches. These identical branches can be combined to form the simple circuit of Figure 7.8d In this case, the total heat generation for all stator teeth is injected into a resistive network with resistances reduced by the number of teeth. Clearly, the solution of Figure 7.8b or 7.8d will be identical.

7.3.8 Radial Heat Flow Over Solid Cylinder with Internal Heat Generation

An electrical machine, being essentially a cylindrical body, generally necessitates the computation of heat flow in cylindrical coordinates. When heat is generated within the cylindrical body, the problem becomes more complicated than the solution for flat plates. The simplest case is where the inner radius is zero in which the body is a simple solid cylinder. This is essentially the case of the rotor shaft which, in some cases has significant losses since it cannot be laminated.

With heat generation, equation (7.1) can be expressed in the cylindrical coordinate form of Poisson's equation as

$$\Lambda_\mathrm{r}\frac{\partial^2\Theta}{\partial r^2}+\frac{\Lambda_\mathrm{r}}{r}\frac{\partial\Theta}{\partial r}+\frac{\Lambda_\Theta}{r^2}\frac{\partial^2\Theta}{\partial\theta^2}+\Lambda_z\frac{\partial^2}{\partial z^2}(\Theta)=-q_\mathrm{h} \tag{7.61}$$

If heat flows only radially, Poisson's equation reduces to the axi-symmetric form

$$\frac{1}{r}\frac{\partial}{\partial r}\left(r\frac{\partial\Theta}{\partial r}\right)=-\frac{q_\mathrm{h}}{\Lambda_\mathrm{r}} \tag{7.62}$$

Multiplying both sides by r and integrating once produces

$$r\frac{\mathrm{d}\Theta}{\mathrm{d}r}=-\frac{q_\mathrm{h}r^2}{2\Lambda_\mathrm{r}}+C_1 \tag{7.63}$$

Dividing this result by r and integrating a second time gives

$$\Theta(r) = -\frac{q_h r^2}{4\Lambda_r} + C_1 \ln(r) + C_2 \tag{7.64}$$

To solve for the constants of integration, it can be noted that the temperature must be maximum at the center of the cylinder, Thus

$$\left.\frac{\partial \Theta}{\partial r}\right|_{r=0} = 0 \tag{7.65}$$

and, on the outer surface of the cylinder

$$\Theta\left(\frac{D_o}{2}\right) = \Theta_o \tag{7.66}$$

Equation (7.65) results in

$$\left.\frac{\partial \Theta}{\partial r}\right|_{r=0} = -\left.\frac{q_h r}{2\Lambda_r}\right|_{r=0} + \left.\frac{C_1}{r}\right|_{r=0} = 0 \tag{7.67}$$

Clearly, $C1$ must be zero to satisfy this constraint. From the second boundary condition

$$C_2 = \Theta_o + \frac{q_h}{16\Lambda_r} D_o^2 \tag{7.68}$$

The temperature distribution is therefore

$$\Theta(r) = \frac{q_h D_o^2}{16\Lambda_r}\left[1 - \left(\frac{2r}{D_o}\right)^2\right] + \Theta_o \tag{7.69}$$

The average value of temperature is found by equating the heat stored in the cylindrical body to an equivalent body in which the temperature is equal throughout, but the heat stored is the same. Equating the heat stored in the two bodies,

$$\int_{\text{vol}} C_h \Theta(r) r (d\theta)(dr) dl = C_h \Theta_{\text{rad,ave}} \frac{\pi}{4} D_o^2 l \tag{7.70}$$

where C_h is the thermal capacitance. Since the heat profile does not vary with θ or axial length, equation (7.70) reduces to

$$\Theta_{\text{rad,ave}} = \frac{8}{D_o^2} \int_0^{\frac{D_o}{2}} \Theta(r) r dr \tag{7.71}$$

Performing the integration

$$\Theta_{\text{rad,ave}} = \frac{q_h D_0^2}{32\Lambda_r} + \Theta_o \tag{7.72}$$

The total heat generated within the cylinder is related to the heat density by

$$\frac{Q_h}{2} = q_h d \frac{\pi}{4} D_o^2 \tag{7.73}$$

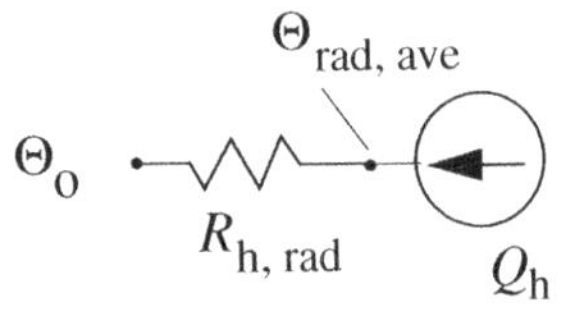

Figure 7.9 Equivalent circuit for radial heat flow in a solid cylinder with internal heat generation.

so that, finally,

$$\Theta_{\mathrm{rad,ave}} = \frac{1}{\Lambda_r} \frac{Q_h}{(16\pi d)} + \Theta_o \tag{7.74}$$

The thermal resistance is, in this case

$$R_{\mathrm{h,rad}} = \frac{1}{\Lambda_r (16\pi d)} \tag{7.75}$$

The equivalent circuit for a cylinder with uniform heat generation within the body is shown in Figure 7.9. It is important to remember that here, the length d corresponds to 1/2 the axial length of the cylinder ($2d = l$) while Q_h is the heating for the entire cylinder.

When heat is also extracted axially from the cylinder the development of Section 7.3.5 clearly applies. Recall that, for this case, the general form is

$$R_{\mathrm{h,rad}} = \frac{d}{24\Lambda_{ax} A} \tag{7.76}$$

For the case of a cylindrical cross section, the length d again corresponds to one-half of the iron stack and A to the cross-sectional area of the cylinder, that is, for cylindrical geometry

$$R_{\mathrm{h,ax}} = \frac{d}{6\pi D_o^2 \Lambda_{ax}} \tag{7.77}$$

An equivalent circuit for both radial and axial heat flow is shown in Figure 7.10a where Θ_{e1}, Θ_{e2} represent the temperatures at each end of the lamination stack and Θ_m is the temperature at the midpoint of the stack. In most cases, it can be assumed that the machine is cooled equally on both sides. In this case $\Theta_{e1} = \Theta_{e2}$ and by symmetry $\Theta_m = \Theta_o$. The equivalent circuit can be reduced to the form of Figure 7.10b. It is also assumed here that the midpoint is fed by a thermal source and that a prediction of the average temperature within the lamination stack is not of interest.

7.3.9 Heat Flow Over Cylindrical Shell with Internal Distributed Heat Generation

A sketch of the general cylindrical geometry was shown in Figure 7.3. It is assumed now that heat flows radially from the inner surface of temperature Θ_i to the outer surface of temperature Θ_o and axially from an inner surface of temperature Θ_m to a

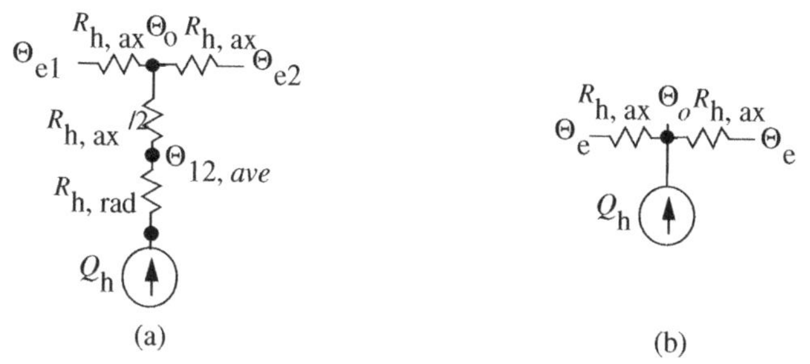

Figure 7.10 (a) Thermal equivalent circuit for a cylinder with both radial and axial heat flow, (b) simplified circuit assuming cooling on both ends of the cylinder are identical.

cooler surface of temperature Θ_e. The cylinder corresponds to either the stator or the rotor core. If the machine has a simple cooling structure with no air ducts, then Θ_m and Θ_e again correspond to the temperature at the center of the core and at the end of the core, respectively.

If the heat first flows only radially outward from the body, then the heat flow is again governed by equation (7.62), that is

$$\frac{\partial^2 \Theta}{\partial r^2} + \frac{1}{r}\frac{\partial \Theta}{\partial r} = -\frac{q_h}{\Lambda_r} \tag{7.78}$$

where Λ_r is the heat conductivity in the radial direction (i.e., in the plane of the laminations).

The resultant solution for temperature has the same form as previously obtained, equation (7.64), that is

$$\Theta(r) = -\frac{q_h r^2}{4\Lambda_r} + C_1 \ln(r) + C_2 \tag{7.79}$$

Let D_i denote either the diameter of the stator core measured at the top of the stator slot or the inner diameter of the rotor core. The diameter D_o then corresponds to either the outer diameter of the stator core or the rotor diameter measured at the bottom of the rotor slots. Heat is assumed to flow only from the inner surface to the outer surface. Thus when $r = D_i/2$, then the heat flow over the inner surface is zero. Mathematically, this means that the gradient of the temperature on the inner surface is zero. In cylindrical coordinates

$$\left.\frac{\partial \Theta}{\partial r}\right|_{r=\frac{D_i}{2}} = 0 \tag{7.80}$$

From equation (7.79)

$$\left.\frac{\partial \Theta}{\partial r}\right|_{r=\frac{D_i}{2}} = -\frac{q_h r}{2\Lambda_r} + \left.\frac{C_1}{r}\right|_{r=\frac{D_i}{2}} = 0 \tag{7.81}$$

Upon solving for C_1

$$C_1 = \frac{q_h D_i^2}{8\Lambda_r} \tag{7.82}$$

Thus,

$$\Theta = -\frac{q_h r^2}{4\Lambda_r} + \frac{q_h D_i^2}{8\Lambda_r} \ln(r) + C_2 \tag{7.83}$$

When $r = D_o/2$, then $\Theta(D_o/2) = \Theta_o$ so that

$$\Theta_o = -\frac{q_h D_o^2}{16\Lambda_r} + \frac{q_h D_i^2}{8\Lambda_r} \ln\left(\frac{D_o}{2}\right) + C_2 \tag{7.84}$$

in which case

$$C_2 = \Theta_o + \frac{q_h D_o^2}{16\Lambda_r} - \frac{q_h D_i}{8\Lambda_r} \ln\left(\frac{D_o}{2}\right) \tag{7.85}$$

The complete solution is therefore

$$\Theta(r) = \frac{q_h}{4\Lambda_r}\left(\frac{D_o^2}{4} - r^2\right) + \frac{q_h D_i^2}{8\Lambda_r} \ln\left(\frac{2r}{D_o}\right) + \Theta_o \tag{7.86}$$

The average value of temperature is again found by equating the heat stored in the cylindrical body to an equivalent body in which the temperature is the same throughout but the heat stored is the same. In this case, equating the heat stored in the two bodies,

$$\int_{\text{vol}} C_h \Theta(r) r(d\theta)(dr)dl = C_h \Theta_{\text{radial,ave}} \frac{\pi}{4}\left(D_o^2 - D_i^2\right) l \tag{7.87}$$

Since the heat profile does not vary with θ or axial length, equation (7.87) reduces to

$$\Theta_{\text{rad,ave}} = \frac{8}{\left(D_o^2 - D_i^2\right)} \int_{\frac{D_i}{2}}^{\frac{D_o}{2}} \Theta(r) r dr \tag{7.88}$$

Utilizing equation (7.79), the solution is

$$\Theta_{\text{rad,ave}} = -\frac{q_h}{32\Lambda_r}\left(D_o^2 + D_i^2\right) + \frac{D_o^2 \ln\left(\frac{D_o}{2}\right) - D_i^2 \ln\left(\frac{D_i}{2}\right)}{D_o^2 - D_i^2} C_1 - \frac{C_1}{2} + C_2 \tag{7.89}$$

where C_1 and C_2 are defined by equations (7.82) and (7.85), respectively.

When the constants C_1 and C_2 are inserted into equation (7.89), the result is

$$\Theta_{\text{rad,ave}} = \frac{q_{\text{h}}}{32\Lambda_{\text{r}}}\left[D_{\text{o}}^2 - 3D_{\text{i}}^2 + \frac{4D_{\text{i}}^4}{D_{\text{o}}^2 - D_{\text{i}}^2}\ln\left(\frac{D_{\text{o}}}{D_{\text{i}}}\right)\right] + \Theta_{\text{o}} \tag{7.90}$$

The total heat generated in half the cylinder formed by half the core is

$$\frac{Q_{\text{h}}}{2} = q_{\text{h}} d \frac{\pi}{4}\left(D_{\text{o}}^2 - D_{\text{i}}^2\right) \tag{7.91}$$

so that the average radial temperature can be expressed in terms of the total heat generated by

$$\Theta_{\text{rad,ave}} = \frac{Q_{\text{h}}}{16\pi\Lambda_{\text{r}} d}\left[1 - \frac{2D_{\text{i}}^2}{D_{\text{o}}^2 - D_{\text{i}}^2} + \frac{4D_{\text{i}}^4}{\left(D_{\text{o}}^2 - D_{\text{i}}^2\right)^2}\ln\left(\frac{D_{\text{o}}}{D_{\text{i}}}\right)\right] + \Theta_{\text{o}} \tag{7.92}$$

Defining the thermal resistance in the radial direction of the core,

$$\Theta_{\text{h,rad}} = \frac{1}{16\pi\Lambda_{\text{r}} d}\left[1 - \frac{2D_{\text{i}}^2}{D_{\text{o}}^2 - D_{\text{i}}^2} + \frac{4D_{\text{i}}^4}{\left(D_{\text{o}}^2 - D_{\text{i}}^2\right)^2}\ln\left(\frac{D_{\text{o}}}{D_{\text{i}}}\right)\right] \tag{7.93}$$

Equation (7.92) can then be written in the form

$$\Theta_{\text{rad,ave}} = \Theta_{\text{o}} + Q_{\text{h}} R_{\text{h,rad}} \tag{7.94}$$

so that the equivalent circuit given in Figure 7.9 again applies to this case.

The heat flow in the axial direction can also be computed by solving Poisson's equation for the axial direction, namely

$$\frac{\partial^2 \Theta}{\partial z^2} = -\frac{q_{\text{h}}}{\Lambda_{\text{z}}} \tag{7.95}$$

where Λ_z is the thermal conductivity in the direction normal to the laminations. The solution is similar to equation (7.46),

$$\Theta(z) = -\frac{q_{\text{h}} z^2}{2\Lambda_z} + \frac{1}{d}(\Theta_{\text{e}} - \Theta_{\text{m}})z + \frac{q_{\text{h}} d}{2\Lambda_z} z + \Theta_{\text{m}} \tag{7.96}$$

Here, the heat source is taken to be at the center of axial length l, that is, at d. The temperature Θ_{m} is taken to be the temperature in the core either at the center of the machine or on the surface of a radial air duct. This temperature can be taken to be the average core temperature at this point, while Θ_{e} is the average temperature at the surface of the core at the two ends of the machine or, if there are more than two air ducts, the temperature on the cooler side of the stator section between air ducts.

The average temperature in the axial direction of the core is, by analogy to equation (7.50)

$$\Theta_{\text{ax,ave}} = \frac{q_{\text{h}} d^2}{12\Lambda_z} + \left(\frac{\Theta_{\text{m}} + \Theta_{\text{e}}}{2}\right) \tag{7.97}$$

or, in terms of the total heat generated in the core Q_{h} is,

$$\Theta_{\text{ax,ave}} = \frac{Q_{\text{h}} d}{6\pi\Lambda_z \left(D_{\text{o}}^2 - D_{\text{i}}^2\right)} + \left(\frac{\Theta_{\text{m}} + \Theta_{\text{e}}}{2}\right) \tag{7.98}$$

Again it is useful to define an equivalent core axial thermal resistance of the form

$$R_{\text{h,ax}} = \frac{d}{6\pi\Lambda_z \left(D_{\text{o}}^2 - D_{\text{i}}^2\right)} \tag{7.99}$$

so that the average temperature in the axial direction becomes

$$\Theta_{\text{axial,ave}} = Q_{\text{h}} R_{\text{h,ax}} + \left(\frac{\Theta_{\text{m}} + \Theta_{\text{e}}}{2}\right) \tag{7.100}$$

Equations (7.60), (7.94), and (7.100) clearly form the same thermal equivalent circuit as for heat flow in a solid cylinder(Figure 7.10). If desired, the cylindrical body could be broken into subsections. Clearly, accuracy improves as the axial length is subdivided into increasingly smaller subsections. However, a reasonable estimate can be obtained by simply dividing the core axial length into two halves.

It should be recalled that the temperature distribution in the radial and the axial direction has been computed independently. This solution is clearly an approximation to a very difficult problem. In reality, the average temperature in the radial direction varies in the axial direction. However, examination of equation (7.92) indicates that this result remains valid for any length l if the *temperature difference* $\Theta_{\text{i}} - \Theta_{\text{o}}$ remains constant as one progresses down the cylinder in the axial direction. This feature is very nearly always met with thermal paths in typical AC machines.

Values of thermal conductivity Λ for typical materials used in electrical machines and transformers are given in Tables 7.4 and 7.5. Typical values of heat flow densities through conducting and magnetic portions of the machine are given in Table 7.6.

TABLE 7.4 Thermal conductivity Λ of some electric insulators

Material	Λ[W/(m-°K)]	Material	Λ[W/(m-°K)]
Glass fiber	0.8–1.2	Mica–Synthetic resins	0.2–0.3
Asbestos	0.2	Bonding epoxy	0.64
Mica	0.4–0.6	Teflon	0.2
Nomex	0.11	Treating varnish	0.26
Kapton	0.12	Typical insulation system	0.2
Water	0.6–0.68	Hydrogen	0.156–0.183
Paper	0.05–0.15	Air	0.025-0.03

TABLE 7.5 Thermal conductivity Λ of metals used in electrical machines

Metal	Λ[W/(m-°K)]	Metal	Λ[W/(m-°K)]
Cast iron	40–46	Copper	360
Steel (Structural)	35–45	Aluminum	200–220
Stainless steel	25–30	Brass/Bronze	100–110
M-22 lamination	22	M-27 lamination	24
M-36 lamination	27–28	M-43 lamination	34
M-45 lamination	40	Normal to lamination	0.6
Ferrite magnet	4.5	Sm–Co magnet	10
NdFeB	9		

TABLE 7.6 Typical heat flow densities p_h in watts/cm^2 in portions of electric machines

Indirectly cooled windings in slots	0.15
Iron laminations	0.4
Solid iron pieces	2.5
Direct air cooled conductors	4.0
Direct water cooled conductors	9.0

7.4 HEAT CONVECTION ON PLANE SURFACES

Typical geometry in which the heat convection takes place in an electric machine is a heated tube through which a cooling fluid is flowing. The heat is transferred from the hot surface to the cooling fluid, creating a temperature increase $\Delta\Theta$ in the cooling fluid according to

$$\Delta\Theta = \frac{Q_c}{c_p \rho_m \left(\frac{dV}{dt}\right)} \tag{7.101}$$

where Q_c (W) is the transferred thermal power from the surface to the cooling fluid, and dV/dt (m^3/s) denotes the volume flow rate, and $c_p\rho_m$ (joules/m^3-°K) is the specific heat content. The temperature rise $\Delta\Theta$ of a cooling fluid in electric machines reaches values between 15 and 40 °C. The gas inlet temperature is usually limited to 40 °C. A table of typical maximum values is shown in Table 7.7.

TABLE 7.7 Maximum temperature rise [°C] for the cooling gas temperature with a 40 °C ambient temperature

Insulation class	A	B	F	H
AC windings	75	80	100	125
DC windings (generally)	75	80	100	125
Cylindrical rotor field windings	—	90	110	—

Heat convection takes place on those surfaces in a machine which are exposed to the cooling fluid. Denoting by Θ_s (°C) the surface temperature, and by Θ_c (°C) the cooling fluid temperature, the power transferred from the hot surface to the cooling fluid is

$$Q_c = \alpha_c A(\Theta_s - \Theta_c) = \alpha_c A \Delta\Theta \quad (7.102)$$

where A(m^2) is the surface area and α_c is the heat convection coefficient. The value of α_c in air varies between seven (W/m^2-°K) for free convection to several thousand for liquid coolants. The value of α_c for free convection can be added to the amount of heat convection coefficient α_r for radiation (Table 7.11), in order to obtain the resulting heat convection coefficient $\alpha_{res} = \alpha_c + \alpha_r$ [approximately 15 (W/m^2-°K)] for naturally cooled casings of electric machines.

The heat convection coefficient for plane surfaces cooled by air at 1 bar in turbulent media can be determined by the following empirical relationship

$$\alpha_{c,a} = 7.8v^{0.78} \quad (7.103)$$

where v denotes the air velocity in meters per second (1 bar = 1.02 kgf/cm^2). Knowing the value of $\alpha_{c,a}$ of air at 1 bar and at a given velocity v, one can express the heat convection coefficient for any gas $\alpha_{c,g}$ at the same velocity v as

$$\alpha_{c,g} = \alpha_{c,a}\left(\frac{\Lambda_g}{\Lambda_a}\right)^{0.22}\left(\frac{c_g\rho_g}{c_a\rho_a}\right)^{0.78} \quad (7.104)$$

where Λ denotes the thermal conductivity and $c\rho$ is the specific heat content (e.g., see Table 7.8).

The heat convection coefficient is dependent on the geometry of surface on which the heat is transferred between the solid medium and cooling fluid, on cooling

TABLE 7.8 Comparison of cooling properties of various media relative to air at 1 bar, α_{ref} is the heat convection coefficient of air at 1 bar

	Relative density ρ_m(pu)	Relative specific heat content $c_p\rho_m$(pu)	Relative heat convection coefficient α_c/α_{ref}	Relative windage losses	Relative fan power
Air at 1 bar	1	1	1	1	1
H2 at 1 bar	0.107	1	1.49	0.148	0.107
H2 at 2 bar	0.214	2	2.56	0.258	0.214
H2 at 4 bar	0.427	4	4.4	0.449	0.427
He at 1 bar	0.138	0.74	1.17	0.209	0.138
CO2 at 1 bar	1.528	1.27	1.08	1.344	1.528
N2 at 1 bar	0.967	1.03	1.02	0.966	0.967
Water	935	3880	43	N/A	N/A
Cooling oil	740	1550	5	N/A	N/A

TABLE 7.9 Ratio between heat convection coefficients for hydrogen and air at various hydrogen pressures

p (bar)	1	2	3	4	5	6
$\alpha_{H2}/\alpha_{air,1\ bar}$	1.49	2.56	3.51	4.40	5.23	6.03

medium thermal parameters and on its velocity. When recalculating the gas thermal parameters from 1 bar to another pressure, one should bear in mind, that

- the gas density ρ is proportional to the pressure,
- the kinematic viscosity ν is proportional to the pressure, the thermal conductivity Λ is independent of pressure,
- the specific heat c is independent of pressure, and
- the specific heat content $c\rho$ is proportional to the pressure.

The behavior of equation (7.104) is illustrated in the following table, in which the ratio between heat convection coefficient for air at 1 bar and hydrogen at various pressures is given in Table 7.9. Note that the heat convection coefficient for hydrogen at a given pressure is about 50% higher than for air at the same pressure.

When a plane surface is cooled by a liquid rather than a gas, the heat convection coefficient can be expressed as

$$\alpha_c = 0.0568(c_p\rho_m)^{0.78}\left(\frac{\Lambda}{l}\right)^{0.22} v^{0.78} \tag{7.105}$$

where l is the surface length in the direction of cooling fluid flow, and v is the coolant velocity.

Since the power passing across a body is related to the temperature by equation (7.11), from equation (7.102) it is possible to define an equivalent resistance for convective heat transfer as,

$$\Delta\Theta = Q_c R_{conv} \tag{7.106}$$

where

$$R_{conv} = \frac{1}{\alpha_c A_{conv}} \tag{7.107}$$

7.5 HEAT FLOW ACROSS THE AIR GAP

Heat transfer occurs across the air gap primarily as a result of both conduction and convection. The influence of the air gap is determined by the nature of the air flow in the gap which can be either laminar or turbulent. In machines designed to operate at normal speeds, the flow is typically laminar. The thermal resistance of the air gap between the stator and rotor in the case of laminar flow can be described by

$$R_{ag} = \frac{g}{\Lambda_{air}\pi D_{is} l_i} \tag{7.108}$$

TABLE 7.10 Equivalent heat coefficient for radiation α_r [W/(m^2°K)]

Ambient temperature Θ_a(°C)	Temperature rise $\Delta\Theta$ (°K)			
	10	30	50	100
10	0.957	3.18	5.88	15.1
20	1.06	3.51	6.47	16.5
30	1.17	3.87	7.1	18.0

where Λ_{air} is obtained from Table 7.4. When turbulent flow begins, the thermal resistance decreases substantially. This case must be considered when designing a high-speed machine (>3600 RPM (min^{-1})). The reader is referred to the literature for further information [4].

7.6 HEAT TRANSFER BY RADIATION

In general, heat is removed from a completely passively cooled machine only by means of the natural air stream and by radiation. This type of cooling has a very poor thermal efficiency. Its application is limited to the smallest machines.

In most cases cooling by means of heat radiation does not play any important role during normal operation of an electric machine. The reason for this is the relatively small temperature difference between the machine surface, from which the heat should be radiated, and its surroundings, to which the heat is radiated. According to Stefan–Boltzmann's law, the heat flow density for radiation is proportional to

$$p_{rad} = c_{rad}\left[\left(\frac{\Theta_s}{100}\right)^4 - \left(\frac{\Theta_a}{100}\right)^4\right] \text{ W/m}^2 \tag{7.109}$$

where Θ_s denotes the absolute temperature (°K) of the surface from which the heat is radiated, and Θ_a is the absolute temperature of its surroundings. The constant c_{rad} is equal to 5.8 W/(°K^4 m^2) for an absolutely black body, and 5 W/(°K^4 m^2) for a surface of a typical electric machine. Zero degree celsius corresponds to 273.1 degrees Kelvin.

The heat flow density due to radiation is sometimes also expressed in terms of a heat convection coefficient for radiation α_r. The dependence of α_r on the temperature of the surroundings Θ_α and the temperature rise $\Delta\Theta$ on the machine surface is given in the Table 7.10. The temperature rise $\Delta\Theta$, the heat convection coefficient α_r, the area A_{rad} of surface from which the heat is radiated, and the radiated power P_{rad} are related to each other as:

$$\Delta\Theta = \frac{Q_r}{\alpha_r A_{rad}} \tag{7.110}$$

The equivalent resistance for radiation can be defined as

$$\Delta\Theta = Q_r R_{rad} \tag{7.111}$$

where

$$R_{rad} = \frac{1}{\alpha_r A_{rad}} \tag{7.112}$$

7.7 COOLING METHODS AND SYSTEMS

It has already been learned from the last chapter that according to the scaling laws, the rated power of an electric machine is proportional to the third power of its dimension. Recall that the total area of all surfaces from which the losses can be taken out from the machine is, however, proportional to the square of machine dimensions. Hence, the larger the machine, the more losses have to be removed from its cooling surface, that is, the higher heat flow density through the cooling surfaces. Thus, machine cooling has to be intensified when its rated power increases. Usual methods for improving the machine cooling capability are [10]

- an increase in the cooling surface area, for example, by introducing radial cooling ducts, and
- direct cooling of conductors with an increase in gas flow (air, hydrogen) or fluid (water, oil).

7.7.1 Surface Cooling by Air

The complete heat generated in a machine with surface cooling is delivered to the surrounding air via the machine's outer surface (frame). A surface-cooled machine can be either cooled with an externally driven fan or self-cooled. The cooling air flow on a surface-cooled machine is shown in Figure 7.11

7.7.2 Internal Cooling

The heat in an internally cooled machine is delivered to the cooling air which flows through it. Fresh cooling air comes from outside of the machine as shown in Figure 7.12.

7.7.3 Cooling in a Circulatory System

The heat in a machine with circulatory system cooling is delivered first to the cooling medium, which, in turn, transports it to a heat exchanger. This cooling mode is applied

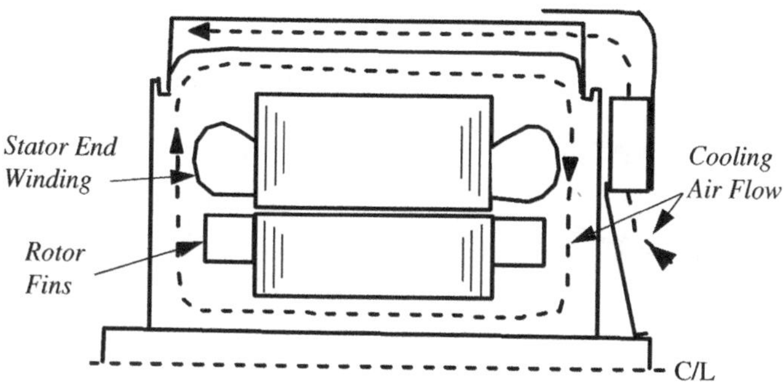

Figure 7.11 A surface cooled (TEFC) squirrel-cage induction machine with cooling ribs on its case.

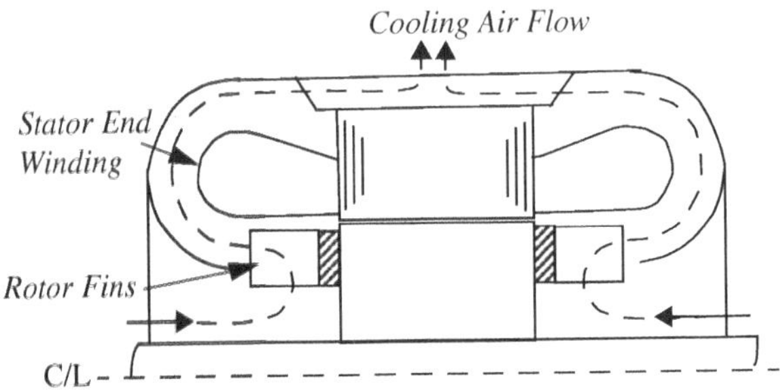

Figure 7.12 Typical drip proof type induction motor with radial internal ventilation showing air flow paths.

in both self and externally cooled machines. A self-cooled turbo-generator with a circulatory system cooling is shown in Figure 7.13.

7.7.4 Cooling with Liquids

The parts of an electric machine which cannot be efficiently cooled by gas are cooled by a liquid: demineralized water or oil similar coolant. Two directly water-cooled stator bars are shown in a cross-sectional view in Figure 7.14.

7.7.5 Direct Gas Cooling

The parts of electrical machines, especially conductors which are directly cooled by gas have holes through which the cooling gas is passed. This cooling mode is used in self and externally cooled machines. A rotor slot of a turbo-generator with directly gas (air or hydrogen) cooled conductors is shown in Figure 7.15. [10]

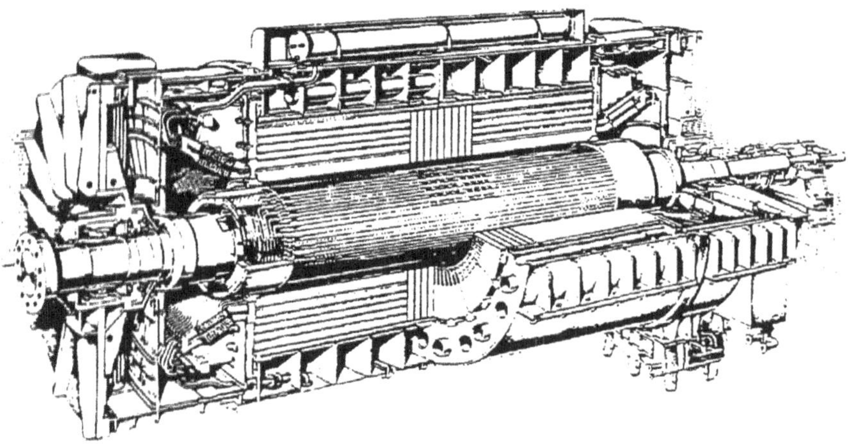

Figure 7.13 Turbo-generator with a circulatory cooling system and direct conductor cooling.

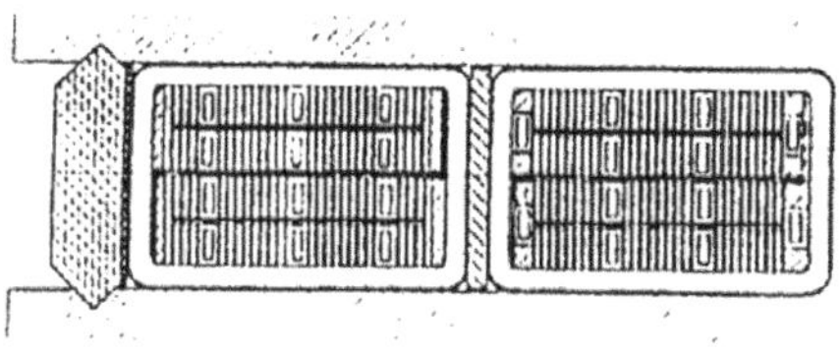

Figure 7.14 Directly water cooled stator bars in a turbo-generator slot.

7.7.6 Gas as a Cooling Medium

Air is the most wide spread gaseous cooling medium for electric machines. It is used for surface, internal, and circulatory system cooling. Cooling air must be free of dust and particles, especially if they are electrically conducting. Cooling air should also not contain dust particles because they can deteriorate voltage strength and decrease cross section of smaller cooling ducts in the machine. Therefore, cooling air has to be filtered in machines with an open cooling system. Cooling air in machines with a closed cooling system (circulatory system cooling) does not leave the cooling system and, therefore, it does not have to be filtered. In very large machines it becomes almost impossible to filter the necessary cooling air. For example, a 100 MVA turbo generator at 3600 rev/min has a mass of 140 metric tons and 1500 kW losses. The generator demands 50 m^3/s air for cooling, which is equivalent to 190 tons/hour. Hence, 1.37 times the generator mass has to be filtered every hour!

The temperature rise of cooling air, as it absorbs energy, is related to the flow of the air by

$$\Delta\Theta = \frac{8.879 \times 10^{-4} P_{diss}}{\text{CMS}} \text{ °C or °K} \tag{7.113}$$

where CMS is the air flow rate in cubic meters per second.

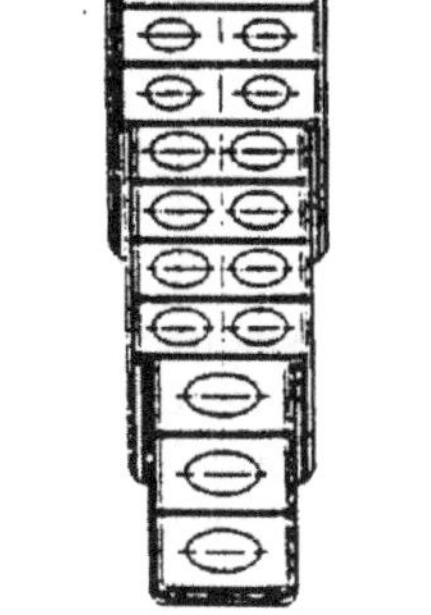

Figure 7.15 Rotor slot of turbo-generator with directly gas cooled conductors.

TABLE 7.11 Thermal properties of cooling liquids at 1 bar pressure

	Temperature (C)	Density $\rho_m(kg/m^3)$	Specific heat $c_p\rho_m$ $(J/m^3-{}^\circ K)$	Kinematic viscosity $v(m^2/s)$	Thermal conductivity $\Lambda(W/m-{}^\circ K)$	Specific electrical resistance (Ωm)
Water	20	998	4173×10^3	1.01×10^{-6}	0.598	$2\text{–}5\times10^3$
Water	40	992	4145×10^3	0.66×10^{-6}	0.627	$2\text{–}5\times10^3$
Water	60	983	4123×10^3	0.48×10^{-6}	0.652	$2\text{–}5\times10^3$
Cooling oil	20	800	1600×10^3	5.0×10^{-6}	0.147	$10^8\text{–}10^{14}$
Cooling oil	40	785	1640×10^3	3.3×10^{-6}	0.143	$10^8\text{–}10^{14}$
Cooling oil	60	770	1670×10^3	2.25×10^{-6}	0.14	$10^8\text{–}10^{14}$
Transformer oil	20	870	1760×10^3	36.5×10^{-6}	0.124	$10^8\text{–}10^{14}$
Transformer oil	40	850	1820×10^3	16.7×10^{-6}	0.123	$10^8\text{–}10^{14}$
Transformer oil	60	840	1860×10^3	8.7×10^{-6}	0.122	$10^8\text{–}10^{14}$

A disadvantage of air as a cooling medium is its relatively high density, which results in high windage losses. Therefore, hydrogen is used as a cooling medium in the largest machines instead of air. A machine cooling system based on hydrogen as a cooling medium must be closed, and the hydrogen pressure goes up to 6 bars. The advantage of hydrogen as compared to air is its higher heat transfer capability, lower windage losses, and higher thermal conductivity. In addition, the coolers and cross-sectional dimensions of gas transport tubes for hydrogen are smaller than those for air. On the other hand, the casing of a hydrogen-cooled machine has to be hermetically closed and made explosion proof. The shaft of such a machine must also be hermetically sealed, and auxiliary devices are needed to produce the hydrogen (e.g., by electrolysis).

Thermal properties of some gases used for machine cooling are given in Table 7.11. Quantities in this table vary as a function of the gas pressure p according to equation $pV = mRT$, which results in:

$$\begin{aligned}\rho(p) &= k_\rho p \\ v(p) &= k_v p\end{aligned} \tag{7.114}$$

whereas the thermal conductivity Λ does not depend on the gas pressure.

7.7.7 Liquids as a Cooling Medium

Oil is used to cool transformer windings and core, as well as stator windings, iron lamination, and press plates in electric machines. Water has better cooling capability and needs lower pressure for circulation than oil. Water is used for cooling of stator and rotor windings, iron lamination, and press plates in electric machines. Cooling water must have electrical conductivity below 500 μS/m (micro-siemens or micro-mhos per meter), and must be demineralized, deionized, and possess a low oxygen content. Water temperature rises as it absorbs energy which can be estimated from

$$\Delta\Theta = \frac{63.09 P_{diss}}{CCS} \quad °C \text{ or } °K \tag{7.115}$$

where CCS is the water flow in cubic centimeters per second (cm^3/s) and P_{diss} is expressed in Watts (W).

Liquid cooling media can interact with the surface of hollow conductors through which they flow, introducing the mechanisms of erosion, cavitation, and corrosion. Therefore, the velocity of liquid coolants in copper cooling tubes is limited to 2 m/s. The maximum allowed fluid velocity in stainless steel hollow conductors is somewhat higher. Thermal properties of typical cooling liquids are shown in Table 7.11.

7.8 THERMAL EQUIVALENT CIRCUIT

A conceptual sketch of a quarter section of a typical induction machine with axial and radial cooling is shown in Figure 7.16. The machine is divided into its basic thermal

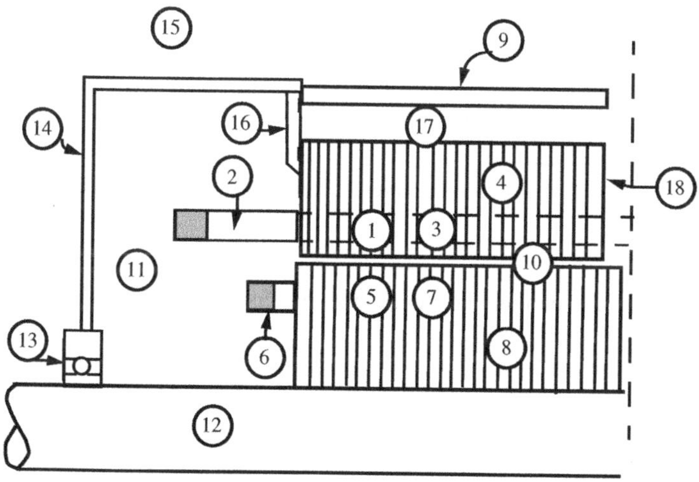

1 & 5 - Stator and Rotor Windings, Slot Portion
2 & 6 - Stator and Rotor End Windings
3 & 7 - Stator and Rotor Teeth
4 & 8 - Stator and Rotor Core

9 - Frame	13 - Bearings	17 - Axial Duct
10 - Air Gap	14 - End Shield	18 - Radial Duct
11 - Overhang Air Space	15 - Ambient Air	
12 - Shaft	16 - Stator End Support	

Figure 7.16 Quarter section of induction machine showing major thermal nodes.

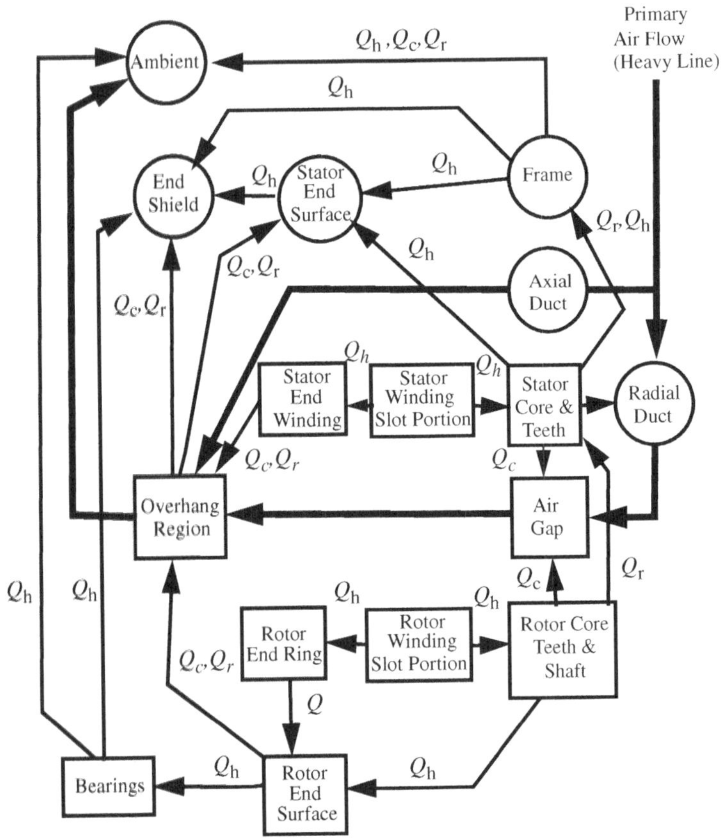

Figure 7.17 Heat flow diagram, Q_h, Q_c, and Q_r represent conduction, convection, and radiation respectively [2].

elements or nodes. The corresponding heat exchange map for this machine showing all heat flow paths is shown in Figure 7.17 [2].

Table 7.3 forms the basis for constructing equivalent circuits for the heat flow within an electrical machine. In its simplest form, radial cooling is accomplished as shown in Figure 7.12. Two fans attached to the rotor push the cool air through the end windings and along the surface of the lamination end portions. In larger machines, part of the fresh air is diverted into axial rotor channels which serve as distribution points for the radial cooling ducts separating the individual lamination stacks on both the rotor and the stator.

The solution of the thermal circuits, implied by Figure 7.17, is clearly a formidable task. However, reasonably accurate answers can be obtained if this thermal network is considerably simplified. In general, the major sources of heat are in the iron and the copper. Transfer of heat between the slot embedded portion of the stator copper $Q_{cu,1}$ and the stator iron $Q_{fe,1}$ is not very significant because of the conductor and slot insulation. In contrast, the heat conduction along the copper conductor in the axial direction is excellent so that most of the heat is taken from the end windings.

Likewise, on the rotor, transfer of heat $Q_{cu,2}$ from the rotor bars to the stator is poor since the air in the gap acts as an insulator. Almost all of the heat pulled from the rotor is taken from the end rings and axial extensions projecting from the end ring used to improve heat transfer. The heat produced in the rotor teeth can be important due to slot ripple but the heat produced within the rotor core can be considered as negligible since the flux in this portion of the machine varies as slip frequency. A simple equivalent circuit which can be used to obtain a reasonable estimate of temperature rise is shown in Figure 7.18. No cooling ducts are assumed. It is also assumed that the heat produced by the stator and rotor conductors is removed solely via the end

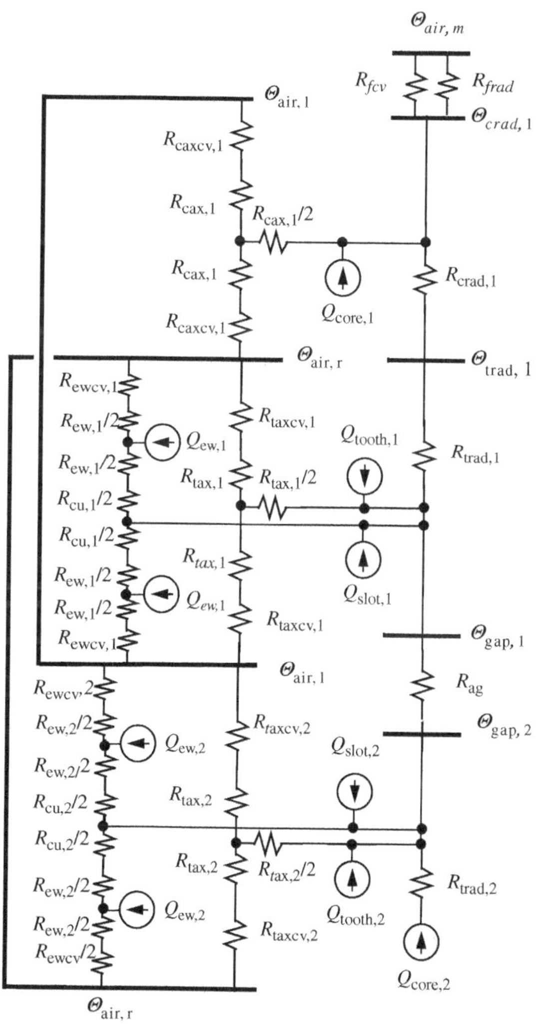

$R_{caxcv,1}$ – Stator Core Surface Convection T.R. (Axial)

R_{frad}, R_{fcv} – Stator Frame Conduction and Convection T.R. (Radial)

$R_{crad,1}$ – Stator Core T.R. (Radial)

$R_{cax,1}$ – Stator Core T.R. (Axial)

$R_{tax,1}$, $R_{tax,2}$ – Stator & Rotor Teeth T.R. (Axial)

$R_{trad,1}$, $R_{trad,2}$ – Stator & Rotor Teeth T.R. (Radial)

$R_{ins,1}$, $R_{ins,2}$ – Stator & Rotor Insulation T.R.

$R_{tcv,1}$ – Stator Tooth Surface Convection T.R.

$R_{cu,1}$, $R_{cu,2}$ – Stator & Rotor Copper T.R.

$R_{ew,1}$, $R_{ew,2}$ – Stator & Rotor End Winding and Conduction T.R

$R_{ewcv,1}$, $R_{ewcv,2}$, Stator & Rotor Convection T.R.

R_{ag} Air Gap T.R.

$Q_{core,1}$, $Q_{core,2}$ – Stator and Rotor Core Iron Loss

$Q_{tooth,1}$, $Q_{tooth,2}$ – Stator & Rotor Tooth Iron Losses

$Q_{slot,1}$, $Q_{slot,2}$ – Stator & Rotor I^2R Losses in Slots

$Q_{ew,1}$, $Q_{ew,2}$ – Stator and Rotor Copper Losses in End Windings

Figure 7.18 Thermal equivalent circuit for an induction machine, subscripts "1" and "2" denote stator and rotor respectively, subscripts "l", "m" and "r" denote air at left, middle, and right side of the machine.

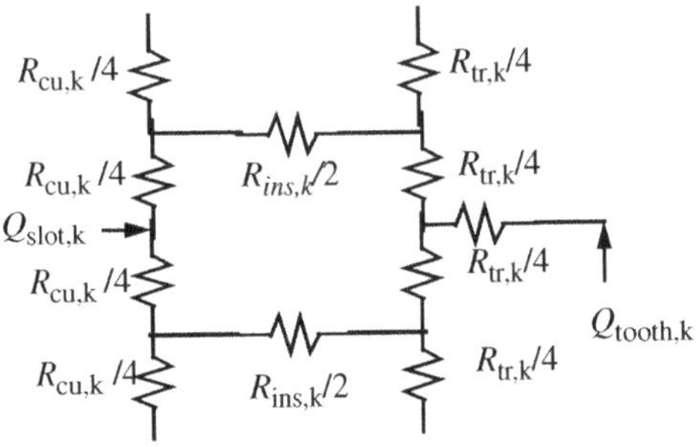

Figure 7.19 Showing coupling between slot-based conductor loss and tooth radial thermal resistance when removal of conductor heat via the laminations is significant. $R_{ins,k}$ denotes the thermal resistance of the slot insulation, where k denotes either 1 or 2 (stator or rotor).

windings. In some cases the heat removed from the conductors via the laminations is significant. In this case, the conductor-based thermal resistances must be "coupled" to the thermal resistances associated with heat removal from the iron in the radial direction. This modification is shown in Figure 7.19.

The thermal resistances of the stator and rotor iron and copper losses have already been treated in Section 7.3. The thermal resistances of the stator and rotor end windings have the form

$$R_{ew} = \frac{1}{S_{ew}}\left(\frac{t_{ins}}{\Lambda_{ins}} + \frac{1}{\alpha_{ew}}\right) \tag{7.116}$$

where

t_{ins} = thickness of the winding insulation
Λ_{ins} = heat conductivity of the winding insulation
α_{ew} = heat transfer coefficient for convection
S_{ew} = surface area of the coil end windng that is exposed to air

To evaluate $Q_{cu,1}$ one can note from Chapter 6, equation (6.84), that the density of power dissipated per unit area of the stator inner surface is

$$p_{diss} = \frac{K_{s(rms)}J_{s(rms)}}{\sigma} \tag{7.117}$$

The total power lost in the stator copper is then

$$Q_{cu,1} = \frac{K_{s(rms)}J_{s(rms)}}{\sigma}(\pi D_{is} l_i)\frac{l_{coil,1}}{l_s} \tag{7.118}$$

where $l_{coil,1}$ is the average total length of a stator coil side and

$$\frac{l_{coil,1}}{l_s} \approx 1 + \frac{\pi D_{is}}{P l_s} \tag{7.119}$$

The surface area of the stator end winding is given by

$$S_{ew,1} = 2N_c p_{coil,1} l_s \left(\frac{l_{coil,1}}{l_s} - 1\right) \tag{7.120}$$

where N_c is the number of stator coils, p_{coil} represents the cross-section perimeter of the end-winding coil bundle. The perimeter of one coil is given by

$$p_{coil,1} = 2(h_s + \tau_s + t_{ave}) \tag{7.121}$$

for one-layer windings and by

$$p_{coil,1} = 2\left(\frac{h_s}{2} + \tau_s - t_{ave}\right) \tag{7.122}$$

for two-layer windings, where

h_s = slot height
τ_s = stator slot pitch
t_{ave} = average stator tooth width

The resistance to heat transfer is therefore,

$$R_{ew,1} = \frac{\left(\dfrac{t_{ins,1}}{\Lambda_{ins,1}} + \dfrac{1}{\alpha_{ew,1}}\right)}{2N_c p_{coil,1} l_s \left(\dfrac{l_{coil,1}}{l_s} - 1\right)} \tag{7.123}$$

where the heat generated by the stator copper windings is

$$Q_{cu,1} = \frac{K_{s(rms)} J_{s(rms)}}{\sigma} (\pi D_{is} l_i) \frac{l_{coil,1}}{l_s} \tag{7.124}$$

A similar expression can be readily derived for the rotor power loss.

The thermal conductivity for insulation is given in Table 7.4. For class B wire insulation $\Lambda_{ins} = 0.15$ W/(°K-m). An estimate of the convection heat transfer coefficient for end windings α_{ew} can be obtained using an empirical formula of G. F. Luke [7–8],

$$\alpha_{ew} \approx 20 v_{air}^{0.6} \text{ (W/°K-m)} \tag{7.125}$$

where v_{air} is the average velocity of the cooling air, typically ranging between 1 and 2 m/s. In the case of the rotor cooling process, the relative velocity between the cooling air and the rotating rotor fins must be used.

One obtains values of heat transfer coefficient ranging from 30 to 50, depending upon v_{air} for open drip proof machines of the type shown in Figure 7.12. In a machine in which cooling air is recirculated such as (totally enclosed fan cooled) TEFC motors, heat transfer coefficients of 20–30 W/(°K-m^2) are common. An additional resistance must be used in the equivalent circuit to account for the resistance to heat flow between the recirculated air and the external air. The transfer process from internal air to external air is a combination of conduction and convection heat transfer. With intense forced air and Class H insulation, the overall heat transfer coefficient α_{ew} may reach 90–100 W/(°K-m^2). Assuming a 140 °C rise above a 40 degree ambient temperature (Class H insulation), this corresponds to a heat flux density of

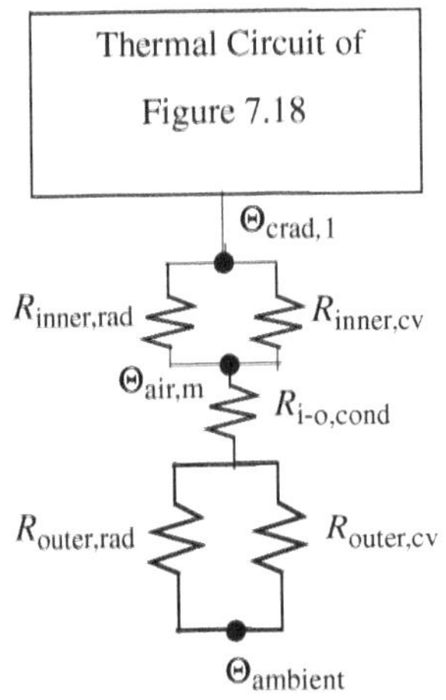

Figure 7.20 Equivalent circuit of totally enclosed fan-cooled machine of Figure 7.11.

1.4 W/cm^2. An equivalent circuit for a machine with recirculated air is shown in Figure 7.20.

In the event that heat is removed from the conductors by way of the laminations, the thermal resistance to heat transfer, R_{ins} must also be computed. By analogy to the above discussion for the end-winding portion, the thermal resistance of the slots is

$$R_{ins} = \frac{t_{ins}}{\Lambda_{ins} S(b_s + 2d_s) l_i} \tag{7.126}$$

where S denotes the number of slots (stator or rotor), Λ_{ins} is the thermal conductivity of the insulation, b_s is the width of the slot bottom and d_s is the total depth of the slot taking into account the actual perimeter of the slot wall.

7.9 EXAMPLE 10—HEAT DISTRIBUTION OF 250 HP INDUCTION MACHINE

Consider again the example of the 250 HP induction machine of Examples 2–8. A sketch of the cooling paths for this machine is shown in Figure 7.21. Ambient air is drawn into the machine on the left side. The air splits into two paths, one of which is used to cool the stator end winding and rotor end ring. This air passes over the outer diameter of the stator through channels. The other portion is guided into the rotor and passes through the cooling ducts and expelled to the outer diameter of the stator at which point it merges with the cooling air from the other path. The cooling air then is forced over the end winding and end ring on the right-hand side of the machine and ejected to the atmosphere.

It will be assumed that the ambient temperature is 30 °C and that the air flow in the end-winding region is 500 cubic feet per minute (0.236 cubic meters per second).

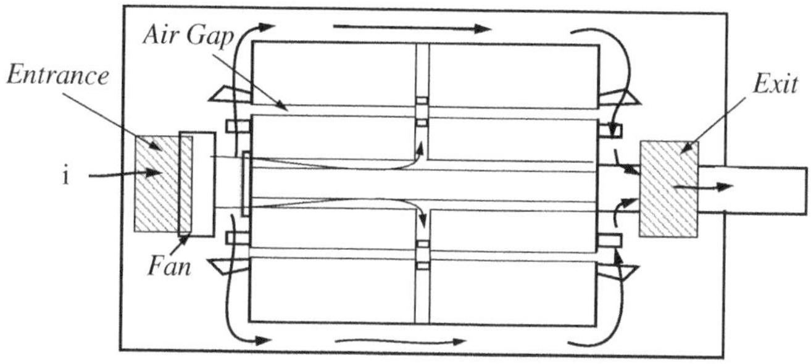

Figure 7.21 Cooling arrangement of 250 HP induction machine.

The velocity of the air flow past the rotor end windings is 3 m/s and past the stator winding is 2 m/s. The velocity in the ducts is 2 m/s and in the convection region (over the outer stator diameter) averages 4 m/s. (The correlation between air flow, air pressure, and air velocity enters the world of thermodynamics which is considered to be outside the scope of this text).

An equivalent electrical circuit of the cooling system is shown in Figure 7.22. Since the thermal circuit is symmetrical about the midpoint of the machine, only one side of the machine need to be modeled and the temperature distribution in the remainder of the machine is obtained by symmetry. The stator copper losses essentially flow through the end-winding region and the cooling ducts. Similarly, the rotor conductor losses flow away from the rotor via the exposed end ring and the air ducts. Heat transfer from the conductors to the iron through the slot insulation has been included by adding the thermal resistor R_{ins}. Since the rotor bars are not insulated, heat can freely pass from the bars to the rotor laminations with negligible intervening thermal resistance. The effect is represented by a short circuit between the rotor bar and rotor tooth heat sources.

7.9.1 Heat Inputs

From the equivalent circuit of Figure 5.33, it can be easily determined that for rated line voltage of 2400 V(rms) and rated output power of 250 HP (186.5 kW), the stator and rotor currents are,

$$|I_1| = 56.6 \text{ A(rms)}$$
$$|I_2| = 50.1 \text{ A(rms)}$$

The stator conductor losses are,

$$P_{cu,1} = 3R_1|I_1|^2 = 3(0.323)56.6^2 = 3104 \text{ W}$$

The portion of the heat introduced in the slots and in the end winding must be determined. From Example 7, it was determined that $l_{ew2} = 1.25''$ and that $l_{ew1} = 1.859''$.

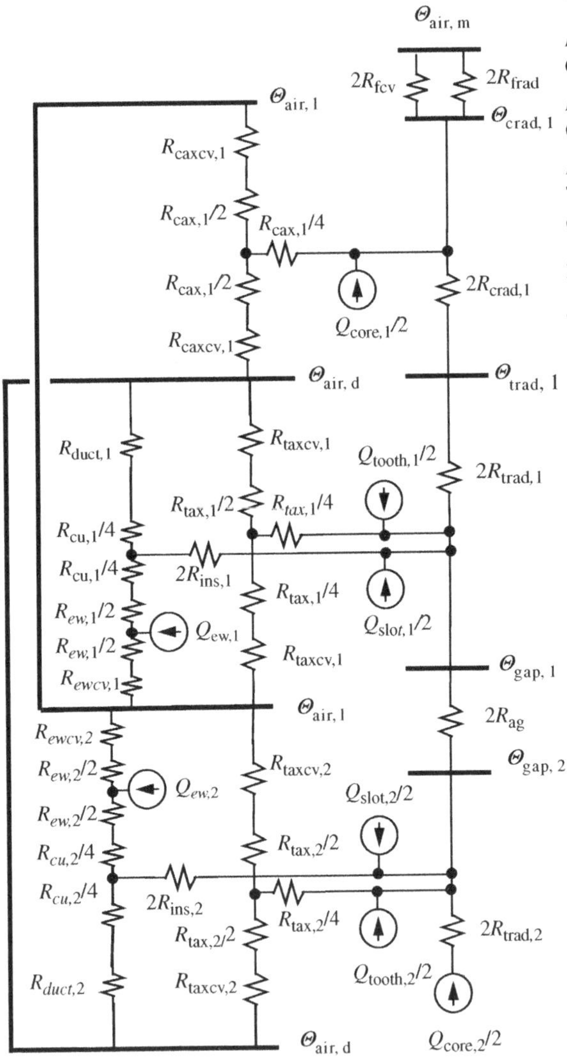

Symbols not defined in Figure 7.18

$R_{caxcv,1d}$ = Stator Core Surface Convection T.R on Duct Wall

$R_{duct,1}$, $R_{duct,2}$ = Stator & Rotor Convection T.R. Over Duct Wall

$R_{taxd,1}$, $R_{taxd,2}$ = Stator & Rotor Tooth Surface Convection T.R. on Duct Wall

$R_{ins,1}$, $R_{ins,2}$ = Stator & Rotor Winding Insulation T.R.

T.R. = Thermal Resistance

Figure 7.22 Thermal equivalent circuit for example 250 HP induction motor, left half of machine is shown, right half can be obtained by symmetry, subscript "d" denotes duct wall.

Since $p\tau_{\mathrm{p(ave)}} = (0.8)10.32 = 8.25''$, the peripheral length of one end-winding portion on one side of the machine is

$$
\begin{aligned}
l_{\mathrm{ew}} &= 2l_{\mathrm{ew2}} + 2\sqrt{l_{\mathrm{ew1}}^2 + \left(\frac{p\tau_{\mathrm{p(ave)}}}{2}\right)^2} \\
&= 2(1.25) + 2\sqrt{1.859^2 + \left(\frac{8.25}{2}\right)^2} = 11.5''
\end{aligned}
\tag{7.127}
$$

The gross length of the core is 9.25 in., so that by simple ratios the power dissipated in the each end-winding side and in the stator slots is

$$P_{\text{slot},1} = P_{\text{cu},1}\frac{9.25}{9.25 + 11.5} = 1380\text{ W}$$

and

$$P_{\text{ew},1} = \frac{P_{\text{cu},1}}{2}\frac{11.5}{9.25 + 11.5} = 860\text{ W}$$

Thus, the heat injected into the equivalent circuit of Figure 7.22 due to stator copper loss is $Q_{\text{slot},1} = 1380$ and $Q_{\text{ew},1} = 860$ W.

From Example 8 of Chapter 5, the fundamental component iron losses in the stator core and teeth are,

$$P_{\text{core},1} = 630\text{ W}$$
$$P_{\text{teeth},1} = 290\text{ W}$$

The fundamental losses in the rotor iron are assumed to be negligible, since only slip frequency exists on the rotor as a result of balanced sinusoidal stator excitation.

Referring to Table 5.1, additional losses exist due to stray no-load and load losses as well as anomalous effects such as stresses. The anomalous losses can be considered as concentrated in the stator core. If a value of 30% for the anomalous loss is selected from the table, the total heat generated within the stator core is

$$Q_{\text{core},1} = P_{\text{core},1} + 0.35(P_{\text{core},1} + P_{\text{teeth},1}) = 950\text{ W}$$

The stray no-load losses occur primarily in the stator and rotor teeth. A 60% value is typical for no-load losses for this machine having semi-open stator and rotor slots. If this loss is split equally between the stator and rotor teeth, the total losses in the stator teeth and therefore one of the heat inputs to Figure 7.22 can be estimated as

$$Q_{\text{teeth},1} = P_{\text{teeth},1} + 0.7(P_{\text{core},1} + P_{\text{teeth},1}) = 290 + 0.7(630 + 290) = 930\text{ W}$$

The remainder of the stray no-load loss is assumed to be in the rotor teeth in which case,

$$Q_{\text{teeth},2} = 0.7(P_{\text{core},1} + P_{\text{teeth},1}) = 0.7(630 + 290) = 644\text{ W}$$

Since the rotor experiences slip frequency during normal operation, the rotor core loss is typically neglected. That is,

$$Q_{\text{core},2} = 0$$

The stray load loss occurs primarily as extra losses in the rotor conductors. Since

$$|I_2| = 50.1\text{ A(rms)}$$

from Example 7 of Chapter 5, the rotor conductor loss due to sinusoidal excitation is 758

$$P_{cu,2} = 3R_2|I_2|^2 = 3(0.786)(50.1)^2 = 5920 \text{ W}$$

Also from Example 7 it was found that the rotor resistance is made up of two components, the bar portion $r_b = 43.7 \times 10^{-6}$ and the end ring portion having a value of 23.40×10^{-6}. By ratioing the rotor loss in proportion to the two resistances, the loss in the end ring is

$$Q_{ew,2} = \frac{23.40}{23.40 + 43.7} 5920 = 2060 \text{ W}$$

In addition to the fundamental components losses in the rotor bars, the extra losses due to stray load losses must be included. Choosing 0.4% of rated output power as the stray load loss, the total heat loss injected into the equivalent circuit by the rotor conductors is

$$Q_{slot,2} = \frac{43.7}{23.40 + 43.7} 5920 + \frac{0.4}{100}(250)(746) = 4600 \text{ W}$$

7.9.2 Thermal Resistances

The thermal resistances must now be calculated. Consider first the thermal resistances of the iron portions of the machine. For the stator tooth, resistance to both axial and radial heat flow must be calculated. For radial flow

$$R_{trad,1} = \frac{d_{ss}}{\Lambda_{i,rad} S_1 l_i \left(\dfrac{t_{ts} + t_{bs}}{2}\right)} \tag{7.128}$$

From Example 2 in Chapter 3,

$$\begin{aligned}
d_{ss} &= 2.2 \text{ in.} \\
S_1 &= 120 \\
S_2 &= 97 \\
l_i &= 9.25 \text{ in.} \\
l_s &= 10 \text{ in.} \\
t_{ts} &= 0.256 \text{ in.} \\
t_{bs} &= 0.369 \text{ in.}
\end{aligned}$$

Also, from Table 7.5, assuming an M-36 grade of silicon steel

$$\Lambda_{i,\,rad} = 28 \text{ W/°K-m}$$

so that

$$R_{trad,1} = \frac{2.2\left(\dfrac{100}{2.54}\right)}{28(120)(8.5)\left(\dfrac{0.256 + 0.369}{2}\right)} = 0.0097 \text{ °K/W} \tag{7.129}$$

For heat flow in the axial direction of the stator tooth

$$R_{tax,1} = \frac{l_i}{\Lambda_{i,ax} S_1 d_{ss} \left(\frac{t_{ts} + t_{bs}}{2} \right)} \tag{7.130}$$

$$= \frac{8.5(39.37)}{0.6(120)(2.2)\left(\frac{0.256 + 0.369}{2} \right)} = 6.76\ ^\circ\text{K/W} \tag{7.131}$$

For the stator core, from equation (7.93), in the radial direction

$$R_{crad,1} = \frac{1}{32\pi \Lambda_{i,rad} l_i} \left\{ D_{os}^2 - 3D_{bs}^2 + \frac{4D_{bs}^4}{\left(D_{os}^2 - D_{bs}^2\right)} \ln\left(\frac{D_{os}}{D_{bs}}\right) \right\} \tag{7.132}$$

where D_{os} is the outer diameter of the stator laminations ($D_{os} = 31.5''$) and D_{bs} is the inner diameter of the stator core measured at the bottom of the stator teeth ($D_{bs} = 31.5 - 2(1.76) = 28$). Equation (7.132) can be readily evaluated as

$$R_{crad,1} = 0.00084\ ^\circ\text{K/W} \tag{7.133}$$

For the stator core in the axial direction

$$R_{cax,1} = \frac{1}{\Lambda_{i,ax}} \left(\frac{l_i}{2} \right) \frac{4}{\pi} \frac{1}{\left(D_{os}^2 - D_{bs}^2\right)} \tag{7.134}$$

$$= \frac{1}{0.6} \left(\frac{8.5}{2} \right) \frac{4}{\pi} \frac{1}{(31.5^2 - 28.48^2)} \times \frac{100}{2.54} = 0.8168\ ^\circ\text{K/W} \tag{7.135}$$

For the rotor tooth, in the radial direction,

$$R_{trad,2} = \frac{d_{s2}}{\Lambda_{i,rad} S_2 l_{ir} \left(\frac{t_{tr} + t_{br}}{2} \right)} \tag{7.136}$$

$$= \frac{0.67\left(\frac{100}{2.54} \right)}{28(97)(8.5)\left(\frac{0.392 + 0.348}{2} \right)} = 0.003\ ^\circ\text{K/W} \tag{7.137}$$

and in the axial direction

$$R_{tax,2} = \frac{l_i}{\Lambda_{i,ax} S_2 d_{s2} \left(\frac{t_{tr} + t_{br}}{2} \right)} \tag{7.138}$$

$$= \frac{8.5\left(\frac{100}{2.54} \right)}{0.6(97)(0.67)\left(\frac{0.392 + 0.348}{2} \right)} = 23.2\ ^\circ\text{K/W} \tag{7.139}$$

It can be noted that the thermal resistances of the iron laminations in the axial direction normal to the plane of the laminations are much greater than their counterpart in the radial direction (a factor of 90 or more). Hence, the flow of heat in the axial direction through the laminations can be safely neglected. The thermal resistance for the rotor core will also be neglected since the heat transfer from the rotor core to the rotor shaft or core end surfaces is assumed to be negligible. Finally, since the rotor bars of a squirrel-cage machine are normally inserted in the rotor slots without the benefit of insulation, the thermal resistance of rotor conductor insulation is neglected.

The thermal equivalent circuit for the 250 HP machine reduces to that shown in Figure 7.23 Additional temperature nodes Θ_1, Θ_2, and Θ_3 have been added to facilitate the eventual solution of the network.

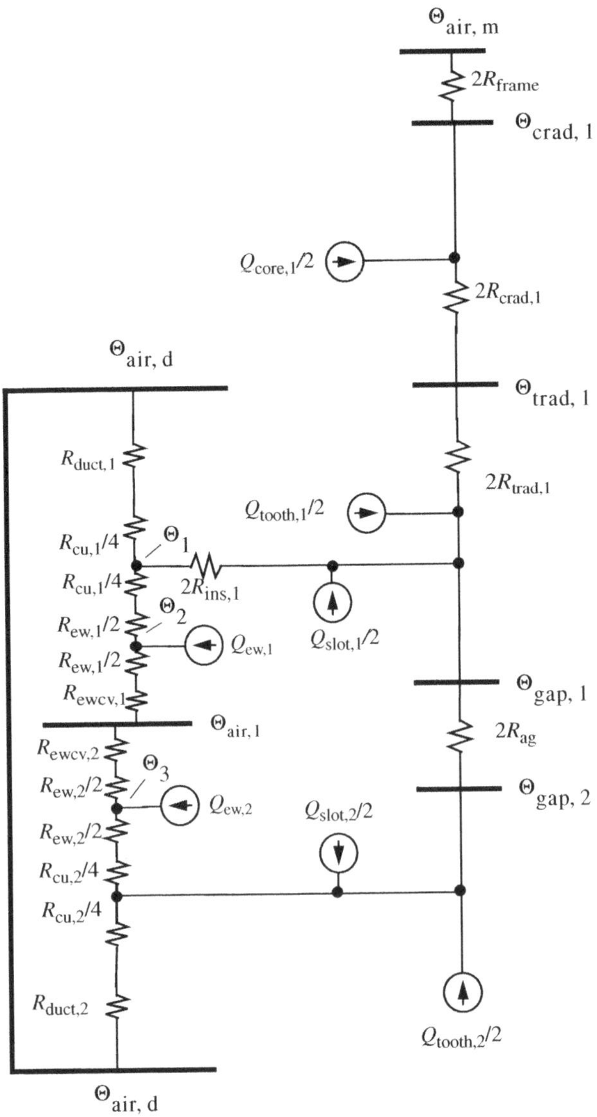

Figure 7.23 Reduced complexity thermal model neglecting heat flow in the laminations in the axial direction.

The remainder of the thermal resistances can now be calculated. For the stator conductor within the slots,

$$R_{cu,1} = \frac{l_s}{\Lambda_{cu} S_1 A_{cu,1}} \tag{7.140}$$

where $A_{cu,1}$ denotes the cross-sectional area of the copper in a slot. From Example 4 and Table 7.5

$$R_{cu,1} = \frac{9.25\left(\frac{100}{2.54}\right)}{360(120)(6)(0.02553)} = 0.0275\ °\text{K/W} \tag{7.141}$$

For the end-winding portion

$$R_{ew,1} = \frac{l_{ew,1}}{\Lambda_{cu} S_1 A_{cu,1}} = \frac{11.5}{9.25} 0.0275 = 0.0342\ °\text{K/W} \tag{7.142}$$

For the rotor bars,

$$R_{cu,2} = \frac{l_{rb}}{\Lambda_{cu} S_2 A_{cu,2}} \tag{7.143}$$

From Example 7, neglecting skew

$$R_{cu,2} = \frac{9.25\left(\frac{100}{2.54}\right)}{360(97)(0.375 \times 0.5625)} = 0.044\ °\text{K/W} \tag{7.144}$$

Also

$$R_{ew,2} = \frac{l_{ew,2}}{\Lambda_{cu} S_2 A_{cu,2}} \tag{7.145}$$

where $l_{ew,2}$ and $A_{cu,2}$ denote the length and cross-sectional area of one rotor end-winding segment. From Example 7, it can be estimated that

$$\frac{l_{ew,2}}{A_{cu,2}} = \frac{1.625}{(0.375)(0.5625)} + \frac{0.71}{(1.0)(0.75)} = 8.65 \tag{7.146}$$

whereupon

$$R_{ew,2} = \frac{8.56\left(\frac{100}{2.54}\right)}{360(97)} = 0.0097\ °\text{K/W} \tag{7.147}$$

The convection heat transfer from the end portions of the winding and core to the circulating air can now be calculated. For one end-winding side, from equation (7.123),

$$R_{ewcv,1} = \frac{\left(\frac{t_{ins,1}}{\Lambda_{ins,1}} + \frac{1}{\alpha_{ew,1}}\right)}{N_c p_{coil,1} l_s \left(\frac{l_{coil,1}}{l_s} - 1\right)} \tag{7.148}$$

The insulation thickness is assumed to be 78 mils or

$$t_{ins,1} = \frac{0.078}{(100/2.54)} = 0.002 \text{ m} \tag{7.149}$$

From Table 7.4, the thermal conductivity for insulation using synthetic resins is

$$\Lambda_{ins,1} = 0.25 \text{ W/(°K-m)} \tag{7.150}$$

The heat transfer coefficient for air passing over the end winding is, from equation (7.125),

$$\alpha_{ew,1} = 20 \times v_{air,1}^{0.6} = 20 \times 2^{0.6} = 30.3 \text{ W/(°K-m}^2) \tag{7.151}$$

The outside perimeter of the coil is

$$p_{coil,1} = 2\frac{(0.984 + 0.374)}{(100/2.54)} = 0.0690 \text{ m} \tag{7.152}$$

The length of one end turn can be directly calculated as

$$l_{end,1} = \frac{2 \times 1.25 + 2 \times 4.527}{(100/2.54)} = 0.2921 \text{ m} \tag{7.153}$$

The thermal resistance for one end-winding side is therefore

$$R_{ewcv,1} = \frac{\left(\frac{t_{ins,1}}{\Lambda_{ins,1}} + \frac{1}{\alpha_{ew,1}}\right)}{N_c p_{coil,1} l_{end,1}} = \frac{\frac{0.002}{0.25} + \frac{1}{30.3}}{120 \times 0.0690 \times 0.2921} = 0.017 \text{ °K/W} \tag{7.154}$$

The thermal resistance for one rotor end ring follows the same procedure, The insulation thickness for the end ring is zero so that the thermal resistance is set totally by the air flow over the end ring. The thermal heat transfer coefficient for the end ring is

$$\alpha_{ew,2} = 20 v_{air,2}^{0.6} = 20 \times 3^{0.6} = 38.66 \text{ °W/K-m}^2 \tag{7.155}$$

The perimeter of the end ring takes into account only the portion exposed to the passing air. The mean diameter of the end ring is

$$D_{ring} = \frac{D_{or} - 2(0.047 + 0.04) - 0.572}{(100/2.54)} = 0.59 \text{ m} \tag{7.156}$$

The average length of the ring is

$$l_{ring} = \pi D_{ring} = 1.853 \text{ m} \tag{7.157}$$

The perimeter of the ring is

$$p_{ring} = \frac{2(0.75 + 1.0)}{(100/2.54)} = 0.0089 \text{ m} \tag{7.158}$$

The length and perimeter of the straight portions corresponding to the ends of the rotor bars projecting into the end region are

$$p_{end,2} = \frac{2(0.5625 + 0.375)}{(100/2.54)} = 0.0476 \text{ m} \tag{7.159}$$

$$l_{ew,2} = \frac{1.625}{(100/2.54)} = 0.0413 \text{ m} \tag{7.160}$$

By analogy to equation (7.143)

$$R_{ewcv,2} = \frac{\left(\frac{1}{\alpha_{ew,2}}\right)}{S_2 p_{end,2} l_{ew,2} + p_{ring} l_{ring}} \tag{7.161}$$

$$= \frac{\frac{1}{38.66}}{97 \times 0.0476 \times 0.0413 + 0.089 \times 1.856} = 0.0727\ {}^\circ\text{K/W} \tag{7.162}$$

The thermal resistances of the stator and rotor ducts must now be obtained. Again,

$$R_{duct,1} = \frac{\frac{t_{ins,1}}{\Lambda_{ins,1}} + \frac{1}{\alpha_{duct}}}{S_1 p_{coil,1} l_{duct}} \tag{7.163}$$

It has already been determined that $t_{ins,1} = 0.002$, $\Lambda_{ins,1} = 0.25$, $S_1 = 120$. Also, while not explicitly defined for heat transfer in a duct, it will be assumed that equation (7.125) remains reasonably valid. Thus,

$$\alpha_{duct} = 20 \times 2^{0.6} = 30.3 \text{ W/}^\circ\text{K-m}^2 \tag{7.164}$$

Also,

$$p_{coil} = \frac{2(2 \times 0.984 + 0.374)}{(100/2.54)} = 0.991 \text{ m} \tag{7.165}$$

$$l_{duct} = \left(\frac{3/8}{(100/2.54)}\right) = 0.0095 \text{ m} \tag{7.166}$$

The stator duct resistance is

$$R_{duct,1} = \frac{\frac{0.002}{0.25} + \frac{1}{30.3}}{120 \times 0.119 \times 0.0095} = 0.302\ {}^\circ\text{K/W} \tag{7.167}$$

Since the rotor bars are not insulated, the duct resistance for the rotor is, in general,

$$R_{duct,2} = \frac{\frac{1}{\alpha_{duct}}}{S_2 p_{bar,2} l_{duct}} \tag{7.168}$$

All quantities are already known except the perimeter of the rotor bar,

$$p_{\text{bar},2} = 2\frac{(0.5625 + 0.375)}{39.37} = 0.0476 \text{ m} \tag{7.169}$$

whereupon the resistance of the rotor duct is

$$R_{\text{duct},2} = \frac{\frac{1}{30.3}}{97 \times 0.0476 \times 0.0095} = 0.7524\ ^\circ\text{K/W} \tag{7.170}$$

It has already been established that the insulation thickness surrounding the stator conductor is 0.078 in. Since an additional air space is inevitable an additional air thickness between the insulation and the iron surface of 0.035 in. will be assumed. The resistances corresponding to these two thicknesses must be added in series. The thermal resistance corresponding to heat passing through the insulation is

$$R_{\text{ins},1a} = \frac{t_{\text{ins},1}}{\Lambda_{\text{ins},1} S_1 \left(b_s + 2d_s\right) l_i} \tag{7.171}$$

$$= \frac{0.078\left(\frac{100}{2.54}\right)}{0.25 \times 120 \times (0.374 + 2(0.984)) \times (8.5)} \tag{7.172}$$

$$= 0.0051\ ^\circ\text{K/W} \tag{7.173}$$

For the heat passing through the air layer between the insulation and the surface of the stator slot,

$$R_{\text{ins},1b} = \frac{t_{\text{air}}}{\Lambda_{\text{air}} S_1 \left(b_s + 2d_s\right) l_i} \tag{7.174}$$

$$= \frac{0.035\left(\frac{100}{2.54}\right)}{0.03 \times 120 \times (0.374 + 2(0.984)) \times (8.5)} \tag{7.175}$$

$$= 0.019\ ^\circ\text{K/W} \tag{7.176}$$

The total resistance to heat flow between the stator iron and stator conductors is

$$R_{\text{ins},1} = R_{\text{ins},1\text{a}} + R_{\text{ins},1\text{b}} = 0.0051 + 0.019 = 0.0241\ ^\circ\text{K/W} \tag{7.177}$$

The heat flow across the air gap is expressed by

$$R_{\text{ag}} = \frac{g}{\Lambda_{\text{air}} \pi D_{\text{is}} l_i} \tag{7.178}$$

whereupon,

$$R_{\text{ag}} = \frac{0.04(39.37)}{0.025(3.14159)(24.08)(8.5)} = 0.0980\ ^\circ\text{K/W} \tag{7.179}$$

The heat generated within the stator iron also rises to the outer surface of the stator, at which point heat is radiated and convected away. For the radiated portion, equation (7.112) gives

$$R_{\text{frad}} = \frac{1}{\alpha_r A_{\text{rad}}} = \frac{1}{\alpha_r \pi D_{\text{os}}(l_i)} \tag{7.180}$$

The equivalent heat convection coefficient depends upon the temperature difference between the iron surface and the surrounding air. Since the air temperature is, at this point, unknown it must be estimated. If a 50 °C temperature difference is assumed, then from Table 7.10, $\alpha_r = 7.1$ W/m^2-°K and

$$R_{\text{frad}} = \frac{1}{7.1 \times \pi \times \left(\dfrac{31.5}{100/2.54}\right) \times \left(\dfrac{8.5}{100/2.54}\right)} = 0.260\ °\text{K/W} \quad (7.181)$$

Finally, the thermal resistance for convection of heat from the stator surface to the air must be calculated. From equation (7.103), assuming turbulence and one bar pressure, the heat convection coefficient is

$$\alpha_c = 7.8v^{0.78} = 7.8 \times 4^{0.78} = 23\ \text{W/m}^2\text{-°K} \quad (7.182)$$

Thus,

$$R_{\text{fcv}} = \frac{1}{\alpha_c A_{\text{conw}}} \quad (7.183)$$

$$= \frac{1}{23 \times \pi \times \left(\dfrac{31.5}{100/2.54}\right) \times \left(\dfrac{8.5}{100/2.54}\right)} = 0.0801\ °\text{K/W} \quad (7.184)$$

For the combination of radiated and convected heat,

$$R_{\text{frame}} = \frac{R_{\text{frad}} R_{\text{fcv}}}{R_{\text{frad}} + R_{\text{fcv}}} = \frac{(0.260)(0.0801)}{0.260 + 0.0801} = 0.0612\ °\text{K/W} \quad (7.185)$$

Using Figure 7.23, the following nodal equations can be defined,

$$\frac{\Theta_{\text{air,d}} - \Theta_1}{R_{\text{duct,1}} + \dfrac{R_{\text{cu,1}}}{4}} + \frac{\Theta_{\text{air,d}} - \Theta_{\text{gap,2}}}{\dfrac{R_{\text{cu,2}}}{4} + R_{\text{duct,2}}} = 0 \quad (7.186)$$

$$\frac{\Theta_1 - \Theta_{\text{air,d}}}{R_{\text{duct,1}} + \dfrac{R_{\text{cu,1}}}{4}} + \frac{\Theta_1 - \Theta_2}{\dfrac{R_{\text{cu,1}}}{4} + \dfrac{R_{\text{ew,1}}}{2}} + \frac{\Theta_1 - \Theta_{\text{gap,1}}}{2R_{\text{ins,1}}} = 0 \quad (7.187)$$

$$\frac{\Theta_2 - \Theta_1}{\dfrac{R_{\text{cu,1}}}{4} + \dfrac{R_{\text{ew,1}}}{2}} + \frac{\Theta_2 - \Theta_{\text{air,l}}}{\dfrac{R_{\text{ew,1}}}{2} + R_{\text{ewcv,1}}} = Q_{\text{ew,1}} \quad (7.188)$$

$$\frac{\Theta_{\text{air,l}} - \Theta_2}{\dfrac{R_{\text{ew,1}}}{2} + R_{\text{ewcv,1}}} + \frac{\Theta_{\text{air,l}} - \Theta_3}{R_{\text{ewcv,2}} + \dfrac{R_{\text{ew,2}}}{2}} = 0 \quad (7.189)$$

$$\frac{\Theta_3 - \Theta_{\text{air,l}}}{R_{\text{ewcv,2}} + \dfrac{R_{\text{ew,2}}}{2}} + \frac{\Theta_3 - \Theta_{\text{gap,2}}}{\dfrac{R_{\text{ew,2}}}{2} + \dfrac{R_{\text{cu,2}}}{4}} = Q_{\text{ew,2}} \quad (7.190)$$

$$\frac{\Theta_{gap,2}-\Theta_{gap,1}}{2R_{ag}}+\frac{\Theta_{gap,2}-\Theta_{3}}{\frac{R_{ew,2}}{2}+\frac{R_{cu,2}}{4}}+\frac{\Theta_{gap,2}-\Theta_{air,d}}{\frac{R_{cu,2}}{4}+R_{duct,2}}=\frac{Q_{slot,2}+Q_{tooth,2}}{2} \quad (7.191)$$

$$\frac{\Theta_{gap,1}-\Theta_{gap,2}}{2R_{ag}}+\frac{\Theta_{gap,1}-\Theta_{trad,1}}{2R_{trad,1}}+\frac{\Theta_{gap,1}-\Theta_{1}}{2R_{ins,1}}=\frac{Q_{slot,1}+Q_{tooth,1}}{2} \quad (7.192)$$

$$\frac{\Theta_{trad,1}-\Theta_{crad,1}}{2R_{crad,1}}+\frac{\Theta_{trad,1}-\Theta_{gap,1}}{2R_{trad,1}}=0 \quad (7.193)$$

$$\frac{\Theta_{crad,1}-\Theta_{air,m}}{2R_{frame}}+\frac{\Theta_{crad,1}-\Theta_{trad1}}{2R_{crad,1}}=\frac{Q_{core,1}}{2} \quad (7.194)$$

It should be noted that some of the eight unknown temperatures could easily be algebraically eliminated since several of the temperatures are the result of summing only two heat flows at a node. However, since the solution of these equations by computer is rapidly done, the full complement of temperatures can be retained.

The correct value for the temperature at the machine midpoint $\Theta_{air,m}$ can be obtained by observing that the heat injected into the air on the inlet side of the machine is

$$\begin{aligned} P_{diss} &= \frac{P_{cu,1}}{2}+\frac{P_{cu,2}}{2}+\frac{Q_{core,1}}{2}+\frac{Q_{teeth,1}}{2}+\frac{Q_{teeth,2}}{2} \\ &= \frac{3104+5920+950+930+644}{2}=5774\ \text{W} \end{aligned} \quad (7.195)$$

From equation (7.113), with CMS in cubic meters per second air flow, the temperature rise of the air from the inlet to the midpoint of the machine is

$$\Delta\Theta=\frac{8.879\times10^{-4}P_{diss}}{\text{CMS}} \quad (7.196)$$

where the air flow is

$$\text{CMS}=\frac{\text{CFM}}{60\times3.28^3}=\frac{500}{60\times3.28^3}=0.236\ \text{m}^3/\text{s} \quad (7.197)$$

where CFM is the air flow in cubic feet per minute. Hence,

$$\Delta\Theta=\frac{8.879\times10^{-4}\times5774}{0.236}=22\ {}^\circ\text{K or }{}^\circ\text{C} \quad (7.198)$$

Thus the temperature of the cooling air at the midpoint of the machine is

$$\Theta_{air,m}=\Theta_{air,l}+\Delta\Theta=30+22=52\ {}^\circ\text{C} \quad (7.199)$$

Note that this value is essentially the same as the 50 °C value assumed to calculate R_{rad} and R_{conv}. In general, an iteration could be performed to converge on a more accurate answer. However, in view of the approximations inherent in the approach taken, it is usually not warranted.

Assuming symmetrical cooling on both sides of the machine, the air at the exit end of the machine will be

$$\Theta_{\text{air,m}} = \Theta_{\text{air,exit}} = 30 + 22 = 52\ °\text{C} \tag{7.200}$$

If the inlet air only enters from the left side of the machine, then the air at the exit end of the machine will be

$$\Theta_{\text{air,r}} = \Theta_{\text{air,exit}} = 30 + 2(22) = 74\ °\text{C} \tag{7.201}$$

Since the air temperature at the midpoint of the machine is algebraically related to the inlet air by equation (7.199), one of the nine temperature equations, equation (7.189), is redundant. Also $\Theta_{\text{air,d}}$ is essentially equal to the inlet temperature, since it is not heated before entering the duct so that equation (7.186) is redundant. In effect, $\Theta_{\text{air,l}}$ becomes the *reference node* of the nodal equations. In this case, all of the temperatures may be calculated in terms of *temperature rise above ambient.* The actual temperatures can simply be found by adding the ambient temperature to the temperature rise at all nodes. It will be assumed here that the air temperature at the inlet to the duct is essentially the same as the ambient air, in which case $\Theta_{\text{air,l}}$ and $\Theta_{\text{air,d}}$ will be set to 30 °C.

The remaining seven nodal equations can be assembled into matrix form for a solution. The matrix equation can be written as

$$[G_{\text{h}}] \cdot [\Theta] = [Q_{\text{h}}] \tag{7.202}$$

where

$$[\Theta]^{\text{t}} = [\Theta_1, \Theta_2, \Theta_3, \Theta_{\text{gap,2}}, \Theta_{\text{gap,1}}, \Theta_{\text{trad,1}}, \Theta_{\text{crad,1}}] \tag{7.203}$$

and

$$[Q_{\text{h}}] = \begin{bmatrix} \dfrac{\Theta_{\text{air,d}}}{R_{\text{duct,1}+\frac{R_{\text{cu,1}}}{4}}} \\ Q_{\text{ew,1}} + \dfrac{\Theta_{\text{air},l}}{\dfrac{R_{\text{ew,2}}}{2} + R_{\text{ewcv,1}}} \\ Q_{\text{ew,2}} + \dfrac{\Theta_{\text{air},l}}{\dfrac{R_{\text{ew,2}}}{2} + R_{\text{ewcv,2}}} \\ \dfrac{Q_{\text{slot,2}} + Q_{\text{tooth,2}}}{2} + \dfrac{\Theta_{\text{air,d}}}{\dfrac{R_{\text{cu,2}}}{4} + R_{\text{duct,2}}} \\ \dfrac{Q_{\text{slot,1}} + Q_{\text{tooth,1}}}{2} \\ 0 \\ \dfrac{Q_{\text{core,1}}}{2} + \dfrac{\Theta_{\text{air,m}}}{2R_{\text{frame}}} \end{bmatrix} \tag{7.204}$$

The matrix defining $[G_h]$ is given below. The elements on the main diagonal $R_1 \ldots R_8$ correspond to the inverse sum of the thermal resistances connected to the eight nodes. For example,

$$\frac{1}{R_1} = \frac{1}{R_{cu,1}/4 + R_{duct,1}} + \frac{1}{R_{cu,1}/4 + R_{ew,1}/2} \tag{7.205}$$

and so forth. The matrix can be easily inverted using any of a number of software programs to solve for the unknown vector $[\Theta]$.

$$[G_h] = \begin{bmatrix} \frac{1}{R_1} & -\frac{1}{\frac{R_{cu,1}}{4}+\frac{R_{ew,1}}{2}} & 0 & 0 & -\frac{1}{2R_{ins,1}} & 0 & 0 \\ -\frac{1}{\frac{R_{cu,1}}{4}+\frac{R_{ew,1}}{2}} & \frac{1}{R_2} & 0 & 0 & 0 & 0 & 0 \\ 0 & 0 & \frac{1}{R_3} & -\frac{1}{\frac{R_{ew,2}}{2}+\frac{R_{cu,2}}{4}} & 0 & 0 & 0 \\ 0 & 0 & -\frac{1}{\frac{R_{ew,2}}{2}+\frac{R_{cu,2}}{4}} & \frac{1}{R_4} & -\frac{1}{2R_{ag}} & 0 & 0 \\ -\frac{1}{2R_{ins,1}} & 0 & 0 & -\frac{1}{2R_{ag}} & \frac{1}{R_5} & -\frac{1}{2R_{trad,1}} & 0 \\ 0 & 0 & 0 & 0 & -\frac{1}{2R_{trad,1}} & \frac{1}{R_6} & -\frac{1}{2R_{crad,1}} \\ 0 & 0 & 0 & 0 & 0 & -\frac{1}{2R_{crad,1}} & \frac{1}{R_7} \end{bmatrix} \tag{7.206}$$

The resulting solution vector is

$$[\Theta] = \begin{bmatrix} \Theta_1 \\ \Theta_2 \\ \Theta_3 \\ \Theta_{gap,2} \\ \Theta_{gap,1} \\ \Theta_{trad,1} \\ \Theta_{crad,1} \end{bmatrix} = \begin{bmatrix} 132.2 \\ 98.0 \\ 423.2 \\ 477.2 \\ 218.6 \\ 205.3 \\ 193.8 \end{bmatrix} \tag{7.207}$$

Thus the internal winding temperature is 132 °C while the end-winding temperature is predicted to be 98 °C, which is within the limit for class F or H insulation material. (see Table 7.1), and could suggest a need for an improved cooling scheme. It is interesting to note that the temperature of rotor is much greater than the stator, a typical result. However, the rotor bars are not equipped with cooling fins which is usually the case. In any case the temperature is far from the melting point of copper (1083 °C).

7.10 TRANSIENT HEAT FLOW

7.10.1 Externally Generated Heat

So far, the stationary heat distribution in electric machines was analyzed which is characteristic for a steady-state operation. When the machine is either in an electrical or in a mechanical transient, the amounts of heat flow and/or the values of heat convection coefficients vary. As a consequence, the temperature distribution in a machine becomes a function of time. The heat source in the Poisson partial differential equation, equation (7.2), becomes time variant and must be used in order to find the temperature distribution. Such a time dependent, three-dimensional partial differential equation (the Fourier partial differential equation) is difficult to solve. However, many transient heat problems in electric machines can be solved by assuming a one-dimensional heat distribution, instead of the three-dimensional given in equation (7.5).

For one-dimensional heat distribution in a homogeneous isotropic body, one can write

$$Q_{\mathrm{h}}dt = (A\alpha_{\mathrm{c}}\Theta)dt + (mc_{\mathrm{p}})d\Theta \tag{7.208}$$

where $Q_{\mathrm{h}}dt$ (Joules) is the heat energy impressed on the surface in time interval dt, m (kg) is the mass of the body, c_{p} (Joules/(kg °K) is its specific heat, and Θ is the temperature rise (°K). The product mc_{p} is called thermal capacitance of the body, and has units Joules/°K. The product $A\alpha_{\mathrm{c}}$ is the thermal conductivity of the body surface on which heat is exchanged with the surroundings. Denoting the thermal capacitance by C_{h}, that is, $C_{\mathrm{h}} = mc_{\mathrm{p}}$, and thermal resistance by $R_{\mathrm{h}} = 1/A\alpha_{\mathrm{c}}$, one can express equation (7.208) as

$$Q_{\mathrm{h}}dt = \frac{\Theta}{R_{\mathrm{h}}}dt + C_{\mathrm{h}}d\Theta \tag{7.209}$$

The first term on the right-hand side of equation (7.209) is the heat transferred to the surroundings at the temperature rise (temperature difference) between the body and the surrounding ambient temperature Θ_{a}. The second term, $C_{\mathrm{h}}\, d\Theta$, is equal to an increase of the accumulated thermal energy in the body.

The body temperature at steady-state Θ_{m} is obtained by setting $d\Theta = 0$ in equation (7.209). This condition gives therefore

$$\Theta_{\mathrm{m}} = Q_{\mathrm{h}}R_{\mathrm{h}} \tag{7.210}$$

The body temperature at steady state is proportional to the power of losses converted into heat and to the thermal resistance of the body. Denoting by

$$\tau_{\mathrm{h}} = R_{\mathrm{h}}C_{\mathrm{h}} = \frac{mc_{\mathrm{p}}}{A\alpha_{\mathrm{c}}} \tag{7.211}$$

the thermal time constant of the body, one can write equation (7.209) as

$$\Theta + \tau_{\mathrm{h}}\frac{d\Theta}{dt} = \Theta_{\mathrm{m}} \tag{7.212}$$

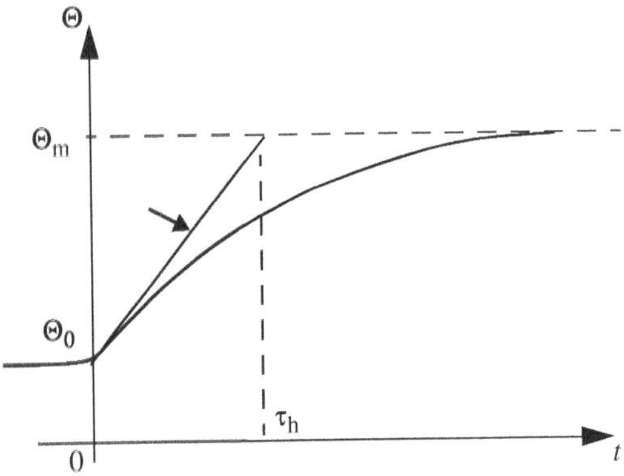

Figure 7.24 Temperature rise in a homogenous body during transient heating.

the solution of which is

$$\Theta = \Theta_m K e^{-\frac{t}{\tau_h}} \tag{7.213}$$

The constant of integration K can be obtained from initial condition $\Theta(0) = \Theta_o$, that is, the temperature rise at time instant $t = 0$ is equal to Θ_o. This results in

$$K = \Theta_o - \Theta_m \tag{7.214}$$

Therefore, the temperature rise during the transient heating phase can be described as

$$\Theta = \Theta_o + (\Theta_m - \Theta_o)\left(1 - e^{-\frac{t}{\tau_h}}\right) \tag{7.215}$$

which is shown in Figure 7.24.

During the cooling phase, the final temperature rise of the body is equal to zero, that is, the value for $\Theta_m = 0$ should be assumed in equation (7.212).

$$\Theta = \tau_h \frac{d\Theta}{dt} = 0 \tag{7.216}$$

Setting for initial condition $\theta(0) = \theta_o$, one obtains

$$\Theta = \Theta_o e^{-\frac{t}{\tau_h}} \tag{7.217}$$

The temperature decrease during cooling is shown in Figure 7.25.

7.10.2 Internally Generated Heat—Stalled Operation

Although the stalled condition of an induction motor is usually an abnormal mode of operation, when it does occur there are many unusual stresses that are involved which must be factored into the design. Failure of many components, both inside and outside the motor can cause such operation. For example, any faulty connection within the motor or in a feed line to the motor which opens a phase will result

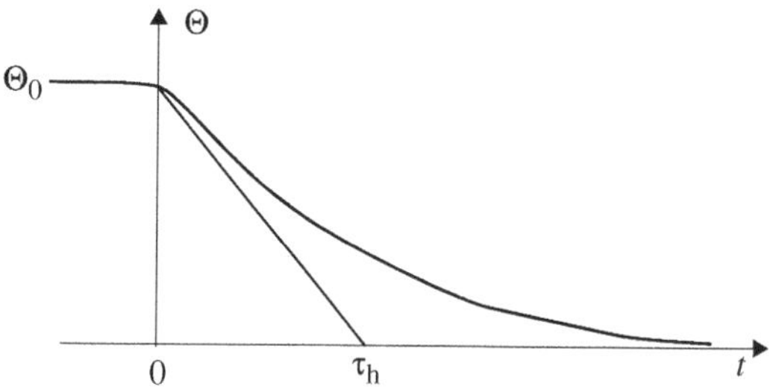

Figure 7.25 Temperature decrease in a homogeneous body during cooling.

in the disappearance of starting torque and high current in the remaining two windings. Broken end rings or rotor bars can sometimes reduce starting torque sufficiently to stall the motor. Since the torque of an induction motor varies as the square of the voltage, performance is sensitive to line voltage. Low line voltage or poor regulation of line voltage due to high starting current is a frequent cause of starting problems.

When a motor becomes stalled, the resulting high currents can quickly cause damage if the protection scheme is not adequate. Stalled motor currents are of the order of four to seven times full-load current. Consequently, depending upon the design, current densities under stalled conditions may be as much as 40,000 amps/in^2. All of the heat generated by the rotor current must be stored within the bars during such a short period of time, very little can transfer by convection from the outer surface. One can write

$$\text{energy stored} = \text{energy dissipated}$$

$$c_p\left(\frac{d\Theta}{dt}\right)V_{bar} = (\rho J^2_{s(rms)})V_{bar} \tag{7.218}$$

where ρ is the resistivity, $J_{s(rms)}$ is the current density in RMS, and c_p is the heat capacity of the conductor material and is equal to the product of the specific heat times the bar density. Note that this quantity is not quite the same but is obviously related to the thermal capacitance C_h which is equal to the specific heat times the mass of the body. This equation can be written as

$$\frac{d\Theta}{dt} = \frac{J^2_{s(rms)}}{c_p/\rho} \tag{7.219}$$

Table 7.12 shows values of c_p/ρ for a number of conductor materials. From this table, it can be determined that if the stator current density is 15,000 amps/in.^{2} (23 amps/mm^2) the rate rise of temperature 3.5 °K/s, resulting in a temperature rise of 150 °C in 35 s. If the current density is increased to 40,000 amps/in.^{2} (62 amps/mm^2), the rate of rise of temperature increases to 25 °K/s producing a temperature rise of 150 °C in only 5 s. Generally speaking, small, lightweight, well-ventilated motors

TABLE 7.12 Transient heating coefficient for several materials

Material	c_p/ρ^*
Copper	63.1×10^6
Cast aluminum	20.0×10^6
Cast zinc	12.9×10^6
Brass	15.6×10^6
Cast bronze	10.6×10^6

* Values of c_p/ρ are for 100C: Units are J/°K-ohm-inch^4

are designed with a relatively high starting current density and the problem of coordination of protection is a major concern.

7.10.3 Thermal Instability

Thermal instability is a phenomenon which can occur in thermally overloaded current-carrying conductors. The specific electrical conductivity σ of a conductor is a function of temperature rise, defined as

$$\sigma(\Delta\Theta) = \frac{\sigma(20\ °\mathrm{C})}{1 + \beta_c \Delta\Theta} \tag{7.220}$$

where $\sigma(20\ °\mathrm{C})$ is the specific electrical conductivity at 20 °C, $\Delta\Theta$ is the temperature rise relative to 20 °C, and β_c is the temperature coefficient equal to 1/255 (°C)−1 for copper. Assuming that the conductor resistance has no influence on the current in the conductor (i.e., the leakage inductance dominates), the I^2R losses increase proportionally to the resistance which is temperature dependent. The effects of this positive feedback between the losses and temperature rise are dependent on the conductor thermal parameters. If the generated losses become greater than the thermal energy transferred to the conductor surroundings, the accumulated thermal energy increases to, theoretically, an infinite amount. This is accompanied by a conductor temperature increase up to the melting point of the conductor. Thermal instability occurs when

$$\frac{P_e(\Delta\Theta)}{A\alpha_c} > \frac{1}{\beta_c} \quad \text{for copper:} \quad \frac{P_e \Delta(20\ °\mathrm{C})}{A\alpha_c} > 255 \tag{7.221}$$

A copper conductor becomes unstable when the current through it leads to a stationary temperature rise greater than 255 °C.

The thermal capacitance has a considerable importance in calculating the heat rise in machines during starting. Often it is this mode of operation rather than steady-state heat flow which determines the ultimate life of the machine. The thermal equivalent circuit for the transient case is essentially the same except that the thermal capacitance for each major component in the machine, that is, stator core, stator teeth, stator copper, etc., includes a thermal capacitance in parallel with the heat loss input node. A typical example for a member with two-dimensional heat flow is shown in Figure 7.26.

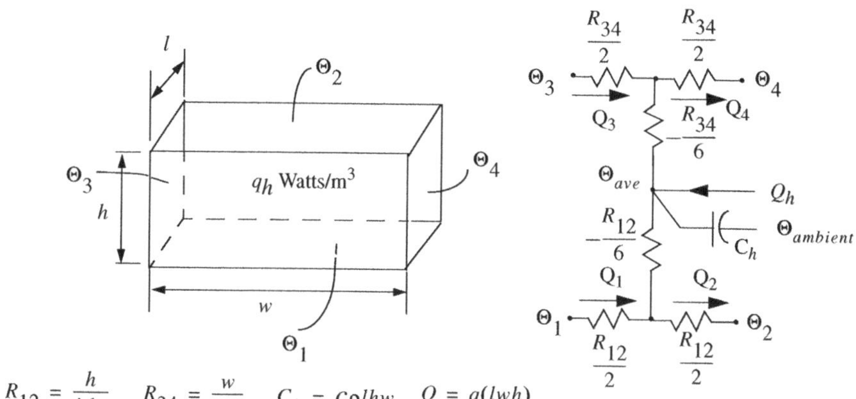

$$R_{12} = \frac{h}{\Lambda lw} \quad R_{34} = \frac{w}{\Lambda lh} \quad C_h = c\rho lhw \quad Q = q(lwh)$$

Figure 7.26 Two-dimensional heat flow including thermal capacitance.

7.11 CONCLUSION

The machine designer can never call himself solely an electrical engineer. Indeed, almost all of the ultimate limiting factors in a good design are outside the branch of electrical engineering and involve mechanical stress, material properties, rotational dynamics, not to mention the thermal issues treated in this chapter. Management and/or resolution of these problems after the electrical design has been completed will ultimately determine whether the machine succeeds in its intended use.

REFERENCES

[1] G. Gotter, *Erwarmung und Kuhlung Elektrischer Maschinen*, Springer-Verlag, 1954.

[2] J. Perez and J. G. Kassakian, "Stationary thermal model for smooth air-gap rotating electric machines," *Electric Machines and Electromechanics*, vol. 3, pp. 285–303, 1979.

[3] V. Ostovic, *Thermal Design of Electric Machines*, UW Short Course, AC Machine Design, August 19–21, 1996.

[4] J. Saari, "Thermal Analysis of High-speed Induction Machines," Ph.D. Thesis, Acta Polytechnica Scandinavica Electrical Engineering Series No. 90, Helsinki University of Technology, Espoo, Finland, 1998.

[5] D. Roberts, "The Application of an Induction Motor Thermal Model to Motor Protection and Other Functions," Ph.D. Thesis, University of Liverpool, September 1986.

[6] C. R. Soderberg, "Steady flow of heat in large turbine generators," *AIEE Transactions*, vol. 50, June 1931, pp. 787–802.

[7] G. F. Luke, "The cooling of electrical machines," *AIEE Transactions*, vol. 42, 1923, pp. 636–652.

[8] G. F. Luke, "Surface Heat Transfer in Electric Machines with Forced Air Flow," *AIEE Transactions*, vol. 45, 1926, p. 1036–1047.

[9] J. Lindstrom, "Thermal Model of a Permanent-Magnet Motor for a Hybrid Vehicle," Technical Report, Chalmers University of Technology, Goteborg, Sweden, April 1999.

[10] B. Eck, *Ventilatoren*, Springer-Verlag, 1962.

CHAPTER 8

PERMANENT MAGNET MACHINES

WITH THE EMERGENCE OF modern ferrite and rare earth magnets, permanent magnet (PM) machines are being increasingly used in applications where efficiency, or more importantly, losses are of concern. One such application is in servo drives in which the motor is commanded to operate for a long period at, or very near, zero speed. In such applications, the heat rejection of a conventional squirrel-cage induction motor drive could be insufficient to keep the machine (usually the rotor) at tolerable levels. Since PM machines use magnets to supply the excitation current (current required to magnetize the iron parts), a substantial reduction of the stator current is possible, thereby greatly reducing the losses. Increased use of such machines will inevitably also become important in conventional operation from the utility supply when the cost of energy increases to a sufficiently high value.

The basic objective of this chapter is to relate the operational requirements of permanent magnet motors to their dimensions, shapes, and material properties. Design is always a combination of science and experience. Some examples will be given, but the numbers should be regarded only as approximations with limited generality. Emphasis should be placed on how motor parameters and characteristics vary with dimensions and, conversely, what latitude is available to the designer to obtain a desired set of motor characteristics.

For simplicity, the initial focus is on the design of three-phase, medium to large rating motors with distributed sinusoidally fed stator windings and surface-mounted neodymium–iron–boron (Nd–Fe–B) magnets on the rotor. Later, the same principles are applied to other magnetic materials (ferrites, Sm–Co) and to rectangular wave-fed PM motors. The subject of buried magnet motor designs (designs in which the magnets are placed inside the rotor body) is introduced, but not pursued in detail. The reader is referred to recent papers on this still evolving topic.

8.1 MAGNET CHARACTERISTICS

Ordinarily, a piece of iron consists not of a single crystal, but of an aggregate of small crystal fragments with axes oriented at random. The situation in a small piece of iron may be represented schematically as in Figure 8.1. Here a number of crystal

Introduction to AC Machine Design, First Edition. Thomas A. Lipo.

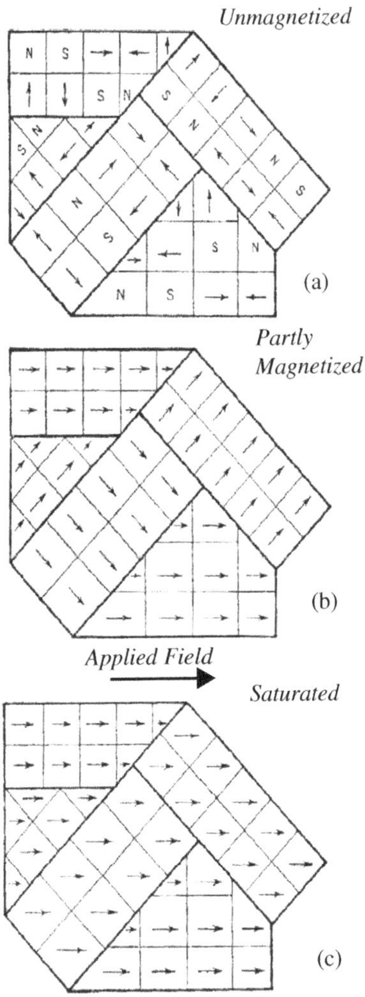

Figure 8.1 Successive stages of magnetization of a polycrystalline specimen with increasing field.

fragments are shown, each with a number of magnetic domains, represented by a small square. The boundaries between crystal fragments are indicated by the heavy lines, and domain boundaries by the light lines which also indicate the direction of the crystal axes. Not only is the piece of iron unmagnetized, but also the individual crystal fragments are unmagnetized. The domains in each crystal are magnetized along the directions of easy magnetization, that is, along the three crystal axes. However, the polarity of adjacent domains is opposite so that the total magnetization of each crystal is negligible.

With the application of a magnetic field $\boldsymbol{H}$ in the direction indicated by the arrows, Figure 8.1b, some of the domains with polarities opposed to or perpendicular to the applied field become unstable and rotate quickly to another direction of easy magnetization in the same direction as the field, or more nearly so. These changes take place on the steep part of the magnetization curve. After all the domains have changed direction with a sufficiently high field, the result is as suggested in Figure 8.1b.

A domain may contain millions of atoms, and since each domain flips from one direction of easy magnetization to another in an interval measured in milliseconds, the magnetization proceeds by steps rather than in a smooth, continuous manner. These steps are called *Barkhausen steps* or jumps. The stepped characteristic can be observed by sensitive measurements. The Barkhausen jumps are largest on the steep part of the magnetization curve.

With further increase in the applied field, the direction of magnetization of the domains not already parallel to the field is rotated gradually toward the direction of $\boldsymbol{H}$. This increase in magnetization is more difficult, and very high fields may be required to reach saturation, where all domains are magnetized parallel to the field, as indicated in Figure 8.1c. This accounts for the flatness of the upper part of the magnetization curve.

This picture of the magnetization process is an oversimplified one, but it accounts qualitatively for many of the important phenomena. Another phenomenon is the change in size of domains during the magnetization process. Not only do domains change in size, but the entire specimen changes in length during magnetization. This effect is called *magnetostriction.*

It has been learned from Chapter 1 that when operating in a magnetic material, magnetic dipoles $\boldsymbol{m}$ are created which line up with the magnetic field resulting in the definition of the magnetic polarization vector

$$\boldsymbol{M} = N\boldsymbol{m} \tag{8.1}$$

where N is the number of magnetic dipoles per unit volume. The magnetic field intensity $\boldsymbol{H}$ inside the magnetic material becomes the result given by equation (1.47), that is

$$\boldsymbol{H} = \frac{\boldsymbol{B}}{\mu_0} - \boldsymbol{M} \tag{8.2}$$

or, alternatively,

$$\boldsymbol{B} = \mu_0(\boldsymbol{H} + \boldsymbol{M}) \tag{8.3}$$

In a soft magnetic material such as common steel, one typically factors out the quantity $\boldsymbol{H}$ from equation (8.3) and writes it as

$$\boldsymbol{B} = \mu_0\left(1 + \frac{\boldsymbol{M}}{\boldsymbol{H}}\right)\boldsymbol{H} \tag{8.4}$$

and, since $\boldsymbol{M}$ and $\boldsymbol{H}$ are collinear, it is then possible to proceed to define,

$$\mu_r = 1 + \frac{\boldsymbol{M}}{\boldsymbol{H}} \tag{8.5}$$

whereupon, the usual expression results

$$\boldsymbol{B} = \mu_0\mu_r\boldsymbol{H} \tag{8.6}$$

Conversely if $\boldsymbol{M}$ is solved in terms of the field intensity $\boldsymbol{H}$, then again using equation (8.5),

$$\boldsymbol{M} = \frac{\mu - \mu_0}{\mu_0}\boldsymbol{H} = (\mu_r - 1)\boldsymbol{H} \tag{8.7}$$

8.2 HYSTERESIS

If the field applied to a specimen is increased to saturation and is then decreased, the flux density B decreases, but not as rapidly as it increased along the initial magnetization curve. Thus, when H reaches zero, there is a residual flux density, or *remanence*, B_r.

In order to reduce B to zero, a negative field $-H_c$ must be applied, see Figure 8.2. This field is called the *coercive force* H_c. As H is further increased in the negative direction, the specimen becomes magnetized with the opposite polarity, the magnetization at first being easy and then difficult as saturation is approached. Bringing the field to zero again leaves a residual magnetization or flux density $-B_r$ and to reduce B to zero, a coercive force $+H_c$ must be applied. With further increase in the field, the specimen again becomes saturated with the original polarity.

The phenomenon which causes B to lag behind H so that the magnetization curve for increasing and decreasing fields is not the same, is termed *hysteresis*, and the loop traced out by the magnetization curve is called a *hysteresis loop* (Figure 8.2). If the material is carried to saturation at both ends of the magnetization curve, the loop is called the saturation, or major hysteresis loop. The residual flux density B_r on the saturation loop is called the *retentivity*, and the coercive force H_c on this loop is

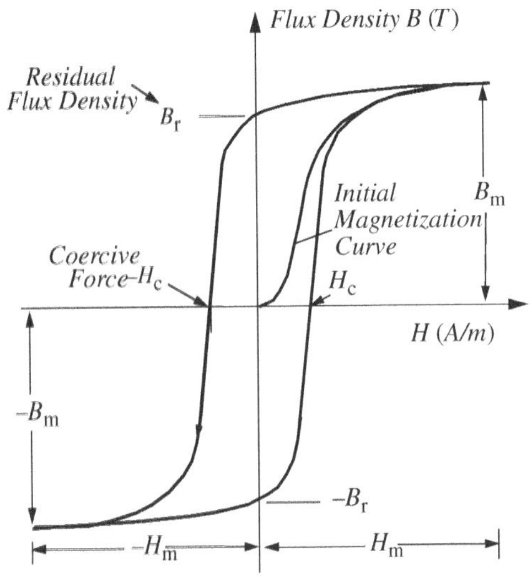

Figure 8.2 Hysteresis loop.

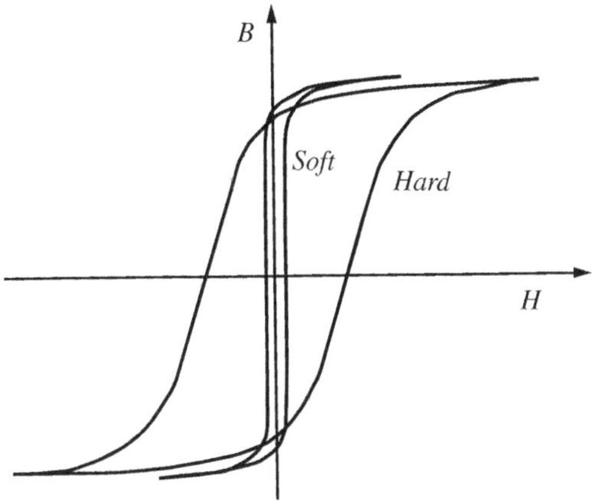

Figure 8.3 Hysteresis loops for soft and hard magnetic materials.

called the *coercivity*. Thus, the retentivity of a material is the maximum value which the residual flux density can attain and the coercivity the maximum value which the coercive force can attain. For a given specimen, no points can be reached on the B–H diagram outside of the saturation hysteresis loop, but any point inside can.

For "soft," or easily magnetized, materials, the hysteresis loop is "thin" as suggested in Figure 8.3, with a small area enclosed. By way of comparison, the hysteresis loop of a hard magnetic material is also shown, the area enclosed in this case being considerably greater. This figure also illustrates that the value of H_c is many times greater than the value of H needed to demagnetize a piece of iron of the same length.

Turning attention to how the permeability varies, consider the hysteresis loop of Figure 8.4a. The corresponding graph of μ as a function of H is as shown in Figure 8.4b. At $H = 0$, it is apparent that μ becomes infinite. On the other hand, when $B = 0$, $\mu = 0$. Under such conditions, the permeability μ becomes meaningless in

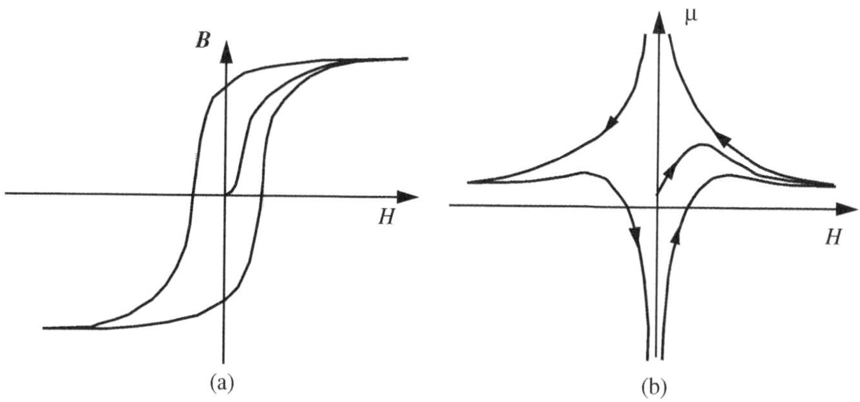

Figure 8.4 (a) Hysteresis loop, (b) corresponding permeability curve.

the context of magnet behavior. Therefore the use of μ must be confined to situations where it has significance, as, for example, to the initial, or DC, magnetization curve. It is to be noted that the term *maximum permeability* signifies specifically the maximum permeability for an initial magnetization curve and not for a hysteresis loop or other type of magnetization curve.

8.3 PERMANENT MAGNET MATERIALS

Rare-earth magnets, in general, and neodymium-based magnets, in particular, have experienced a remarkable improvement in the few decades as can be seen in Figure 8.5. This material is an alloy which actually has an exact composition of $Nd_2Fe_{14}B$ and can also involve even more complex compositions. This material is significantly cheaper than samarium cobalt, but nevertheless has a higher flux density, coercivity, and maximum energy produce than $SmCo_5$. Unfortunately, it generally has poor corrosion resistance and the coercivity decreases rapidly with increasing temperature. Small amounts of other materials such as cobalt is often added to improve its high temperature characteristics Nd–Fe–B exists in a variety of grades and formulations with energy product ranging from 200 to 400 kJ/m^3 (26–50 MegaGauss-Oersteds). Two different techniques exist for manufacturing this type of magnet including the melt-spun Magnaquench® process originally developed by General Motors and the sintering technique originated by Sumitomo.

The demagnetization characteristic of Nd–Fe–B material is typically of the form shown in Figure 8.6 where the residual flux density B_r is typically in the range 1.1–1.2 teslas, but as high as 1.4 teslas and the slope B–H is plotted with relative recoil

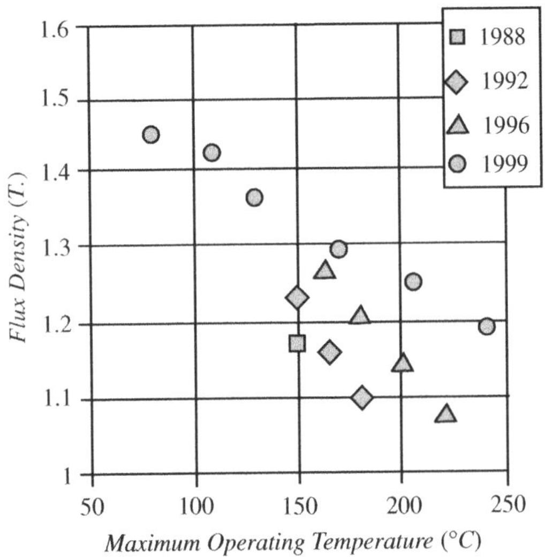

Figure 8.5 Historical improvement of neodymium rare earth permanent magnets over a ten year period of rapid development.

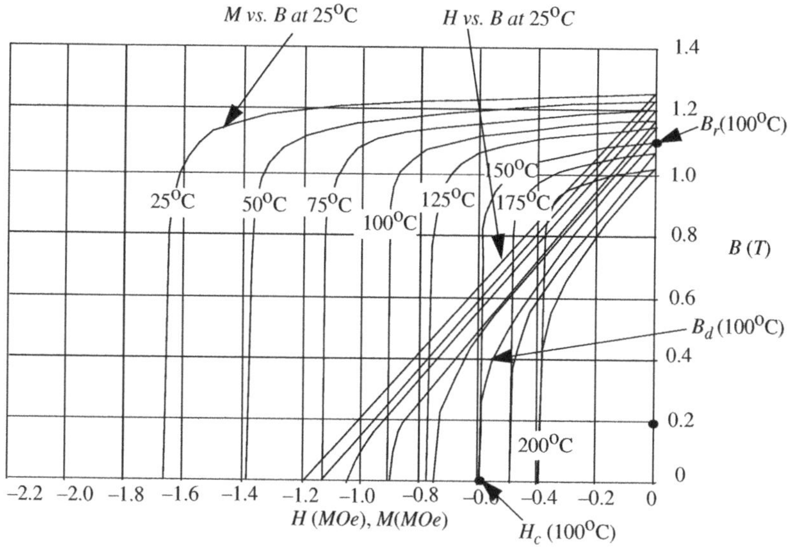

Figure 8.6 *B–H* characteristic for a typical neodymium–iron–boron magnetic material [79.6**H* (Oe) = *H* (A/m)].

relative permeability μ_r of about 1.05. The characteristic is linear and repeatable for negative values of field intensity H up to the value H_d at which point the magnetic polarization vector $\boldsymbol{M}$ collapses. This value is typically about −1.5 MA/m at 60°C reducing to about −1 MA/m at 100°C. The corresponding value of flux density is termed B_d.

Neodymium, sometimes termed "Neo" magnets, comes in a wide variety of grades to match different performance requirements and cost objectives. In order to assess the suitability of a particular magnet grade, figure of merits have been established. The most common figure of merit is the magnet's "energy product" which can be expressed as

$$\text{Energy product} = \frac{B_r^2}{2\mu_{rm}\mu_0} \tag{8.8}$$

This figure of merit is most useful for evaluating the magnet's effectiveness in producing gap field energy. An alternative figure of merit has been suggested by researchers [1] for applications which must operate in the presence of large demagnetizing fields.

$$\text{Demagentization energy product} = \frac{B_r H_c}{4} \tag{8.9}$$

Table 8.1 provides a listing of current Neo magnet properties and their respective figure of merits. The temperature sensitivities of the various magnet grades are represented by their reversible temperature coefficients: α for the residual induction and β for the intrinsic coercivity. Figure 8.7 graphs B_r versus H_c values for these

TABLE 8.1 Typical high performance neodymium magnet data

	B_r (T)	α (%°C)	H_c (kA/m)	β (%/°C)	Energy product at 120°C (kJ/m³)	Demagnetization energy product at 20°C (kJ/m³)	Energy product at 120°C (kJ/m³)	Demagnetization energy product at 20°C (kJ/m³)
N5O	1.4	−0.1	−0.995	−0.61	390	349	316	123
N48M	1.37	−0.1	−1.17	−0.6	373	400	302	144
N45H	1.35	−0.1	−1.274	−0.59	363	430	294	158
N4SH	1.30	−0.1	−1.672	−0.55	337	544	272	220
N36UH	1.22	−0.09	−2.149	−0.5	296	656	245	298
N34Z	1.16	−0.1	−2.574	−0.45	267	739	216	365

grades at two temperature points. Overall, the best magnet grade for developing maximum shear stress is not simply the one with the highest B_r. Generally the grade with the highest coercivity is favored to help reduce the magnet volume. A small amount of the rare-earth element dysprosium is frequently added to raise the coercivity for demanding applications such as permanent magnet motors.

8.4 DETERMINATION OF MAGNET OPERATING POINT

In contrast to iron, permanent magnet materials are "hard" resulting in nearly a square magnetization characteristic. When placed in a magnetic circuit, the magnetic polarization M acts as equivalent MMF source, forcing the flux to flow around a magnetic circuit. Consider, for example, the simple circuit shown in Figure 8.8 in which the magnet is assumed to be connected in series with a number of materials having no (negligible) hysteresis. Note that the B–H curve corresponding to the magnet can be considered as the sum of two curves, one produced by the magnetization vector M

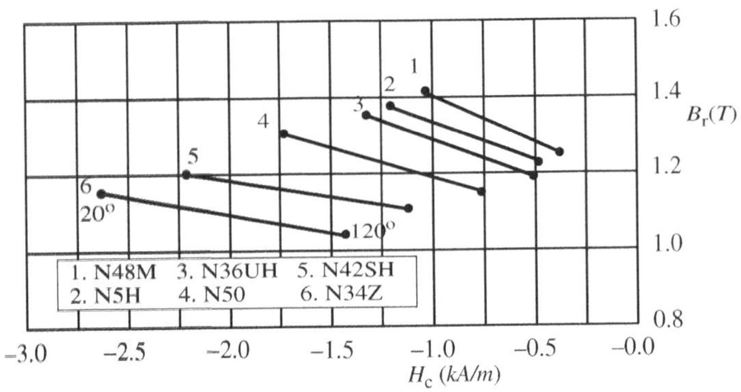

Figure 8.7 Variation of corner points H_c–B_r for Nd magnets over the temperature range 20–120°C.

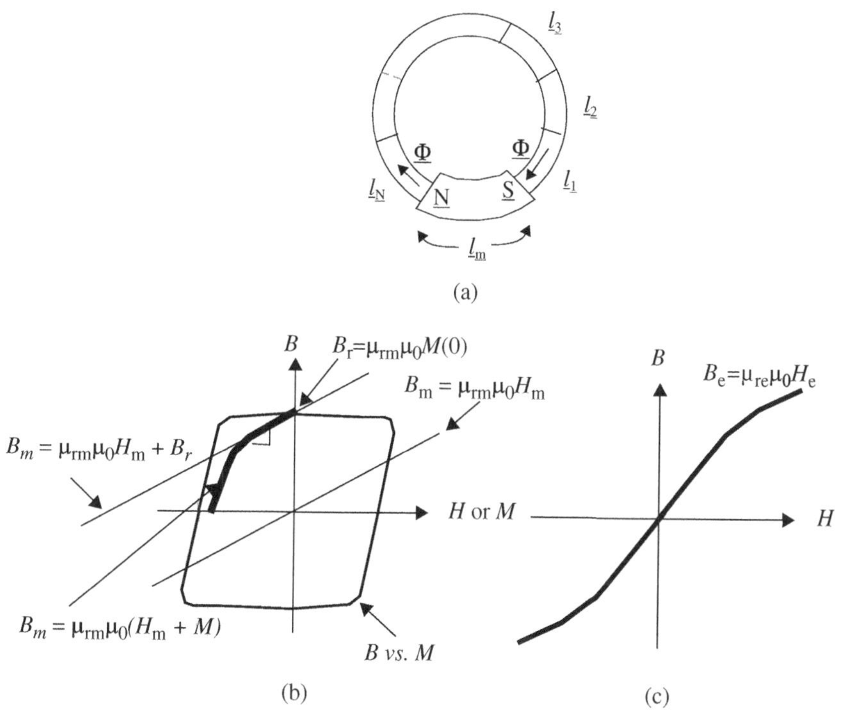

Figure 8.8 (a) Simple flux path representing flux path of a typical PM machine, (b) *B* versus *H* in the permanent magnet material, (c) *B* versus *H* in the non-permanent magnet portion of the magnetic circuit.

versus *B* and the other simply produced by a linear field intensity *H* versus *B* corresponding essentially to air although mrm is slightly greater than unity ($\mu_r \cong 1.05$–1.1).

As done in previous chapters, the length g_e can be considered as an equivalent air gap length as determined by the flux density and total MMF drop in the iron material as well as the air gap, that is,

$$\mathcal{F}_e = H_e l_e = \sum_{n=1}^{N} H_n(B_n) l_n \tag{8.10}$$

where $\mathcal{F}_e$ is the total MMF drop around the external path *excluding* the magnet thickness and $H_n(B_n)$ denotes the non-linear dependency of the field intensity on the flux density.

The operating point of the magnetic circuit can be determined by examining the consequences of Gauss' law and Ampere's law. Since the flux over any cross section of the toroid is constant, assuming no leakage and that the toroidal cross-sectional areas over the permanent magnet and non-permanent magnet portions remain constant, the flux in the magnetic circuit is

$$\Phi = B_m A_m = B_e A_e \tag{8.11}$$

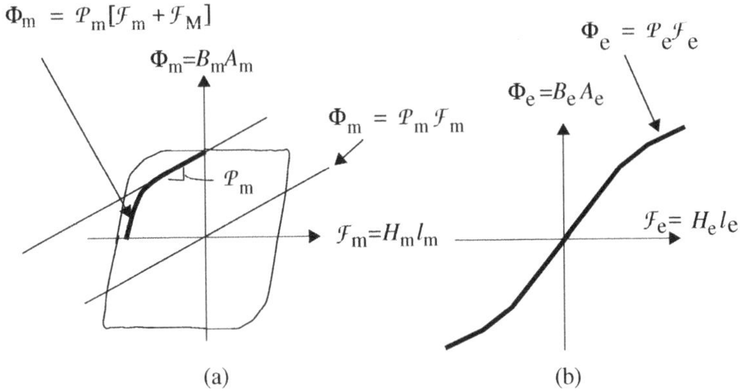

Figure 8.9 Modified B vs. H curves for (a) the magnet and (b) non-magnet portion of the toroid of Figure 8.8.

where B_m and B_e represent the magnet and external flux densities. One can ensure this equality is maintained by multiplying the ordinate of Figures 8.8b and 8.8c by A_m and A_e, respectively.

While the proper relationship between the flux densities within and outside the magnet has been established, the precise value is yet to be determined. This can be accomplished by also working out the constraint relating the abscissa axes of the two B–H curves. From Ampere's law, integrating around the center of the magnetic circuit, since no physical current is enclosed around the path,

$$\oint_{\text{circuit}} H \bullet dl = H_m l_m + H_e l_e = 0 \tag{8.12}$$

Hence, if one multiplies the abscissa of the magnet and non-magnet B–H curves by l_m and l_e respectively, the sum of the two modified abscissas must add to zero. One can now construct the two modified B–H curves shown in Figure 8.9. The non-linear relationship between flux density and field intensity can be changed by first multiplying equation (8.3) by A_m whereupon,

$$B_m A_m = \mu_{rm}\mu_0 A_m [H_m + M] = \Phi_m \tag{8.13}$$

Also, defining the MMF drops,

$$H_m l_m = \mathcal{F}_M \tag{8.14}$$

$$M l_m = \mathcal{F}_m \tag{8.15}$$

The MMF $\mathcal{F}_M$ can be considered to be the equivalent of the electromotive source voltage in an electric circuit while the MMF $\mathcal{F}_m$ can be considered as the equivalent of the Thevenin internal impedance. Note that the slope of the line corresponding to the Thevenin voltage drop becomes

$$\frac{\Phi_m}{\mathcal{F}_m + \mathcal{F}_M} = \mu_{rm}\mu_0 \frac{A_m}{l_m} = \mathcal{P}_m \tag{8.16}$$

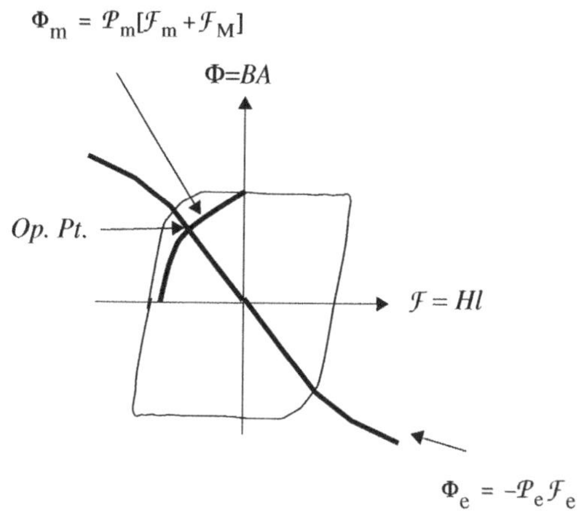

Figure 8.10 Construction to determine the operating point for the magnetic circuit of Figure 8.8.

which is simply the permeance of the magnet portion of the magnetic circuit assuming the permeability of what is essentially air. A similar situation exists for the non-magnet portion except that μ_{rm} is replaced by μ_{re}. That is,

$$\frac{\Phi_e}{\mathcal{F}_e} = \mu_{re}\mu_0\frac{A_e}{l_e} = \mathcal{P}_e \tag{8.17}$$

The solution for the problem corresponds to the conditions where $\Phi_m = \Phi_e$ and $\mathcal{F}_m + \mathcal{F}_M + \mathcal{F}_e = 0$ and is most conveniently obtained by setting

$$\mathcal{F}_m + \mathcal{F}_M = -\mathcal{F}_e \tag{8.18}$$

The solution can now be obtained graphically by plotting Φ_e versus the negative of $\mathcal{F}_e$ on the same coordinates as Φ_m versus $\mathcal{F}_m + \mathcal{F}_M$, as shown in Figure 8.10.

8.5 SINUSOIDALLY FED SURFACE PM MOTOR

Figure 8.11 shows sections of a two-pole, surface-magnet PM motor indicating its major dimensional parameters. The stator is assumed to be fitted with a three-phase winding with its turns of each phase distributed approximately sinusoidally around the periphery. Since the machine has no starting means from a fixed frequency supply, it is assumed to operate from an inverter supply. Assuming the general case of a P-pole machine, the stator windings are provided with a set of balanced sinusoidal currents I_s giving an MMF distribution per pole around the stator periphery of the form

$$\mathcal{F}_s(\theta) = \mathcal{F}_{s1}\sin\left(\frac{P\theta}{2} - \omega_e t\right) \quad \text{A-turns/m} \tag{8.19}$$

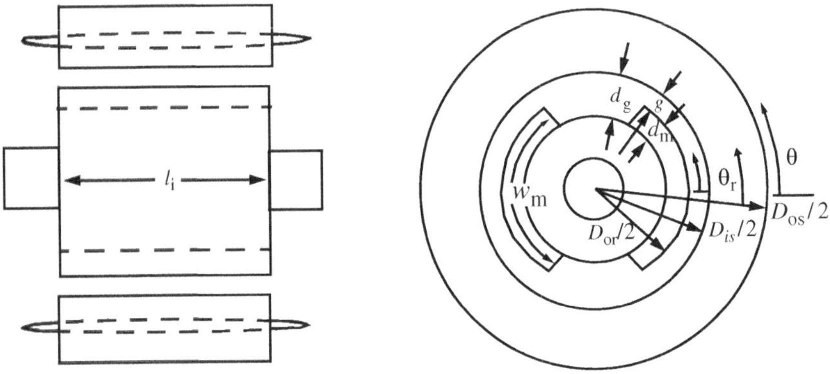

Figure 8.11 Cross sections of a 2-pole PM motor with surface magnets.

where, from equation (2.73)

$$\mathcal{F}_{s1} = \frac{3}{2}\left(\frac{\pi}{4}\right)\left(\frac{k_1 N_s}{P}\right) I_s \tag{8.20}$$

and ω_e is the electrical supply angular frequency (rad/s).

From equation (2.13)

$$B_{gs}(\theta) = \mu_0 \frac{\mathcal{F}_s(\theta)}{g_e} \tag{8.21}$$

The stator surface current density $K_s(\theta)$ which produces the gap flux density $B_{gs}(\theta)$ is in phase quadrature with the flux density $B_{gs}(\theta)$ and consequently $\mathcal{F}_s(\theta)$. The equivalent gap g_e again includes Carter's coefficient and the effects of saturation. Since the MMF distribution per pole is proportional to the space integral of the surface current density,

$$\mathcal{F}_s(\theta) = \int_0^{2\pi} K_s(\theta) \frac{D_{is}}{2} d\theta \tag{8.22}$$

one can then relate $K_s(\theta)$ to $\mathcal{F}_s(\theta)$ by,

$$K_s(\theta) = \frac{2}{D_{is}} \frac{d\mathcal{F}_s(\theta)}{d\theta} \tag{8.23}$$

whereupon

$$K_s(\theta) = \left(\frac{P}{D_{is}}\right) \mathcal{F}_{s1} \cos\left(\frac{P\theta}{2} - \omega_e t\right) \tag{8.24}$$

The maximum value of the sheet current density fundamental component is therefore proportional to the peak value of MMF per pole or,

$$K_{s1} = \left(\frac{P}{D_{is}}\right) \mathcal{F}_{s1} = \frac{6}{\pi} \frac{k_1 N_s}{D_{is}} I_s \ \text{A/m} \tag{8.25}$$

which is the same as the quantity K_s obtained in Chapter 6, equation (6.7). While the same value K_{s1} represents the peak value of the rotating, sinusoidally distributed surface current, K_s simply corresponds to the peak value of current that exists in each slot of a multiphase machine.

The rotor has radially directed permanent magnets mounted on the surface of an iron rotor core. Only the fundamental space component B_{gm1} of air gap flux density produced by the magnet can interact with K_{s1} to produce torque, that is,

$$B_{gm}(\theta) = B_{gm1} \sin\left(\frac{P}{2}\theta + \beta - \omega_e t\right) \quad \text{tesla} \tag{8.26}$$

where it is assumed that the magnet rotates synchronously with the stator MMF, but is spatially displaced with respect to the MMF by the angle β expressed in electrical degrees.

Recall that the force on a current-carrying conductor in the presence of a magnetic field is

$$\boldsymbol{F} = (\boldsymbol{B} \times \boldsymbol{l})I \quad \text{N} \tag{8.27}$$

In the context of this problem, the magnetic flux density is radially directed and the stator current is axially directed, making the force for each conductor circumferentially directed. As usual, l_e represents the axial effective length of the stator and is roughly equal to $l_e = l_i + 2g$ assuming no cooling ducts in the machine. In terms of current density, the stator current can be expressed as

$$dI_s = K_s(\theta)\frac{D_{is}}{2}d\theta \tag{8.28}$$

The total force is therefore found by integrating over the entire surface of the machine,

$$F = \int_0^{2\pi} \left[B_{gm1} \sin\left(\frac{P}{2}\theta + \beta - \omega_e t\right)\right]\left[K_{s1} \cos\left(\frac{P}{2}\theta - \omega_e t\right)\right]\frac{D_{is}}{2} l_e d\theta \tag{8.29}$$

Equation (8.29) works out to

$$F = \frac{\pi}{2} D_{is} l_e B_{gm1} K_{s1} \sin\beta \tag{8.30}$$

Equation (8.30) can be written alternatively as

$$F = \frac{\pi}{2} D_{is} l_e B_{gm1} K_{s1} \cos\varepsilon \tag{8.31}$$

where $\beta - \varepsilon = 90$. Since the angle ε locates the position of the MMF with respect to the air gap voltage (time rate of change of air gap flux), it is the equivalent of the power factor angle measured at the air gap so that equation (8.31) is sometimes preferred.

In terms of MMF, equation (8.31) can be written as

$$F = \frac{\pi}{2} P l_e B_{gm1} \mathcal{F}_{s1} \cos\varepsilon \tag{8.32}$$

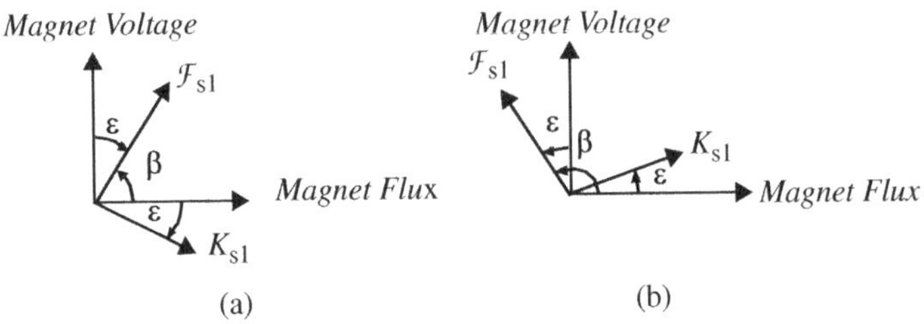

Figure 8.12 Vector diagrams (a) magnet flux insufficient to magnetize magnetic path, $\mathcal{F}_{s1}$ contributes to magnetization, (b) magnet flux more than sufficient to magnetize the magnetic path, $\mathcal{F}_{s1}$ contributes to demagnetization.

or, in terms of current as

$$F = 3l_e B_{gm1}(k_1 N_s) I_s \cos\varepsilon \tag{8.33}$$

Vector diagrams showing cases in which the magnet is insufficient to magnetize the machine and when the magnet is more than sufficient are given in Figure 8.12. When the magnet strength is insufficient, additional MMF must be provided to completely magnetize the iron path in which case lagging current flows. When the magnet strength is more than sufficient, stator MMF is provided to reduce the magnetization to the amount dictated by the voltage across the air gap (leading current operation).

Maximum force exerted on the rotor is reached when the peak values of the magnetic flux density and surface current waves are coincident, that is $\varepsilon = 0$ (or alternatively, when $\beta = \pi/2$).

The motor torque is clearly given by

$$T_e = F\frac{D_{is}}{2} \tag{8.34}$$

or

$$T_e = \frac{\pi}{4} D_{is}^2 l_e B_{gm1} K_{s1} \cos\varepsilon \quad \text{N-m} \tag{8.35}$$

The coefficient $(\pi/4)D_{is}^2 l_e$ can be recognized as essentially the rotor volume so that one figure of merit often used for designing a permanent magnet machine is

$$\text{Torque}/(\text{Rotor volume}) = B_{gm1} K_{s1} \cos\varepsilon \tag{8.36}$$

Alternatively, equation (8.35) can be written as

$$T_e = \frac{\pi}{4} P D_{is} l_e B_{gm1} \mathcal{F}_{s1} \cos\varepsilon \tag{8.37}$$

or,

$$T_e = \frac{3}{2}(k_1 N_s I_s) B_{gm1}(D_{is} l_e) \cos\varepsilon \tag{8.38}$$

Setting $\varepsilon = 0$, the maximum average force F per unit area A of gap surface is then

$$\left(\frac{F}{A}\right)_{avg} = \frac{B_{gm1} K_{s1}}{2} \quad \text{N/m}^2 \tag{8.39}$$

or, equivalently, (8.38)

$$\left(\frac{F}{A}\right)_{\text{avg}} = B_{\text{gm1(rms)}}K_{\text{s(rms)}} \tag{8.40}$$

The product $B_{\text{gm1}}K_{\text{s1}}/2$ is thus the *magnetic shear stress* σ_{m} already identified in Chapter 6 and is another common figure of merit for comparing machines. Normal, totally enclosed PM motors using rare-earth magnets typically have values of σ_{m} ranging from 10 to 20 kN/m^2 while ferrite magnet machines have values from 3 to 7 kN/m^2. The value of magnetic shear stress for a rare-earth permanent magnet machine is generally limited to 25 kN/m^2 without resorting to extraordinary cooling techniques.

8.6 FLUX DENSITY CONSTRAINTS

From Figure 8.8, the straight line portion of the demagnetization curve is described by

$$B_{\text{m}} = \mu_0\mu_{\text{rm}}H_{\text{m}} + B_{\text{r}} \tag{8.41}$$

whereas in the external circuit

$$B_{\text{g}} = \mu_0 H_{\text{g}} \tag{8.42}$$

From Ampere's law, assuming that the external circuit can be replaced by an equivalent gap,

$$H_{\text{m}}d_{\text{m}} + H_{\text{g}}g_{\text{e}} = 0 \tag{8.43}$$

where d_{m} is the radial thickness (depth) of the magnet.

Using equations (8.41) and (8.42),

$$\frac{B_{\text{m}} - B_{\text{r}}}{\mu_0\mu_{\text{rm}}}d_{\text{m}} + \frac{B_{\text{g}}}{\mu_0}g_{\text{e}} = 0 \tag{8.44}$$

Since the flux produced by the magnet must equal the flux entering the gap

$$\Phi_{\text{m}} = \Phi_{\text{e}} \tag{8.45}$$

or

$$B_{\text{m}}A_{\text{m}} = B_{\text{g}}A_{\text{p}} \tag{8.46}$$

where $A_{\text{p}} = \tau_{\text{p}}l_{\text{e}}$ or the area of one pole measured along the rotor surface. Eliminating B_{g} using equation (8.46), equation (8.44) can now be written

$$B_{\text{m}}\left(\frac{d_{\text{m}}}{\mu_0\mu_{\text{rm}}} + \frac{A_{\text{m}}}{A_{\text{p}}}\frac{g_{\text{e}}}{\mu_0}\right) = B_{\text{r}}\frac{d_{\text{m}}}{\mu_0\mu_{\text{rm}}} \tag{8.47}$$

which can be written as

$$B_{\text{m}} = B_{\text{r}}\left(\frac{\dfrac{d_{\text{m}}}{\mu_{\text{rm}}}}{\dfrac{d_{\text{m}}}{\mu_{\text{rm}}} + \dfrac{A_{\text{m}}g_{\text{e}}}{A_{\text{p}}}}\right) \tag{8.48}$$

Assuming that the external flux density B_e is essentially the same as the flux density in the gap B_g, then from equation (8.46)

$$B_g = B_r \left(\frac{\dfrac{d_m}{\mu_{rm}}}{\dfrac{A_p d_m}{A_m \mu_{rm}} + g_e} \right) \tag{8.49}$$

If the cross-sectional areas of the magnet and external circuit are assumed to be equal, the flux densities B_m and B_g also become equal so that

$$B_g = \frac{\dfrac{d_m}{\mu_{rm}}}{g_e + \dfrac{d_m}{\mu_{rm}}} B_r \quad \text{tesla} \tag{8.50}$$

where g_e = the effective air gap length which is again somewhat greater than the actual gap g due to stator slotting and the effects of stator core saturation.

In choosing an appropriate value for B_g, it should be noted that the torque depends on the product $B_{g1}K_{s1}$ in equation (8.39). The gap flux density $B_g(\theta)$ is limited by magnetic saturation in the stator teeth width while the linear current density is limited by the slot width. As was the case for induction machines, the product $B_{g1}K_{s1}$ is maximized when about half of the stator periphery is used for teeth and half for slots.

Since the maximum tooth flux density is in the range 1.7–1.8 teslas, a reasonable value for the peak gap flux density B_{gm} is 0.85–0.9 tesla, that is, about 70–80% of B_r for Nd–Fe–B. The radial length (thickness or depth) d_m of the magnet to achieve this flux density is then obtained from equation (8.49) as

$$\frac{d_m}{g_e} = \frac{\mu_{rm}}{\dfrac{B_r}{B_{gm}} - \dfrac{A_g}{A_m}} \approx 3.5 - 7 \tag{8.51}$$

The minimum value of the air gap length g is set by mechanical constraints and is similar to the values encountered in induction motors. A typical empirical expression for permanent magnet machines is given in Reference [2].

$$g = 0.2 + 0.003\sqrt{D_{or} l_i / 2} \quad \text{mm} \tag{8.52}$$

It may be preferable to use a minimum gap length of about 1 mm to permit easy insertion of a magnetized rotor into the stator. This gap may have to be increased further if banding is required around the rotor for high-speed operation. Surface magnets are frequently fixed to the rotor core using a thermosetting epoxy adhesive. This adhesive is considered to be adequate for most operating speeds. As an added insurance, glass or Kevlar® tape can be wound around the rotor under tension.

In a surface-magnet machine, the air gap flux density $B_{gm}(\theta)$ is maintained over the magnetic width w_m or over an angle α_m as shown in Figure 8.13, where, for P poles,

$$\frac{\alpha_m}{\pi} = \frac{P w_m}{\pi D_{is}} \tag{8.53}$$

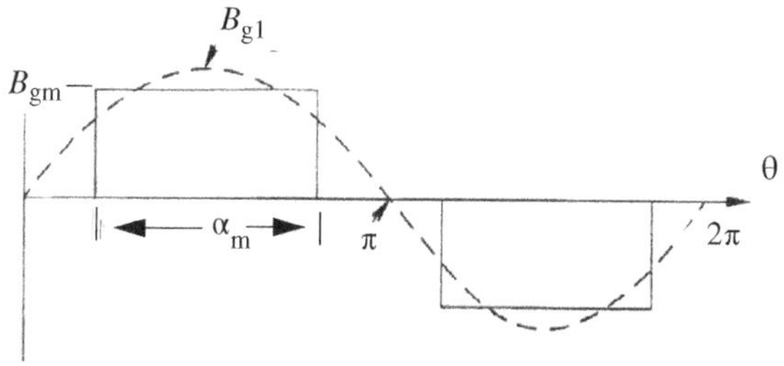

Figure 8.13 Actual and effective gap density assuming surface magnets.

or

$$\alpha_m = \frac{Pw_m}{D_{is}} \quad \text{radians} \tag{8.54}$$

The effective or fundamental space component B_{g1} of gap magnet flux density is then readily determined to be

$$B_{g1} = \frac{4}{\pi} B_{gm} \sin \frac{\alpha_m}{2} \quad \text{tesla} \tag{8.55}$$

In choosing the magnet width w_m, it should be noted that the effective flux density increases only 15% as α_m is increased from 120° to 180° while the magnet volume and cost increase by 50%. Also, the yoke flux, and therefore the rotor and stator yoke thicknesses d_{cs} and d_{cr} increase in proportion to α_m.

From the point of view of minimum iron losses, an optimum value of a magnet span can be obtained by minimizing the weighted total harmonic distortion (WTHD) created by the rectangular wave. The WTHD is, in general,

$$\text{WTHD} = \frac{\sqrt{\sum_{h=2}^{\infty} \left(\frac{B_{gh}}{h}\right)^2}}{B_{g1}} \tag{8.56}$$

where B_{gh} represents the amplitudes of the harmonic components of the waveform. Since the waveform of Figure 8.13 is half-wave symmetric, the even harmonic components are zero. The odd harmonics are

$$B_{gh} = \left(\frac{1}{h}\right)\left(\frac{4}{\pi}\right) B_{gm} \sin\left(h\frac{\alpha_m}{2}\right) \quad h = 3, 5, \ldots \tag{8.57}$$

Thus

$$\text{WTHD} = \frac{\sqrt{\left[\frac{1}{3}\sin\left(3\frac{\alpha_m}{2}\right)\right]^2 + \left[\frac{1}{5}\sin\left(5\frac{\alpha_m}{2}\right)\right]^2 + \cdots}}{\sin\left(\frac{\alpha_m}{2}\right)} \tag{8.58}$$

The function is minimized roughly where $\alpha_m = 130°$.

It is also logical to simply minimize the RMS value of the unweighted total harmonic content of the flux density wave. In this case

$$\text{THD} = \sqrt{\left[\sin\left(3\frac{\alpha_m}{2}\right)\right]^2 + \left[\sin\left(5\frac{\alpha_m}{2}\right)\right]^2 + \cdots} \tag{8.59}$$

This function is minimized at roughly $\alpha_m = 125°$.

Typically, α_m is in the range 110–160° for sinusoidally fed motors. The value of 120° is frequently used which tends to minimize torque ripple for frequently used slot choices, that is, 24, 36, 48, … slots. Assuming NdFeB magnets, the peak fundamental value of magnet flux density in the gap is generally in the range $B_{gm1} = 0.92–1.2$ teslas.

As an example, consider a motor with an air gap of $g = 1$ mm or $g_e = 1.05$ and having $A_g/A_m = 1$ and $\mu_{rm} = 1$. With a residual flux density of 1.2 teslas, an air gap flux density of 0.9 tesla can be achieved with

$$\frac{d_m}{d_m + g_e} = \frac{B_{gm}}{B_r} = \frac{0.9}{1.2} = 0.75$$

in which case the magnet thickness should be

$$d_m = \left[\frac{0.75(1.2)}{1 - 0.75}\right] \approx 3.6 \text{ mm}$$

If the magnet width is two-thirds of the maximum possible value, $\alpha_m = 120°$ and the peak fundamental gap flux density will be

$$B_{g1} = \left(\frac{4}{\pi}\right)(0.9)\sin(\pi/3) = 0.99 \text{ teslas}$$

It is important to note that the thickness of the magnet is independent of the size of the motor. As the motor torque rating is increased, the volume of the magnet increases as the square of the dimension (k^2). However, the overall volume increases as the cube of the dimension (k^3) so that the amount of magnet as a per of the unit volume will, therefore, decrease as the machine size increases.

8.7 CURRENT DENSITY CONSTRAINTS

The normalized current loss is an important figure of merit for machine design and has already been discussed in Section 6.10. In the case of permanent magnet machines, the relevant expressions are

$$P_{mech} = T_{e,rated}\frac{2\omega_e}{P} = \frac{\pi}{2P}B_{g1}K_{s1}D_{is}^2 l_e \omega_e \cos\varepsilon \text{ W} \tag{8.60}$$

where

$$K_{s1} = \sqrt{2}K_{s(rms)} \tag{8.61}$$

and is assumed to be at its rated value, and

$$P_{\mathrm{cu(pu)}} = \frac{\sqrt{2}P\rho_{\mathrm{cu}}\left[l_{\mathrm{s}} + \upsilon_{\mathrm{e}}\frac{(\pi D_{\mathrm{is}})}{P}\right]K_{\mathrm{s(rms)}}}{k_1 D_{\mathrm{is}} l_{\mathrm{e}} k_{\mathrm{cu}} d_{\mathrm{s}} \omega_{\mathrm{e}} B_{\mathrm{g1}} \cos\varepsilon} \tag{8.62}$$

The result is the same as equations (6.101) and (6.102) except that the quantity $\eta_{\mathrm{gap}} \cos \phi_{\mathrm{gap}}$ has been replaced by $\cos \varepsilon$. Estimates of initial values of parameters can be determined by solving equations (8.60) and (8.62) in terms of two of the parameters, for example D_{is} and $K_{\mathrm{s(rms)}}$, making reasonable assumptions for the remainder. An example of this technique was shown in Section 6.9.2.

8.8 CHOICE OF ASPECT RATIO

The basic shape of the rotor is related to the ratio of rotor length to motor diameter (aspect ratio). This choice may be based on several alternate criteria. One would be the shape giving minimum loss in the stator winding which essentially is the one in which the end winding length as a fraction of the pole pitch is a minimum. From Chapter 6 by definition

$$\upsilon_{\mathrm{e}} = \frac{l_{\mathrm{ew}}}{\tau_{\mathrm{p}}} = \frac{P l_{\mathrm{ew}}}{\pi D_{\mathrm{is}}} \tag{8.63}$$

High-performance drives are a typical application of PM motors where often high acceleration is required. The shape should, in this case, be chosen to maximize the torque to inertia ratio. Consider for example a solid rotor of diameter D_{or}, length l_{s}, and iron density ρ_{i}, assuming that the lower density of the magnets and spacers between magnets are compensated by the inertia of the shaft ends and fans. In this case, the polar moment of inertia of the rotor is

$$J_{\mathrm{r}} = \rho_{\mathrm{i}} \pi \frac{D_{\mathrm{or}}^2}{4} l_{\mathrm{r}} \left(\frac{D_{\mathrm{or}}^2}{8}\right) \quad \text{kg-m}^2 \tag{8.64}$$

Assuming $D_{\mathrm{is}} \cong D_{\mathrm{or}}$ and $l_{\mathrm{r}} = l_{\mathrm{s}}$, the acceleration capability of the motor with rated torque is then given, from equations (8.35) and (8.64), as

$$\frac{T_{\mathrm{e,rated}}}{J_{\mathrm{r}}} = \frac{8 B_{\mathrm{gm1}} K_{\mathrm{s1}}}{\rho_{\mathrm{i}} D_{\mathrm{or}}^2} \quad \text{rad/s}^2 \tag{8.65}$$

By choosing a value of D_{or} as small as is mechanically practical, the torque to inertia ratio can be maximized. However, removal of heat becomes increasingly difficult so that the minimum value of D_{or} is ultimately limited by thermal issues.

8.9 EDDY CURRENT IRON LOSSES

To determine the total heat loss in the stator, an estimate is required of the core losses. Data on core loss for stator steel at normal line frequencies are readily available as discussed in Chapter 5. They can be approximated over a reasonable range of flux density and frequency by the expression.

$$p_{cs} = k_c B^2 \omega_e^2 \quad \mathrm{W/m^3} \tag{8.66}$$

For 3.25% Si steel, 0.63-mm thick (25 mils), a typical value of k_c is 0.11.

Core loss data are given for sinusoidally varying flux density. However, in the PM motor, the flux density in the stator teeth rises rapidly as the magnet edge crosses the tooth and then remains relatively constant while the tooth is covered. It is, therefore, preferable to express the core loss as a function of the time rate of change of flux density. For a sine wave of flux density,

$$\left(\frac{dB}{dt}\right)_{\mathrm{rms}} = \frac{B\omega_e}{\sqrt{2}} \tag{8.67}$$

Thus, it can be inferred that another expression for the instantaneous iron loss is

$$p_c = 2k_c\left(\frac{dB}{dt}\right)^2 \quad \mathrm{W/m^3} \tag{8.68}$$

8.9.1 Eddy Current Tooth Iron Losses

Approximately, the tooth flux density B_{ts} may be considered to rise linearly to $\hat{B}_{ts}$ over the time ΔT required for the magnet edge to travel the top of the tooth width τ_t as shown in Figure 8.14.

$$\Delta T = \frac{\tau_t}{v} = \frac{P\tau_t}{\omega_e D_{is}} \tag{8.69}$$

so that, over this interval

$$\frac{dB_{ts}}{dt} = \frac{D_{is}\omega_e B_{ts}}{P\tau_t} \tag{8.70}$$

This rate of change occurs four times per period, that is, for a fraction of the time of

$$\frac{4\Delta T}{T} = \frac{2P\tau_t}{\pi D_{is}} \tag{8.71}$$

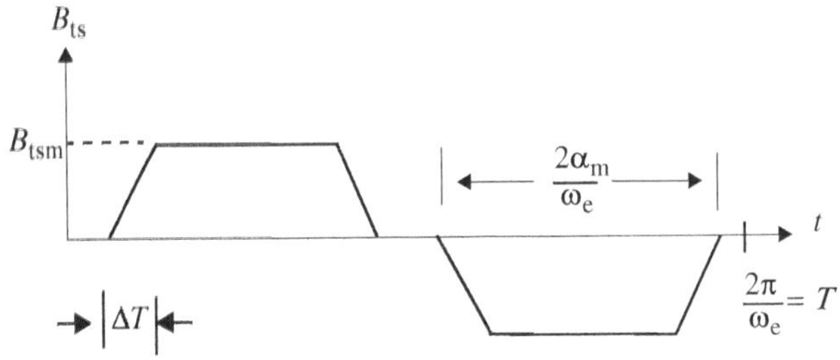

Figure 8.14 Tooth flux density as a function of time.

Over the remaining fraction of time, the eddy current iron loss can be neglected. Thus, combining equations (8.68), (8.70), and (8.71), the iron loss per unit of tooth volume can be expressed as

$$p_{\mathrm{t}} = 2k_{\mathrm{c}}\left(\frac{4\Delta T}{T}\right)\left(\frac{dB_{\mathrm{ts}}}{dt}\right)^2 = \frac{4k_{\mathrm{c}}D_{\mathrm{is}}}{\pi P t_{\mathrm{t}}}(B_{\mathrm{tsm}}\omega_{\mathrm{e}})^2 \quad \mathrm{W/m^3} \tag{8.72}$$

For simplicity, if the widths of the tooth and slot are assumed equal so that $t_{\mathrm{t}} = b_{\mathrm{o}}$, and if the number of slots in the stator is S_1

$$S_1 = \frac{\pi D_{\mathrm{is}}}{t_{\mathrm{t}} + b_{\mathrm{o}}} = \frac{\pi D_{\mathrm{is}}}{2t_{\mathrm{t}}} \tag{8.73}$$

Thus,

$$p_{\mathrm{t}} = \frac{8k_{\mathrm{c}}S_1}{\pi^2 P}(B_{\mathrm{tsm}}\omega_{\mathrm{e}})^2 \quad \mathrm{W/m^3} \tag{8.74}$$

The tooth volume is

$$V_{\mathrm{ts}} = l_{\mathrm{i}}\left[\frac{\pi}{4}(D_{\mathrm{is}} + 2d_{\mathrm{s}})^2 - \frac{\pi}{4}D_{\mathrm{is}}^2 - \pi D_{\mathrm{is}}d_{\mathrm{s}}\frac{b_{\mathrm{s}}}{b_{\mathrm{s}} + t_{\mathrm{r}}}\right] \tag{8.75}$$

Assuming parallel-sided slots and $t_{\mathrm{r}} = b_{\mathrm{s}}$,

$$V_{\mathrm{ts}} = \frac{\pi}{2}l_{\mathrm{i}}d_{\mathrm{s}}(D_{\mathrm{is}} + 2d_{\mathrm{s}}) \quad \mathrm{m^3} \tag{8.76}$$

The maximum flux density in the teeth occurs near the stator surface and has a value

$$B_{\mathrm{tsm}} = \frac{t_{\mathrm{t}} + b_{\mathrm{o}}}{t_{\mathrm{t}}}B_{\mathrm{gm}} \approx 2B_{\mathrm{gm}} \tag{8.77}$$

If the reduction in this density along the tapered tooth is ignored, the core losses in the teeth may be expressed as

$$P_{\mathrm{ts}} = \frac{32}{\pi P}k_{\mathrm{c}}S_1 d_{\mathrm{s}} l_{\mathrm{i}}\left(\frac{D_{\mathrm{is}}}{2} + d_{\mathrm{s}}\right)(B_{\mathrm{gm}}\omega_{\mathrm{e}})^2 \quad \mathrm{W} \tag{8.78}$$

Note that the loss increases linearly with the number of slots since reducing the tooth width increases the rate of flux density change.

8.9.2 Eddy Current Yoke Iron Losses

In the yoke, the flux density due to a magnet can be considered to change linearly from $-B_{\mathrm{cm}}$ to $+B_{\mathrm{cm}}$ as the magnet moves through a distance w_{m}, that is the width of the magnet. Thus, over this interval

$$\frac{dB_{\mathrm{cs}}}{dt} = \frac{2B_{\mathrm{cm}}}{(w_{\mathrm{m}}/v)} \tag{8.79}$$

$$= \frac{2B_{\mathrm{cm}}}{\left(\dfrac{Pw_{\mathrm{m}}}{\omega_{\mathrm{e}}D_{\mathrm{is}}}\right)} = \frac{2\omega_{\mathrm{e}}B_{\mathrm{cm}}}{\alpha_{\mathrm{m}}} \quad \mathrm{T/s} \tag{8.80}$$

The fraction of the time that this change occurs is α_m/π. Thus, from equations (8.68) and (8.80),

$$p_c = 2k_c\left(\frac{2\omega_e B_{cm}}{\alpha_m}\right)^2 \frac{\alpha_m}{\pi} \tag{8.81}$$

$$= \frac{8}{\pi}\frac{k_c}{\alpha_m}(\omega_e B_{cm})^2 \quad \text{W/m}^3 \tag{8.82}$$

The yoke volume is, roughly

$$V_y \approx \pi(D_{is} + 2d_s + d_{cs})d_{cs}l_i \ \text{m}^3 \tag{8.83}$$

The core loss in the stator yoke is then

$$P_{cs} = \frac{8k_c}{\alpha_m}(D_{is} + 2d_s + d_{cs})d_{cs}l_i\omega_e^2 B_{cm}^2 \quad \text{W} \tag{8.84}$$

or, in terms of the air gap flux density B_{gm}, since

$$B_{cm}d_{cs}l_i = B_{gm}\frac{\alpha_m D_{is}}{2P}l_i \tag{8.85}$$

then

$$P_{cs} = \left(\frac{2k_c}{P^2}\right)\alpha_m\frac{(D_{is} + 2d_s + d_{cs})}{d_{cs}}D_{is}^2 l_i\omega_e^2 B_{gm}^2 \quad \text{W} \tag{8.86}$$

This analysis of core losses contains a number of approximations, but has been found to match reasonably well to measurements. For small machines with parallel-sided slots, the tooth loss is overestimated because of the significant drop in tooth flux density along the tooth depth.

The major part of the motor losses consists of the stator winding loss (equation 5.42) and the eddy current core loss (equations 8.78 and 8.86). The stray-load loss which is significant in induction machines is negligible in the PM motor because of the large effective air gap and the lack of rotor slotting. With a reasonably good distribution of the stator winding, the rotor losses due to space harmonics are also negligible, again because of the large effective air gap and because of the relatively high resistivity of the rotor magnets (approximately 1.5×10^{-6} ohm-m) [3]. The windage and friction losses are similar to those encountered in induction motors.

8.10 EQUIVALENT CIRCUIT PARAMETERS

In this section, the major parameters of the motor circuit model are summarized. Figure 8.15 shows the conventional steady-state equivalent circuit for a PM motor [4]. In this representation, the MMF produced by the magnet is represented as a current source I_f. The open circuit peak voltage per phase when operated at speed $\omega_r = 2\omega_e/P$ rad/s can be calculated from the results of Chapter 3. The stator leakage inductance L_{ls} has slot, end winding, harmonic and tooth top leakage components as already described in Chapter 4 and Section 8.6. The magnetizing inductance L_{ms} is

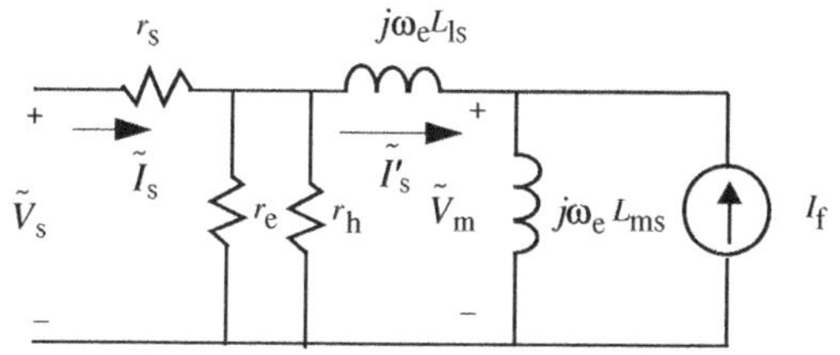

Figure 8.15 Per phase equivalent circuit of a surface magnet PM motor.

greatly dependent on the rotor structure, but in the case of a simple surface magnet, it is readily obtained as will be shown. The resistance r_s is the stator copper loss resistance and is again given by equation (5.42). The resistance r_e represents the stator iron eddy current loss which must also be treated in more detail in Section 8.10.3. The hysteresis loss, represented by r_h, if significant can be calculated in the same manner as for an induction motor (Section 5.12).

8.10.1 Magnetizing Inductance

The PM motor with surface-mounted Nd–Fe–B or Sm–Co magnets is essentially non-salient since the incremental permeability of the magnet material is approximately the same as that of air. Thus, the magnetizing inductance of the motor has the same form as for an induction motor. From equations (3.76) and (8.25), L_{ms} is expressed in terms of gap flux density and the stator current sheet as

$$L_{ms} = \left(\frac{12}{\pi}\right)\frac{(k_1 N_s)^2 B_{g1} l_e}{P K_{s1}} \quad \text{H} \tag{8.87}$$

and in terms of physical dimensions as given by equation (3.83)

$$L_{ms} = \left(\frac{3}{2}\right)\left(\frac{8}{\pi}\right)\left(\frac{k_1 N_s}{P}\right)^2 \mu_0 \frac{D_{is} l_e}{d_g} \quad \text{H} \tag{8.88}$$

where the effective gap g_e is replaced by the gap depth which includes the thickness of the magnet d_m.

Note that the magnetizing inductance is inversely proportional to the effective length d_g between the stator inner radius and the surface of the iron rotor core and the bottom surface of the magnet. Typically, this may be in the range of 5–10 mm for a PM motor in contrast with 0.5–1 mm for an induction motor. While the per unit magnetizing inductance of the induction motor is 1–3 per unit, that of the PM motor is typically in the range of 0.1–0.5 per unit.

The magnetizing inductance decreases as the number of poles is increased, a feature which is familiar with induction machines. As the motor torque rating is increased, the value of d_m remains essentially constant while the diameter D_{is} will increase. Also, the current density $K_{s(rms)}$ will increase while flux density B_{g1} will remain reasonably constant. Thus, the magnetizing inductance, as well as its per unit value, can be expected to increase with increased rating.

8.10.2 Current Source

In the equivalent circuit of Figure 8.15, the magnet may be represented by a peak current source I_f, where, from equation (6.36)

$$V_{gap} = \frac{2}{P} k_1 N_s B_{gm1} D_{is} l_e \omega_e \quad \text{V pk} \tag{8.89}$$

and B_{gm1} again denotes the fundamental component of air gap flux density produced by the magnet. Thus, the field current source equivalent to the magnet is

$$I_f = \frac{V_{gap}}{\omega_e L_{ms}} = \left(\frac{\pi}{6}\right) \frac{P d_g B_{gm1}}{\mu_0 k_1 N_s} \quad \text{A pk} \tag{8.90}$$

8.10.3 Eddy Current Iron Loss Resistance

Since the iron loss has been assumed to be proportional to the square of the $B\omega_e$ product, the loss can be represented in the equivalent circuit by a constant resistance r_m. Note that this resistance has been connected across the total induced stator voltage in Figure 8.15 rather than across the inductance L_m. This choice takes into consideration the fact that the flux densities in the teeth and yoke are proportional to the total stator flux linkage which in turn is the spatial sum of the air gap and leakage flux distributions.

The eddy current iron loss resistance which will produce the required losses in the equivalent circuit of Figure 8.15 is

$$r_e = \frac{V_{l-l}^2}{P_{ts} + P_{cs}} \quad \Omega \tag{8.91}$$

where

$$V_{l-l} = \left(\frac{L_{ls} + L_{ms}}{L_{ms}}\right) V_{l-l,gap} = \sqrt{3}\left(\frac{L_{ls} + L_{ms}}{L_{ms}}\right) V_{gap} \tag{8.92}$$

$$= \frac{2\sqrt{3}}{P} k_1 N_s B_{g1} D_{is} l_e \omega_e \left(\frac{L_{ls} + L_{ms}}{L_{ms}}\right) \quad \text{V pk} \tag{8.93}$$

in which

$$B_{g1} = \frac{4}{\pi} B_{gm} \sin\frac{\alpha_m}{2} \quad \text{T} \tag{8.94}$$

and P_{ts} and P_{cs} are given by equations (8.78) and (8.86), respectively. The result is

$$r_e = \frac{(48)(k_1 N_s)^2 \left(\sin\frac{\alpha_m}{2}\right)^2 D_{is}^2 l_e}{16 p\pi k_c S_1 d_s \left(\frac{D_{is}}{2} + d_s\right) + \pi^2 k_c \alpha_m \frac{(D_{is} + 2d_s + d_{cs})}{d_{cs}} D_{is}^2} \left(\frac{L_{ls} + L_{ms}}{L_{ms}}\right)^2 \quad \Omega \tag{8.95}$$

Since the per unit value of voltage is again unity, the value of the eddy current resistance in per unit is the inverse of the per unit core loss.

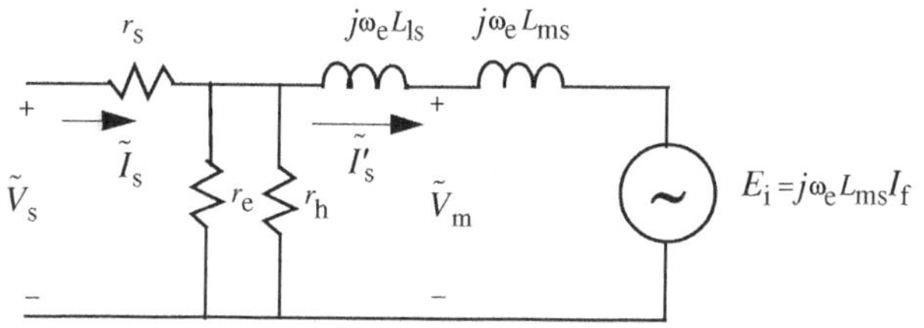

Figure 8.16 Per phase equivalent circuit of a surface magnet PM motor employing induced emf concept.

8.10.4 Alternate Equivalent Circuit

If one expresses the voltage drop across the magnetizing reactance defined by Figure 8.15,

$$\tilde{V}_m = j\omega_e L_{ms}\left(\tilde{I}'_s + \tilde{I}_f\right) \tag{8.96}$$

defining

$$\tilde{E}_i = j\omega_e L_{ms}\tilde{I}_f \tag{8.97}$$

so that

$$\tilde{V}_m = j\omega_e L_{ms}\tilde{I}'_s + \tilde{E}_i \tag{8.98}$$

Thus equivalent circuit of Figure 8.16 will produce the same defining equations as Figure 8.15. The voltage $\tilde{E}_i$ is defined as the emf induced in the stator windings due to rotor rotation. While conceptually the difference seems trivial, when saturation occurs, Figure 8.15 is preferable to Figure 8.16 since only one quantity (L_{ms}) varies with saturation while two quantities (L_{ms} and E_i) vary in Figure 8.16.

8.11 TEMPERATURE CONSTRAINTS AND COOLING CAPABILITY

The temperature limits for the PM motor stator are similar to those conventionally used with induction motors. As mentioned in Chapter 6, the maximum temperature for the winding insulation should not exceed 130°C for Class B and 155°C for Class F insulation. In the rotor, the critical component is the magnet material. For Nd–Fe–B magnets, the value of maximum reverse field H_d reduces rapidly with increased temperature. To allow the magnet flux density to be reduced to zero under transient load conditions, that is, with field intensities $H \approx B_r/\mu_{rm}\mu_0$, the magnet temperature should not exceed roughly 100°C with currently available materials. Thus, the maximum temperature in the stator winding insulation should normally be limited to about 110–115°C.

Ventilation techniques are similar to those used with induction machines. The major differences are (1) although the losses in the rotor are the losses in a solid rotor shaft, supporting the magnets can be appreciable if the stator slots are open, and (2) the rotor temperature limit is generally lower than that of the stator.

8.12 MAGNET PROTECTION

As noted in Section 8.5, the magnet must be protected against reverse fields exceeding the value H_d, that is, the magnet flux density must not be reduced below the value B_d which corresponds to the knee of magnetization vector $\boldsymbol{M}$ where its value begins to rapidly decrease. Typically, the value of B_d is about −0.2 tesla at 100°C for currently available Nd–Fe–B materials. The magnet flux density can be reduced either by applying too large a stator current from the inverter or by short circuiting of the stator windings.

8.12.1 Magnet Protection for Maximum Steady-State Current

Figure 8.17 shows the normal operating conditions of the motor with the magnet field centered on the peak of the stator current surface density. The effect of the resulting stator field on the magnet is to increase its flux density on the leading edge and decrease it on the trailing edge.

The maximum value of MMF acting in the air gap as a result of stator current is, from equation (3.63)

$$\mathcal{F}_{s1} = \left(\frac{3}{2}\right)\left(\frac{4}{\pi}\right)\left(\frac{k_1 N_s}{P}\right) I_s \quad \text{A-t} \tag{8.99}$$

The maximum value of the flux density B_{g1} produced in the air gap or in the magnet by the stator current acting alone is, from equation (3.77)

$$B_{g1} = \left(\frac{3}{2}\right)\left(\frac{4}{\pi}\right)\frac{\mu_0 k_1 N_s}{P g_e} I_s \quad \text{T} \tag{8.100}$$

To protect the magnet at any value of the angle β between the stator and magnet fields

$$B_{g1} \leq B_m - B_d \quad \text{T} \tag{8.101}$$

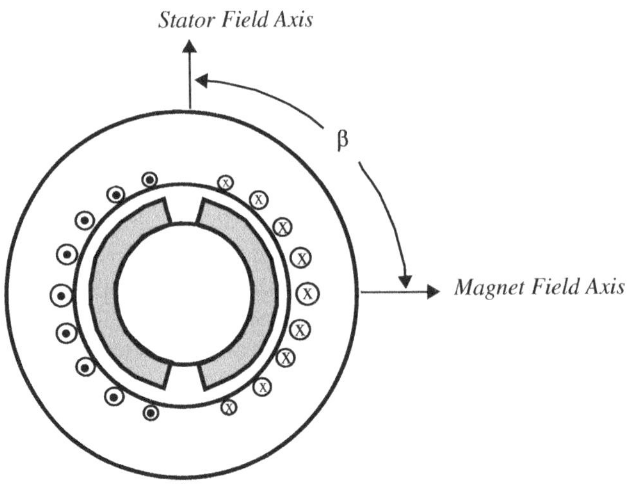

Figure 8.17 Relation of rotor and stator fields with $\beta = 90°$

From equation (8.48), since the areas A_m and A_g are, in this case, equal

$$B_m = B_r\left(\frac{d_m}{d_m + \mu_{rm}g_e}\right) \tag{8.102}$$

Combining equations (8.101) and (8.102),

$$B_{g1} \le \frac{B_r d_m - B_d(d_m + \mu_{rm}g_e)}{(d_m + \mu_{rm}g_e)} \tag{8.103}$$

Combining equations (8.100) and (8.103), the maximum value of steady-state RMS stator current is

$$I_{s(max)} \le \frac{\pi P g_e}{6\mu_0 k_1 N_s}\left[\frac{B_r d_m - B_d(d_m + \mu_{rm}g_e)}{(d_m + \mu_{rm}g_e)}\right] \tag{8.104}$$

When saturation is ignored and it is assumed that the permeability of the magnet is essentially that of air, $d_m + \mu_{rm}g_e = d_m + g = d_g$ and equation (8.104) reduces to

$$I_{s(max)} \le \frac{P\pi}{6\mu_0(k_1N_s)}(B_r d_m - B_d d_g) \quad \text{A} \tag{8.105}$$

The result can also be expressed in per unit format using equation (6.38) as a base quantity

$$I_{s(max)} \le \frac{P}{\mu_0}\frac{(B_r d_m - B_d d_g)}{k_1 D_{is} K_s} \quad \text{per unit} \tag{8.106}$$

where $K_s = \sqrt{2}K_{s(rms)}$ and is assumed to be at the rated steady-state value.

The most important factor in these expressions is the direct proportionality to the number of poles. With K_s constant in the per unit expression, increasing P decreases the magnet width and thus decreases the reverse armature field. It is also noted that the maximum current is inversely proportional to D_{is} and K_s both of which increase as the torque rating is increased. Thus, the per unit overcurrent limit may be expected to reduce with large motors. The overcurrent limit is approximately proportional to the magnet length, since $d_m \approx d_g$.

One criterion that may be important for motor design is the peak torque available under brief overcurrent conditions since this quantity establishes the maximum available acceleration of the rotor and its load. With the angle β set at 90°, the per unit value of maximum torque will be equal to the per unit value of maximum current from equation (8.106) since the per unit values of air gap voltage and speed are assumed to be unity. This is also based on an assumption that the stator teeth are not unduly saturated under such overcurrent conditions in the region of the leading edge of the magnet.

To maximize the maximum torque capability, a small value of magnetizing inductance is required to minimize the reverse field produced by stator current. This can be accomplished by use of a relatively large value of magnet length and a relatively large number of poles. For example, a 5 kW PM motor has been reported with a linear relation between torque and stator current up to about 6 per unit current [5].

8.12.2 Magnet Protection for Transient Conditions

The other danger to the magnet arises from a possible short circuit of the stator terminals [5]. A three-phase short circuit is considered to be the typical defining case. Consider the equivalent circuit of Figure 8.15, ignoring the resistances. The steady-state short circuit current is

$$I_s = \frac{L_{ms}}{L_{ls} + L_{ms}} I_f \tag{8.107}$$

The transient short circuit current will include the sinusoidal component of equation (8.107) plus a unidirectional offset having a peak value kI_s, $0 \le k \le 1$ and decaying at a time constant

$$\tau_1 = \frac{L_{ms} + L_{ls}}{r_s} \tag{8.108}$$

The total current wave in a fully offset phase is shown in Figure 8.18. If the time constant τ_1 is long in comparison with a half period, the peak stator current can approach $2I_s$. In this case, the maximum steady-state current must be limited to one-half the value of equation (8.104), or

$$I_{s(max)} = \frac{\pi P d_g}{12\mu_0 k_1 N_s} \left[\frac{B_r d_m - B_d(d_m + \mu_{rm} g_e)}{(d_m + \mu_{rm} g_e)} \right] \quad \text{A} \tag{8.109}$$

From equations (8.107) and (8.90)

$$I_{s(max)} = \frac{L_{ms}}{L_{ls} + L_{ms}} I_f = \frac{L_{ms}}{L_{ls} + L_{ms}} \frac{\pi P B_{gm1} d_g}{6k_1 N_s \mu_0} \tag{8.110}$$

Thus, assuming for simplicity equation (8.105)

$$\frac{P\pi}{6\mu_0(k_1 N_s)}(B_r d_m - B_d d_g) \ge \frac{L_{ms}}{L_{ls} + L_{ms}} \frac{\pi P B_{gm1} d_g}{6k_1 N_s \mu_0} \tag{8.111}$$

where, from equation (8.55)

$$B_{gm1} = \frac{4}{\pi} B_{gm} \sin \frac{\alpha_m}{2} \tag{8.112}$$

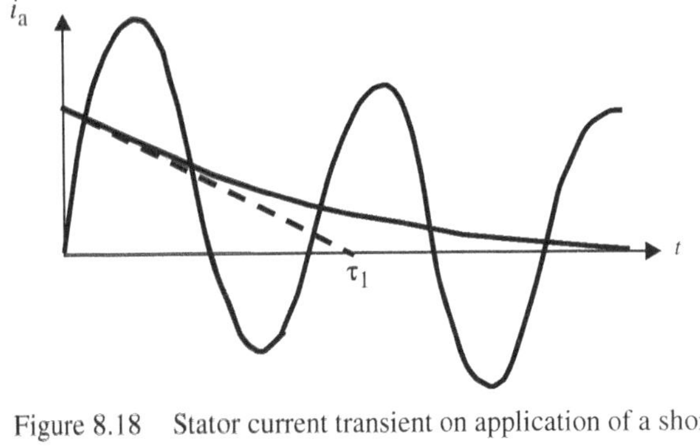

Figure 8.18 Stator current transient on application of a short circuit.

Upon solving for L_{ls}/L_{ms}, setting $B_{gm} = B_r(d_m/d_g)$, the necessary value of leakage inductance L_{ls} to protect the magnet from damage can be determined to be

$$\frac{L_{ls}}{L_{ms}} \geq \left(\frac{\frac{4}{\pi} B_r d_m \sin \frac{\alpha_m}{2}}{B_r d_m - B_d d_g} \right) - 1 \tag{8.113}$$

From the above, it is seen that to ensure magnet integrity, design means may be required to increase the leakage inductance and/or to decrease the magnetizing inductance. The main means of increasing the inductance ratio is increasing the effective gap length l_g, that is, a longer air gap or a larger magnet. Also, use of deep slots and a relatively short rotor length will increase the leakage. Partial closing of the top of the stator slots or designing intentionally deep slots provides a means of increasing leakage in small motors with random-wound coils, but is difficult in larger motors with form-wound coils. Increasing the number of poles also increases the slot leakage component.

The design of a PM motor to withstand a short circuit may be a critical consideration for large ratings, but is of lesser importance for small motors. For the latter, the magnetizing inductance can be readily made small and the leakage inductance can be made relatively large. Also, the per unit stator resistance increases as the rating is decreased (equation 8.88). This reduces the decay time constant τ_1 of equation (8.108) to considerably less than a half period. The expression for inductance ratio becomes

$$\frac{L_{ls}}{L_{ms}} \geq \left(\frac{\frac{4}{\pi} B_r d_m \sin \frac{\alpha_m}{2} e^{-\frac{\pi}{2w\tau_1}}}{B_r d_m - B_d d_g} \right) - 1 \tag{8.114}$$

Before concluding, it should be mentioned that it has been determined that other faults such as a line-to-line short-circuit can produce peak currents greater than the two per unit values assumed above. A good design should prepare for 3–4 per unit current if single-phase short circuits are to be anticipated.

8.13 DESIGN FOR FLUX WEAKENING

For some applications, a permanent magnet motor operating from an inverter supply is required to operate at constant power over its upper speed range. The surface-mounted PM motor has only limited capability in this region. The following analysis shows the design factors which can optimize this capability [6].

Ignoring the loss elements r_s, r_h, and r_e, the equivalent circuit of Figure 8.15 can be simplified using Norton's theorem. The equivalent reactance is found by measuring the reactance with the magnet current source open circuited. Clearly,

$$L'_{ms} = L_{ls} + L_{ms} \tag{8.115}$$

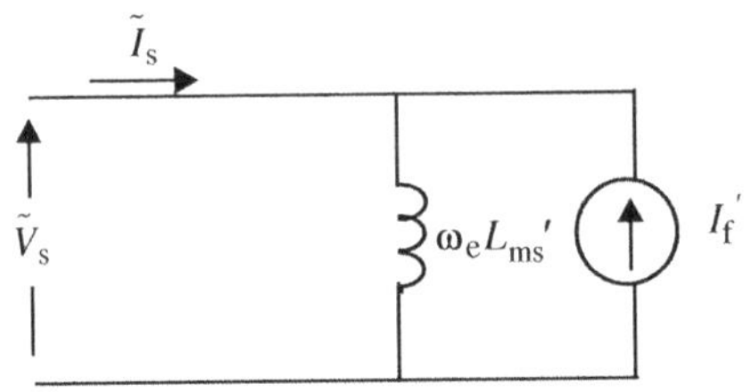

Figure 8.19 Simplified equivalent circuit.

The equivalent current source is found by short-circuiting the stator terminals and measuring the current in the short circuit. First, the voltage across the current source during a short circuit is

$$V_f = \frac{\omega_e L_{ls} L_{ms}}{L_{ls} + L_{ms}} I_f \tag{8.116}$$

The current in the inductor L_{ls} is the current in the short-circuited input and is defined as I_f',

$$I_f' = \frac{V_f}{\omega_e L_{ls}} = \frac{L_{ms}}{L_{ls} + L_{ms}} I_f \tag{8.117}$$

The resulting equivalent circuit is shown in Figure 8.19.

Maximum torque is obtained with the current angle β set at 90°. This condition can apply up to the mechanical speed Ω_r or angular frequency ω_e at which point the limiting available inverter voltage is reached. Since the currents I_f' and I_s are orthogonal, the speed at which this point occurs (base speed) is

$$\omega_{e,base} = \frac{V_{s(max)}}{L_m' \sqrt{\left(T_f^2 + I_s^2\right)}} \quad \text{rad/s} \tag{8.118}$$

To operate at increased speed, the current angle β must be increased beyond 90° (Figure 8.12b) and the motor enters leading power factor operation. Assuming peak sinusoidal values of currents, the torque and frequency relations are then given by the equations

$$T_e = \frac{3}{2} \omega_e L_{ms}' I_f' I_s \sin\beta \quad \text{N-m} \tag{8.119}$$

$$\omega_e = \frac{V_{s(max)}}{L_{ms}' \left[\left(I_f' + I_s \cos\beta\right)^2 + \left(I_s \sin\beta\right)^2 \right]^{1/2}} \quad \text{rad/s} \tag{8.120}$$

At $\beta = 180°$, the torque will be zero and the frequency will be

$$\omega_{e,max} = \frac{V_{s(max)}}{L_{ms}'(I_f' - I_s)} \quad \text{rad/s} \tag{8.121}$$

The only effective design means of increasing the available constant power speed ratio is to increase the inductance $L_{ms}' = L_{ls} + L_{ms}$. Means for increasing its components $L_{ms}$and L_{ls}have been discussed in Section 8.12.2. Alternatively, an external series inductance can be inserted between the inverter and the motor which increases the field-weakening range but reduces the maximum torque capability of the motor thereby reducing the value of the base speed corner point (equation 8.118). The leakage inductance can also be increased by utilizing fractional slot windings with slots/pole/phase < 1 [7, 8]. This approach can generally be used only if the number of motor poles is relatively high. This is not necessarily a limitation, however, since the speeds on the order of 1200–3600 RPM can be reached by simply increasing the inverter frequency. The approach is ultimately limited by iron losses.

The denominator of equation (8.121) is related to the unprimed variables by

$$L'_{ms}(I'_f - I_s) = L_{ms}I_f - (L_{ms} + L_{ls})I_s \tag{8.122}$$

Setting this quantity to zero would extend $\omega_{e,max}$ to infinity and produce an infinite field-weakening range at rated power. This point is reached where [9]

$$I_f = \left(\frac{L_{ms} + L_{ls}}{L_{ms}}\right) I_{s,rated} \tag{8.123}$$

or, in terms of magnet emf (Figure 8.16)

$$E_i = X_s I_{s,rated} \tag{8.124}$$

For machines with air gap magnets, this value of voltage is reached when E_i is relatively low since the synchronous reactance of such a machine is only 0.1–0.5 per unit. Hence, the torque-producing capability of such a machine is poor and the machine will be large and heavy. However, this difficulty can be considerably improved with machines equipped with either inset or buried magnets since the per unit synchronous reactance of such machines is considerably larger. Also buried machines can be designed to possess a considerable amount of reluctance torque approaching or even more than 40% of its rated torque.

8.14 PM MOTOR WITH INSET MAGNETS

The surface-magnet machine is only one of a wide variety of magnet configurations used for PM machines. A magnet arrangement in an inset magnet motor is shown in Figure 8.20. Structurally, this construction can provide a more secure magnet setting. The motor can also produce somewhat more maximum torque and its overspeed range is larger due to its saliency [6].

An equivalent circuit for a permanent magnet machine with saliency is shown in Figure 8.21 Neglecting the resistances for simplicity, the equivalent circuit corresponds to the two equations,

$$V_{qs} = \omega_e(L_{ls} + L_{md})I_{ds} + \omega_e L_{md} I_f \tag{8.125}$$

$$V_{ds} = -\omega_e(L_{ls} + L_{mq})I_{qs} \tag{8.126}$$

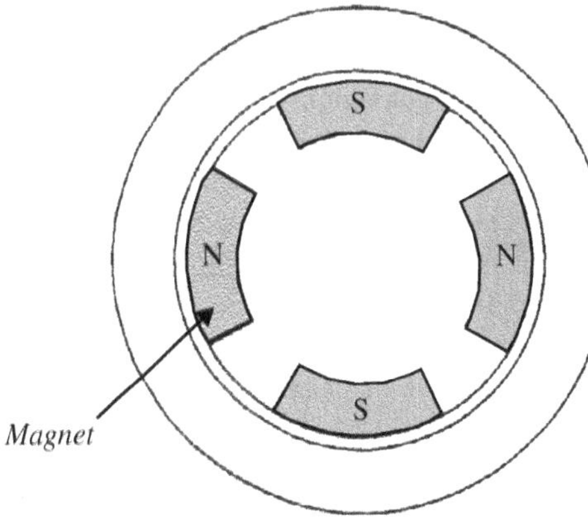

Figure 8.20 A 4-pole inset magnet PM motor

Neglecting losses and assuming peak values, the power input is equal to the power output so that

$$P_{in} = P_{out} = \frac{3}{2}(V_{qs}I_{qs} + V_{ds}I_{ds}) \tag{8.127}$$

$$P_{out} = \frac{3}{2}[\omega_e(L_{ls} + L_{md})I_{ds}I_{qs} - \omega_e(L_{ls} + L_{mq})I_{ds}I_{qs} + \omega_e L_{md} I_{qs} I_f] \tag{8.128}$$

$$= \frac{3}{2}\omega_e[(L_{md} - L_{mq})I_{ds}I_{qs} + L_{md}I_{qs}I_f] \tag{8.129}$$

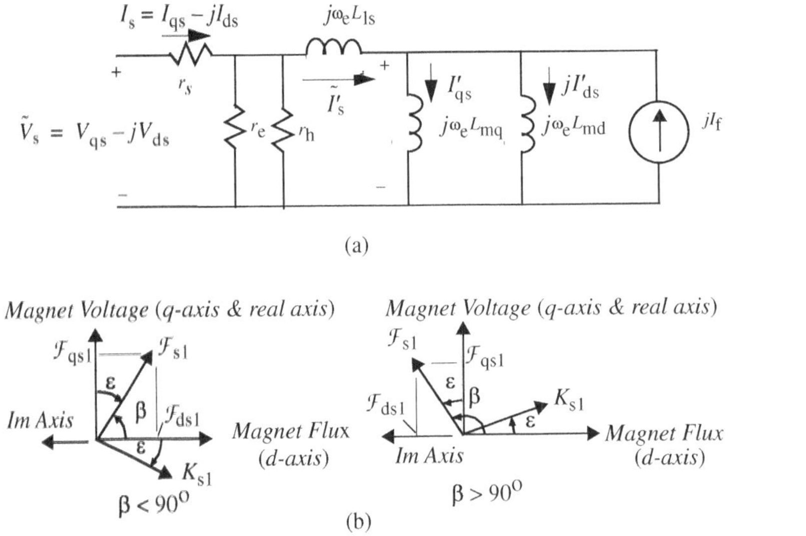

Figure 8.21 (a) Per phase steady state equivalent circuit and (b) phasor diagrams for a permanent magnet machine with saliency.

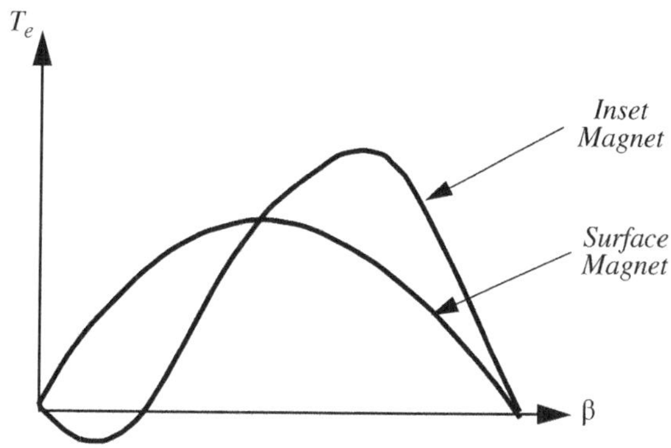

Figure 8.22 Comparison of inset and surface mounted magnet torque-angle relationships.

The phasor direction of the MMF $\mathcal{F}_{s1}$ is the same as that of the stator current so that

$$I_{qs} = I_s \sin\beta \tag{8.130}$$

$$I_{ds} = I_s \cos\beta \tag{8.131}$$

The electromagnetic torque is therefore

$$T_e = \frac{P_{out}}{\omega_e} = \frac{3}{4}(L_{md} - L_{mq})I_s^2 \sin 2\beta + \frac{3}{2}L_{md}I_s I_f \sin\beta \quad \text{N-m} \tag{8.132}$$

A typical comparison of torque-angle characteristics for surface mounted and inset magnet motors is shown in Figure 8.22. Since q-axis inductance is clearly larger than the d-axis inductance, the first torque (saliency torque) in the equation is negative for small positive values of β but then becomes positive for $90° < \beta < 180°$. Maximum torque is then achieved at an angle greater than 90° where the stator field tends to produce positive torque-producing flux in the iron section leading each magnet. The flux density in these steel interpolar sections is limited by magnetic saturation in the stator teeth to a value which is not much greater than the flux density above the magnets. Thus, the effect is to produce the equivalent of a magnet with maximum possible width, that is, $\alpha_m = 180°$. The ratio of the maximum torques for inset and surface magnets is then approximately

$$\frac{T_{inset}}{T_{surface}} \approx \frac{1}{\sin\left(\frac{\alpha_m}{2}\right)} \tag{8.133}$$

The maximum torque will occur at an angle of approximately $\beta = 180 - \alpha_m/2$. The above expressions apply reasonably well down to a minimum value of $\alpha_m =$ 90° beyond which point the torque contribution due to saliency ceases. Thus, more torque can be obtained with the use of less magnet material. For example, replacing a surface-mounted magnet with $\alpha_m = 120°$ by an inset magnet with an optimum case of $\alpha_m = 90°$ would result in an increase in maximum torque of about 15%.

To obtain maximum benefit from the inset design, the air gap around the machine should be uniform and as small as possible. In designing the thickness of the stator yoke of this type of motor, the flux should be considered to be equivalent to that of a magnet with $\alpha_m = 180°$. Otherwise, most design constraints are similar to those of surface-mounted PM motors. The principle disadvantage of this type of construction is the stator iron losses due to the steeply rising flux at the front and back ends of the magnet which are accentuated by the smaller air gap of this machine.

8.14.1 Short-Circuit Protection

When the inset magnet motor is short-circuited, the stator field is again aligned in direct opposition to the magnetic field. The effect of the stator field is to produce a reverse flux density in the steel sections on each side of each magnet. This reduces the net air gap flux. With a short circuit, the stator flux linkage due to the air gap flux must be equal and opposite to the leakage flux linkage.

For the most severe case in which the stator time constant τ_1 is considered as large, the criterion for magnet protection from a short circuit is given by the following expression [5],

$$\frac{L_{ls}}{L_{ms}} \geq \left[\frac{8\left[\sin\alpha_m - \dfrac{B_s d_g}{B_r d_m}(1-\sin\alpha_m)\right]}{\pi\left(1 - \dfrac{B_d d_g}{B_r d_m}\right)} - \left(\frac{2\alpha_m + \sin 2\alpha_m}{\pi}\right) \right] \tag{8.134}$$

In this expression, the inductance L_{ms} can be calculated as for a surface-mounted magnet using equation (8.88). The quantity B_s is the value of flux density which is established in the interpolar air gap when the teeth above it are effectively saturated.

8.14.2 Flux Weakening

The use of the inset magnet construction also increases the range of constant power operation. An increase in the current angle β beyond $180° - \alpha_m/2$ causes a reduction in the effective air gap flux and thus allows operation at a higher speed. Ignoring the effect of leakage inductance, the maximum frequency at which zero torque is reached can be related to the limit frequency $\omega_{e,base}$ for maximum torque by Reference [6],

$$\frac{\omega_{e,max}}{\omega_{e,base}} = \frac{1}{2\sin\left(\dfrac{\alpha_m}{2}\right) - 1} \tag{8.135}$$

For $\alpha_m = 90°$, this ratio is about 2.4. A magnet width of 60° will produce an infinite field-weakening range. However, reduction of the magnet width effectively reduces the magnet emf which consequently requires a large stator MMF to achieve a given torque requirement. A larger machine will result. The situation is much the same as for the air gap magnet machine (Section 8.13).

The presence of leakage inductance will again increase the potential speed range. Also, the range can again be extended through the use of external series inductance with the penalty of reducing $\omega_{e,base}$ unless the inverter voltage rating is increased.

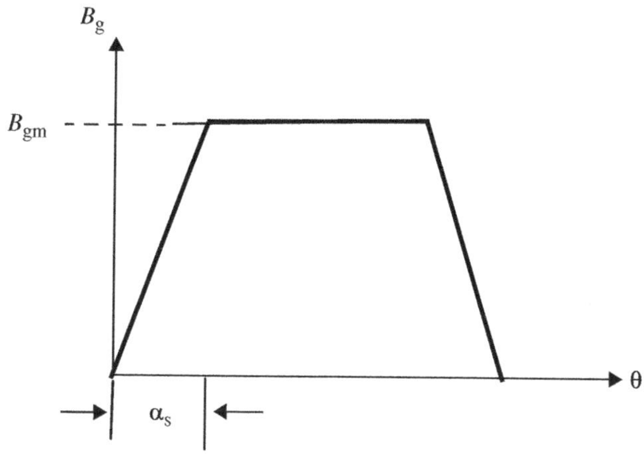

Figure 8.23 Effective magnet gap flux density with one slot skew

8.15 COGGING TORQUE

Cogging torque arises from the variation of the magnetic reluctance of the stator teeth as seen by the magnets as the rotor rotates. Unless precautions are taken, this torque can be very large, that is, up to 0.25 per unit. One means of controlling the cogging torque is to skew either the stator slots or the rotor magnets by one slot pitch τ_s as indicated in Figure 8.23 [10]. However, the construction of either the stator or the rotor is simpler if skewing is avoided. An alternate approach is to choose the magnet width w_m for minimum cogging. It has been experimentally demonstrated [11] that for equal tooth and slot widths, the magnet width w_m should be related to the slot pitch τ_s by

$$w_m \approx (m + 0.14)\tau_s \tag{8.136}$$

where m is an integer, being the appropriate number of teeth under a pole. For example, consider a sine wave fed motor with $q = 3$ slots per pole/phase. Appropriate choices for the pole width would be 6.14, 7.14, or 8.14 representing values of the angle $\alpha_m = 122.8$, 142.8, or 162.8°.

This approach to cogging control is most effective with $w_s \approx 0.5\,\tau_s$. In this region, the ripple torque produced at each end of the magnet varies nearly sinusoidally and appropriate choice of the width allows cancellation of the fundamental component of torque variation. The frequency of this component is $S_1\omega_e$ rad/s for S_1 stator slots.

In some instances, the cogging torque can be further reduced by displacing one pair of magnets with respect to an adjoining pair of a 4, 8, 12, ... pole machine by an electrical angle of $90°/S_1$ to cancel the second harmonic of the cogging torque.

8.16 RIPPLE TORQUE

In addition to the cogging torque, a ripple torque may be produced by the interaction of the rotor magnets with the stator current MMF. For a sine wave fed motor, an oscillating torque can be produced at frequency $6\omega_e$ by the interaction of the fifth and seventh space harmonics of the air gap flux density with the fifth and seventh harmonics respectively of the stator current density distribution, that is,

$$T_{e6} = \frac{\pi}{4} D_{is}^2 l_e B_{gh} K_{sh} \quad \text{N-m} \tag{8.137}$$

where $h = 5$ or 7. The RMS harmonic flux density is related to the fundamental by

$$\frac{B_{gh}}{B_{g1}} = \frac{1}{h} \sin\left(h\frac{\alpha_m}{2}\right) \tag{8.138}$$

The RMS value of the harmonic linear current density is related to its fundamental by

$$\frac{K_{sh}}{K_{s1}} = \left(\frac{1}{h}\right)\frac{k_h}{k_1} \tag{8.139}$$

where k_h is the winding factor (distribution, pitch, and skew) for the winding. It is noted that both the above ratios decay at least with $1/h$ and thus, for any component, the per unit value of the sixth harmonic torque is

$$\frac{T_{e6}}{T_1} \leq \frac{1}{h^2} \tag{8.140}$$

Usually, the pitch and distribution of the winding can be arranged to reduce this ratio substantially.

8.17 DESIGN USING FERRITE MAGNETS

Ferrite magnet material has a much lower cost per unit volume than neodymium or samarium–cobalt material. However, its residual flux density B_r is typically in the range 0.3–0.4 tesla as shown in Figure 8.24, that is, roughly one-third of that of Nd–Fe–B. Its relative recoil permeability is about 1.05. The temperature coefficient of B_r is typically about −0.2%/°C, that is, about twice that of Nd–Fe–B. Its limiting value of magnetic field H_d is typically about −200 kA/m at 20°C. In contrast with Nd–Fe–B, the value of H_d increases with an increase in temperature typically by about 0.4% per degree celsius up to about 100°C. However, the remnant flux density continues to decrease with increasing temperature.

PM motors may be designed with surface ferrite magnets arranged as shown in Figure 8.25. The air gap flux density resulting from the magnets is then typically in the range $B_{gm} = 0.25 - 0.3$ tesla using a magnet thickness $l_m = 5 - 10$ mm. The required tooth width w_t is then considerably less than half of the slot pitch τ_s with the increased space w_s available for the slots. Thus, the values of the surface current

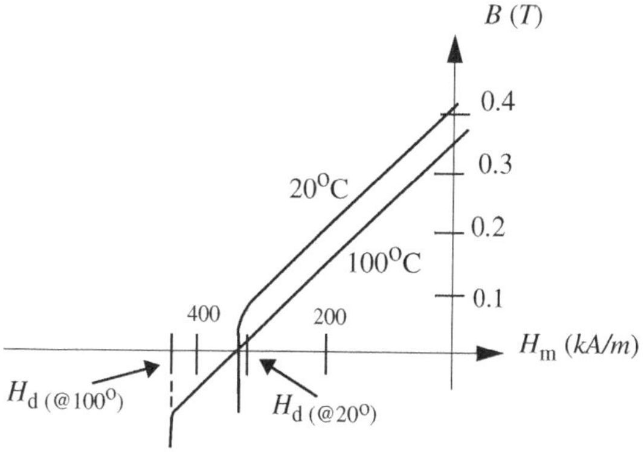

Figure 8.24 Demagnetization characteristics for a typical ferrite magnet material.

density K_{s1} can be increased. It is, however, the machine, of course, remains subject to the constraints of adequate cooling as described in Section 8.11.

In order to keep the air gap flux as large as possible, it is necessary to provide as low a reluctance in the air gap as possible. The slots are normally semi-closed with tooth tips providing as small a slot opening as possible as shown in Figure 8.25. This type of stator construction is applicable for small motors using random-wound coils which can be inserted turn by turn through the slot opening. The yoke of motors using radial ferrite magnets is relatively thin because of the low average air gap flux density B_{gm}. Small machines can, therefore, be constructed with two poles without excessive yoke thickness.

8.18 PERMANENT MACHINES WITH BURIED MAGNETS

The largest family of permanent magnet machine involves those with buried magnets. Figure 8.26 shows some of the possibilities. Clearly, an exhaustive treatment of such machines is beyond the scope of this book. However, several typical examples of these machines can be discussed to illustrate the issues involved.

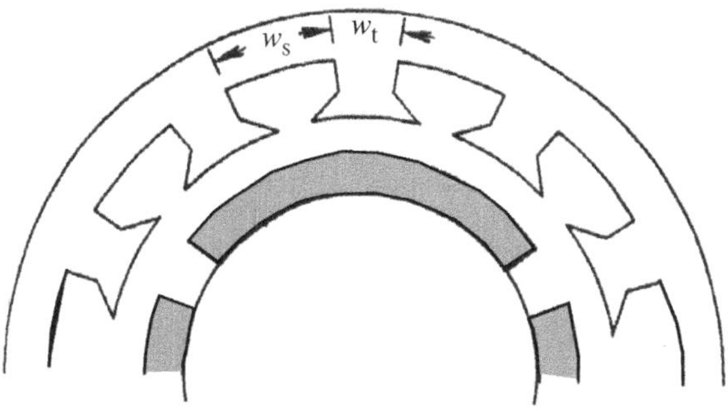

Figure 8.25 PM motor with ferrite surface magnets.

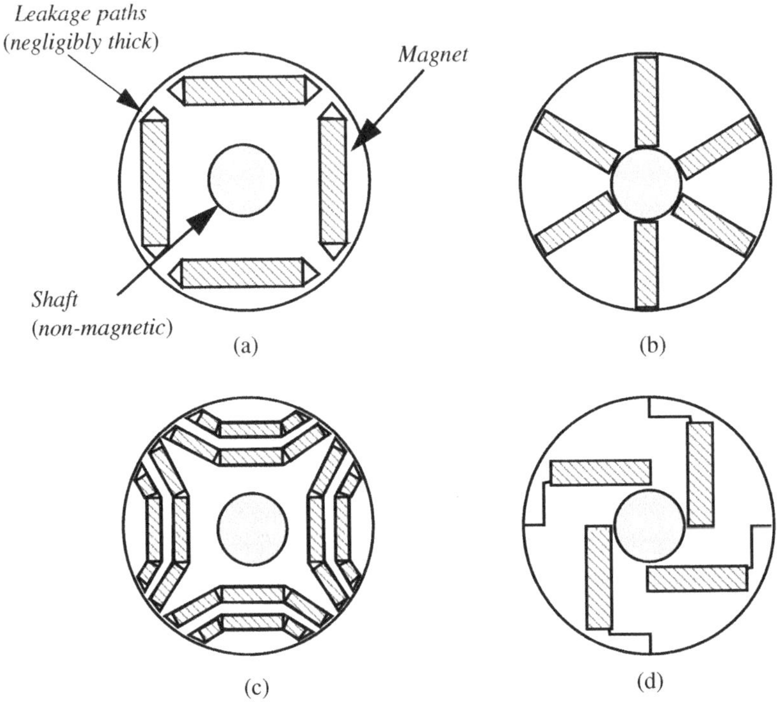

Figure 8.26 Typical four pole buried magnet configurations (a) radially magnetized, (b) circumferentially magnetized, (c) two layer radial magnetization (d) asymmetrical distribution.

8.18.1 PM Machines with Buried Circumferential Magnets

The air gap flux density can be substantially increased by arranging either ferrite or rare-earth magnets located in the spoke type form shown in Figure 8.27. Since the inner portion of the rotor is utilized to position magnets, flux density in the gap can be raised to a value greater than that existing in the magnets themselves. This structure, however, imposes several constraints on the rotor dimensions and numbers of poles.

Consider a design for a specified magnet air gap flux density B_g over each pole arc of length $(\pi D_{or} - Pd_m)/P$. The flux over this arc is equal to the flux produced by two adjacent magnets (ignoring fringing). Thus

$$B_g\left(\frac{\pi D_{or} - Pd_m}{P}\right) = 2w_m B_m \tag{8.141}$$

where B_m is the flux density within the magnet.

When the magnets just touch at the nonmagnetic core radius r_{ir}

$$r_{ir} = Pd_m/2\pi \tag{8.142}$$

Thus,

$$w_m = r_{or} - r_{ir} = r_{or} - Pd_m/2\pi \tag{8.143}$$

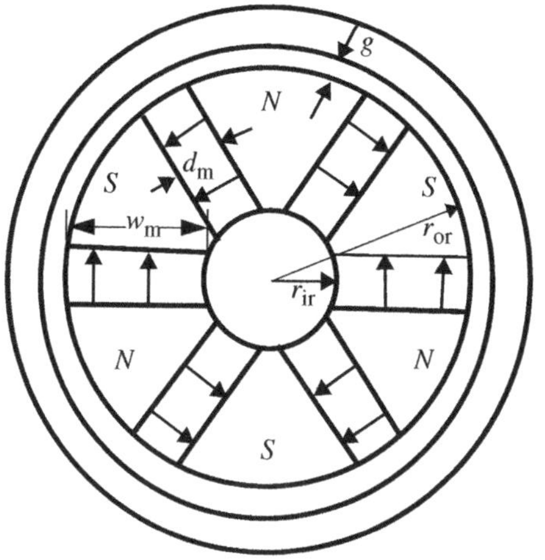

Figure 8.27 PM motor with spoke magnets (idealized).

Combining equations (8.141) and (8.143) gives

$$\frac{B_g}{B_m} = \frac{P}{\pi} \tag{8.144}$$

Thus the gap flux density will be raised above the magnet flux density if the number of poles is four or more. This effect can be exploited to increase the effectiveness of ferrite magnets when the number of poles is large, typically eight or more.

Consider a closed flux path across one magnet and two air gaps each of effective length g_e. For this path,

$$H_m d_m + 2g_e \frac{B_g}{\mu_0} = 0 \tag{8.145}$$

From Section 5.4, the permanent magnet material can be characterized by the expression

$$B_m = B_r + \mu_{rm}\mu_0 H_m \tag{8.146}$$

Combining equations (8.141), (8.142), and (8.143) yields

$$\frac{B_g}{B_r} = \frac{P}{\pi + 2\mu_{rm} P g_e / d_m} \tag{8.147}$$

and

$$\frac{B_m}{B_r} = \frac{\pi}{\pi + 2\mu_{rm} P g_e / d_m} \tag{8.148}$$

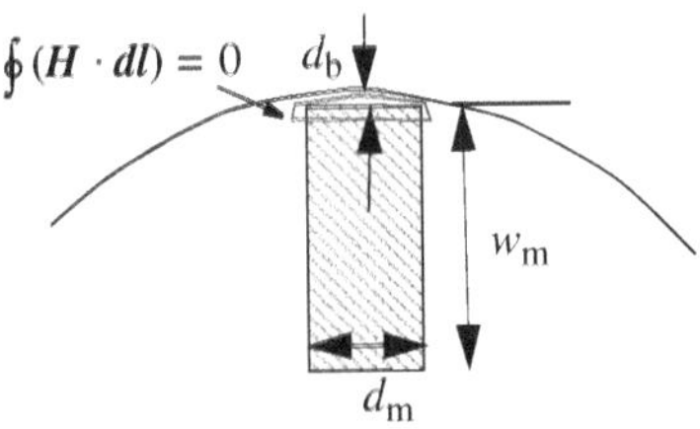

Figure 8.28 Magnetic bridge for retaining a magnet.

Thus, for a material with residual density B_r, P poles, and an effective air gap g'_e the magnet length w_m required to produce a specified gap flux density B_g can be found.

$$w_m = \frac{2P\mu_{rm}g_e}{\left(P\frac{B_r}{B_g} - \pi\right)} \quad (8.149)$$

For the configuration of Figure 8.27, the proportion of the rotor periphery which has the radial flux density B_{gm} is

$$\frac{\text{Pole arc}}{\text{Pole pitch}} = 1 - \frac{P}{\pi}\frac{d_m}{D_{or}} \quad (8.150)$$

In general, it is desirable to have this quantity close to 2/3. Depending upon the value of P, D_{or} must be considerably greater than d_m. As the rotor diameter is reduced, the length of the magnet must be correspondingly reduced making the air gap flux density smaller if a reasonable value of pole arc is to be maintained.

Magnet retention is an important issue in the design of permanent magnet motors. When the magnets are on the surface of the rotor, the magnets are generally fixed by high-tech fixatives (glues) or by wrapping the magnet with a retentive band. When the magnet is buried, other means are often adopted to keep the air gap as small as possible and thus raise the air gap flux density to as large a value as possible. Often the magnets are fixed by use of thin magnetic bridges as shown in Figure 8.28. These bridges should be made as small as possible since a small amount of the magnet flux is diverted through this short- circuiting path, reducing the effectiveness of the magnet.

The flux passing through the bridge causes some of the magnet flux to be sacrificed in order to secure the magnet to the rotor. Evaluating Ampere's law around a path passing through the magnet and through the bridge, as shown in Figure 8.28, and assuming that H is directed along the path l

$$\oint \boldsymbol{H} \bullet \boldsymbol{dl} = H_m d_m + H_b d_b = 0 \quad (8.151)$$

where H_m and H_b are the magnetic field intensities encountered inside and in the bridge outside of the magnet respectively. Since d_m and d_b are essentially the same length

$$H_b \cong -H_m \quad (8.152)$$

Since

$$B_{\mathrm{m}} = B_{\mathrm{r}} + \mu_{\mathrm{rm}}\mu_0 H_{\mathrm{m}} \tag{8.153}$$

then

$$H_{\mathrm{m}} = -H_{\mathrm{b}} = \frac{B_{\mathrm{m}} - B_{\mathrm{r}}}{\mu_{\mathrm{rm}}\mu_0} \tag{8.154}$$

and from equation (8.148),

$$H_{\mathrm{b}} = \frac{B_{\mathrm{r}}\left(1 - \dfrac{\pi}{\pi + 2\mu_{\mathrm{rm}}Pg_{\mathrm{e}}/w_{\mathrm{m}}}\right)}{\mu_{\mathrm{rm}}\mu_0} \tag{8.155}$$

$$= \frac{B_{\mathrm{r}}(2\mu_{\mathrm{rm}}Pg_{\mathrm{e}}/w_{\mathrm{m}})}{\mu_0\mu_{\mathrm{rm}}(\pi + 2\mu_{\mathrm{rm}}Pg_{\mathrm{e}}/w_{\mathrm{m}})} \tag{8.156}$$

The value of H_{b} determines the MMF drop across the bridge and thus the flux level in the bridge. The flux in the bridge is equal to a sacrificial amount of magnet flux such that

$$B_{\mathrm{b}}w_{\mathrm{t}}l_{\mathrm{e}} = B_{\mathrm{m}}w_{\mathrm{x}}l_{\mathrm{i}} \tag{8.157}$$

Again if $l_{\mathrm{e}} \cong l_{\mathrm{i}}$,

$$w_{\mathrm{x}} = \frac{B_{\mathrm{b}}}{B_{\mathrm{m}}}d_{\mathrm{b}} = \frac{B_{\mathrm{b}}}{B_{\mathrm{r}}}\left(\frac{\pi + 2\mu_{\mathrm{rm}}Pg_{\mathrm{e}}/w_{\mathrm{m}}}{\pi}\right)t \tag{8.158}$$

The value of B_{b} to be used in this expression is the value of the flux density corresponding to H_{b}, equation (8.155) above, obtained from the *B–H* characteristic of the steel. As a rough approximation, it can be assumed that $B_{\mathrm{b}} \cong 2$ tesla for common iron.

In most cases, the magnet is usually provided with a second bridge at the other end of the magnet width. Assuming that the thickness of the two bridges is identical, the bridges cause the peak flux density in the air gap to decrease to

$$B_{\mathrm{gm}} = \frac{PB_{\mathrm{r}}}{\pi + 2\mu_{\mathrm{rm}}Pg_{\mathrm{e}}/d_{\mathrm{m}}}\left(\frac{d_{\mathrm{m}} - 2d_{\mathrm{x}}}{d_{\mathrm{m}}}\right) \tag{8.159}$$

After a rotor has been designed for a specific air gap density B_{gm}, the fundamental component of the flux density should be calculated as shown in Section 8.6. The stator design then proceeds as discussed previously.

8.19 CONCLUSION

This chapter has served as only an introduction into the design of permanent magnet machines. The purpose of the chapter is to allow the reader to gain insight into how the physical dimensions and parameters are influenced by permanent magnet excitation. It should have become apparent to the reader that the subject is rich in detail and any attempt to cover the design of PM motors in a single chapter will often not be applicable for a specific situation of interest. For more information on this subject, the reader is referred to several books devoted to the design of permanent magnet machines [12–15].

ACKNOWLEDGMENT

This chapter has relied quite heavily on the work of G. R. Slemon and T. Sebastian as presented in several workshops sponsored by the IEEE Industry Applications Society. The author gratefully acknowledges this important body of work which formed a major portion of his source material.

REFERENCES

[1] R. J. Parker, *Advances in Permanent Magnetism*, John Wiley & Sons, 1990, pp. 37–42.

[2] M. G. Say, *The Performance and Design of Alternating Current Machines*, Pitman, 1948.

[3] T. Sebastian and G. R. Slemon, "Transient modelling and performance of variable speed permanent magnet motors," *IEEE Transactions on Industry Applications*, vol. IA-25, no. 1, January/February 1989, pp. 101–107.

[4] T. Sebastian, G. R. Slemon, and M. A. Rahman, "Modelling of permanent magnet synchronous motors," *IEEE Transactions on Magnetics*, vol. MAG-22, no. 5, September 1986, pp. 1069–1071.

[5] T. Sebastian and G. R. Slemon, "Transient torque and short circuit capabilities of variable speed permanent magnet motors," *IEEE Transactions on Magnetics*, vol. MAG-23, no. 5, September 1987, pp. 3619–3621.

[6] T. Sebastian and G. R. Slemon, "Operating limits of inverter driven permanent magnet motor drives," *IEEE Transactions on Industry Applications*, vol. IA-23, no. 2, March/April 1987, pp. 327–333.

[7] J. Cros and P. Viarouge, "Synthesis of high performance PM motors with concentrated windings," *IEEE Transactions on Energy Conversion*, vol. 17, no. 2, June 2002, pp. 248–253.

[8] F. Magnussen and C. Sadarangani, "Winding Factors and Joule Losses of Permanent Magnet Machines with Concentrated Windings," in Proceedings of the IEEE International Electric Machines and Drives Conference (IEMDC), Madison, WI, March 2003, pp. 333–339.

[9] R. Schiferl and T. A. Lipo, "Power capability of salient pole permanent magnet synchronous motors in variable speed drive applications," *IEEE Transactions on Industry Applications*, vol. 26, no. 1, January/February 1990, pp. 115–123.

[10] J. De La Ree and N. Boules, "Torque production in permanent magnet synchronous motors," *IEEE Industry Applications Society Annual Meeting Conference Record*, October 1987, pp. 15–20.

[11] T. Li and G. R. Slemon, "Reduction of cogging torque in permanent magnet motors," *IEEE Transactions on Magnetics*, vol. MAG-24, no. 6, November 1988, pp. 2901–2903.

[12] J. C. Gieras and M. Wing, *Permanent Magnet Motor Technology – Design and Applications*, Marcel Dekker, New York, 1997.

[13] D. C. Hanselman, *Brushless Permanent-Magnet Motor Design*, McGraw Hill, New York, 1994.

[14] J. R. Hendershot and T. J. E. Miller, "Design of Brushless Permanent-Magnet Machines," Motor Design Books. 2010.

[15] S. A. Nasar, I. Boldea, and L. E. Unnewehr, *Permanent Magnet, Reluctance, and Self-Synchronous Motors*, CRC Press, Boca Raton, 1993.

CHAPTER 9

ELECTROMAGNETIC DESIGN OF SYNCHRONOUS MACHINES

THE DESIGN OF INDUCTION MACHINES essentially concerns the analysis of two distinct structures namely squirrel cage and wound rotor configurations. On the other hand, the family of synchronous machines is more extensive encompassing round rotor and salient pole synchronous machine structures, reluctance machines, and permanent magnet machines to name only the four major categories. While there is a certain amount of similarity to the design of each of these machines, they are sufficiently different that they cannot be treated as a group but must be considered separately. In this chapter, the design of salient pole synchronous machines will be considered in some detail. Fortunately, the effort devoted to the study of induction machines can be put to good use, so that it will not be necessary to begin the analysis of synchronous machines from first principles.

9.1 CALCULATION OF USEFUL FLUX PER POLE

The computation of fluxes and MMF in the induction motor magnetic circuit was considered in detail in Chapter 3. Since the machine was symmetrical, that is, since the stator and rotor had symmetrically placed windings and the air gap length was constant, the computation of the parameters of only a single circuit was necessary since the equivalent circuit viewed from each of the three phases was identical. Hence, a "per phase" equivalent circuit was possible. In the case of a salient pole synchronous machine, however, the air gap is anything but uniform. In addition, windings are placed on the rotor in an unsymmetrical manner since the field windings are wound around salient poles. Hence, the equivalent circuit viewed from each phase depends on the instantaneous position of the rotor. Nonetheless, an analysis of a synchronous machine can proceed if one resolves the equivalent circuit representation of the stator into two circuit components. One of the circuits portrays the behavior of the fluxes and currents in the field axis, the *direct axis*, and the other circuit represents the fluxes and currents in the interpolar space, the *quadrature axis*. Since these two circuits are fixed relative to the rotor axes, they are said to rotate with the rotor or that the reference frame rotates with the rotor.

Introduction to AC Machine Design, First Edition. Thomas A. Lipo.

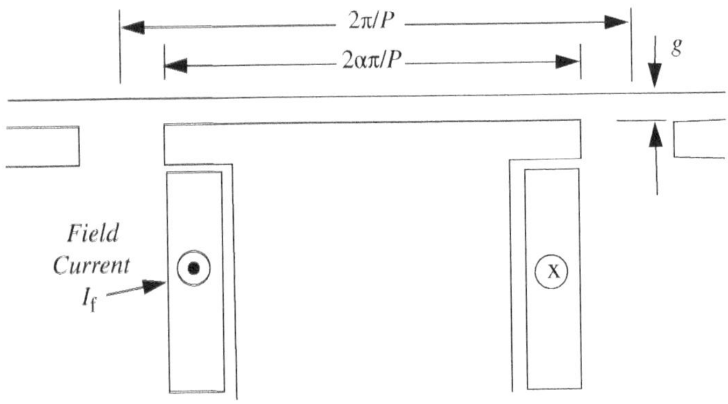

Figure 9.1 Simplified geometry of pole face for a salient pole synchronous machine.

One important difference between induction and synchronous machines is that power is brought to or taken from the induction machine via only one port, the stator, while the synchronous machine is a two-port system since the field winding is also excited. In addition, synchronous machines are nearly always designed so that the magnetic circuit, particularly the rotor poles, is rather highly saturated. Hence, in order to obtain useful values for machine parameters, a simple superposition scheme based on computing the fluxes produced first by the field and then by the stator MMF independently and adding the result is usually inapplicable. A solution must be obtained with the machine at a representative operating condition, usually as close to rated load and power factor as can be estimated. Fortunately, a very good estimate of a full-load condition can be obtained as an extension of the results obtained from an open-circuit condition. Hence, it will be necessary to concentrate on this condition initially.

It can be recalled from the analysis of induction machines that the procedure began with an analysis commencing at the gap of the machine and worked one's way outward to the terminals. This will again be the preferred approach. However, in this case, the possibility of rotor saliency of the synchronous machine will introduce some additional complications. The analysis begins by assuming that an ideal pole shape exists in which the gap is uniform and that the flux in the interpolar space can be neglected. Such a geometry is shown in Figure 9.1.

Because of the shape of the field poles, the air gap flux cannot be assumed to be sinusoidal whether it arises from the field winding current alone, the stator winding current alone, or some combination. The analysis begins by calculating the magnetizing inductance associated with the direct axis component of stator current.

9.2 CALCULATION OF DIRECT AND QUADRATURE AXIS MAGNETIZING INDUCTANCE

It will first be assumed that the stator of the machine is excited with three-phase balanced sinusoidal voltages resulting in a uniformly rotating stator MMF wave. The winding configuration of the stator of the synchronous machine is essentially the same as an induction motor, and the stator MMF distribution is as described in Chapter 2. If one neglects the higher harmonics, the stator MMF rotates synchronously in the gap according to

$$\mathcal{F}_s = \mathcal{F}_{s1} \cos\left(\frac{P\theta}{2} - \omega_e t\right) \tag{9.1}$$

where $\mathcal{F}_{s1}$ is the amplitude of the fundamental component "1" of the stator MMF, that is, from equation (2.73),

$$\mathcal{F}_{s1} = \left(\frac{3}{2}\right)\left(\frac{4}{\pi}\right)\left(\frac{k_1 N_t}{P}\right)\frac{I_s}{C_s} = \left(\frac{3}{2}\right)\left(\frac{4}{\pi}\right)\left(\frac{k_1 N_s}{P}\right) I_s \tag{9.2}$$

where I_s is the amplitude of the current in each of the three stator phases and N_s is the number of series-connected turns. The subscript "s" has also been added to the number of stator parallel circuits C for eventual clarity when rotor field circuits are considered.

In terms of *d–q* components as defined by equations (2.76) and (2.77), the peak stator MMF for each of P poles is, for balanced conditions, given by the results of equations, (2.78) to (2.79). Namely

$$\mathcal{F}_{ds1} = \frac{3}{2}\left(\frac{4}{\pi}\right)\frac{(k_1 N_t)}{P}\left(\frac{I_s}{C_s}\cos\varepsilon\right) = \frac{3}{2}\left(\frac{4}{\pi}\right)\left(\frac{k_1 N_s}{P}\right) I_d \tag{9.3}$$

in the *d*-axis circuit and as

$$\mathcal{F}_{qs1} = \frac{3}{2}\left(\frac{4}{\pi}\right)\frac{(k_1 N_t)}{P}\left(\frac{I_s}{C_s}\sin\varepsilon\right) = \frac{3}{2}\left(\frac{4}{\pi}\right)\left(\frac{k_1 N_s}{P}\right) I_q \tag{9.4}$$

in the *q*-axis circuit and where I_d and I_q are equivalent to the two-phase currents I_x and I_y of Section 2.10 but expressed in the synchronously rotating reference frame as

$$I_d = I_s \cos\varepsilon \tag{9.5}$$

and

$$I_q = I_s \sin\varepsilon \tag{9.6}$$

The angle ε is the spatial location of the peak value of stator MMF relative to the *d*-axis and is termed the *MMF angle*. Each of the two components of MMF is assumed to produce a corresponding flux density so that the gap flux density $B_g(\theta)$ of Section 3.10 can be considered as being comprised of two components namely $B_{gd}(\theta)$ and $B_{gq}(\theta)$.

In order to calculate the direct axis air gap self-inductance, it will be assumed that the machine is excited only by the direct axis component of stator MMF. The issue of leakage inductances will be set aside for the moment. Initially, the simple geometry shown in Figure 9.1 can be adopted in which the gap can be assumed to be uniform under the pole and essentially infinite in the interpolar space. It will be assumed that the flux goes straight across the gap and that fringing in the interpolar region other types of leakage flux are neglected. The resulting stator air gap flux linkages or "magnetizing" flux linkages for one pole can be found by integrating the gap flux density over one pole, properly weighting the value by a sinusoidal distribution of turns (i.e., effectively equation (1.146)) or, for the direct axis stator flux linkage per pole,

$$\lambda_{\text{mdp}} = l_{\text{e}} \frac{D_{\text{is}}}{2} \int_{-\pi/P}^{\pi/P} N_{\text{ds}}(\theta) B_{\text{gd}}(\theta) d\theta \tag{9.7}$$

Neglecting spatial harmonics, the winding function corresponding to the current I_{ds} is

$$N_{\text{ds}}(\theta) = \left(\frac{4}{\pi}\right)\left(\frac{k_1 N_{\text{t}}}{P}\right) \cos\left(\frac{P\theta}{2}\right) \tag{9.8}$$

Since the gap is assumed as uniform over the pole arc $2\alpha\pi/P$, the gap flux density $B_{\text{gd}}(\theta)$ produced by $\mathcal{F}_{\text{ds}}$ is sinusoidal over this region with amplitude B_{gd1} and zero elsewhere. Equation (9.7) becomes

$$\lambda_{\text{mdp}} = \frac{4}{\pi}\left(\frac{k_1 N_{\text{t}}}{P}\right) \int_{-(\alpha\pi)/P}^{\alpha\pi/P} B_{\text{gd1}} \cos^2\left(\frac{P\theta}{2}\right) l_{\text{e}} \frac{D_{\text{is}}}{2} d\theta \tag{9.9}$$

$$= \frac{2}{\pi}\left(\frac{k_1 N_{\text{t}}}{P}\right) B_{\text{gd1}} \tau_{\text{p}} l_{\text{e}} \left[\frac{\alpha\pi + \sin\alpha\pi}{\pi}\right] \tag{9.10}$$

where the pole pitch

$$\tau_{\text{p}} = \frac{\pi D_{\text{is}}}{P}$$

and B_{gd1} is the peak value of the fundamental component of gap flux density produced by the d-axis MMF for the case where $\alpha = 1$, that is, the case of the smooth air gap machine.

The total flux linkages produced in the d-axis are equal to the number of poles connected in series or

$$\begin{aligned}\lambda_{\text{md}} &= \frac{P}{C_{\text{s}}} \lambda_{\text{mdp}} \\ &= \frac{2}{\pi} k_1 N_{\text{s}} B_{\text{gd1}} \tau_{\text{p}} l_{\text{e}} k_{\text{d}}\end{aligned} \tag{9.11}$$

where $N_{\text{s}} = N_{\text{t}}/C_{\text{s}}$ and where k_{d} is defined as

$$k_{\text{d}} = \frac{\alpha\pi + \sin\alpha\pi}{\pi} \tag{9.12}$$

By definition, the direct axis magnetizing inductance is equal to the flux linkages produced by the stator currents divided by the d-axis component of stator current. Hence,

$$L_{\text{md}} = \frac{\lambda_{\text{md}}}{I_{\text{d}}} \tag{9.13}$$

Therefore, from equations (9.11) and (9.3),

$$L_{\text{md}} = \left(\frac{12}{\pi^2}\right)\frac{(k_1^2 N_s^2)}{P}\left(\frac{B_{\text{gd1}}}{\mathcal{F}_{\text{ds1}}}\right)\tau_{\text{p}} l_{\text{e}} k_{\text{d}} \tag{9.14}$$

This equation is sufficiently general to also include the effects of MMF drop in the iron paths if $\mathcal{F}_{\text{ds1}}$ includes this effect. The peak value of the d-axis MMF $\mathcal{F}_{\text{ds}}$ is then related to the peak fundamental flux density in the gap by equation (3.77). That is,

$$\mathcal{F}_{\text{ds1}} = \frac{B_{\text{gd1}}}{\mu_0} g_{\text{e}} \tag{9.15}$$

where g_{e} can now be considered as the effective gap for the d-axis of the machine including the effects of both fringing (Carter factor) and saturation so that one obtains, finally

$$L_{\text{md}} = \left(\frac{12}{\pi^2}\right)\left(\frac{k_1^2 N_s^2}{P}\right)\mu_0 \frac{\tau_{\text{p}} l_{\text{e}}}{g_{\text{e}}} k_{\text{d}} \tag{9.16}$$

Note that this result is the same as equation (3.80) except for the extra factor k_{d}. The function k_{d} is plotted in Figure 9.2 as a function of the per unit pole arc α.

In a like manner, the quadrature axis magnetizing inductance can be found. In this case, it is necessary to align the quadrature axis with the synchronously rotating stator MMF or, alternatively, assume excitation only by q-axis stator MMF. The

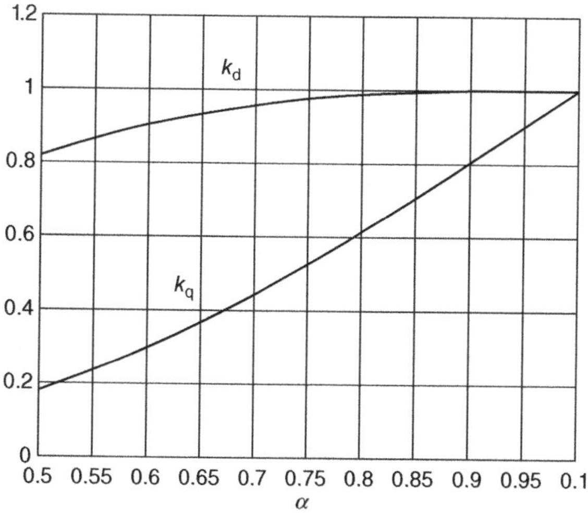

Figure 9.2 Idealized values of k_{d} and k_{q} as a function of per unit pole arc α.

equation which describes the total air gap flux linkages for all P poles in the quadrature axis is

$$\lambda_{\mathrm{mq}} = P\frac{D_{\mathrm{is}}}{2}l_{\mathrm{e}} \int_{-\alpha\pi/P}^{\alpha\pi/P} N_{\mathrm{qs}}(\theta)B_{\mathrm{gq}}(\theta)d\theta \tag{9.17}$$

where the winding function $N_{\mathrm{qs}}(\theta)$ is defined by the fundamental component of equation (2.83). Equation (9.17) becomes

$$\lambda_{\mathrm{mq}} = \left(\frac{2}{\pi}\right) k_1 N_{\mathrm{s}} B_{\mathrm{gq1}} \tau_{\mathrm{p}} l_{\mathrm{e}} k_{\mathrm{q}} \tag{9.18}$$

where again $\tau_{\mathrm{p}} = (\pi D_{\mathrm{is}}/P)$ and

$$k_{\mathrm{q}} = \left[\frac{\alpha\pi - \sin\alpha\pi}{\pi}\right] \tag{9.19}$$

Therefore, using equation (9.4), the quadrature axis magnetizing inductance is

$$L_{\mathrm{mq}} = \left(\frac{12}{\pi}\right)\left(\frac{k_1^2 N_{\mathrm{s}}^2}{P}\right)\frac{B_{\mathrm{gq1}}}{\mathcal{F}_{\mathrm{qs1}}}\tau_{\mathrm{p}} l_{\mathrm{e}}\left[\frac{\alpha\pi - \sin\alpha\pi}{\pi}\right] \tag{9.20}$$

and, by equation (3.78),

$$L_{\mathrm{mq}} = \left(\frac{12}{\pi^2}\right)\left(\frac{k_1^2 N_{\mathrm{s}}^2}{P}\right)\mu_0\frac{\tau_{\mathrm{p}} l_{\mathrm{e}}}{g_{\mathrm{e}}}k_{\mathrm{q}} \tag{9.21}$$

where k_{q} is defined in the ideal case as

$$k_{\mathrm{q}} = \frac{\alpha\pi - \sin\alpha\pi}{\pi} \tag{9.22}$$

The function k_{q} is also plotted in Figure 9.2. The equivalent gap g_{e} is again taken to be the gap g as measured at any point along the surface of the pole modified by saturation and by the Carter factor for the stator slot openings as well as the effect of the pole curvature. Since the value is referred to the value of a smooth round-rotor machine, it is essentially the same as the d-axis equivalent gap. Saturation of the iron paths is still a factor in computing the q-axis inductance since this q-axis flux encounters saturated iron parts caused by d-axis excitation (the so-called cross saturation effect). If the applicable value is clearly not the same as for the direct axis and it would be necessary to use a different subscript to denote this difference.

The functions that have been computed are, of course, only an approximation to the actual case since rather radical assumptions have been made concerning the flux distribution along the pole face. In practice, the problem to be solved is much more complicated since the flux density is not uniform over the surface of the rotor pole. In reality, the pole face is frequently tapered as shown in Figure 9.3 having a minimum gap $g_{\min}$ at the center line of the pole and a maximum gap $g_{\max}$ at the edges of the pole.

Two factors, pole taper and pole fringing combine to make the calculation of flux linkages a much more difficult task.

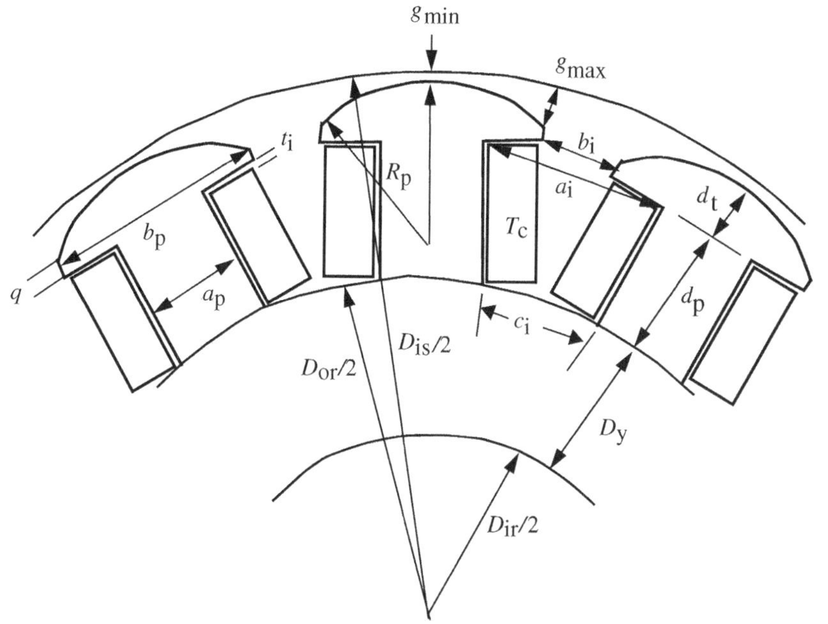

Figure 9.3 Practical geometry for a salient pole structure. Contour of the rotor poles is exaggerated.

Figure 9.4 shows a flux density plot from a classic study by Wieseman [1] who did an exhaustive study of flux density waveforms for over more than 150 pole geometries by using curvilinear square flux plotting techniques. Other detailed analytical solutions are available [2] but without much improvement in accuracy. However, modern finite element algorithms allow for more accurate computation of such solutions.

Following Wieseman, the fundamental component of flux produced by d-axis armature current can be obtained for more complicated topologies than that of Figure 9.1 by treating the pole geometry as a function of two shape factors, pole arc/pole pitch and maximum gap/minimum gap. Figure 9.5 shows two curves equivalent to the work of Wieseman which allows for the direct computation of flux linkage and consequently inductance. While Wieseman used planar geometry (no curvature), this figure includes the effect of curvature resulting from the cylindrical shape of the machine. Specifically, a typical, point of reference four-pole machine was assumed with a diameter $D_{is} = 200$ cm and minimum gap of 2.2 mm. Machines of other diameters and pole numbers were determined to have shape factors which differ only marginally from the reference machine.

In terms of the shape factor k_{dA}, the d-axis air gap flux linkage for a machine with an arbitrary pole arc and pole shape is

$$\lambda_{md} = \frac{2}{\pi} k_1 N_s B_{gd1} l_e (\tau_p k_{dA}) \tag{9.23}$$

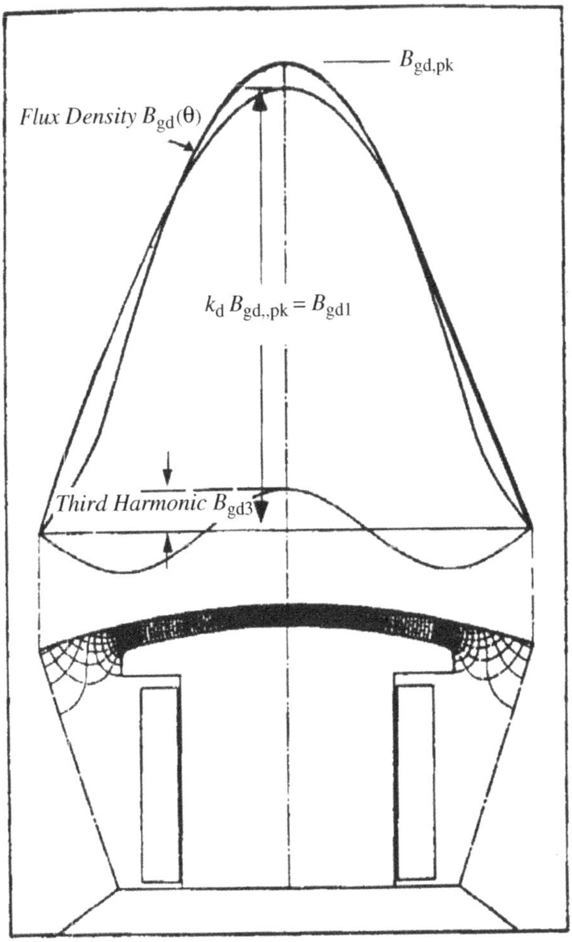

Figure 9.4 Air gap flux density with d-axis armature current and zero field current [1].

Since the factor k_{dB} refers to the minimum gap of the reference machine,

$$L_{md,ref} = \left(\frac{12}{\pi^2}\right)\left(\frac{k_1^2 N_s^2}{P}\right)\mu_0(\tau_p k_{dA}) l_e \frac{k_{dB}}{g_{ref}} \tag{9.24}$$

where g_{ref} is the air gap of the reference machine, namely 2 mm. The inductance of the machine being evaluated can be determined by scaling the gap by

$$L_{md} = \frac{g_{ref}}{g_{min}} L_{md,ref} = \left(\frac{12}{\pi^2}\right)\left(\frac{k_1^2 N_s^2}{P}\right)\mu_0(\tau_p k_{dA}) l_e \frac{k_{dB}}{g_{min}} \tag{9.25}$$

It should be noted that the Carter factor k_c must again be included in the calculation of this value of the gap. Also, when saturation is taken into account, the effective

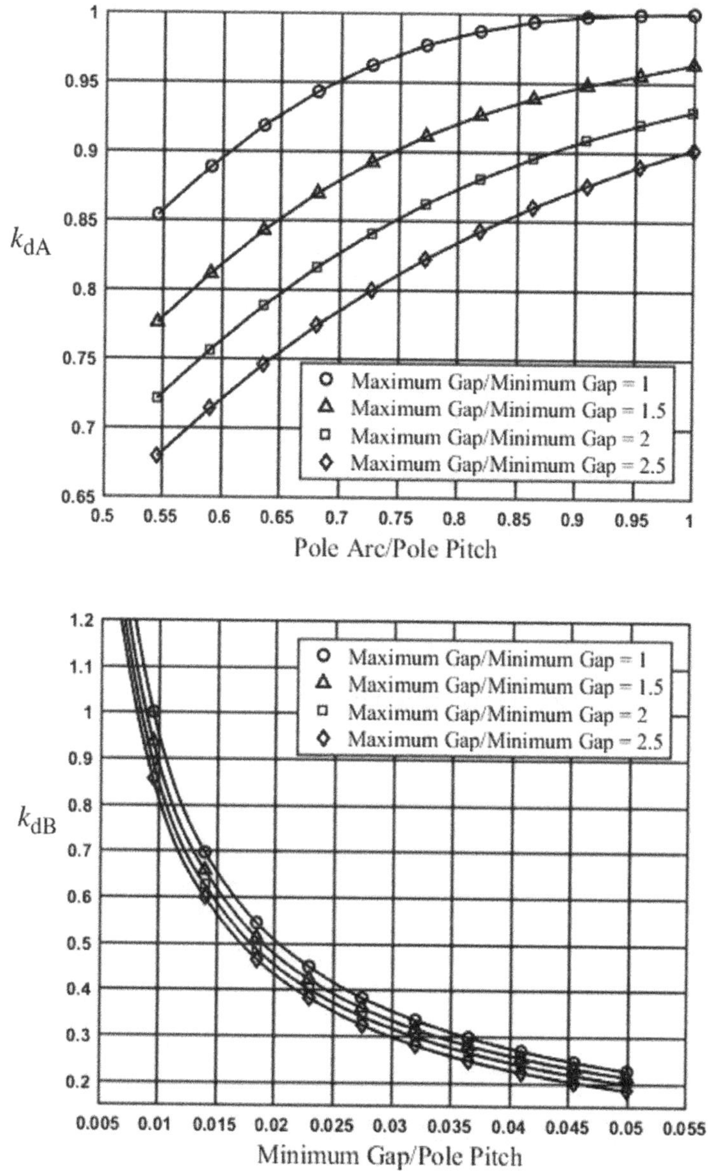

Figure 9.5 Curves for calculating k_d obtained by flux plotting. $k_d = k_{dA}\, k_{dB}$, $B_{gd1} = k_d\, B_{gdm}$ (from Wieseman [1]).

gap must again be modified by a saturation factor k_{sat} to account for the MMF drop in the iron. Thus, finally

$$L_{md} = \left(\frac{12}{\pi^2}\right)\left(\frac{k_1^2 N_s^2}{P}\right)\mu_0\left(\frac{\tau_p l_e}{g_e}\right)k_{dA}k_{dB} \tag{9.26}$$

where

$$g_e = k_c k_{sat} g_{min} \tag{9.27}$$

In a similar manner, the fundamental component of flux density produced by q-axis armature current can be computed. Figure 9.6 shows similar curves for computation of k_q.

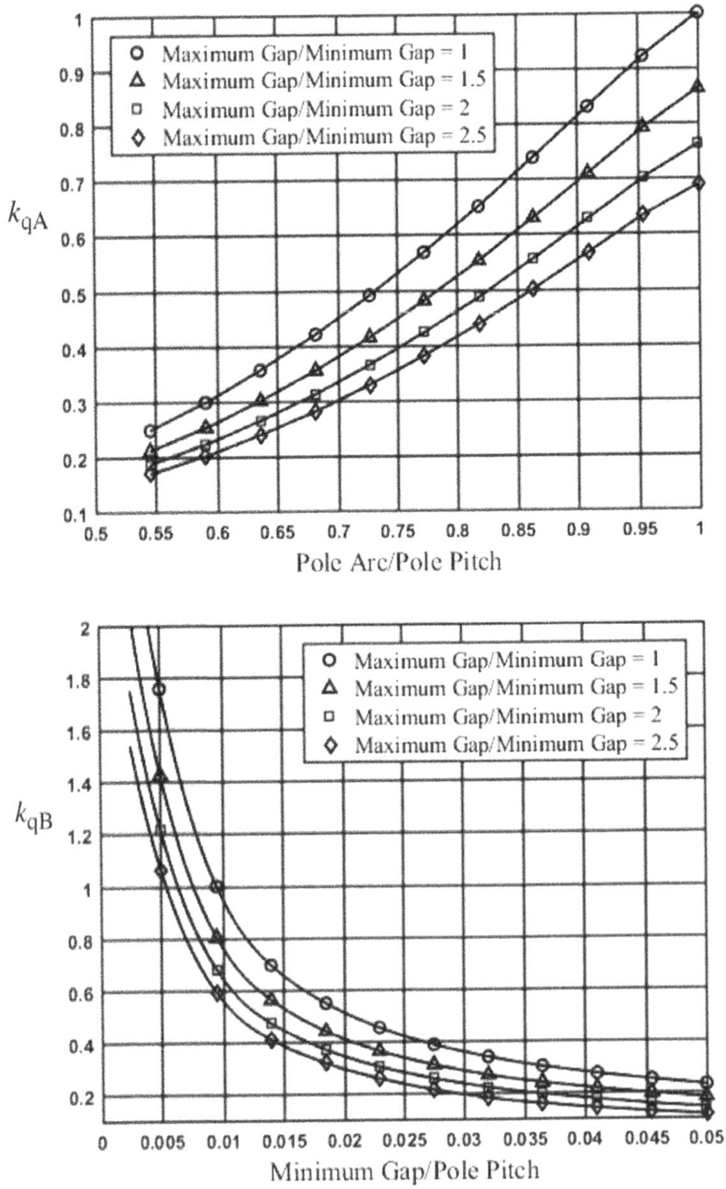

Figure 9.6 Curves for calculating k_q obtained by flux plotting $k_q = k_{qA}\, k_{qB}$, $B_{gq1} = k_q\, B_{gqm}$ (from Wieseman [1]).

In terms of the shape factor k_{qA}, the q-axis air gap flux linkage for a machine with arbitrary pole arc and pole shape is

$$\lambda_{mq} = \frac{2}{\pi} k_1 N_s B_{gq1} l_e (\tau_p k_{qA}) \tag{9.28}$$

Since the factor k_{dB} refers to the minimum gap of the reference machine

$$L_{mq,ref} = \left(\frac{12}{\pi^2}\right)\left(\frac{k_1^2 N_s^2}{P}\right) \mu_0 (\tau_p k_{qA}) l_e \frac{k_{qB}}{g_{ref}} \tag{9.29}$$

The inductance of the machine being evaluated can be determined by scaling the gap by

$$L_{mq} = \frac{g_{ref}}{g_{min}} L_{mq,\,ref} = \left(\frac{12}{\pi^2}\right)\left(\frac{k_1^2 N_s^2}{P}\right) \mu_0 \left(\frac{\tau_p l_e}{g_{min}}\right) k_{qA} k_{qB} \tag{9.30}$$

Finally, with saturation and fringing included,

$$L_{mq} = \left(\frac{12}{\pi^2}\right)\left(\frac{k_1^2 N_s^2}{P}\right) \mu_0 \left(\frac{\tau_p l_e}{g_e}\right) k_{qA} k_{qB} \tag{9.31}$$

where, as previously defined

$$g_e = k_c k_{sat} g_{min} \tag{9.32}$$

9.3 DETERMINATION OF FIELD MAGNETIZING INDUCTANCE

Consider next the determination of the magnetizing portion of the field inductance. A plot of the field pattern for this case is shown in Figure 9.7. If it is first assumed that $B_{gf}(\theta)$ is a constant equal to $B_{gf,pk}$ over the entire pole. When expressed in terms of the fundamental component of square wave, the useful magnetizing (air gap) portion of the field flux linkages per pole is

$$\lambda_{mfp1} = \left(\frac{N_{tf}}{P}\right) l_e \frac{D_{is}}{2} \left(\frac{4}{\pi}\right) \int_{-\alpha\pi/P}^{\alpha\pi/P} B_{gf,pk} \cos\left(\frac{P\theta}{2}\right) d\theta \tag{9.33}$$

where N_{tf} is the total number of field turns. Equation (9.33) reduces to

$$\lambda_{mfp1} = \frac{8}{\pi}\left(\frac{N_{tf}}{P^2}\right) B_{gf,pk} \sin\left(\frac{\alpha\pi}{2}\right) l_e D_{is} \tag{9.34}$$

or

$$\lambda_{mfp1} = \frac{8}{\pi^2}\left(\frac{N_{tf}}{P}\right) B_{gf,pk} k_f l_e \tau_p \tag{9.35}$$

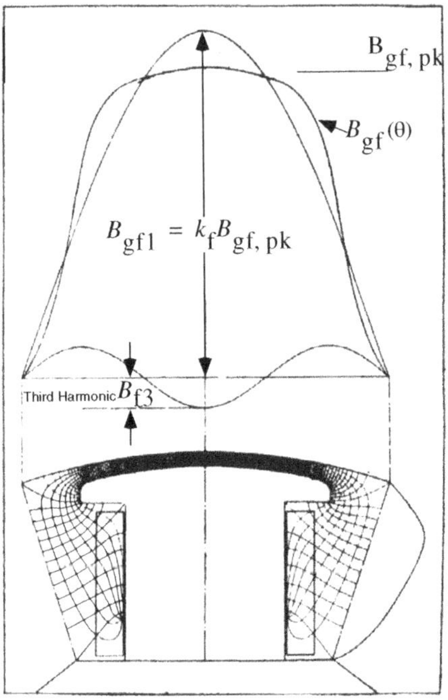

Figure 9.7 Air gap flux density plot with field excited and zero armature current [1].

and, for all poles in series

$$\lambda_{\mathrm{mf1}} = \frac{P}{C_{\mathrm{f}}}\lambda_{\mathrm{mfp1}} = \frac{8}{\pi^2}N_{\mathrm{f}}B_{\mathrm{gf,pk}}k_{\mathrm{f}}l_{\mathrm{e}}\tau_{\mathrm{p}} \tag{9.36}$$

where C_{f} is the number of parallel field circuits, $N_{\mathrm{f}} = N_{\mathrm{tf}}/C_{\mathrm{f}}$ and

$$k_{\mathrm{f}} = \sin\left(\frac{a\pi}{2}\right) \tag{9.37}$$

The coefficient $(8/\pi)k_{\mathrm{f}}B_{\mathrm{gf,pk}}$ can be considered as the term which corresponds to the amplitude of the fundamental component of the flux density waveform or B_{gf1}. Note that a subscript "1" has been specifically added to equations (9.33)–(9.36) to designate the fundamental component of magnetizing field flux linkage.

A more accurate estimate of the field flux can again be obtained by use of finite elements. Extending the work of Wieseman, the fundamental peak air gap field flux linkage on open circuit for a non-uniform air gap under the pole can be more accurately obtained by multiplying the peak flux density by k_{f} where k_{f} is defined as the product of k_{fA} times k_{fB} given in Figure 9.8. From this figure, the field flux linkage of each of the C_{f} circuits is

$$\lambda_{\mathrm{mf1}} = \frac{8}{\pi^2}N_{\mathrm{f}}B_{\mathrm{gf,pk}}l_{\mathrm{e}}\tau_{\mathrm{p}}k_{\mathrm{fA}}k_{\mathrm{fB}} \tag{9.38}$$

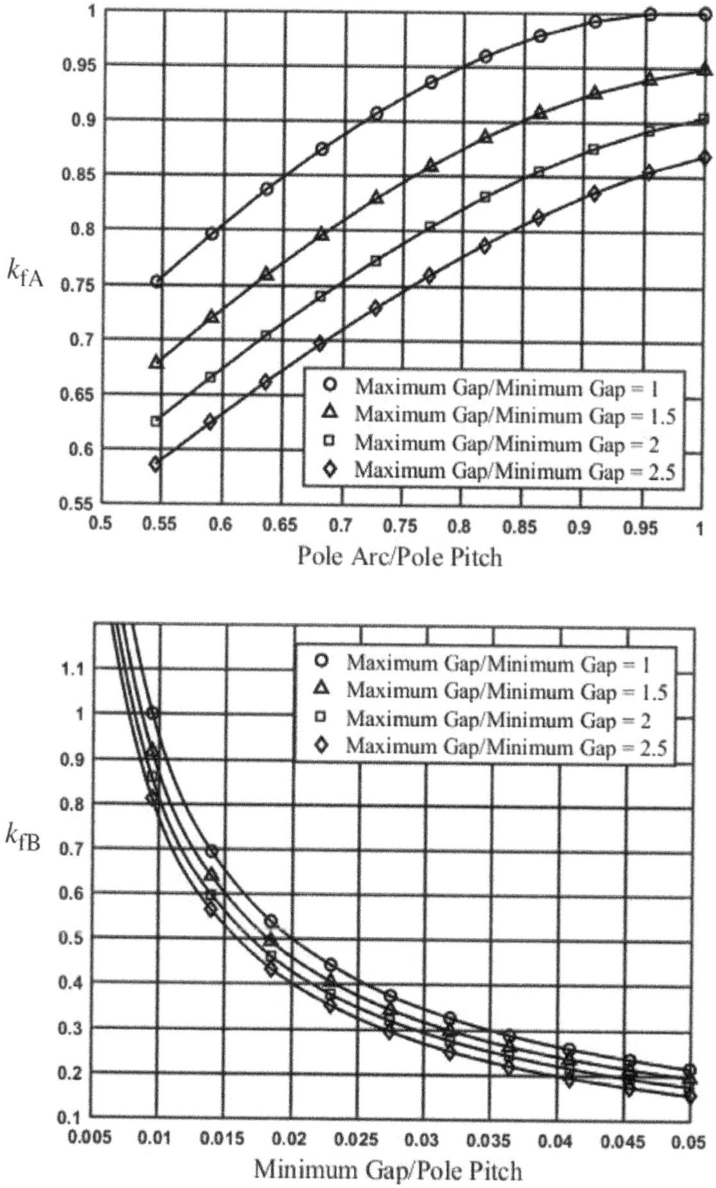

Figure 9.8 Curves for calculating k_f obtained from finite element plots, $k_f = k_{fA}k_{fB}$.

and, since the MMF drop over each of the circuits is

$$\mathcal{F}_{fp} = \frac{N_f(I_f/C_f)}{(P/C_f)} = \frac{N_f I_f}{P} \tag{9.39}$$

the magnetizing portion of the field winding inductance corresponding to C_{f} parallel circuits is

$$L_{\mathrm{mf}} = \frac{8}{\pi^2} N_{\mathrm{f}}^2 \left(\frac{B_{\mathrm{gf,pk}}}{\mathcal{F}_{\mathrm{fp}}} \right) l_{\mathrm{e}} \tau_{\mathrm{p}} k_{\mathrm{fA}} k_{\mathrm{fB}} \tag{9.40}$$

$$= \frac{8}{\pi^2} \left(\frac{N_{\mathrm{f}}^2}{P} \right) \mu_0 \left(\frac{l_{\mathrm{e}} \tau_{\mathrm{p}}}{g_{\mathrm{e}}} \right) k_{\mathrm{fA}} k_{\mathrm{fB}} \tag{9.41}$$

Equation (9.38) represents the useful, fundamental component of flux produced by the field since only this component links the stator. However, one cannot use this expression to solve the magnetic equivalent circuit since a considerable portion of the field leakage flux also crosses the gap. That is, non-fundamental components of field flux density cross the gap and, while they do not link the stator windings (which are assumed to be sinusoidally distributed), they contribute to the saturation of the stator teeth and core, and form a part of the field leakage inductance. Hence, it is necessary to solve instead for the total flux per pole crossing the gap. When the idealized geometry of Figure 9.1 is used it is evident that the total field flux crossing the gap to the stator is simply

$$\Phi_{\mathrm{fp}} = \alpha l_{\mathrm{e}} \tau_{\mathrm{p}} B_{\mathrm{gf,pk}} \quad 0 < \alpha < 1 \tag{9.42}$$

The total air gap field flux linkages per pole based on an ideal uniform gap beneath the pole is

$$\lambda_{\mathrm{mfp}} = \frac{N_{\mathrm{tf}}}{P} (\alpha B_{\mathrm{gf,pk}}) \tau_{\mathrm{p}} l_{\mathrm{e}} \tag{9.43}$$

where N_{tf} is again the total number of field winding turns.

Wieseman has also developed curves for obtaining the total air gap field flux per pole in terms of the fundamental component of field flux per pole as given in Figure 9.9. That is,

$$\Phi_{\mathrm{mfp}} = k_{\phi} \Phi_{\mathrm{mfp1}} \tag{9.44}$$

where Φ_{mfp1} is the fundamental component of flux in the air gap. Since the flux is the spatial integral of the flux density,

$$\Phi_{\mathrm{mfp1}} = \left(\frac{2}{\pi} \right) \tau_{\mathrm{p}} l_{\mathrm{e}} B_{\mathrm{gf1}} \tag{9.45}$$

where B_{f1} is the amplitude of the fundamental component of the field flux density in the air gap.

From the previous discussion concerning k_{f},

$$B_{\mathrm{gf1}} = k_{\mathrm{f}} B_{\mathrm{gf,pk}} \tag{9.46}$$

so that equation (9.44) becomes, finally,

$$\Phi_{\mathrm{mfp}} = k_{\phi} k_{\mathrm{f}} \left(\frac{2}{\pi} \right) \tau_{\mathrm{p}} l_{\mathrm{e}} B_{\mathrm{gf,pk}} \tag{9.47}$$

where $B_{\mathrm{f,pk}}$ is the peak field flux density at the center line of the field pole.

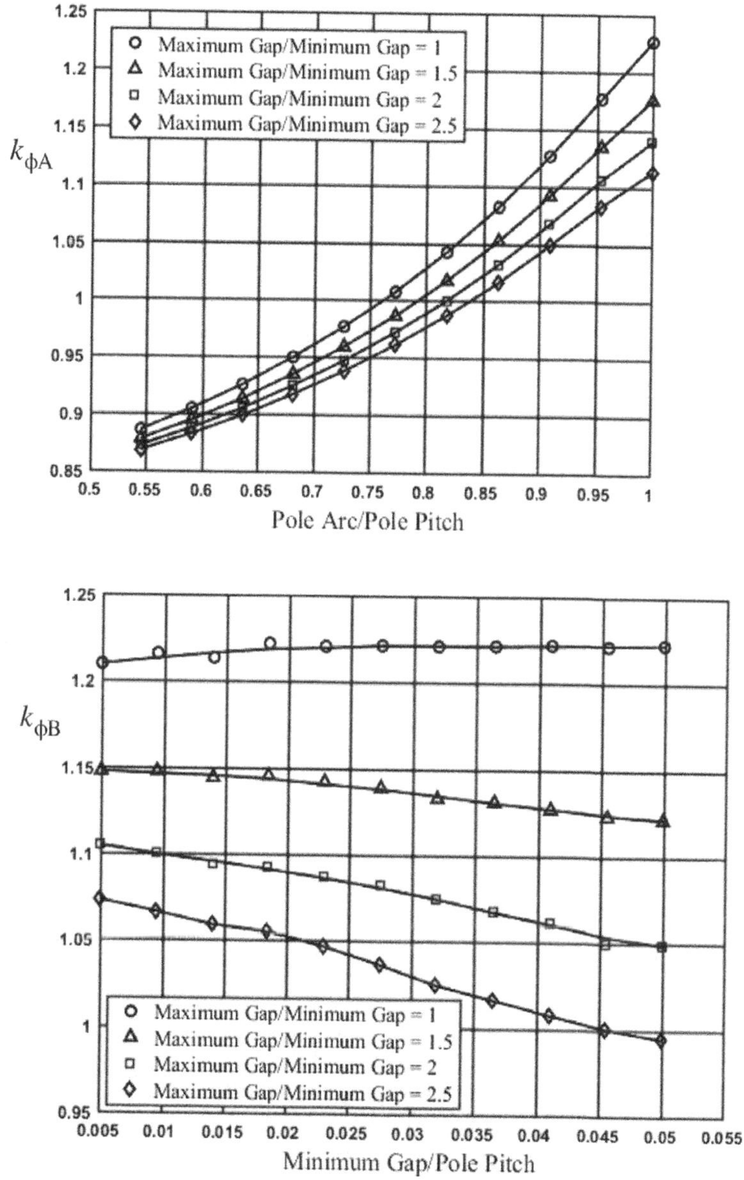

Figure 9.9 Curves for $k_\phi = k_{\phi A} k_{\phi B}$ including the effects of pole taper where $\phi_{fp} = k_\phi \phi_{fp1}$.

The MMF per pole which creates this flux density is defined by the excitation curve, and in general, one can write that

$$\mathcal{F}_{fp} = \frac{N_{tf}}{P} \frac{I_f}{C_f} \tag{9.48}$$

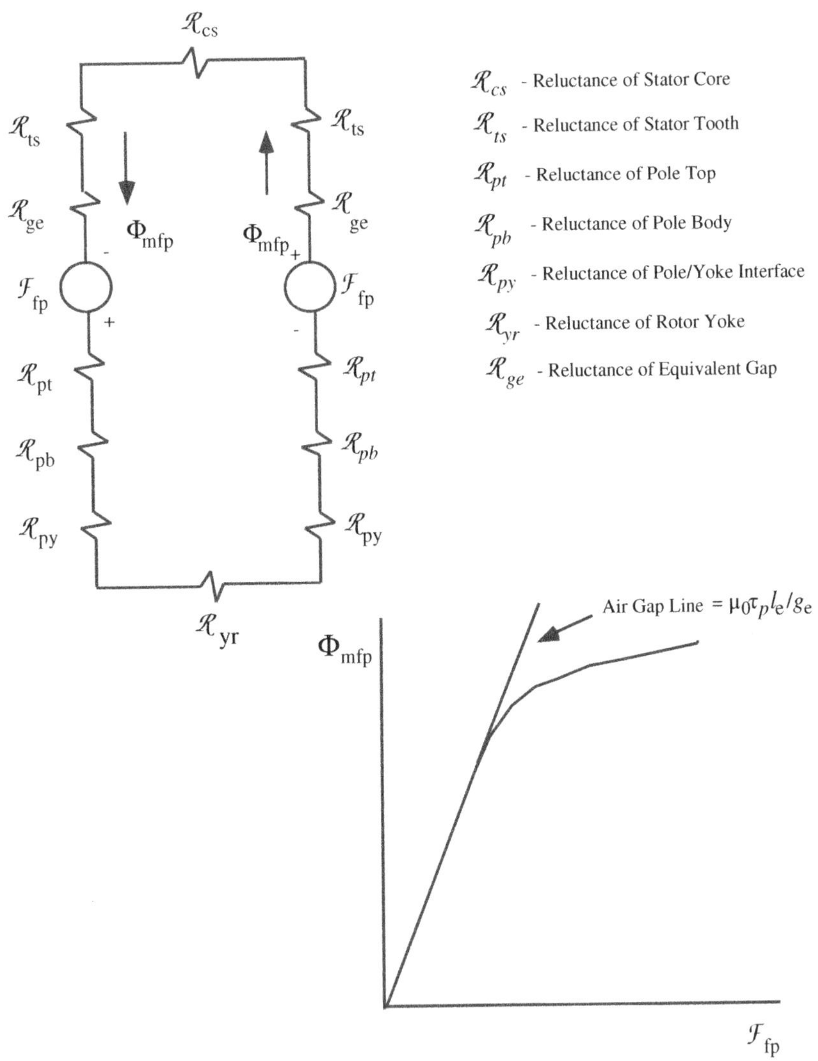

Figure 9.10 Magnetic circuit for two pole pairs assuming field excitation and zero armature current and resulting equivalent magnetizing curve.

where N_{tf} is the total number of field turns for all P poles and C_f is the number of parallel field circuits. Note that the pitch and distribution factor are unity for this case since the field winding is effectively concentrated in one equivalent "slot" per pole.

Equation (9.47) can now be employed to solve the magnetic circuit for two stator poles to obtain the MMF drop around the magnetic circuit for field excitation without load. Figure 9.10 shows a simplified representation of the magnetic circuit to be solved. In this case, it is necessary to choose the stator tooth and slot corresponding to the point of maximum flux density in this circuit instead of a point 30° from the maximum as used for induction machines. This is done because of the uncertain nature of the exact flux density distribution in the gap and the effect saturation has on the flux density distribution. Note that the equivalent circuit contains reluctances for portions of the rotor which will be discussed in a future section.

The resulting flux linkages per pole corresponding to the field flux Φ_{mfp} is

$$\lambda_{\mathrm{mfp}} = \frac{N_{\mathrm{tf}}}{P}\Phi_{\mathrm{mfp}} = k_{\phi}\lambda_{\mathrm{mfp1}} \tag{9.49}$$

from which the total d-axis magnetizing (air gap) field flux linkages are expressed as

$$\lambda_{\mathrm{mf}} = \lambda_{\mathrm{mfp}}\frac{P}{C_{\mathrm{f}}} = \left(\frac{8}{\pi^2}\right)\frac{N_{\mathrm{tf}}}{C_{\mathrm{f}}}k_{\mathrm{f}}k_{\phi}\tau_{\mathrm{p}}l_{\mathrm{e}}B_{\mathrm{gf,pk}} \tag{9.50}$$

in which C_{f} is the total number of parallel field circuits.

Since

$$I_{\mathrm{f}} = \frac{\mathcal{F}_{\mathrm{fp}}PC_{\mathrm{f}}}{N_{\mathrm{tf}}} \tag{9.51}$$

the magnetizing portion of the field inductance is

$$L_{\mathrm{mf}} = \frac{\lambda_{\mathrm{mf}}}{I_{\mathrm{f}}} = \left(\frac{8}{\pi^2}\right)k_{\mathrm{f}}k_{\phi}\tau_{\mathrm{p}}l_{\mathrm{e}}\frac{B_{\mathrm{gf,pk}}}{\mathcal{F}_{\mathrm{fp}}}\frac{N_{\mathrm{tf}}^2}{PC_{\mathrm{f}}^2} \tag{9.52}$$

However, the total field turns N_{tf} are related to the number of series-connected turns N_{f} by $N_{\mathrm{f}} = N_{\mathrm{tf}}/C_{\mathrm{f}}$ so that the field magnetizing inductance is then,

$$L_{\mathrm{mf}} = \left(\frac{8}{\pi^2}\right)k_{\mathrm{f}}k_{\phi}\tau_{\mathrm{p}}l_{\mathrm{e}}\frac{B_{\mathrm{gf,pk}}}{\mathcal{F}_{\mathrm{fp}}}\frac{N_{\mathrm{f}}^2}{P} \tag{9.53}$$

Also, the fundamental component of flux density produced by either the stator or field MMF is the same (Figure 9.10) so that

$$B_{\mathrm{gd1}} = k_{\mathrm{f}}B_{\mathrm{gf,pk}} = k_{\mathrm{d}}B_{\mathrm{gd,pk}} \tag{9.54}$$

The MMF needed to produce B_{gd1} is the same no matter whether it originates from the stator or the field MMF. Therefore,

$$\mathcal{F}_{\mathrm{fp}} = \mathcal{F}_{\mathrm{dp}} \tag{9.55}$$

in which case

$$L_{\mathrm{mf}} = \left(\frac{8}{\pi^2}\right)(\tau_{\mathrm{p}}l_{\mathrm{e}})k_{\phi}k_{\mathrm{d}}\left(\frac{B_{\mathrm{gd,pk}}}{\mathcal{F}_{\mathrm{dp}}}\right)\frac{N_{\mathrm{f}}^2}{P} \tag{9.56}$$

and from equation (9.15) the air gap component of the field inductance can then be written as

$$L_{\mathrm{mf}} = \left(\frac{8}{\pi^2}\right)\left(\mu_0\frac{\tau_{\mathrm{p}}l_{\mathrm{e}}}{g_{\mathrm{e}}}\right)k_{\phi}k_{\mathrm{d}}\frac{N_{\mathrm{f}}^2}{P} \tag{9.57}$$

The process for calculating magnetizing inductance can be repeated for the case where only the fundamental component of field flux, equation (9.45), is considered. In this case, the field magnetizing inductance becomes, for the useful flux which links the stator winding

$$L_{\mathrm{mf1}} = \left(\frac{8}{\pi^2}\right)\left(\mu_0 \frac{\tau_{\mathrm{p}} l_{\mathrm{e}}}{g_{\mathrm{e}}}\right) k_{\mathrm{d}} \frac{N_{\mathrm{f}}^2}{P} \tag{9.58}$$

The fact that the total amount of the flux produced by the field does not link the stator suggests that the difference between equations (9.52) and (9.58) must correspond to a leakage component. One can now formally define the *pole face field leakage inductance* as

$$L_{\mathrm{l,pf}} = L_{\mathrm{mf}} - L_{\mathrm{mf1}} = \left(\frac{8}{\pi^2}\right) \mu_0 \frac{\tau_{\mathrm{p}} l_{\mathrm{e}}}{g_{\mathrm{e}}} k_{\mathrm{d}}(k_{\phi} - 1) \frac{N_{\mathrm{f}}^2}{P} \tag{9.59}$$

It should be mentioned here that a term similar to equation (9.59) clearly exists for the stator gap flux since all of the flux produced by the stator in the air gap does not correspond to a fundamental component. Even with an ideal uniform air gap where the flux density distribution over one pole pitch is a sine wave, zero intervals also exist in the interpolar space between rotor poles resulting in non-fundamental harmonic components. Wieseman did not provide a set of curves to determine this effect. However, these flux components are essentially odd harmonics of the fundamental which are relatively small. The stator pole face leakage inductance will be calculated in a different manner in a later section.

9.4 DETERMINATION OF *d*-AXIS MUTUAL INDUCTANCES

It is now necessary to relate the excitation curve, Figure 9.10, to circuit quantities. That is, the mutual inductance relating the field current excitation to the resulting open circuit stator voltage must be calculated. The useful flux per pole produced by the field winding has already been defined in terms of the peak fundamental gap flux density by equation (9.45).

$$\Phi_{\mathrm{mfp1}} = \left(\frac{2}{\pi}\right) \tau_{\mathrm{p}} l_{\mathrm{e}} B_{\mathrm{f1}} \tag{9.60}$$

or from equation (9.46)

$$\Phi_{\mathrm{mfp1}} = \left(\frac{2}{\pi}\right) \tau_{\mathrm{p}} l_{\mathrm{e}} k_{\mathrm{f}} B_{\mathrm{f,pk}} \tag{9.61}$$

This flux links one pole of the sinusoidally distributed stator winding having $k_1 N_s/P$ effective turns so that, for one pole, the flux from the field linking a stator winding is

$$\lambda_{\mathrm{dfp}} = \int_{\pi/P}^{\pi/P} N_{\mathrm{ds}}(\theta) B_{\mathrm{f}}(\theta) d\theta \tag{9.62}$$

The fundamental component of the stator winding distribution $N_{\mathrm{ds}}(\theta)$ is expressed in terms of equivalent two-phase components as discussed in Section 2.10. Thus, because of the scaling convention for transforming *a,b,c* components to *d,q* components a 3/2 term again appears so that

$$N_{\rm ds}(\theta) = \frac{3}{2}\left(\frac{4}{\pi}\right)\left(\frac{k_1 N_{\rm t}}{P}\right)\cos\left(\frac{P\theta}{2}\right) \tag{9.63}$$

The field gap flux density is a constant over the pole arc $(\alpha\pi)/P$ so that, per pole,

$$\lambda_{\rm dfp} = \left(\frac{6}{\pi}\right)\left(\frac{k_1 N_{\rm t}}{P}\right) B_{\rm f,pk} \int_{-\alpha\pi/P}^{\alpha\pi/P} \cos\left(\frac{P\theta}{2}\right) l_{\rm e} \frac{D_{\rm is}}{2} d\theta \tag{9.64}$$

$$= \left(\frac{12}{\pi^2}\right)\left(\frac{k_1 N_{\rm t}}{P}\right)\sin\left(\frac{\alpha\pi}{2}\right) B_{\rm f,pk} l_{\rm e} \tau_{\rm p} \tag{9.65}$$

The corresponding flux linkages for all $P/C_{\rm s}$ series-connected stator poles is

$$\lambda_{\rm df} = \left(\frac{12}{\pi^2}\right)\left(\frac{k_1 N_{\rm t}}{C_{\rm s}}\right) \tau_{\rm p} l_{\rm e} k_{\rm f} B_{\rm f,pk} \tag{9.66}$$

$$= \left(\frac{12}{\pi^2}\right)(k_1 N_{\rm s})\tau_{\rm p} l_{\rm e} k_{\rm f} B_{\rm f,pk} \tag{9.67}$$

where

$$k_{\rm f} = \sin\left(\frac{\alpha\pi}{2}\right) \tag{9.68}$$

The mutual inductance between the field and the *d*-axis of the armature, $L_{\rm df}$ is defined as

$$L_{\rm df} = \frac{\lambda_{\rm df}}{I_{\rm f}} \tag{9.69}$$

and from equations (9.67) and (9.48),

$$L_{\rm df} = \left(\frac{12}{\pi^2}\right)\left(\frac{k_1 N_{\rm s}}{P}\right)\left(\frac{N_{\rm tf}}{C_{\rm f}}\right) k_f \left(\frac{B_{\rm f,pk}}{\mathcal{F}_{\rm fp}}\right) \tau_{\rm p} l_{\rm e} \tag{9.70}$$

The ratio $N_{\rm tf}/C_{\rm f}$ is the total number of series-connected field turns defined as $N_{\rm f}$ whereupon one can write,

$$L_{\rm df} = \left(\frac{12}{\pi^2}\right)\left(\frac{k_1 N_{\rm s} N_{\rm f}}{P}\right) k_{\rm f} \left(\frac{B_{\rm f,pk}}{\mathcal{F}_{\rm fp}}\right) \tau_{\rm p} l_{\rm e} \tag{9.71}$$

In Section 8.3, it was shown that since the fundamental flux density produced by either the stator MMF or field MMF remains unchanged, or

$$B_{\rm gd1} = k_{\rm f} B_{\rm f,pk} = k_{\rm d} B_{\rm gd,pk} \tag{9.72}$$

and that

$$\mathcal{F}_{\rm dp} = \mathcal{F}_{\rm fp}$$

in which case equation (9.71) can be expressed as

$$L_{\rm df} = \left(\frac{12}{\pi^2}\right)\left(\frac{k_1 N_{\rm s} N_{\rm f}}{P}\right)\left(\frac{\mu_0 \tau_{\rm p} l_{\rm e}}{g_{\rm e}}\right) \tag{9.73}$$

Again note the similarity of this equation with equation (3.80).

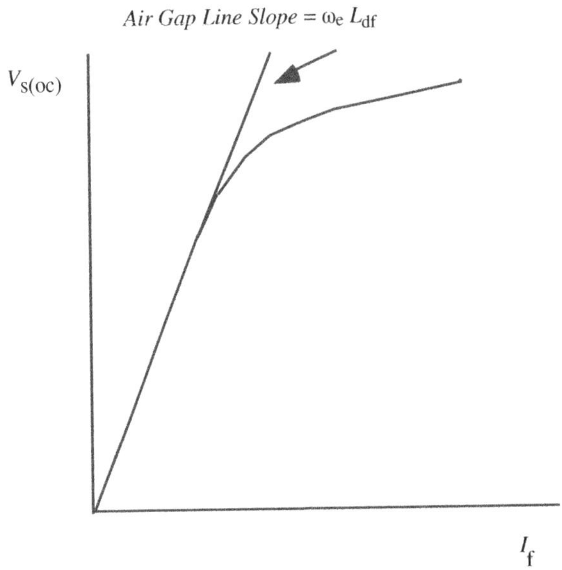

Figure 9.11 Typical open circuit saturation curve.

It is apparent that the ratio of gap flux density to the MMF per pole is once again a key parameter in the determination of synchronous machine inductances. In general, for more accuracy, this ratio must again be computed by solving a magnetic circuit as was done in Section 9.2. If one neglects saturation, it is possible to obtain an expression for mutual inductance which is independent of the characteristics of the magnetic material. The inductance so obtained defines the inductance along the *air gap line*.

Having obtained the unsaturated mutual inductance between the field and the stator armature winding, it is now possible to convert Figure 9.10 to a characteristic involving terminal quantities. If the machine is rotating at a synchronous speed ω_e then the open circuit voltage is

$$V_{s(oc)} = \omega_e L_{df} I_f \tag{9.74}$$

The actual saturated value of L_{df} is obtained from equation (9.70). An idealized unsaturated value of L_{df} along the air gap line which neglects MMF drop in the iron is derived from equation (8.58) with g_e replaced by an unsaturated value of g_e, namely $g_e = k_c g_{min}$ where k_c is the Carter coefficient calculated for the stator slot located over the center of the field pole. The open circuit saturation curve of terminal voltage versus field current for the example of Section 9.3 is shown in Figure 9.11. Note that it is essentially proportional to the magnetization curve of Figure 9.10.

9.5 CALCULATION OF ROTOR POLE LEAKAGE PERMEANCES

Although field pole leakage components that cross the gap have already been included, other components of leakage also exist which have their paths entirely on

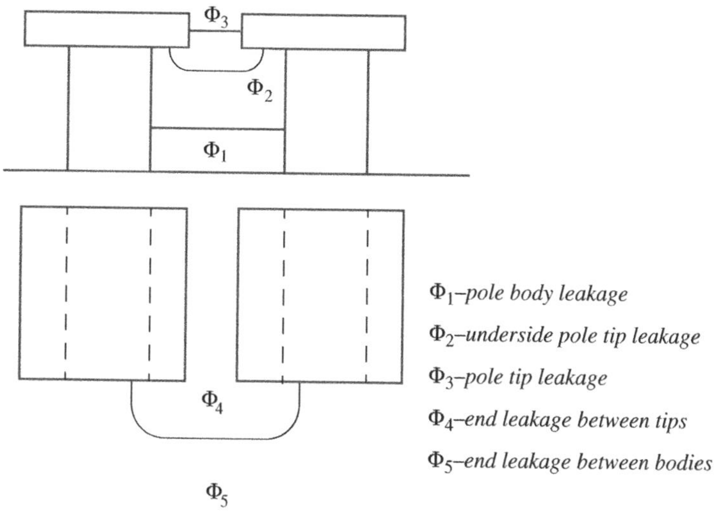

Figure 9.12 Illustrating pole body leakage.

the rotor side of the air gap. While leakage fluxes are usually sufficiently small that their effect on iron saturation can be neglected, such is not possible for the field poles since the leakage fluxes are appreciable in this case. Figure 9.12 shows a typical field pole for the purpose of calculating leakage flux. Three major components of leakage can be identified corresponding to pole body leakage, pole tip leakage, and pole end leakage. The fourth major component, pole top leakage has already been incorporated by the use of the flux coefficient k_ϕ.

Consider first the pole body leakage permeance. This component will be evaluated by simply assuming that the interpolar space is a large rotor slot containing $2N_{\text{tf}}/P$ turns. Here, cylindrical geometry has been approximated in rectangular coordinates by slanting the rotor poles away from each other. Figure 9.13 illustrates the problem to be solved. Note that since the field winding is wrapped around the field poles the interpolar space is not filled but the distribution of turns in the "slot" can be considered as uniform from top to bottom. The differential specific permeance of a strip dx at a height x is

$$dp_{\text{pb}} = \frac{\mu_0 dx}{a_i + \dfrac{c_i - a_i}{d_{\text{p}}}x} \tag{9.75}$$

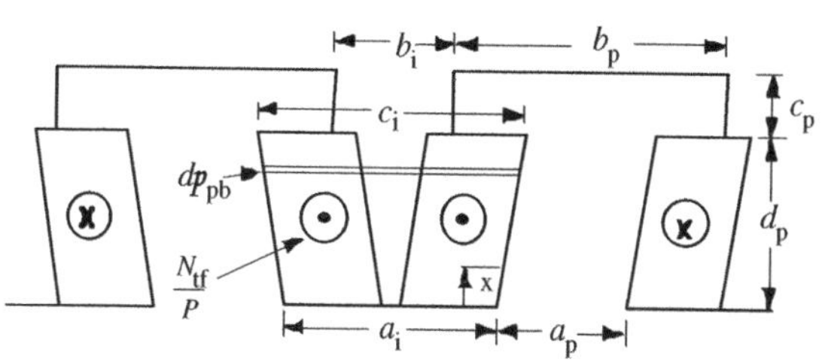

Figure 9.13 Magnetic circuit to be solved.

The differential flux through this strip is

$$d\Phi_{\mathrm{pb}} = \mathcal{F}_{\mathrm{f}} l_{\mathrm{e}} dp_{\mathrm{pb}}$$

$$= 2\frac{N_{\mathrm{tf}}}{P}\left(\frac{I_{\mathrm{f}}}{C_{\mathrm{f}}}\right)\left(\frac{x}{d_{\mathrm{p}}}\right) l_{\mathrm{e}} dp_{\mathrm{pb}} \tag{9.76}$$

$$= \left(\frac{2N_{\mathrm{tf}}I_{\mathrm{f}}}{C_{\mathrm{f}}P}\right)\left(\frac{x}{d_{\mathrm{p}}}\right) l_{\mathrm{e}} dp_{\mathrm{pb}} \tag{9.77}$$

The flux linkages produced by $d\Phi_{\mathrm{pb}}$ are

$$d\lambda_{\mathrm{pb}} = \frac{\mu_0\left(\frac{N_{\mathrm{tf}}}{P}\right)^2 2\left(\frac{I_{\mathrm{f}}}{C_{\mathrm{f}}}\right)\left(\frac{x}{d_{\mathrm{p}}}\right)^2}{a_i + \frac{c_i - a_i}{d_{\mathrm{p}}}x} l_{\mathrm{e}} dx \tag{9.78}$$

Upon integrating from 0 to d_{p} one obtains,

$$\lambda_{\mathrm{pb}} = 2\mu_0\left(\frac{I_{\mathrm{f}}}{C_{\mathrm{f}}}\right)\left(\frac{N_{\mathrm{tf}}}{P}\right)^2 l_{\mathrm{e}}\left(\frac{d_{\mathrm{p}}}{c_i - a_i}\right)\left[\frac{1}{2} - \frac{a_i}{(c_i - a_i)} + \frac{a_i^2}{(c_i - a_i)^2}\ln\left(\frac{c_i}{a_i}\right)\right] \tag{9.79}$$

The permeance per unit axial length associated with pole body leakage is therefore

$$p_{\mathrm{pb}} = 2\mu_0\left(\frac{d_{\mathrm{p}}}{c_i - a_i}\right)\left[\frac{1}{2} - \frac{a_i}{(c_i - a_i)} + \frac{a_i^2}{(c_i - a_i)^2}\ln\left(\frac{c_i}{a_i}\right)\right] \tag{9.80}$$

The second component of leakage involves the flux which goes across the interpolar space from pole tip to pole tip. If it is again assumed that the flux goes straight across the "slot" then the pole tip leakage is easily found to be

$$p_{\mathrm{pt}} = 2\mu_0\frac{c_{\mathrm{p}}}{b_i} \tag{9.81}$$

Because of the "bowing out" of the flux lines from the pole tips, the permeance of equation (9.81) is too small. As a means of correcting for this error, it is possible to add an additional term corresponding to flux which emanates from the bottom of the pole overhang as shown in Figure 9.12. The corrected permeances per unit length can be written as

$$p_{\mathrm{pt}} = 2\mu_0\frac{c_{\mathrm{p}}}{b_i} + 2\int_0^{c_i/2}\frac{\mu_0 dr}{\pi r + b_i} \tag{9.82}$$

$$p_{\mathrm{pt}} = 2\mu_0\left\{\frac{c_{\mathrm{p}}}{b_i} + \frac{1}{\pi}\ln\left[1 + \frac{\pi}{2}\frac{(c_i - b_i)}{b_i}\right]\right\} \tag{9.83}$$

It is assumed that all of the flux on the top surface of the pole reaches the stator surface and therefore has already been accounted for by the flux coefficient k_ϕ.

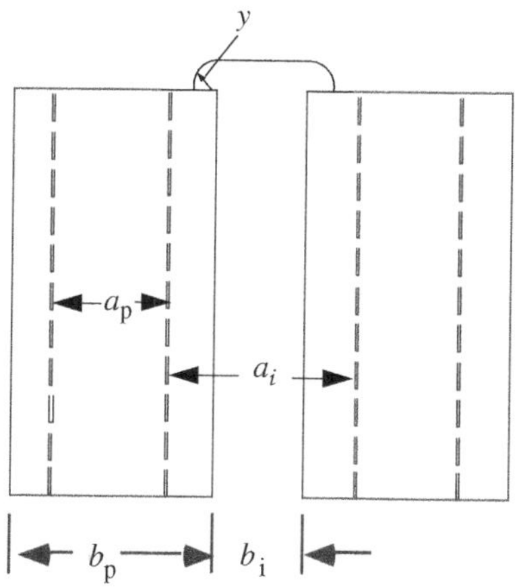

Figure 9.14 Pole dimensions for calculation of pole end leakage.

The final leakage component to be considered is the pole end leakage in which the flux bows out of the machine proper in the axial direction. A plot of the pertinent geometric dimensions is given Figure 9.14.

For the pole end leakage from the region of the pole tips considering both sides of the pole, one can write,

$$\mathcal{P}_{\mathrm{pet}} = 2\mu_0 \int_0^{b_p/2} \frac{2c_{\mathrm{p}}dy}{b_i + \pi y} \tag{9.84}$$

So that, after some simplification

$$\mathcal{P}_{\mathrm{pet}} = \mu_0 \frac{4c_{\mathrm{p}}}{\pi} \ln\left(1 + \frac{\pi}{2}\frac{b_p}{b_i}\right) \tag{9.85}$$

In a similar manner, for the pole end leakage from the region of the pole body,

$$\mathcal{P}_{\mathrm{peb}} = \mu_0 \int_0^{(a_{\mathrm{p}}+b_{\mathrm{p}})/2} \frac{2d_{\mathrm{p}}}{\left[\frac{(a_i + c_i)}{2} + \pi y\right]} dy \tag{9.86}$$

upon which

$$\mathcal{P}_{\mathrm{peb}} = \mu_0 \frac{2d_{\mathrm{p}}}{\pi} \ln\left[1 + \pi \frac{a_{\mathrm{p}} + b_{\mathrm{p}}}{a_i + c_i}\right] \tag{9.87}$$

The total pole end leakage permeance per pole is

$$\mathcal{P}_{\mathrm{pel}} = \mathcal{P}_{\mathrm{pet}} + \mathcal{P}_{\mathrm{peb}} \tag{9.88}$$

The total leakage permeance per pole is

$$\mathcal{P}_{\mathrm{pl}} = \mathcal{P}_{\mathrm{pel}} + (\mathcal{P}_{\mathrm{bp}} + \mathcal{P}_{\mathrm{pt}}) l_{\mathrm{e}} \tag{9.89}$$

The overall field leakage inductance per pole can now be written as

$$\lambda_{\text{plf}} = \frac{I_{\text{f}}}{C_{\text{f}}}\left(\frac{N_{\text{tf}}}{P}\right)^2 \mathcal{P}_{\text{pl}} \tag{9.90}$$

The total flux linkages for all series-connected poles is then

$$\frac{P\lambda_{\text{plf}}}{C_{\text{f}}} = \frac{I_{\text{f}}}{P}\left(\frac{N_{\text{tf}}}{C_{\text{f}}}\right)^2 \mathcal{P}_{\text{pl}} = \frac{I_{\text{f}}}{P} N_{\text{f}}^2 \mathcal{P}_{\text{pl}} \tag{9.91}$$

The leakage inductance of the field circuit pole body is, finally,

$$L_{\text{lfpb}} = \frac{P\lambda_{\text{plf}}}{I_{\text{f}}} = \frac{N_{\text{f}}^2}{P} \mathcal{P}_{\text{pl}} \tag{9.92}$$

The total field leakage inductance is obtained by adding the inductance obtained for the pole face, equation (9.59), to the inductance for the pole body, equation (9.92). Formally,

$$L_{\text{lf}} = L_{\text{lfpb}} + L_{\text{lfpf}} \tag{9.93}$$

9.6 STATOR LEAKAGE INDUCTANCES OF A SALIENT POLE SYNCHRONOUS MACHINE

Since the armature of a synchronous machine is wound in an identical fashion to the armature of an induction machine, the stator leakage inductances of both machines are very similar. In particular, the inductances of a synchronous machine can be categorized into slot, end-winding, belt, and zigzag components in much the same manner as an induction machine. Sections 4.4–4.8 relating to slot and end-winding leakage apply equally for synchronous machine armatures. However, modifications are required for the zigzag and belt leakage components due to the effects of saliency.

9.6.1 Zigzag or Tooth-Top Leakage Inductance of Salient Pole Machines

One key difference between the synchronous machine and the induction machine is the fact that the synchronous machine carries its own excitation winding and the need for small air gaps to minimize the magnetizing current is not required. Hence, the synchronous machine air gap is relatively large compared to the induction machine. The increased air gap causes a variety of changes, some good, and some undesirable. Clearly, the excitation losses in the field increase since the same flux density in the stator iron must be maintained for good utilization of the stator iron and copper. The large field current also causes the rotor poles to become more saturated. However, an increased air gap also tends to reduce the stray losses so that some of the increased excitation losses are overcome. Also, the increased air gap limits the steady-state short-circuit current to a value not much different than its rated value.

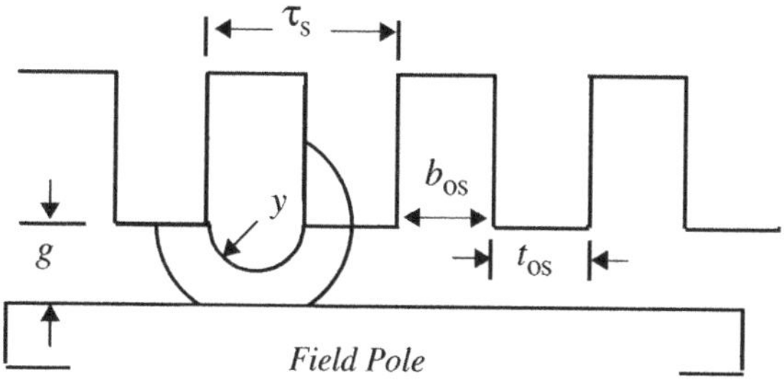

Figure 9.15 Configuration for calculation of zigzag leakage of salient pole machine.

The large air gap also causes some of the flux lines that previously crossed the gap to take paths which bypass the rotor surface. Since the rotor cage of a salient pole machine (amortisseur winding) is typically buried below the rotor surface, the effects of rotor slots are not important and the picture for computation of zigzag leakage changes to that of Figure 9.15. Note that the flux components can be segregated into two components, an approximate semicircular component which is completely in the gap and another component which crosses the gap twice before enclosing the slot of interest. The flux component which does not reach the rotor surface is frequently called the *tooth top leakage*. However, both of these components will be assigned to zigzag leakage inductance. It should also be noted that this computation accounts for the stator-side equivalent of the pole face leakage flux of Section 9.3 accounted for by the factor k_ϕ.

The semicircular component can be found by taking an elementary flux tube and then integrating over all possible differential flux tubes. An expression for the tooth top portion of zigzag leakage permeance can be expressed

$$\mathcal{P}_1 = \int_{b_{os}/2}^{g} \frac{\mu_0 l_e dr}{\pi r} \tag{9.94}$$

which becomes, after integration

$$\mathcal{P}_1 = \frac{\mu_0}{\pi} l_e \ln\left(\frac{2g}{\tau_s - t_{os}}\right) \tag{9.95}$$

If it is assumed that the remaining flux lines continue to go straight across the gap, one has, for the reluctance to flux going from stator to rotor

$$\mathcal{R}_2 = \frac{g}{\mu_0 \left(\frac{\tau_s}{2} - g\right) l_e} \tag{9.96}$$

The reluctance in going from stator to rotor and back again is simply twice $\mathcal{R}_2$. The corresponding permeance is therefore the reciprocal of twice $\mathcal{R}_2$ or

$$\mathcal{P}_2 = \mu_0 \frac{l_e(\tau_s - 2g)}{4g} \tag{9.97}$$

The total zigzag permeance per unit length is, therefore

$$p_{zzq} = \mu_0 \left[\frac{\tau_s - 2g}{4g} + \frac{1}{\pi} \left(\ln \frac{2g}{\tau_s - t_{os}} \right) \right] \tag{9.98}$$

Equation (9.98) can be used to find the inductance for any particular slot by inserting the gap distance g corresponding to the position of the slot in question along the pole face. However, the pole face of most salient pole machines is tapered so that it is difficult to use this equation directly to form the inductance per phase.

In general, two limiting alignments of the rotor are possible with respect to a given phase. When the rotor pole is aligned with the magnetic axis of the stator phase, the parameters of the phase corresponds to the parameters of the d-axis equivalent circuit. When the rotor pole is in quadrature with the magnetic axis of a particular phase the parameters of the q-axis equivalent circuit are obtained. Figure 9.16 shows the position of the rotor when lined up and in quadrature with the magnetic axis of one of the phases. Sixty degree phase belts are assumed. Since the magnetizing inductance associated with this phase corresponds to the direct and quadrature axes magnetizing inductances, the associated leakage of this phase can also be considered as forming the direct or quadrature axis leakage inductance depending upon the alignment of the rotor.

Note from Figure 9.16 that the conductors of the phase are centered over the rotor pole when the q-axis circuit is energized. This case will be considered first and the case where the center line of the conductors is located on d-axis case second. Although many variations are possible, it will be assumed that the pole is constructed so that the gap varies linearly from a maximum value to a minimum value. In practice, the pole is usually constructed so that the pole face forms an arc having a radius slightly smaller than the maximum radius of the rotor measured from the center line of the machine. The difference between two offset circles (stator inner surface and

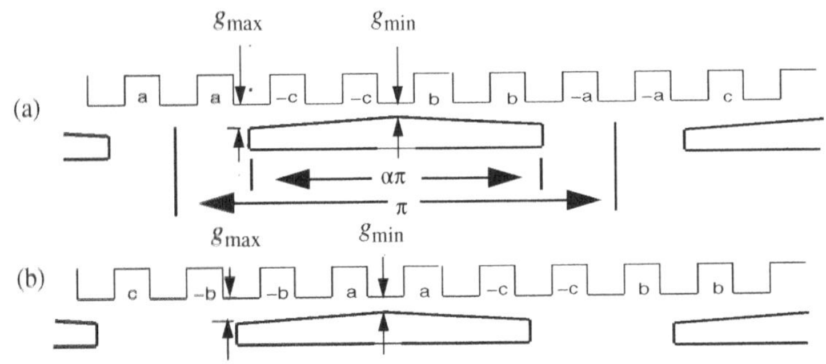

Figure 9.16 Location of the conductors of phase a when the rotor poles are aligned with the magnetic axis of the phase (d–axis, upper sketch) and 90 electrical degrees from alignment (q–axis, lower sketch), 60° phase belts are assumed.

rotor pole face) of different radii is very nearly linear. The gap can, therefore, be expressed in terms of the peripheral angle θ as

$$g = g_{\text{mim}} + (g_{\max} - g_{\min})\frac{2\theta}{\alpha\pi} \tag{9.99}$$

Since it is assumed that the winding has 60° phase belts, the average gap over the 60° intervals formed by the phase belt is

$$g_{\text{e(ave)}} = \frac{6}{\pi}\int_0^{\pi/6}\left[g_{\min} + (g_{\text{e(max)}} - g_{\text{e(min)}})\frac{2\theta}{\alpha\pi}\right]d\theta \tag{9.100}$$

where $g_{\text{e(max)}}$ and $g_{\text{e(min)}}$include the effects of slotting on the maximum and minimum physical gap. Equation (9.100) ultimately becomes

$$g_{\text{e(ave)}} = g_{\text{e(min)}} + \frac{1}{6\alpha}[g_{\text{e(max)}} - g_{\text{e(min)}}] \tag{9.101}$$

Hence, from Section 4.10, the q-axis zigzag leakage inductance for windings with 60° phase belts and full pitch is

$$L_{\text{zzq}} = 12N_{\text{s}}^2 l_{\text{e}}\frac{p_{\text{zzq}}}{S_1} \tag{9.102}$$

Where p_{zzq} is given by equation (9.98) and $g_{\text{e(ave)}}$ is used for the effective gap. If necessary, the zigzag leakage inductance must be reduced to account for the effects of the pitch. The discussion in Section 4.10 again applies. The result becomes

$$L_{\text{zzq}} = 12\frac{N_{\text{s}}^2}{S_1}l_{\text{e}}\frac{(21p-5)}{16}p_{\text{zzq}} \quad 2/3 \le p \le 1 \tag{9.103}$$

A magnetic alignment of the rotor pole with the axis of the phase as in Figure 9.16a results in the d-axis equivalent circuit. In this case, the stator conductors are in the interpolar space and all of the zigzag leakage flux becomes tooth top leakage. Examination of Figure 9.15a indicates that it is possible to quickly obtain the permeance for this case if one simply sets $2g$ equal to τ_{s} in equation (9.98). The result is

$$p_{\text{zzd}} = \frac{\mu_0}{\pi}\ln\left(\frac{\tau_{\text{s}}}{\tau_{\text{s}} - t_{\text{os}}}\right) \tag{9.104}$$

With pitch, the corresponding d-axis zigzag leakage inductance is

$$L_{\text{zzq}} = 12\frac{N_{\text{s}}^2}{S_1}l_{\text{e}}\frac{(21p-5)}{16}p_{\text{zzq}} \quad 2/3 \le p \le 1 \tag{9.105}$$

Note that the result is independent of air gap length. When the pitch is involved, it is apparent that a portion of the conductors of the phase now begins to encroach upon the edges of the rotor poles. However, the gap of the pole is a maximum at this point and may be assumed that the leakage flux lines continue to use the air path rather than leap to the rotor surface. In any case, the effect is probably small. Since the rotor slots are assumed to be imbedded under the surface of the rotor pole, the corresponding zigzag leakage for the rotor cage is zero. If a machine with rotor slotting is used the zigzag leakage for the rotor bars is easily obtained in an analogous fashion.

It is interesting to observe that due to zigzag leakage effects, the leakage inductance for the d– and the q–axes are different. This causes little difficulty insofar as analysis is concerned. However, the large air gap of a synchronous machine results in a zigzag leakage which is only on the order of 20% of the total stator leakage. Therefore, the effect is small for most machines and an average value is typically used to calculate both the d– and the q–axes leakage inductances. That is,

$$L_{\text{zzd}} = L_{\text{zzq}} = 12\frac{N_{\text{s}}^2}{S_1} l_{\text{e}} \frac{(21p-5)}{16} p_{\text{zzq}} \quad 2/3 \le p \le 1 \tag{9.106}$$

where

$$p_{\text{zz}} = \mu_0 \left\{ \frac{\tau_{\text{s}} - 2g_{\text{e(ave)}}}{8g_{\text{e(ave)}}} + \left(\frac{1}{2\pi}\right) \ln\left(\frac{2g_{\text{e(ave)}}\tau_{\text{s}}}{\left(\tau_{\text{s}} - t_{\text{os}}\right)^2} \right) \right\} \tag{9.107}$$

The remaining leakage inductances of the stator are essentially the same as for the induction motor. Again, the stator resistance can again be calculated by equation (5.42). The total stator leakage inductance for the synchronous machine is consequently,

$$L_{\text{ls}} = L_{\text{sl}} + L_{\text{ew}} + L_{\text{zz}} + L_{\text{lk}} \tag{9.108}$$

where L_{sl}, L_{ew}, L_{zz}, and L_{lk} are the slot, end-winding, zigzag, and belt leakage respectively. As with the induction motor, the belt leakage term is negligibly small if the synchronous machine is equipped with an amortisseur winding (the equivalent of a squirrel-cage winding in an induction motor). The complications imposed by the presence of an amortisseur winding is the next subject to be considered.

9.7 THE AMORTISSEUR WINDING PARAMETERS

The *amortisseur winding* is sometimes called the "killer" winding since the word is derived from the French "mort" which means "death". This term is appropriate since this winding was (and is) typically used to "deaden" oscillations in rotor speed when the load of the machine is suddenly changed. The equivalent expression often used is *damper winding*, which has the same meaning regarding suppressing oscillations. The amortisseur winding has essentially the same features as the squirrel-cage induction machine except for the fact that it is embedded only on the salient poles. Figure 9.17 shows a sketch of the two major types of damper winding construction: *continuous* and *interrupted*. In this book, the interrupted type (which is the more analytically challenging of the two) will be examined in detail and it will be left to the reader to develop the corresponding results for the continuous type.

A simplified representation of the placement of the bars along the pole face is shown in Figure 9.18. Note that all bars are generally equidistantly placed on the rotor. The small slot openings above the bars are not shown on this figure. Since the spacing between bars is measured in electrical degrees while the spatial angle θ is

Figure 9.17 (a) Continuous and (b) interrupted amortisseur windings.

measured in mechanical degrees, the flux linking the 1–1′ pair of bars is

$$\Phi_{1-1'} = B_{gd1} \int_{-\gamma/p}^{\gamma/p} \cos\left(\frac{P\theta}{2}\right) (\sin \omega_e t) l_e \left(\frac{D_{or}}{2}\right) d\theta \tag{9.109}$$

$$= \frac{2B_{gd1} l_e D_{or}}{P} \sin\left(\frac{\gamma}{2}\right) \sin \omega_e t \tag{9.110}$$

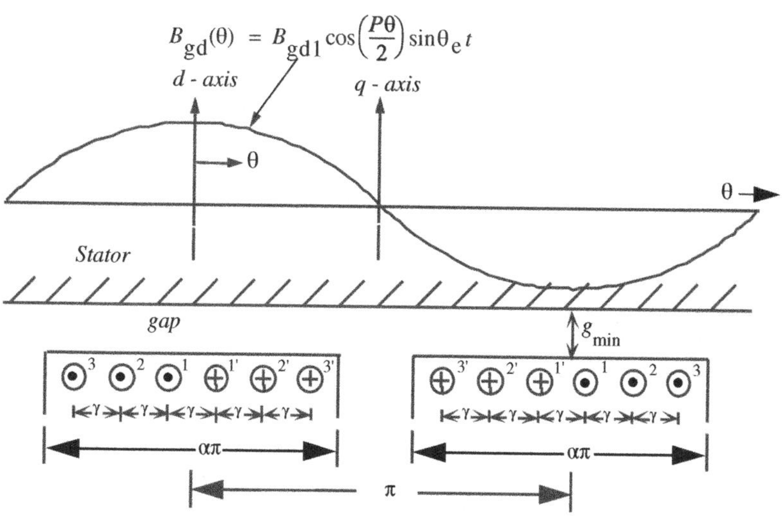

Figure 9.18 Amortisseur bar positions on the rotor pole face assuming six bars per pole. Bars are assumed to be impressed with an alternating air gap flux density.

where D_{or} is the rotor outer diameter measured at the d-axis. From Section 9.2, it is recalled that,

$$B_{\mathrm{gd1}} = k_{\mathrm{d}} B_{\mathrm{gd,pk}} \tag{9.111}$$

and

$$D_{\mathrm{or}} = \frac{P\tau_{\mathrm{p}}}{\pi} \tag{9.112}$$

so that equation (9.110) can be written as

$$\Phi_{1-1'} = \frac{2}{\pi} k_{\mathrm{d}} B_{\mathrm{gd,pk}} l_{\mathrm{e}} D_{\mathrm{or}} \sin\left(\frac{\gamma}{2}\right) \sin \omega_{\mathrm{e}} t \tag{9.113}$$

The voltage induced in this mesh is

$$\frac{d\Phi_{1-1'}}{dt} = e_1 - e_1' = \frac{2}{\pi} \omega_e k_{\mathrm{d}} B_{\mathrm{gd,pk}} l_{\mathrm{e}} \tau_{\mathrm{p}} \sin\left(\frac{\gamma}{2}\right) \cos(\omega_{\mathrm{e}} t) \tag{9.114}$$

where e_1 and e_1' are the voltage induced in bars 1 and $1'$ respectively. It is apparent from the symmetry that $e_1' = -e_1$. Similarly, for the circuit formed from bars 1–$1'$ and the corresponding portion of the end rings,

$$\frac{d\Phi_{2-2'}}{dt} = e_2 - e_2' = \frac{2}{\pi} \omega_{\mathrm{e}} k_{\mathrm{d}} B_{\mathrm{gd,pk}} l_{\mathrm{e}} \tau_{\mathrm{p}} \sin\left(\frac{3\gamma}{2}\right) \cos(\omega_{\mathrm{e}} t) \tag{9.115}$$

In general, for the nth pair of bars,

$$\frac{d\Phi_{n-n'}}{dt} = e_n - e_n' = \frac{2}{\pi} \omega_{\mathrm{e}} k_{\mathrm{d}} B_{\mathrm{gd,pk}} l_{\mathrm{e}} \tau_{\mathrm{p}} \sin\left(\frac{(2n-1)\gamma}{2}\right) \cos(\omega_{\mathrm{e}} \mathrm{t}) \tag{9.116}$$

If R_{b} and L_{b} denote the resistance and leakage inductance of one bar, respectively, and R_{e} and L_{e} that of one end ring segment, then for mesh 1–$1'$,

$$\frac{d\Phi_{1-1'}}{dt} = 2e_1 = 2R_{\mathrm{b}} i_{\mathrm{b1}} + 2L_{\mathrm{b}} \frac{di_{\mathrm{b1}}}{dt} + 2R_{\mathrm{e}} i_{\mathrm{e1}} + 2L_{\mathrm{e}} \frac{di_{\mathrm{e1}}}{dt} \tag{9.117}$$

where i_{b1} and i_{e1} are the currents in the bar and end ring respectively.

For mesh 2–$2'$, one has

$$\frac{d\phi_{2-2'}}{dt} = 2e_2 = 2R_{\mathrm{b}} i_{\mathrm{b2}} + 2L_{\mathrm{b}} \frac{di_{\mathrm{b2}}}{dt} + 4R_{\mathrm{e}} i_{\mathrm{e2}} + 4L_{\mathrm{e}} \frac{di_{\mathrm{e2}}}{dt} + 2R_{\mathrm{e}} i_{\mathrm{e1}} + 2L_{\mathrm{e}} \frac{di_{\mathrm{e1}}}{dt} \tag{9.118}$$

and so forth until the nth mesh.

The currents in the end ring can be expressed in terms of the bar currents by the following constraints,

$$i_{\mathrm{e1}} = i_{\mathrm{b1}} + i_{\mathrm{b2}} + i_{\mathrm{b3}} + \cdots + i_{\mathrm{b}n} \tag{9.119}$$

$$i_{\mathrm{e2}} = i_{\mathrm{b2}} + i_{\mathrm{b3}} + i_{\mathrm{b4}} + \cdots + i_{\mathrm{b}n} \tag{9.120}$$

$$i_{\mathrm{e3}} = i_{\mathrm{b3}} + i_{\mathrm{b4}} + i_{\mathrm{b5}} + \cdots + i_{\mathrm{b}n} \tag{9.121}$$

and so forth until, for the last end ring current,

$$i_{\mathrm{e}n} = i_{\mathrm{b}n} \tag{9.122}$$

The expression for the mesh voltage of the circuit involving bars 1–1′ becomes

$$\begin{aligned}\frac{d\Phi_{1-1'}}{dt} &= 2e_1 = 2R_b i_{b1} + 2L_b\frac{di_{b1}}{dt} + 2R_e(i_{b1} + i_{b2} + i_{b3} + \cdots + i_{bn}) \\ &\quad + 2L_e\frac{d(i_{b1} + i_{b2} + i_{b3} + \cdots + i_{bn})}{dt}\end{aligned} \tag{9.123}$$

and so forth.

The resulting equations can be accumulated to form the matrix equation,

$$\frac{d}{dt}\begin{bmatrix}\Phi_{1-1'}\\ \Phi_{2-2'}\\ \Phi_{3-3'}\\ \cdots\\ \Phi_{n-n'}\end{bmatrix} = 2\begin{bmatrix}e_1\\ e_2\\ e_3\\ \cdots\\ e_n\end{bmatrix} = \begin{bmatrix}2(R_b+R_e) & 2R_e & 2R_e & 2R_e & \cdots & 2R_e\\ 2R_e & 2R_b+6R_e & 6R_e & 6R_e & \cdots & 6R_e\\ 2R_e & 6R_e & 2R_b+10R_e & 10R_e & \cdots & 10R_e\\ \cdots & \cdots & \cdots & \cdots & \cdots & \cdots\\ 2R_e & 6R_e & 10R_e & 14R_e & \cdots & 2R_b+2(N_b-1)R_e\end{bmatrix}\begin{bmatrix}i_{b1}\\ i_{b2}\\ i_{b3}\\ \cdots\\ i_{bn}\end{bmatrix}$$
$$+\begin{bmatrix}2(L_b-L_e) & 2L_e & 2L_e & 2L_e & \cdots & 2L_e\\ 2L_e & 2L_b+6L_e & 6L_e & 6L_e & \cdots & 6L_e\\ 2L_e & 6L_e & 2L_b+10L_e & 10L_e & \cdots & 10L_e\\ \cdots & \cdots & \cdots & \cdots & \cdots & \cdots\\ 2L_e & 6L_e & 10L_e & 14L_e & \cdots & 2L_b+2(N_b-1)L_e\end{bmatrix}\times\frac{d}{dt}\begin{bmatrix}i_{b1}\\ i_{b2}\\ i_{b3}\\ \cdots\\ i_{bn}\end{bmatrix} \tag{9.124}$$

where N_b equals the (even) number of bars in each pole. In simplified form,

$$\frac{d}{dt}[\Phi] = [e] = [R][i] + [L]\frac{d}{dt}[i] \tag{9.125}$$

Equation (9.125) is clearly a coupled nth order set of differential equations making each of the bar currents an independent state variable. However, a simplification is possible if equation (9.124) is written in the form,

$$\frac{1}{2}\frac{d}{dt}\begin{bmatrix}\phi_{1-1'}\\ \phi_{2-2'}\\ \phi_{3-3'}\\ \cdots\\ \phi_{n-n'}\end{bmatrix} = \begin{bmatrix}e_1\\ e_2\\ e_3\\ \cdots\\ e_n\end{bmatrix} = R_b\begin{bmatrix}1+\frac{R_e}{R_b} & \frac{R_e}{R_b} & \frac{R_e}{R_b} & \frac{R_e}{R_b} & \cdots & \frac{R_e}{R_b}\\ \frac{R_e}{R_b} & 1+\frac{3R_e}{R_b} & \frac{3R_e}{R_b} & \frac{3R_e}{R_b} & \cdots & \frac{3R_e}{R_b}\\ \frac{R_e}{R_b} & \frac{3R_e}{R_b} & 1+\frac{5R_e}{R_b} & \frac{5R_e}{R_b} & \cdots & \frac{5R_e}{R_b}\\ \cdots & \cdots & \cdots & \cdots & \cdots & \cdots\\ \frac{R_e}{R_b} & \frac{3R_e}{R_b} & \frac{5R_e}{R_b} & \frac{7R_e}{R_b} & \cdots & 1+(N_b-1)\frac{R_e}{R_b}\end{bmatrix}\begin{bmatrix}i_{b_1}\\ i_{b_2}\\ i_{b_3}\\ \cdots\\ i_{b_n}\end{bmatrix}$$
$$+L_b\begin{bmatrix}1+\frac{L_e}{L_b} & \frac{L_e}{L_b} & \frac{L_e}{L_b} & \frac{L_e}{L_b} & \cdots & \frac{L_e}{L_b}\\ \frac{L_e}{L_b} & 1+3\frac{L_e}{L_b} & 3\frac{L_e}{L_b} & 3\frac{L_e}{L_b} & \cdots & 3\frac{L_e}{L_b}\\ \frac{L_e}{L_b} & 3\frac{L_e}{L_b} & 1+5\frac{L_e}{L_b} & 5\frac{L_e}{L_b} & \cdots & 5\frac{L_e}{L_b}\\ \cdots & \cdots & \cdots & \cdots & \cdots & \cdots\\ \frac{L_e}{L_b} & 3\frac{L_e}{L_b} & 5\frac{L_e}{L_b} & 7\frac{L_e}{L_b} & \cdots & 1+(N_b-1)\frac{L_e}{L_b}\end{bmatrix}\times\frac{d}{dt}\begin{bmatrix}i_{b_1}\\ i_{b_2}\\ i_{b_3}\\ \cdots\\ i_{b_n}\end{bmatrix} \tag{9.126}$$

In practice, it is approximately true that

$$\frac{R_e}{R_b} \approx \frac{L_e}{L_b} \tag{9.127}$$

in which case the two large matrices in equation (9.126) are essentially equal. Letting the two $n \times n$ matrices be denoted by $[S]$, equation (9.125) can be written as

$$[e_b] = R_b[S][i_b] + L_b[S]\frac{d}{dt}[i_b] \tag{9.128}$$

Upon multiplying this equation by the inverse of $[S]$

$$L_b[S]^{-1}[S]\frac{d}{dt}[i_b] + R_b[S]^{-1}[S][i_b] = [S]^{-1}[e_b] \tag{9.129}$$

which reduces to, simply,

$$L_b\frac{d}{dt}[i_b] + R_b[i_b] = [S]^{-1}[e_b] \tag{9.130}$$

Since all n bar voltages are now impressed on the same resistance R_b and inductance L_b, the n currents are all in phase. Hence, once the current in one of the bars is known, all of the others are fixed by the proportionality implied in equation (9.130).

While this derivation has been carried out for an even number of bars, the result will be the same for an odd number with only the voltage vector $[e_b]$ changed appropriately. The odd numbered bar located in the center of the pole is, in this case, designated as the "b0" bar. Since current will not flow in this bar as a result of d-axis air gap flux, equation (9.130) remains essentially unchanged. The "S" matrix, in this case, can be determined to be,

$$[S]_{odd} = \begin{bmatrix} 1+2\frac{L_e}{L_b} & 2\frac{L_e}{L_b} & 2\frac{L_e}{L_b} & 2\frac{L_e}{L_b} & \cdots & 2\frac{L_e}{L_b} \\ 2\frac{L_e}{L_b} & 1+4\frac{L_e}{L_b} & 4\frac{L_e}{L_b} & 4\frac{L_e}{L_b} & \cdots & 4\frac{L_e}{L_b} \\ 2\frac{L_e}{L_b} & 4\frac{L_e}{L_b} & 1+6\frac{L_e}{L_b} & 6\frac{L_e}{L_b} & \cdots & 6\frac{L_e}{L_b} \\ \cdots & \cdots & \cdots & \cdots & \cdots & \cdots \\ 2\frac{L_e}{L_b} & 4\frac{L_e}{L_b} & 6\frac{L_e}{L_b} & 8\frac{L_e}{L_b} & \cdots & 1+(N_b-1)\frac{L_e}{L_b} \end{bmatrix} \tag{9.131}$$

Defining now a peak or base bar voltage

$$E_b = \frac{\omega_e k_d B_{gd1} l_e D_{or}}{P} \tag{9.132}$$

the system of equations can be written also as

$$L_b\frac{d}{dt}[i_b] + R_b[i_b] = E_b\cos(\omega_e t)[S]^{-1}[e_{b,pu}] \tag{9.133}$$

where

$$[e_{\mathrm{b,pu}}] = \begin{bmatrix} \sin\left(\frac{\gamma}{2}\right) \\ \sin\left(\frac{3\gamma}{2}\right) \\ \sin\left(\frac{5\gamma}{2}\right) \\ \cdots \\ \sin\left[\left(\frac{(2n-1)\gamma}{2}\right)\right] \end{bmatrix} \tag{9.134}$$

In the steady-state,

$$[i_{\mathrm{b}}] = \frac{E_{\mathrm{b}}}{Z_{\mathrm{b}}}\cos(\omega_e t - \phi_{\mathrm{b}})[S]^{-1}[e_{\mathrm{b,pu}}] \tag{9.135}$$

where

$$Z_{\mathrm{b}} = \sqrt{R_{\mathrm{b}}^2 + (\omega_{\mathrm{e}} L_{\mathrm{b}})^2} \tag{9.136}$$

and

$$\phi_{\mathrm{b}} = \mathrm{a}\tan\left(\frac{\omega_{\mathrm{e}} L_{\mathrm{b}}}{R_{\mathrm{b}}}\right) \tag{9.137}$$

From equation (9.135), it is apparent that all of the bar currents oscillate sinusoidally in phase with each other. It can be observed that the basic nature of the current distribution in the bars is contained in the matrix product $[S]^{-1}[e_{\mathrm{b,pu}}]$ which can be viewed as a per unit representation of the bar currents, that is,

$$[i_{\mathrm{b,pu}}] = [S]^{-1}[e_{\mathrm{b,pu}}] \tag{9.138}$$

If the bar currents are regarded as consisting of Kronecker delta or "impulse" functions, the resulting d-axis MMF can be expressed analytically, over one pole, as

$$\mathcal{F}_{\mathrm{bd,pu}}(\theta) = \int_0^{\pi} \left[\sum_{k=1}^{N_{\mathrm{b}}} i_{\mathrm{b,pu}}(k)\delta(k)\right] d\theta \tag{9.139}$$

where N_{b} is the number of bars per pole. A typical plot of the per unit bar currents and the resulting per unit air gap MMF is shown in Figure 9.19.

The harmonics of the MMF of Figure 9.19 including the fundamental component can be readily extracted and plotted for various ratios of $R_{\mathrm{e}}/R_{\mathrm{b}} = L_{\mathrm{e}}/L_{\mathrm{b}}$. This MMF acts in the air gap to produce a component of gap flux density which links the field and d-axis stator winding. Defining $I_{\mathrm{b}} = E_{\mathrm{b}}/Z_{\mathrm{b}}$, the d-axis rotor bar MMF can be expressed as

$$\mathcal{F}_{\mathrm{bd}}(\theta) = I_{\mathrm{b}}\mathcal{F}_{\mathrm{bd,pu}}(\theta) \tag{9.140}$$

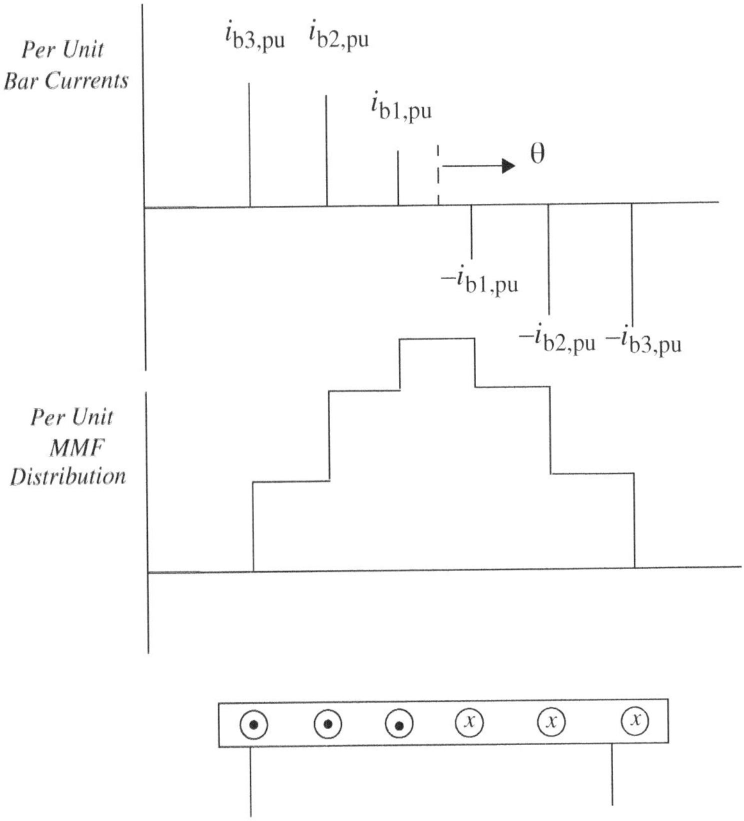

Figure 9.19 Per unit bar currents and resulting MMF distribution. Note that these currents are not sinusoidally distributed.

The fundamental component of $I_b\mathcal{F}_{b,pu}(\theta)$ represents the useful portion of MMF which produces flux linking the field and stator windings and, assuming the distribution of the rotor bars is the same for all the rotor poles, this expression can be defined as

$$\mathcal{F}_{bd,pu1} = \frac{2}{\pi} \int_{-(\alpha\pi)/P}^{(\alpha\pi)/P} \mathcal{F}_{bd,pu}(\theta) \cos\left(\frac{P\theta}{2}\right) d\theta \tag{9.141}$$

The origin for θ is again assumed as the pole center line. The winding factor for the amortisseur bars is related to $\mathcal{F}_{bd,pu1}$ by

$$\mathcal{F}_{bd,pu1} = \frac{4}{\pi} k_{bd} \tag{9.142}$$

The $4/\pi$ appears since the winding factors are defined to be the harmonic components of a square wave winding distribution and not a sine wave distribution.

The higher odd harmonics correspond to a leakage component.

$$k_{\mathrm{bd,h}} = \frac{2}{\pi} \int_{-(\alpha\pi)/P}^{(\alpha\pi)/P} \mathcal{F}_{\mathrm{bd,pu}}(\theta) \cos\left(\frac{hP}{2}\theta\right) d\theta \quad h = 3, 5, 7, \ldots \tag{9.143}$$

The leakage inductance associated with energy stored in the higher harmonic field set up by the bars will be related to the inductance of the bar by defining,

$$L_{\mathrm{lkd}} = k_{\mathrm{lkd}} L_{\mathrm{b}} \tag{9.144}$$

where

$$k_{\mathrm{lkd}} = \sqrt{\sum_{h=3,5,7,\ldots}^{\infty} k_{\mathrm{bd,h}}} \tag{9.145}$$

9.8 MUTUAL AND MAGNETIZING INDUCTANCES OF THE AMORTISSEUR WINDING

The magnetizing inductance of the amortisseur winding is calculated in much the same manner as for the field winding except that in this case, the distribution of MMF across the pole is assumed to be sinusoidal. The result is

$$\lambda_{\mathrm{mkd}} = B_{\mathrm{kd,pk}} \left(\frac{2}{\pi}\right) \int_{-(\alpha\pi)/P}^{(\alpha\pi)/P} N_{\mathrm{b}} \cos\left(\frac{P\theta}{2}\right) d\theta$$

$$L_{\mathrm{mkd}} = \left(\frac{8}{\pi^2}\right) \frac{(k_{\mathrm{bd}} N_{\mathrm{b}})^2}{P} k_{\mathrm{d}} \left(\frac{\mu_0 \tau_{\mathrm{p}} l_{\mathrm{e}}}{g_{\mathrm{e}}}\right) \tag{9.146}$$

Since the current flow in each pole is independent of the others, it is the same for all poles and therefore the effective number of circuits is equal to the number of poles. Likewise, the mutual inductances between the bar circuit and the stator and field windings are

$$L_{\mathrm{dkd}} = \left(\frac{12}{\pi^2}\right) \left(\frac{k_1 N_{\mathrm{s}} k_{\mathrm{bd}} N_{\mathrm{b}}}{P}\right) k_{\mathrm{d}} \left(\frac{\mu_0 \tau_{\mathrm{p}} l_{\mathrm{e}}}{g_{\mathrm{e}}}\right) \tag{9.147}$$

$$L_{\mathrm{fkd}} = \left(\frac{8}{\pi^2}\right) \left(\frac{k_{\mathrm{f}} N_{\mathrm{f}} k_{\mathrm{bd}} N_{\mathrm{b}}}{P}\right) k_{\mathrm{d}} \left(\frac{\mu_0 \tau_{\mathrm{p}} l_{\mathrm{e}}}{g_{\mathrm{e}}}\right) \tag{9.148}$$

9.9 DIRECT AXIS EQUIVALENT CIRCUIT

Having now expressed equations for all of the direct axis magnetizing and mutual inductances, the *d*-axis equivalent circuit can now be determined. In most machines the number of parallel circuits of the stator and field are different. Moreover, the amortisseur winding current circuits are not connected in series or parallel so that it is desirable to simply refer the rotor circuits to the stator by their turns ratio as was the case for the induction machine.

It is now assumed that currents exist in all d-axis windings so that the flux linkage λ_{md} now contains flux linkage components from all three d-axis windings. In order to avoid confusion, it is useful to derive the circuit first based on one pole of the machine. In this case, one can write, from equations (9.16), (9.69), and (9.147), for the air gap portion of the stator flux linkages,

$$\lambda_{mdp} = \frac{\lambda_{md}}{(P/C_s)} = \frac{L_{md}}{(P/C_s)} i_d + \frac{L_{df}}{(P/C_s)} i_f + \frac{L_{dkd}}{(P/C_s)} i_{kd} \tag{9.149}$$

Or, with some manipulation, this expression can be arranged to form

$$\lambda_{mdp} = \left(\frac{12}{\pi^2}\right)\frac{(k_1N_t)^2}{P^2}k_d\mathcal{P}_p\left(\frac{i_d}{C_s}\right) + \left(\frac{12}{\pi^2}\right)\frac{(k_1N_t)(N_{tf})}{P^2}k_d\mathcal{P}_p\left(\frac{i_f}{C_f}\right) + \left(\frac{12}{\pi^2}\right)\frac{(k_1N_t)(k_{bd}N_b)}{P^2}k_d\mathcal{P}_p i_{kd} \tag{9.150}$$

where, from equation (3.82),

$$\mathcal{P}_p = \frac{\mu_0 \tau_p l_e}{g_e} \tag{9.151}$$

The "$12/\pi^2$" factor for the stator-based terms rather than "$8/\pi^2$" which appears in rotor-based terms in equations (9.159) and (8.150) occurs because of the non-reciprocal nature of the d–q transformation.

Equation (9.150) can be rearranged to form,

$$\lambda_{mdp} = \frac{12}{\pi^2}\frac{(k_1N_t)^2}{C_sP^2}k_d\mathcal{P}_p i_d + \frac{12}{\pi^2}\frac{(k_1N_t)^2}{C_sP^2}k_d\mathcal{P}_p\left[\frac{2}{3}\left(\frac{C_s}{C_f}\right)\frac{N_{tf}}{(k_1N_t)}i_f\right] + \frac{12}{\pi^2}\frac{(k_1N_t)^2}{C_sP^2}k_d\mathcal{P}_p\left[\frac{2}{3}C_s\frac{k_{bd}N_b}{(k_1N_t)}i_{kd}\right] \tag{9.152}$$

Defining

$$L_{mdp} = \frac{12}{\pi^2}\frac{(k_1N_t)^2}{P}k_d\mathcal{P}_p \tag{9.153}$$

$$i'_f = \frac{2}{3}\left(\frac{C_s}{C_f}\right)\frac{(N_{tf})}{(k_1N_t)}i_f = \frac{2}{3}\frac{(N_f)}{(k_1N_s)}i_f \tag{9.154}$$

$$i'_{kd} = \frac{2}{3}C_s\frac{(k_{bd}N_b)}{(k_1N_t)}i_{kd} = \frac{2}{3}\frac{(k_{bd}N_b)}{(k_1N_s)}i_{kd} \tag{9.155}$$

Equation (8.133) can be written as

$$\lambda_{mdp} = \frac{L_{mdp}}{C_sP}\left(i_d + i'_f + i'_{kd}\right) \tag{9.156}$$

or, for the P/Cs poles connected in series,

$$\lambda_{md} = \frac{P}{C_s}\lambda_{mdp} = L_{md}\left(i_d + i'_f + i'_{kd}\right) \tag{9.157}$$

where L_{md} is defined by equation (9.16) namely

$$L_{md} = \left(\frac{12}{\pi^2}\right)\frac{(k_1N_s)^2}{P}\mu_0\frac{\tau_p l_e}{g_e}k_d$$

Similarly, for the field circuit and for the amortisseur circuit,

$$\lambda_{mfp} = \frac{\lambda_{mf}}{(P/C_f)} = \frac{L_{mf1}}{(P/C_f)}i_f + \frac{L_{df}}{(P/C_f)}i_d + \frac{L_{fkd}}{(P/C_f)}i_{kd} \tag{9.158}$$

or

$$\lambda_{mfp} = \frac{8}{\pi^2}\frac{N_{tf}^2}{P^2}k_d\mathcal{P}_p\left(\frac{i_f}{C_f}\right) + \frac{12}{\pi^2}\frac{(k_1N_t)(N_{tf})}{P^2}k_d\mathcal{P}_p\left(\frac{i_d}{C_s}\right) + \frac{8}{\pi^2}\frac{(N_{tf})(k_{bd}N_b)}{P^2}k_d\mathcal{P}_p i_{kd} \tag{9.159}$$

In a very similar manner, equation (9.159) can first be multiplied by $(k_1N_t)/(N_{tf})$ to form,

$$\left(\frac{k_1N_t}{N_{tf}}\right)\lambda_{mfp} = \frac{8}{\pi^2}\frac{(N_{tf})(k_1N_t)}{P^2}k_d\mathcal{P}_p\left(\frac{i_f}{C_f}\right) + \frac{12}{\pi^2}\frac{(k_1N_t)^2}{P^2}k_d\mathcal{P}_p\left(\frac{i_d}{C_s}\right) + \frac{8}{\pi^2}\frac{(k_1N_t)(k_{bd}N_b)}{P^2}k_d\mathcal{P}_p i_{kd} \tag{9.160}$$

Defining

$$\lambda'_{mfp} = \frac{(k_1N_t)}{N_{tf}}\lambda_{mfp} \tag{9.161}$$

and further rearranging,

$$\lambda'_{mfp} = \frac{12}{\pi^2}\frac{(k_1N_t)^2}{C_sP^2}k_d\mathcal{P}_p\left[\frac{3}{2}\left(\frac{C_s}{C_f}\right)\left(\frac{N_{tf}}{k_1N_t}\right)i_f\right] + \frac{12}{\pi^2}\frac{(k_1N_t)}{C_sP^2}k_d\mathcal{P}_p i_d + \frac{12}{\pi^2}\frac{(k_1N_t)^2}{C_sP^2}k_d\mathcal{P}_p\left[\frac{2}{3}C_s\frac{(k_{bd}N_b)}{(k_1N_t)}i_{kd}\right] \tag{9.162}$$

after which, using equations (9.153) to (9.155),

$$\lambda'_{mfp} = \frac{12}{\pi^2}\frac{(k_1N_t)^2}{C_sP^2}k_d\mathcal{P}_p\left(i'_f + i_d + i'_{kd}\right) \tag{9.163}$$

or simply

$$\lambda'_{mfp} = \frac{L_{mdp}}{C_sP}\left(i'_f + i_d + i'_{kd}\right) \tag{9.164}$$

where i'_f, i'_{kd}, and L_{mdp} are defined by equations (9.154), (9.155), and (9.153) respectively.

Referring now the field air gap flux linkages to a complete stator phase,

$$\lambda'_{mf} = \left(\frac{P}{C_s}\right)\lambda'_{mfp} \tag{9.165}$$

whereupon

$$\left(\frac{P}{C_s}\right)\frac{L_{mfp}}{C_s P} = \left(\frac{P}{C_s}\right)\left(\frac{12}{\pi^2}\right)\frac{(k_1 N_t)^2}{C_s P^2} k_d \mathcal{P}_p = \frac{3}{\pi}\frac{(k_1 N_s)}{P} k_d \mathcal{P}_p = L_{md} \tag{9.166}$$

so that

$$\lambda'_{mf} = L_{md}\left(i'_f + i_d + i'_{kd}\right) \tag{9.167}$$

Finally, for the d-axis amortisseur circuit

$$\lambda_{mkdp} = \lambda_{mkd} = L_{mkd} i_{kd} + L_{dkd} i_d + L_{fkd} i_f \tag{9.168}$$

or

$$\lambda_{mkdp} = \frac{8}{\pi^2}(k_{bd} N_b)^2 k_d \mathcal{P}_p i_{kd} + \frac{12}{\pi^2}\frac{(k_1 N_t)(k_{bd} N_b)}{P} k_d \mathcal{P}_p \frac{i_d}{C_s} + \frac{8}{\pi^2}\frac{(N_{tf})(k_{bd} N_b)}{P} k_d \mathcal{P}_p \frac{i_f}{C_f} \tag{9.169}$$

Equation (8.150) can be referred to one stator phase in exactly the same manner.

$$\begin{aligned}\left(\frac{k_1 N_t}{k_{bd} N_b}\right)\lambda_{mkdp} &= \frac{8}{\pi^2}(k_{bd} N_b)(k_1 N_t) k_d \mathcal{P}_p i_{kd} + \frac{12}{\pi^2}\frac{(k_1 N_t)^2}{P} k_d \mathcal{P}_p \frac{i_d}{C_s} \\ &+ \frac{8}{\pi^2}\frac{(N_{tf})(k_1 N_t)}{P} k_d \mathcal{P}_p \frac{i_f}{C_f}\end{aligned} \tag{9.170}$$

The resulting flux linkage for all series-connected poles

$$\lambda'_{mkd} = L_{md}(i'_{kd} + i_d + i'_f) \tag{9.171}$$

where

$$\lambda'_{mkd} = \left(\frac{k_1 N_t}{k_{bd} N_b}\right)\left(\frac{1}{C_s}\right)\lambda_{mkdp} \tag{9.172}$$

Note from equations (9.167) and (9.172) that $\lambda'_{mf} = \lambda'_{mkd} = \lambda_{md}$.

9.10 REFERRAL OF ROTOR PARAMETERS TO THE STATOR

The final task in the process of deriving the d-axis equivalent circuit is to refer the field and amortisseur parameters in terms of stator turns and phases. One pole of the field circuit can be described by the differential equation

$$v_{fp} = r_{fp} i_{fp} + L_{1fdp}\frac{di_{fp}}{dt} + \frac{d\lambda_{mfp}}{dt} \tag{9.173}$$

where the "p" appended as the last subscript of each variable again denotes "per pole". Multiplying this equation by $(k_1 N_t)/(N_{tf})$ results in

$$\left(\frac{k_1 N_t}{N_{tf}}\right) v_{fp} = \left(\frac{k_1 N_t}{N_{tf}}\right) r_{fp}\left(\frac{i_f}{C_f}\right) + \left(\frac{k_1 N_t}{N_{tf}}\right) L_{lfp}\frac{d}{dt}\left(\frac{i_f}{C_f}\right) + \frac{d\left(\frac{k_1 N_t}{N_{tf}}\right)\lambda_{mfp}}{dt} \tag{9.174}$$

Recalling equation (9.161), and in terms of the total field voltage

$$\frac{C_f}{P}\left(\frac{k_1 N_t}{N_{tf}}\right) v_f = \left(\frac{k_1 N_t}{N_{tf}}\right) r_{fp}\frac{i_f}{C_f} + \left(\frac{k_1 N_t}{N_{tf}}\right) L_{lfp}\frac{d}{dt}\left(\frac{i_f}{C_f}\right) + \frac{d\lambda'_{mfp}}{dt} \tag{9.175}$$

Rearranging, paying attention to the definition of equation (9.154),

$$\begin{aligned}\frac{C_f}{P}\left(\frac{k_1 N_t}{N_{tf}}\right) vf &= \frac{3}{2}\frac{1}{C_s}\left(\frac{k_1 N_t}{N_{tf}}\right)^2 r_{fp}\left[\frac{2}{3}\left(\frac{C_s}{C_f}\right)\frac{(N_{tf})}{(k_1 N_t)} i_f\right] \\ &+ \frac{3}{2}\frac{1}{C_s}\left(\frac{k_1 N_t}{N_{tf}}\right)^2 L_{lfp}\frac{d}{dt}\left[\frac{2}{3}\left(\frac{C_s}{C_f}\right)\frac{(N_{tf})}{k_1 N_t} i_f\right] + \frac{d\lambda'_{mfp}}{dt}\end{aligned} \tag{9.176}$$

The resistance and inductance per pole are related to the actual field resistance and leakage inductance, r_f and L_{lf}, as viewed from the field terminals by

$$r_{fp} = \frac{C_f^2}{P} r_f \tag{9.177}$$

$$L_{lfp} = \frac{C_f^2}{P} L_{lf} \tag{9.178}$$

Multiplying by P/C_s, and utilizing equations (9.177) and (9.178), equation (9.176) becomes

$$\begin{aligned}\frac{C_f}{C_s}\left(\frac{k_1 N_t}{N_{tf}}\right) v_f &= \frac{3}{2}\frac{C_f}{C_s}\left[\left(\frac{k_1 N_t}{N_{tf}}\right)^2 r_f\right]\left[\frac{2}{3}\left(\frac{C_s}{C_f}\right)\frac{(N_{tf})}{k_1 N_t} i_f\right] \\ &+ \frac{3}{2}\frac{C_f}{C_s}\left(\frac{k_1 N_t}{N_{tf}}\right)^2 L_{lf}\left(\frac{d}{dt}\left[\frac{2}{3}\left(\frac{C_s}{C_f}\right)\frac{(N_{tf})}{(k_1 N_t)} i_f\right]\right) + \frac{d}{dt}\left(\frac{P}{C_s}\lambda_{mfp}\right)\end{aligned} \tag{9.179}$$

Noting equation (9.154) and defining

$$r'_f = \frac{3}{2}\left(\frac{C_f}{C_s}\right)^2\left(\frac{k_1 N_t}{N_{tf}}\right)^2 r_f = \frac{3}{2}\left(\frac{k_1 N_s}{N_f}\right)^2 r_f \tag{9.180}$$

$$L'_{lf} = \frac{3}{2}\left(\frac{C_f}{C_s}\right)^2\left(\frac{k_1 N_t}{N_{tf}}\right)^2 L_{lf} = \frac{3}{2}\left(\frac{k_1 N_s}{N_f}\right)^2 L_{lf} \tag{9.181}$$

$$v'_f = \frac{C_f}{C_s}\left(\frac{k_1 N_t}{N_{tf}}\right) v_f = \left(\frac{k_1 N_s}{N_f}\right) v_f \tag{9.182}$$

$$i'_f = \frac{2}{3}\left(\frac{C_s}{C_f}\right)\frac{(N_{tf})}{(k_1 N_t)} i_f = \frac{2}{3}\left(\frac{N_f}{k_1 N_s}\right) i_f \tag{9.183}$$

results in, finally,

$$v'_{\rm f} = r'_{\rm f} i'_{\rm f} + L'_{\rm lf} \frac{di'_{\rm fr}}{dt} + \frac{d\lambda'_{\rm mf}}{dt} \tag{9.184}$$

where $\lambda'_{\rm mf}$ is defined by (9.167).

The direct axis amortisseur winding can clearly be referred to the stator by the same technique. The result of this transformation is

$$v'_{\rm kd} = 0 = r'_{\rm kd} i'_{\rm kd} + L'_{\rm lkd} \frac{di'_{\rm kd}}{dt} + \frac{d\lambda'_{\rm mkd}}{dt} \tag{9.185}$$

where

$$r'_{\rm kd} = \frac{3}{2} \frac{1}{C_{\rm S}^2} \left(\frac{k_1 N_{\rm t}}{k_{\rm bd} N_{\rm b}} \right)^2 r_{\rm b} = \frac{3}{2} \left(\frac{k_1 N_{\rm s}}{k_{\rm bd} N_{\rm b}} \right)^2 r_{\rm b} \tag{9.186}$$

$$L'_{\rm lkd} \frac{3}{2} \frac{1}{C_{\rm S}^2} \left(\frac{k_1 N_{\rm t}}{k_{\rm bd} N_{\rm b}} \right)^2 L_{\rm be} = \frac{3}{2} \left(\frac{k_1 N_{\rm s}}{k_{\rm bd} N_{\rm b}} \right)^2 L_{\rm be} \tag{9.187}$$

The quantity $L_{\rm be}$ should take into account the effect of the gap leakage defined as the higher harmonic components of the air gap flux density produced by the amortisseur bars, equation (9.143). That is, $L_{\rm be} = (1 + k_{\rm bar,h})L_{\rm b}$. The complete equivalent circuit for the d-axis of the wound field salient pole synchronous machine is shown in Figure 9.20.

In this circuit, the effect of the pole face leakages has been incorporated into the field leakage inductance term. That is, formally,

$$L''_{\rm lf} = L'_{\rm lf,pf} + L'_{\rm lf} \tag{9.188}$$

The pole face leakage inductance $L'_{\rm lf,pf}$ is assumed to incorporate all field leakage inductances including both pole body and tooth top leakages of Sections 9.5 and 9.6. The inductance $L'_{\rm lkd}$ includes both the bar inductance and the harmonic leakage discussed in Section 9.7. Both of these inductances must be referred to stator turns in a manner similar to

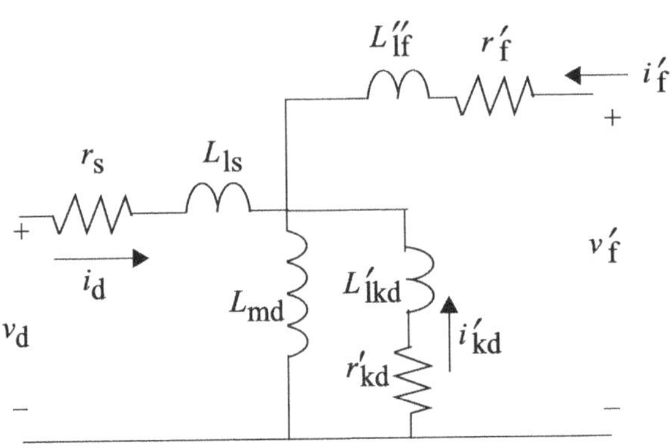

Figure 9.20 Final direct axis equivalent circuit for a wound field, salient pole synchronous machine. "Speed voltages" arising from Park's transformation are not shown.

9.11 QUADRATURE AXIS CIRCUIT

It should now be apparent that the parameters of the quadrature axis circuit can be obtained by analogy to what has been considered thus far. Again assuming an interrupted end ring with six bars per rotor pole, the amortisseur circuit can again be excited with flux from the stator, this time exciting the q-axis rotor circuit as shown in Figure 9.21. Note that the bars have been renumbered when compared with Figure 9.18.

It can be shown that the rotor bar circuits for this case result in a system of equations as follows:

$$\frac{d}{dt}\begin{bmatrix}\Phi_{12}\\ \Phi_{23}\\ \Phi_{34}\\ \cdots\\ \Phi_{n-1,n}\end{bmatrix}=\begin{bmatrix}e_1-e_2\\ e_2-e_3\\ e_3-e_4\\ \cdots\\ e_{n-1}-e_n\end{bmatrix}=R_b\begin{bmatrix}1+\frac{2R_e}{R_b} & -1 & 0 & 0 & \cdots & 0 & 0\\ \frac{2R_e}{R_b} & 1+\frac{2R_e}{R_b} & -1 & 0 & \cdots & 0 & 0\\ \frac{2R_e}{R_b} & \frac{2R_e}{R_b} & 1+\frac{2R_e}{R_b} & -1 & \cdots & 0 & 0\\ \cdots & \cdots & \cdots & \cdots & \cdots & \cdots & 0\\ \frac{2R_e}{R_b} & \frac{2R_e}{R_b} & \frac{2R_e}{R_b} & \frac{2R_e}{R_b} & \cdots & 1+\frac{2R_e}{R_b} & -1\end{bmatrix}\begin{bmatrix}i_{b1}\\ i_{b2}\\ i_{b3}\\ \cdots\\ i_{bn}\end{bmatrix}$$
$$+L_b\begin{bmatrix}1+\frac{2L_e}{L_b} & -1 & 0 & 0 & \cdots & 0 & 0\\ \frac{2L_e}{L_b} & 1+\frac{2L_e}{L_b} & -1 & 0 & \cdots & 0 & 0\\ \frac{2L_e}{L_b} & \frac{2L_e}{L_b} & 1+\frac{2L_e}{L_b} & -1 & \cdots & 0 & 0\\ \cdots & \cdots & \cdots & \cdots & \cdots & \cdots & 0\\ \frac{2L_e}{L_b} & \frac{2L_e}{L_b} & \frac{2L_e}{L_b} & \frac{2L_e}{L_b} & \cdots & 1+\frac{2L_e}{L_b} & -1\end{bmatrix}\times\frac{d}{dt}\begin{bmatrix}i_{b1}\\ i_{b2}\\ i_{b3}\\ \cdots\\ i_{bn}\end{bmatrix} \tag{9.189}$$

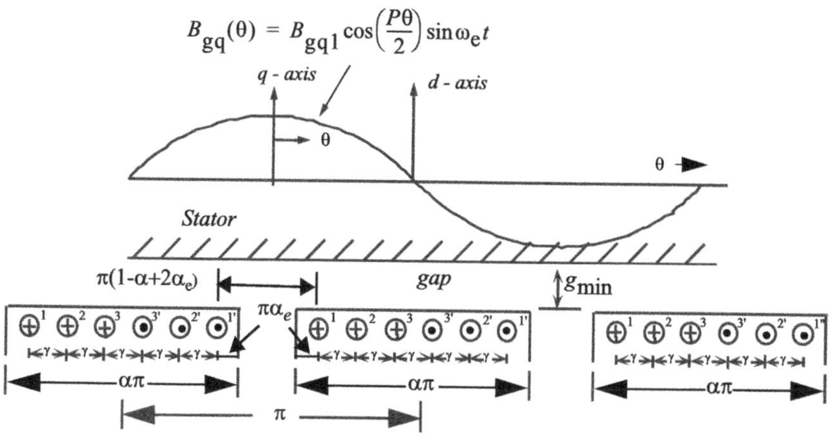

Figure 9.21 Amortisseur bar excitation with interrupted end ring impressed with alternating gap flux density located in the interpolar space (q–axis).

Observe again the related symmetry of the two matrices. One can again proceed to solve for the time derivative of current in much the same manner as equation (9.129). However, in this case, the inverse of the inductance matrix cannot be obtained since the matrix is not square. This difficulty can be eliminated if one introduces the obvious constraint that

$$i_{bn} = -(i_{b1} + i_{b2} + \cdots + i_{b,n-1}) \tag{9.190}$$

Equation (9.189) becomes

$$\frac{d}{dt}\begin{bmatrix} \Phi_{12} \\ \Phi_{23} \\ \Phi_{34} \\ \cdots \\ \Phi_{n,n-1} \end{bmatrix} = \begin{bmatrix} e_1 - e_2 \\ e_2 - e_3 \\ e_3 - e_4 \\ \cdots \\ e_{n-1} - e_n \end{bmatrix} = R_b \begin{bmatrix} 1+\frac{2R_e}{R_b} & -1 & 0 & 0 & \cdots & 0 & 0 \\ \frac{2R_e}{R_b} & 1+\frac{2R_e}{R_b} & -1 & 0 & \cdots & 0 & 0 \\ \frac{2R_e}{R_b} & \frac{2R_e}{R_b} & 1+\frac{2R_e}{R_b} & -1 & \cdots & 0 & 0 \\ \cdots & \cdots & \cdots & \cdots & \cdots & \cdots & 0 \\ \frac{2R_e}{R_b} & \frac{2R_e}{R_b} & \frac{2R_e}{R_b} & \frac{2R_e}{R_b} & \cdots & 1+\frac{2R_e}{R_b} & -1 \\ 1+\frac{2R_e}{R_b} & 1+\frac{2R_e}{R_b} & 1+\frac{2R_e}{R_b} & 1+\frac{2R_e}{R_b} & \cdots & 1+\frac{2R_e}{R_b} & 2+\frac{2R_e}{R_b} \end{bmatrix} \begin{bmatrix} i_{b1} \\ i_{b2} \\ i_{b3} \\ \cdots \\ i_{b,n-1} \end{bmatrix}$$
$$+ L_b \begin{bmatrix} 1+\frac{2L_e}{L_b} & -1 & 0 & 0 & \cdots & 0 & 0 \\ \frac{2L_e}{L_b} & 1+\frac{2L_e}{L_b} & -1 & 0 & \cdots & 0 & 0 \\ \frac{2L_e}{L_b} & \frac{2L_e}{L_b} & 1+\frac{2L_e}{L_b} & -1 & \cdots & 0 & 0 \\ \cdots & \cdots & \cdots & \cdots & \cdots & \cdots & 0 \\ \frac{2L_e}{L_b} & \frac{2L_e}{L_b} & \frac{2L_e}{L_b} & \frac{2L_e}{L_b} & \cdots & 1+\frac{2L_e}{L_b} & -1 \\ 1+\frac{2L_e}{L_b} & 1+\frac{2L_e}{L_b} & 1+\frac{2L_e}{L_b} & 1+\frac{2L_e}{L_b} & \cdots & 1+\frac{2L_e}{L_b} & 2+\frac{2L_e}{L_b} \end{bmatrix} \times \frac{d}{dt}\begin{bmatrix} i_{b1} \\ i_{b2} \\ i_{b3} \\ \cdots \\ i_{b,n-1} \end{bmatrix} \tag{9.191}$$

Again, it is now possible to write that this matrix equation is equivalent to

$$[e_b] = R_b[S_1][i_b] + L_b\frac{d}{dt}[S_2][i_b] \tag{9.192}$$

where

$$[S_1] = [S_2] = [S]$$

The system of equations becomes

$$L_b\frac{d}{dt}[i_b] + R_b[i_b] = [S]^{-1}[e_b] \tag{9.193}$$

Again, the final expression relating the bar currents is not coupled and therefore the currents which flow in the bars are all in phase and differ only in amplitude.

In this case, $[e_b]$ can be determined as follows. The flux linking the circuit defined by bars 1 and 2 is, referring to Figure 9.21,

$$\Phi_{12} = \frac{k_q B_{gq,pk} l_e D_{or}}{2} \sin \omega_e t \int_{\frac{[(1-\alpha+2\alpha_e)]\pi/2}{P}}^{\frac{[(1-\alpha+2\alpha_e)]\pi/2+\gamma}{P}} \cos \frac{P\theta}{2} d\theta \tag{9.194}$$

where $B_{gq,pk}$ is the maximum value of gap flux density when the machine is excited in the q-axis and in which

$$k_q B_{gq,pk} = B_{gq1} \tag{9.195}$$

The quantity D_{or} is taken to be the diameter as measured from the rotor d-axis. Upon working out the integral, and differentiating the result with respect to time,

$$\begin{aligned} \frac{d\Phi_{12}}{dt} = e_1 - e_2 = &\frac{\omega_e k_q B_{gq,pk} l_e D_{or}}{2} \cos \omega_e t \\ &\times \left\{ \sin \frac{1}{2} \left(\left[\frac{1-\alpha}{2} + \alpha_e \right] \pi + \gamma \right) - \sin \frac{1}{2} \left(\left[\frac{1-\alpha}{2} + \alpha_e \right] \pi \right) \right\} \end{aligned} \tag{9.196}$$

Similarly,

$$\begin{aligned} \frac{d\Phi_{23}}{dt} = e_2 - e_3 = &\frac{\omega_e k_q B_{gq,pk} l_e D_{or}}{2} \cos \omega_e t \\ &\times \left\{ \sin \frac{1}{2} \left(\left[\frac{1-\alpha}{2} + \alpha_e \right] \pi + 2\gamma \right) - \sin \frac{1}{2} \left(\left[\frac{1-\alpha}{2} + \alpha_e \right] \pi + \gamma \right) \right\} \end{aligned} \tag{9.197}$$

and so forth until $(d\Phi_{n-1,n})/(dt)$. Equation (9.193) can again be normalized by defining

$$E_b = \frac{\omega_e k_q B_{gq,pk} l_e D_{or}}{2} \tag{9.198}$$

in which case, one can write that,

$$L_b \frac{d}{dt}[i_b] + R_b[i_b] = E_b \cos \omega_e t [S]^{-1} [e_{b,pu}] \tag{9.199}$$

The solution to equation (9.199) is

$$[i_b] = \frac{E_b}{Z_b} \cos(\omega_e t - \phi_b) [S]^{-1} [e_{b,pu}] \tag{9.200}$$

where

$$[e_{\mathrm{b,pu}}] = \begin{bmatrix} \sin\frac{1}{2}\left(\left[\frac{1-\alpha}{2}+\alpha_e\right]\pi+\gamma\right)-\sin\left(\frac{1}{2}\left[\frac{1-\alpha}{2}+\alpha_e\right]\pi\right) \\ \sin\frac{1}{2}\left(\left[\frac{1-\alpha}{2}+\alpha_e\right]\pi+2\gamma\right)-\sin\left(\frac{1}{2}\left[\frac{1-\alpha}{2}+\alpha_e\right]\pi+\gamma\right) \\ \cdots \\ \cdots \\ \sin\frac{1}{2}\left(\left[\frac{1-\alpha}{2}+\alpha_e\right]\pi+(n-1)\gamma\right)-\sin\left(\frac{1}{2}\left[\frac{1-\alpha}{2}+\alpha_e\right]\pi+(n-2)\gamma\right) \end{bmatrix} \tag{9.201}$$

A plot of a typical quadrature axis current distribution and corresponding MMF for the case of six bars per pole is shown in Figure 9.22. Note that the inner bar currents near the center of the pole are, in fact, reversed in polarity compared to their assumed positive direction as denoted by the dots and xs. It is apparent that the outermost bar is stressed most severely when excited with quadrature axis currents. This behavior has important ramifications when the machine is excited with a solid-state power supply and the machine is controlled to operate at high power factor (high quadrature axis component of stator current. Again it can be noted that the MMF produced by the quadrature axis rotor bar currents is related to the bar currents by,

$$\mathcal{F}_{\mathrm{bq}}(\theta) = N_{\mathrm{b}}\left(\int_0^{2\pi}\left[\sum_{k=1}^{N_b} i_{\mathrm{b,pu}}(k)\delta(k)\right]\right)d\theta \tag{9.202}$$

where $\delta(k)$ is the impulse function.

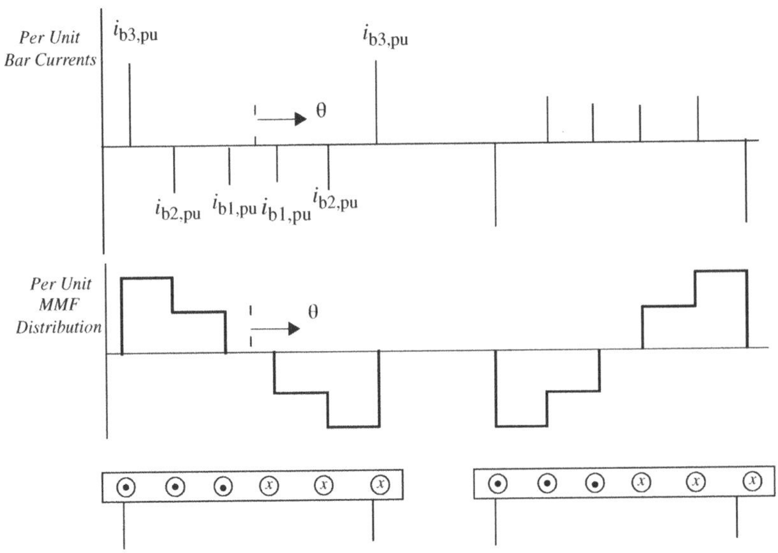

Figure 9.22 Per unit bar currents and resulting MMF distribution for quadrature axis excitation with interrupted end ring.

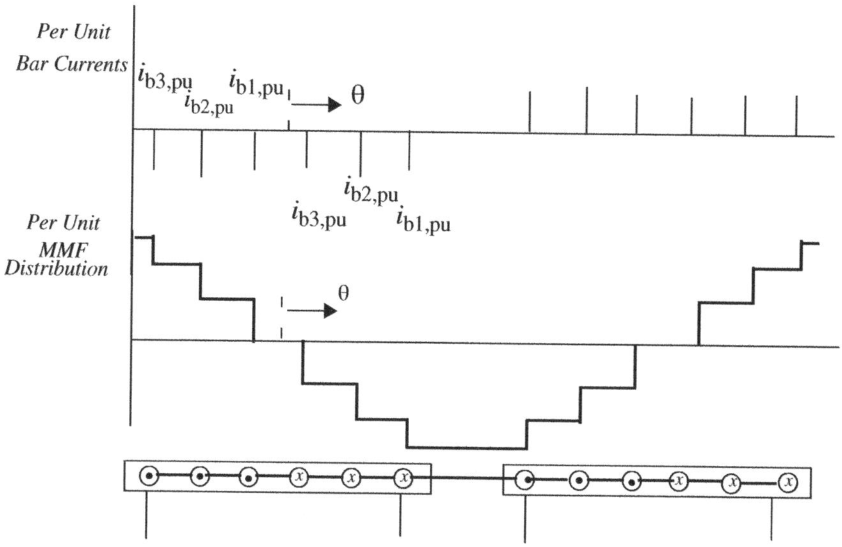

Figure 9.23 Per unit bar currents and resulting MMF distribution for quadrature axis excitation with continuous end ring.

While the specific case of an even number of bars has been chosen, the solution for an odd number of bars is essentially identical, except for the fact that one must set $e_{b,n} = 0$ where $n = (N_b + 1)/2$ denotes the center bar. Another important amortisseur winding configuration is the continuous ring case. Whereas this feature was shown to be unimportant for the d-axis equivalent circuit, this is not the case for the q-axis. Figure 9.23 shows a typical bar current and MMF distribution for this case.

Again, this MMF can be expressed as

$$\mathcal{F}_{bq}(\theta) = I_b \mathcal{F}_{bq,pu}(\theta) \tag{9.203}$$

The fundamental component of $\mathcal{F}_{bq,pu}(\theta)$ is defined as the quadrature axis bar factor k_{bq},

$$\mathcal{F}_{bq,pu1} = \frac{2}{\pi} \int_{-(\alpha\pi)/P}^{(\alpha\pi)/P} \mathcal{F}_{bq,pu}(\theta) \sin\left(\frac{P}{2}\theta\right) d\theta \tag{9.204}$$

where

$$k_{bq} = \left(\frac{\pi}{4}\right) \mathcal{F}_{bq,pu1} \tag{9.205}$$

The non-fundamental component is again considered as a leakage component where

$$L_{qb,h} = k_{lkq} L_b \tag{9.206}$$

and where

$$k_{lkb} = \sqrt{\sum_{h=3,5,7}^{\infty} k_{bq,h}} \tag{9.207}$$

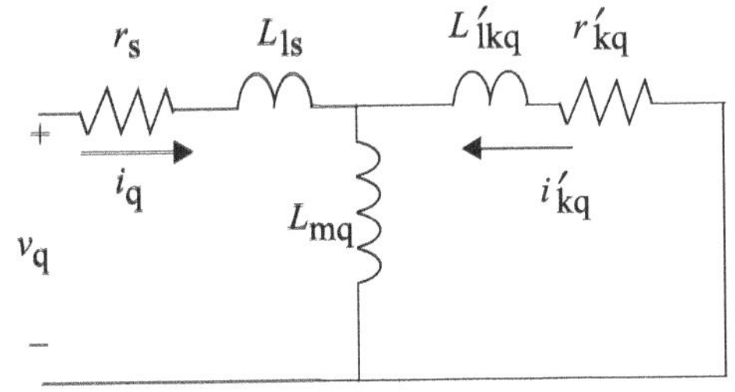

Figure 9.24 Final q–axis equivalent circuit (speed voltages not shown).

Again, the bar parameters can be referred to the stator by the same technique as Section 9.10. The final result is

$$v_q = r_s i_q + L_{ls}\frac{di_q}{dt} + \frac{d\lambda_{mq}}{dt} \tag{9.208}$$

$$0 = r'_{kq} i'_{kq} + L'_{lkq}\frac{di'_{kq}}{dt} + \frac{d\lambda_{mq}}{dt} \tag{9.209}$$

where

$$\lambda_{mq} = L_{mq}\left(i_q + i'_{kq}\right) \tag{9.210}$$

and, in which,

$$r'_{kq} = \frac{3}{2}\left(\frac{k_1 N_s}{k_{bq} N_b}\right)^2 r_b \tag{9.211}$$

$$L'_{lkq} = \frac{3}{2}\left(\frac{k_1 N_s}{k_{bq} N_b}\right)^2 L_{be} \tag{9.212}$$

$$i'_{kq} = \frac{2}{3}\left(\frac{k_{bq} N_b}{k_1 N_s}\right) i_{bq} \tag{9.213}$$

$$L_{mq} = \left(\frac{12}{\pi^2}\right)\frac{(k_1 N_s)^2}{P}\mu_0\frac{\tau_p l_e}{g_e}k_q \tag{9.214}$$

The completed equivalent circuit for the q-axis is shown in Figure 9.24 where again L_{be} should take into account gap leakage derived from the higher harmonic components of air gap flux density as well as leakage flux in the rotor slots and end ring.

9.12 POWER AND TORQUE EXPRESSIONS

Assuming steady-state conditions the power flow into the synchronous machine can be expressed as

$$P_e = \frac{3}{2}(v_q i_q + v_d i_d) \tag{9.215}$$

where the "3/2" term occurs as a result of the transformation from three-phase to two-phase variables as discussed in Section 2.10. Neglecting the stator resistance of

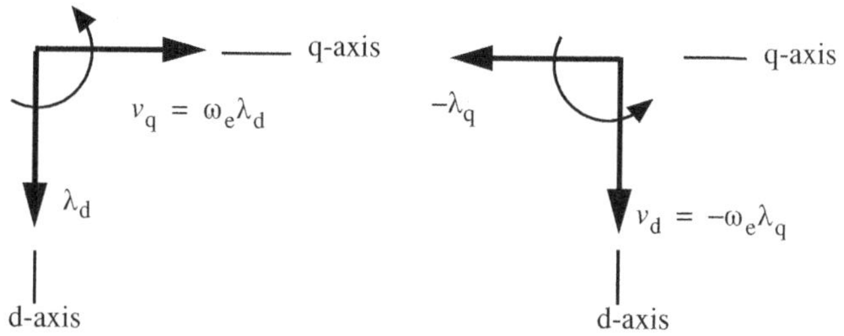

Figure 9.25 Synchronously rotating d-q voltage vectors and corresponding flux linkage vectors.

the *d*- and *q*-axis stator circuits for the moment, and assuming that the axes defining the *d–q* quantities are not rotating, the voltages v_d and v_q become equal to the time rate of change of flux linkage in the d- and q-axes respectively.

Assuming three-phase sinusoidal excitation, the time rate of change of a sinusoidal flux linkage (voltage) leads the corresponding flux linkage quantity by 90 electrical degrees as shown in Figure 9.25.

$$P_e = \frac{3}{2}(\omega_e\lambda_d i_q - \omega_e\lambda_q i_d) \tag{9.216}$$

Since the electrical and mechanical angular frequency are related by $\omega_e = (P/2)\omega_r$,

$$P_e = \frac{3}{2}\left(\frac{P}{2}\right)\omega_r(\lambda_d i_q - \lambda_q i_d) \tag{9.217}$$

In the steady-state, since

$$\lambda_q = (L_{ls} + L_{mq})I_q \tag{9.218}$$

$$\lambda_d = (L_{ls} + L_{md})(I_d + I_f') \tag{9.219}$$

the torque can be written as

$$T_e = \frac{P_e}{\omega_r} = \frac{3}{2}\left(\frac{P}{2}\right)[L_{md}(I_d + I_f')I_q - L_{mq}I_qI_d] \tag{9.220}$$

The torque consists of two terms corresponding to the reaction torque (field torque) and the reluctance torque, or

$$T_e = T_{e,fd} + T_{e,rel} \tag{9.221}$$

wherein

$$T_{e,fd} = \frac{3}{2}\left(\frac{P}{2}\right)L_{md}I_qI_f' \tag{9.222}$$

$$T_{e,rel} = \frac{3}{2}\left(\frac{P}{2}\right)(L_{md} - L_{mq})I_dI_q \tag{9.223}$$

in which

$$L_{md} = \frac{12}{\pi}\left(\frac{k_1N_s}{P}\right)^2\left(\mu_0\frac{D_{is}l_e}{g_e}\right)k_d \tag{9.224}$$

and

$$L_{mq} = \frac{12}{\pi}\left(\frac{k_1 N_s}{P}\right)^2 \left(\mu_0 \frac{D_{is} l_e}{g_e}\right) k_q \tag{9.225}$$

From equation (8.25), the amplitude of the sheet current for a three-phase system is

$$K_{s1} = \left(\frac{P}{D_{is}}\right)\mathcal{F}_{s1} = \frac{6}{\pi}\left(\frac{k_1 N_s}{D_{is}}\right) I_s \tag{9.226}$$

The corresponding sheet current for a two-phase equivalent system, is, for the q component

$$K_{q1} = \frac{6}{\pi}\left(\frac{k_1 N_s}{D_{is}}\right) I_q \tag{9.227}$$

The two sheet currents are related by

$$K_{q1} = K_{s1}\cos\varepsilon = \frac{6}{\pi}\left(\frac{k_1 N_s}{D_{is}}\right) I_s \cos\varepsilon \tag{9.228}$$

where ε is the MMF angle defined in Chapter 8 as the spatial angle between the amplitude of the stator current sheet K_{s1} and the magnet (or in this case the field winding) flux.

Substituting equations (9.224) and (9.227) into equation (9.222), the field torque is

$$T_{e,fd} = \frac{3}{2}\left(\frac{P}{2}\right)\left[\frac{12}{\pi}\left(\frac{k_1 N_s}{P}\right)^2 \left(\mu_0 \frac{D_{is} l_e}{g_e}\right) k_d\right]\left(\frac{K_{q1} I'_f}{\frac{6}{\pi}\left(\frac{k_1 N_s}{D_{is}}\right)}\right) \tag{9.229}$$

Utilizing equation (9.228)

$$T_{e,fd} = \frac{3}{2}\left(\frac{P}{2}\right)\left[\frac{12}{\pi}\left(\frac{k_1 N_s}{P}\right)^2 \left(\mu_0 \frac{D_{is} l_e}{g_e}\right) k_d\right]\left(\frac{K_{s1} I'_f \cos\varepsilon}{\frac{6}{\pi}\left(\frac{k_1 N_s}{D_{is}}\right)}\right) \tag{9.230}$$

Upon simplifying

$$T_{e,fd} = \frac{3}{2}\left(\frac{k_1 N_s}{P}\right)\left(\mu_0 \frac{D_{is}^2 l_e}{g_e}\right) k_d (K_{s1} I'_f \cos\varepsilon) \tag{9.231}$$

The air gap flux density produced by the field current is

$$B_{gf1} = \frac{\mu_0 \mathcal{F}_{f1}}{g_e} = \mu_0 \frac{3}{2}\left(\frac{4}{\pi}\right)\left(\frac{k_1 N_s}{P}\right) k_d \left(\frac{I'_f}{g_e}\right) \tag{9.232}$$

Using this expression to replace I'_f in equation (9.230) yields

$$T_{e,fd} = \left(\frac{\pi}{4}\right)(D_{is}^2 l_e) K_{s1} B_{gf1} \cos\varepsilon \tag{9.233}$$

It is important to note that this result is essentially the same as for the induction machine, Equation (6.16) and for the permanent magnet machine, Equation (9.55) and as such forms the basis for commencing the design of almost any AC machine.

The second term in the torque expression, equation (9.221), is the torque produced by the saliency of the machine and is typically much smaller than the reaction torque that it is frequently neglected. The reluctance torque expression can be expanding by noting that

$$L_{\mathrm{md}} - L_{\mathrm{mq}} = \frac{12}{\pi}\left(\frac{k_1 N_{\mathrm{s}}}{P}\right)^2 \mu_0 \left(\frac{D_{\mathrm{is}} l_{\mathrm{e}}}{g_{\mathrm{e}}}\right)(k_{\mathrm{d}} - k_{\mathrm{q}}) \tag{9.234}$$

so that

$$T_{\mathrm{e,rel}} = \frac{3}{2}\left(\frac{P}{2}\right)\frac{12}{\pi}\left(\frac{k_1 N_{\mathrm{s}}}{P}\right)^2 \mu_0 \left(\frac{D_{\mathrm{is}} l_{\mathrm{e}}}{g_{\mathrm{e}}}\right)(k_{\mathrm{d}} - k_{\mathrm{q}}) I_{\mathrm{d}} I_{\mathrm{q}} \tag{9.235}$$

Furthermore

$$K_{\mathrm{q1}} = K_{\mathrm{s1}} \sin\varepsilon = \frac{6}{\pi}\left(\frac{k_1 N_{\mathrm{s}}}{D_{\mathrm{is}}}\right) I_{\mathrm{q}} \tag{9.236}$$

$$B_{gd1} = B_{g1} \cos\varepsilon = \mu_0 \frac{6}{\pi}\left(\frac{k_1 N_{\mathrm{s}}}{P}\right)\frac{I_{\mathrm{d}}}{g_{\mathrm{e}}} \tag{9.237}$$

from which it is readily determined that

$$T_{\mathrm{e,rel}} = \frac{\pi}{4}\left(D_{\mathrm{is}}^2 l_{\mathrm{e}}\right)(k_{\mathrm{d}} - k_{\mathrm{q}}) B_{\mathrm{g1}} K_{\mathrm{s1}} \left(\frac{\sin 2\varepsilon}{2}\right) \tag{9.238}$$

Here, the gap flux density B_{g1} arises from the d-axis component of the armature current itself and not the field current. This result shows that the reluctance torque is also not dependent on the number of poles nor the number of turns.

An alternative method to specify the reluctance torque is to express it solely in terms of the current sheet current density. In this case, in addition to Equation (9.236),

$$K_{\mathrm{d1}} = K_{\mathrm{s1}} \cos\varepsilon = \frac{6}{\pi}\left(\frac{k_1 N_{\mathrm{s}}}{D_{\mathrm{is}}}\right) I_{\mathrm{d}} \tag{9.239}$$

in which case the reluctance torque can be expressed as

$$T_{\mathrm{e,rel}} = \frac{\pi}{4}\left(\frac{1}{P}\right)\left(\mu_0 \frac{D_{\mathrm{is}}^3 l_{\mathrm{e}}}{g_{\mathrm{e}}}\right)(k_{\mathrm{d}} - k_{\mathrm{q}}) K_{\mathrm{s1}}^2 \left(\frac{\sin 2\varepsilon}{2}\right) \tag{9.240}$$

While the reluctance torque can often be neglected in the design of a synchronous machine, these equations form the starting point for the design of a synchronous reluctance machine which develops its torque solely as a result of its saliency.

9.13 MAGNETIC SHEAR STRESS

In general, the design of a wound field synchronous machine proceeds in a very similar manner to induction motor design with limits on surface current and current density set by the values discussed in Sections 6.8 and 6.9. Curves for estimates of

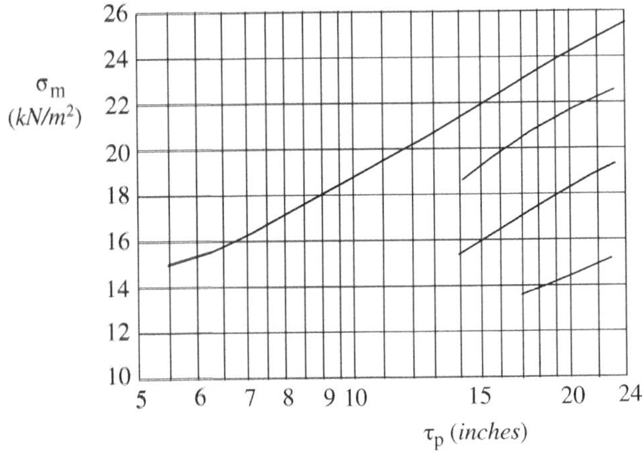

Figure 9.26 Magnetic shear stress σ_m of wound field salient pole synchronous motors in *kilo-newtons/square meter* as a function of pole pitch τ_p (Adapted from [3]).

the magnetic shear stress and aspect ratio for wound field synchronous machines have also been provided by Liwschitz-Garik and are plotted in Figures 9.26 and 9.27. The magnetic shear stress varies from 11 to 22 kN/m^2 whereas the variation was only 5–19 for a squirrel-cage induction machine. It is mentioned by Liwschitz-Garik that σ_m may reach 34 kN/m^2 for large salient pole synchronous machines without water cooling. Most often, synchronous machines have smaller aspect ratios meaning that the machine has a larger diameter and shorter length than a similarly rated induction machine. A partial answer to this trend is the need to provide room for the pole structure and associated field winding on the rotor of the synchronous machine.

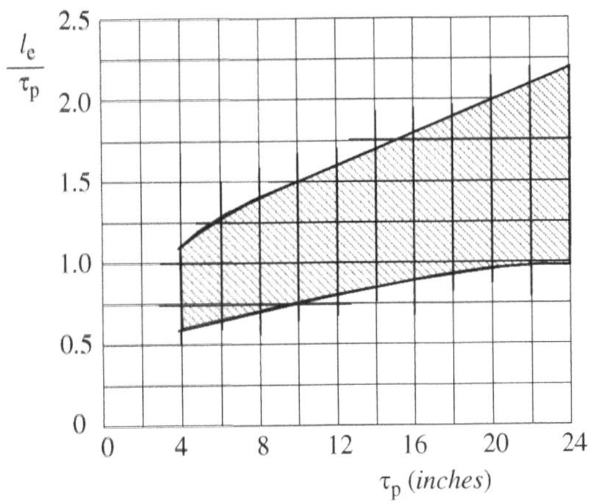

Figure 9.27 Ratio of equivalent core length to pole pitch (aspect ratio) of a wound field salient-pole synchronous machine as a function of the pole pitch [3].

9.14 FIELD CURRENT PROFILE

As opposed to induction and permanent magnet machines, the wound field synchronous machine is a two-port, rather than a single-port, system. If the machine is designed to develop a specified torque, as is generally the case, the desired value can be reached by a multitude of combinations of armature and field current since the torque is a result of the product of these two variables. In general, the d-axis component of armature current is determined by the power factor requirements. Synchronous generators are typically designed to develop rated terminal voltage at unit power factor to completely satisfy the magnetizing requirement of the machine or even a leading power factor of 0.9 or 0.8 to help provide magnetizing current for inductive loads.

The phasor equivalent circuit of Figure 9.28 can be used to solve for the necessary components of current (and thus sheet current density) to regulate the terminal voltage and power factor. In this circuit, the power flow into (or out of) the voltage source $\tilde{E}_i = \tilde{X}_{md} i'_f$ represents the field power (field or reaction torque times speed) and the power dissipated (or produced) by the fictitious resistor R_{rel} denotes the power corresponding to the reluctance torque. Upon solving the circuit, the component of stator current in phase with $\tilde{E}_i$ corresponds to I_q and the orthogonal component to I_d. The derivation of this circuit is contained in Reference [4].

When operating as a motor from a fixed frequency and amplitude voltage source such as the utility grid, the requirements are much the same as for the generator, normally requiring unity power factor operation along with a specified electromagnetic torque. However, for modern day motor drives, the issue of power factor is of lesser importance than motor efficiency. As the motor drive operates at various speeds and torque conditions, it is desirable to control the power sources (inverter and field supply) to minimize the losses. Assume for simplicity that the reaction torque is expressed by

$$T_{e,fd} = K_T I_q I'_f \tag{9.241}$$

and that the corresponding copper loss is

$$P_{cu} = \frac{3}{2}[I_q^2 R_s + (I'_f)^2 R'_f] \tag{9.242}$$

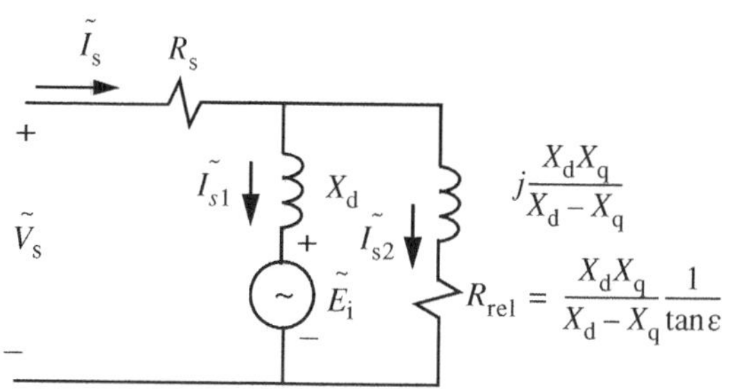

Figure 9.28 Per phase equivalent circuit of a salient pole synchronous machine.

In order to determine the minimum value of copper loss

$$\frac{\partial P_{cu}}{\partial I_q} = 0 = \frac{\partial}{\partial I_q}\left[I_q^2 R_s + \left(\frac{T_{e,fd}}{K_T I_q}\right)^2 R_f'\right] \tag{9.243}$$

$$0 = 2I_q R_s + \left(\frac{T_{e,fd}}{K_T}\right)^2\left(-\frac{2}{I_q^3}\right)R_f' \tag{9.244}$$

Upon solving

$$I_q = \left(\sqrt{\frac{T_{e,fd}}{K_T}}\right)\left(\sqrt[4]{\frac{R_f'}{R_s}}\right) \tag{9.245}$$

$$I_f' = \left(\sqrt{\frac{T_{e,fd}}{K_t}}\right)\left(\sqrt[4]{\frac{R_s}{R_f'}}\right) \tag{9.246}$$

Thus, if I_d is set to zero, minimum copper losses will occur if the following relationship is maintained for any value of torque or speed,

$$\frac{I_q}{I_f'} = \sqrt{\left(\frac{R_f'}{R_s}\right)} \tag{9.247}$$

Recalling that the prime quantities are referred to stator turns, it is important to remember that the actual values of field resistance and current are, from equations (9.180) and (9.183)

$$R_f = \frac{2}{3}\left(\frac{N_f}{k_1 N_s}\right)^2 R_f' \tag{9.248}$$

and

$$I_f = \frac{3}{2}\frac{(k_1 N_s)}{(N_f)} I_f' \tag{9.249}$$

Thus, in terms of actual values, equation (9.247) can be written as

$$\frac{I_q}{I_f} = \frac{3}{2}\left(\frac{k_1 N_s}{N_f}\right)^2 \sqrt{\frac{R_f}{R_s}} \tag{9.250}$$

The d-axis component of current can, of course, also be increased to enhance the torque by utilizing the relatively small reluctance component but, if efficiency is to be maximized, the rotor speed must be sufficiently great such that

$$\frac{3}{2} I_d^2 R_s < \omega_r T_{e,rel} \tag{9.251}$$

9.15 CONCLUSION

This lengthy chapter has again only touched the surface concerning the design of synchronous machines. Clearly, the trade-offs between power density and cooling capability will again play a key role in arriving at a satisfactory design. The material provided in Chapter 9 will hopefully serve as a reasonable starting point in this trade-off. A proper treatment of this topic is beyond the scope of this book. However, the reader is referred to valuable information in the English language in References [5] and [6]. Regrettably, the definitive information on this subject is in the German literature [7], and others.

REFERENCES

[1] R. W. Wieseman, "Graphical determination of magnetic fields. practical applications to salient-pole synchronous machine design", *Trans. AIEE*, vol. 46 1927, pp. 141–148.

[2] D. Ginsberg, A. L. Jokl, and Blum "Calculation of no-load wave shape of salient-pole a-c generators", *Trans. AIEE*, part III Power Apparatus and Systems, October 1953, vol. 72, no. 2, pp. 974–980.

[3] M. Liwschitz-Garik and C. C. Whipple, *Electric Machinery*, vol. 2 *AC Machines*, 2nd edition, Van Nostrand Publishers, 1961.

[4] T. A. Lipo, *Analysis of Synchronous Machines*, 2nd edition, CRC Press, 2012.

[5] M. G. Say, *The Performance and Design of Alternating Current Machines*, 3rd edition, Pitman, 1958.

[6] J. H. Walker, *Large Synchronous Machines*, Clarendon Press, Oxford, 1981.

[7] W. Schuisky, *Berechnung Electrischer Maschinen*, Springer-Verlag, Vienna, 1960 (in German).

CHAPTER 10

FINITE-ELEMENT SOLUTION OF MAGNETIC CIRCUITS

APPROXIMATE FLUX PATH calculations such as those used thus far throughout this book are indispensable for rapid, reasonably accurate calculations for machine design. Since the permeances of each cell are related directly to geometrical constants of the machine (slot height, tooth width, core depth, and so forth) a good understanding of the relationship between the machine dimensions and their effect on the motor parameters (resistances and inductances) is obtained. Another method of analysis which has now become automated to the point of being useful in machine design is the method of finite elements. The method offers improved accuracy and almost an unlimited flexibility in the geometrical shape, boundary conditions, and material properties in different regions of the problem. However, it still remains the task of the analyst to relate this field plot to the circuit parameters of the machine.

A detailed description of the finite-element method is a demanding subject which is, in reality, beyond the scope of this book. However, for dedicated students of machine design, a solution simply provided by a "black box" in the form of a finite-element computation package is not very intellectually satisfying. Such students should be interested in what is going on "under the hood" of a modern finite-element program. In this spirit, the basic method of first-order triangular finite elements will be presented to illustrate its use in both two-dimensional linear and nonlinear magnetostatic problems.

10.1 FORMULATION OF THE TWO-DIMENSIONAL MAGNETIC FIELD PROBLEM

Consider first the magnetic field region shown in Figure 10.1. This region is comprised of air, iron, and current-carrying conductors in some specific but generalized planar (two-dimensional) region. The plane will be assumed to have x and y coordinates. The "input" to the field is assumed to be a component of current (current density) normal to the plane (z-direction). In most electric machine applications the domain may be subdivided into three principal sub- regions as shown, for which different field equations apply, corresponding to air, iron, and conductor sub-regions.

Introduction to AC Machine Design, First Edition. Thomas A. Lipo.

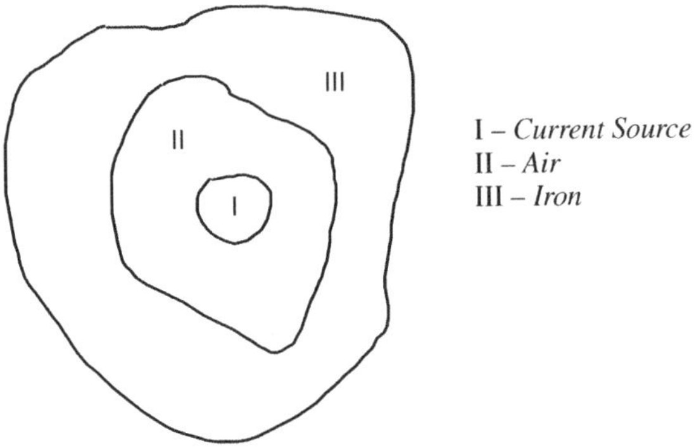

Figure 10.1 Two-dimensional region containing air, current sources and magnetic material.

The relevant field formulation equations are, from Maxwell's equations

$$\nabla \times \boldsymbol{H} = \boldsymbol{J} \tag{10.1}$$

$$\nabla \bullet \boldsymbol{B} = 0 \tag{10.2}$$

and, from the properties of the material in the sub-regions,

$$\boldsymbol{B} = \mu_r \mu_0 \boldsymbol{H} \tag{10.3}$$

where:

$\boldsymbol{H}$: Field intensity vector in ampere-turns / meter

$\boldsymbol{J}$: Current density vector in amperes / square meter

$\boldsymbol{B}$: Magnetic flux density vector in webers / square meter (teslas)

μ_r: Relative permeability

μ_0: Permeability of "free space"

It is useful to reintroduce the magnetic vector potential, $\boldsymbol{A}$ having the dimensions of webers/meter and which satisfies the relation:

$$\boldsymbol{B} = \nabla \times A \tag{10.4}$$

In cartesian coordinates, this equation may be written in the alternate form

$$B = det\begin{bmatrix} \boldsymbol{u}_x & \boldsymbol{u}_y & \boldsymbol{u}_z \\ \dfrac{\partial}{\partial x} & \dfrac{\partial}{\partial y} & \dfrac{\partial}{\partial z} \\ A_x & A_y & A_z \end{bmatrix} \tag{10.5}$$

where A_x, A_y, A_z are the x, y, z components of $\boldsymbol{A}$ in the right-handed Cartesian Coordinate System, and $\boldsymbol{u}_x$, $\boldsymbol{u}_y$, $\boldsymbol{u}_z$ the respective unit vectors.

Recall from Chapter 1, equation (1.24), that the vector potential is defined as

$$\boldsymbol{A} = \frac{\mu_0}{4\pi}\int_V \frac{\boldsymbol{J}}{R}dV \tag{10.6}$$

If the current density is assumed to be only z-directed, then, clearly, the vector potential A can only have a z-directed component as well.

Substituting for $\boldsymbol{H}$ in equation (10.1), in terms of $\boldsymbol{B}$, $\mu_r\mu_0$, and $\boldsymbol{A}$

$$\nabla \times \boldsymbol{H} = \nabla \times \left(\frac{1}{\mu_r\mu_0} \boldsymbol{B} \right) = \nabla \times \left(\frac{1}{\mu_r\mu_0} \nabla \times \boldsymbol{A} \right) = \boldsymbol{J} \tag{10.7}$$

Expanding this equation, one has

$$\boldsymbol{J} \Leftarrow \nabla \times \left(\frac{1}{\mu_r\mu_0} det \begin{bmatrix} \boldsymbol{u}_x & \boldsymbol{u}_y & \boldsymbol{u}_z \\ \frac{\partial}{\partial x} & \frac{\partial}{\partial y} & \frac{\partial}{\partial z} \\ 0 & 0 & A_z \end{bmatrix} \right) \tag{10.8}$$

or

$$\boldsymbol{J} = \nabla \times \left\{ \frac{1}{\mu_r\mu_0} \left[\frac{\partial A_z}{\partial y} \boldsymbol{u}_x - \frac{\partial A_z}{\partial x} \boldsymbol{u}_y \right] \right\} \tag{10.9}$$

Equation (10.9) can again be expanded as

$$\boldsymbol{J} = det \begin{bmatrix} \boldsymbol{u}_x & \boldsymbol{u}_y & \boldsymbol{u}_z \\ \frac{\partial}{\partial x} & \frac{\partial}{\partial y} & \frac{\partial}{\partial z} \\ \frac{1}{\mu_r\mu_0} \left(\frac{\partial A_z}{\partial y} \right) & \frac{-1}{\mu_r\mu_0} \left(\frac{\partial A_z}{\partial x} \right) & 0 \end{bmatrix} \tag{10.10}$$

$$= -\left[\frac{\partial}{\partial x} \left(\upsilon \frac{\partial A_z}{\partial x} \right) + \frac{\partial}{\partial y} \left(\upsilon \frac{\partial A_z}{\partial y} \right) \right] \boldsymbol{u}_z + \frac{\partial}{\partial z} \left(\upsilon \frac{\partial A_z}{\partial x} \right) \boldsymbol{u}_x + \frac{\partial}{\partial z} \left(\upsilon \frac{\partial A_z}{\partial y} \right) \boldsymbol{u}_y \tag{10.11}$$

where ν, the reluctivity, replaces $1/(\mu_r\mu_0)$ for convenience.

In equation (10.11) above, it is evident that the last two terms are zero since under these conditions, the current density $\boldsymbol{J}$ has only a z-directed component $J_z\ \boldsymbol{u}_z$. Dropping the unit vector $\boldsymbol{u}_z$ for simplicity, the two-dimensional field equation can now be written in scalar form as

$$\frac{\partial}{\partial x} \left(\upsilon \frac{\partial A_z}{\partial x} \right) + \frac{\partial}{\partial y} \left(\upsilon \frac{\partial A_z}{\partial y} \right) = -J_z \tag{10.12}$$

Equation (10.12) is called the quasi-Poisson equation for the linear or the nonlinear field problem since υ can either be a constant or a function of A_z. Henceforth, the subscript z will be dropped for this particular case and equation (10.12) written simply as

$$\frac{\partial}{\partial x} \left(\upsilon \frac{\partial A}{\partial x} \right) + \frac{\partial}{\partial y} \left(\upsilon \frac{\partial A}{\partial y} \right) = -J \tag{10.13}$$

The reluctivity υ can be, in some cases, orthotropic, that is, have two orthogonal components. This can be considered as a special case of anisotropic reluctivity. In this case, equation (10.13) can be written

$$\frac{\partial}{\partial x}\left(\upsilon_x \frac{\partial A}{\partial x}\right) + \frac{\partial}{\partial y}\left(\upsilon_y \frac{\partial A}{\partial y}\right) = -J \tag{10.14}$$

In the various regions shown in Figure 10.1, the following field equations uniquely define the magnetic field subject to the appropriate boundary conditions.

Region I: (Source or current region). Here the permeability is that of free space and the Poisson equation applies directly, that is,

$$\upsilon_o \frac{\partial^2 A}{\partial x^2} + \upsilon_o \frac{\partial^2 A}{\partial y^2} = -J \tag{10.15}$$

Region II: (Laplacian region or current-free air region)

$$\upsilon_o \frac{\partial^2 A}{\partial x^2} + \upsilon_o \frac{\partial^2 A}{\partial y^2} = 0 \tag{10.16}$$

Region III: (Current-free iron region) In this case the so-called quasi-Laplace equation applies, so that

$$\frac{\partial}{\partial x}\left(\upsilon \frac{\partial A}{\partial x}\right) + \frac{\partial}{\partial y}\left(\upsilon \frac{\partial A}{\partial y}\right) = 0 \tag{10.17}$$

Equations (10.1), (10.2), and (10.3) can clearly be expressed by the single equation defined by equation (10.12).

10.2 SIGNIFICANCE OF THE VECTOR POTENTIAL

For a two-dimensional magnetostatic problem, the vector potential concept is a useful approach which reduces the number of unknown functions from two (B_x, and B_y) to one (A_z). The significance of A can be illustrated by considering the magnetic flux crossing a surface S. From Gauss' law,

$$\Phi_s = \int_s \boldsymbol{B} \bullet d\boldsymbol{S}$$

and from equation (10.4)

$$\Phi_s = \int_s \nabla \times \boldsymbol{A} \bullet d\boldsymbol{S}$$

Applying Stokes theorem,

$$\Phi_s = \int_s \nabla \times \boldsymbol{A} \bullet d\boldsymbol{S} = \int_l \boldsymbol{A} \bullet d\boldsymbol{l}$$

where the line integral is along the path l which encloses S. In the two-dimensional case, $\boldsymbol{A}$ has only a z component. Therefore, the flux Φ_s crossing a surface of length l_z perpendicular to the x,y plane and spanning from point x_1, y_1 to x_2, y_2 in the two-dimensional case is simply

$$\Phi_s = l_z[A(x_1, y_1) - A(x_2, y_2)] \tag{10.18}$$

where l_z represents the depth of the body in the direction normal to the x,y plane. Hence, the vector potential itself physically represents the flux per unit length crossing a given area.

It is important to also take note that since

$$\boldsymbol{B} = \nabla \times \boldsymbol{A}$$

and

$$\boldsymbol{A} = A_z(x, y)\boldsymbol{u}_z$$

then, if one lets $A_z\ (x,y) = A$ for the two-dimensional case

$$B_x = \frac{\partial A}{\partial y} \tag{10.19}$$

And

$$B_y = -\frac{\partial A}{x} \tag{10.20}$$

so that

$$|\boldsymbol{B}| = B = \sqrt{\left(\frac{\partial A}{\partial x}\right)^2 + \left(\frac{\partial A}{\partial y}\right)^2} \tag{10.21}$$

10.3 THE VARIATIONAL METHOD

Numerous methods can be used for the solution of the quasi-Poisson equation. For example, it is possible to use

1. numerical relaxation,
2. direct discretization of the differential operator by the finite difference method,
3. variational method,
4. integral equation solution method.

In the present discussion, only the variational approach will be discussed and related to the finite-element discretization scheme. The variational method consists of constructing a set of approximate functions which minimize the energy stored in the system. It is possible to show that the approximating function is the required solution to the differential equation which satisfies the given boundary conditions.

The advantages of the variational method are

1. The approximating function can be a higher order polynomial which includes more terms of a Taylor's series than the corresponding finite difference approximation.
2. It can be shown that the boundary conditions are implicit in the energy formulation used in the variational method.
3. It is mathematically a rigorous method which leads to a convergent, unique solution.
4. Since the energy stored in the system is the prime quantity, any set of approximating functions or a linear combination of functions which minimize the stored energy will necessarily lead to a unique solution.

10.4 NONLINEAR FUNCTIONAL AND CONDITIONS FOR MINIMIZATION

In nearly every engineering application, there exists a variational formulation corresponding to the partial differential equations of the corresponding field problem. That is to say that there exist certain scalar quantities, for example, energy, which must be minimized if a given field is to exist and the field differential equations are the conditions for minimization. The energy function can be expressed in terms of the functions representing the vector potential and the current density. Hence, the energy representation is termed a *functional*. The functional representation of energy is clearly a scalar since it is not direction-dependent.

From Chapter 1, it has been stated that if the permeability of the material is constant, the energy density corresponding to the magnetic field at any point in a system can be written as

$$\begin{aligned} w_k &= \frac{1}{2}\boldsymbol{B}\bullet\boldsymbol{H} \\ &= \frac{1}{2}\mu_r\mu_0|\boldsymbol{H}|^2 \\ &= \frac{1}{2}\frac{|\boldsymbol{B}|^2}{\mu_r\mu_0} \end{aligned}$$

In the general case, if the material is nonlinear

$$w_k = \int_0^B \boldsymbol{H}\bullet d\boldsymbol{B} \tag{10.22}$$

The energy expressed by equation (10.22) has units of J/m^3. The total energy stored over a volume V would be obtained by integrating equation (10.22) over the volume of interest.

In general, equation (10.22) represents only that portion of the system energy which is stored in the magnetic field. A more accurate expression for the energy can be derived from Maxwell's equations. Consider the expression $\nabla \bullet (\boldsymbol{E} \times \boldsymbol{H})$, which may be expanded by means of a vector identity to form,

$$\nabla \bullet (\boldsymbol{E} \times \boldsymbol{H}) = \boldsymbol{H} \bullet (\nabla \times \boldsymbol{E}) - \boldsymbol{E} \bullet (\nabla \times \boldsymbol{H}) \tag{10.23}$$

However, from Maxwell's equations, neglecting displacement currents

$$\nabla \times \boldsymbol{H} = \boldsymbol{J} \tag{10.24}$$

and

$$\nabla \times \boldsymbol{E} = -\frac{\partial \boldsymbol{B}}{\partial t} \tag{10.25}$$

Equation (10.23) can be put in the form

$$\nabla(\boldsymbol{E} \times \boldsymbol{H}) = -\left(\boldsymbol{H} \bullet \frac{\partial \boldsymbol{B}}{\partial t}\right) - \boldsymbol{J} \bullet \boldsymbol{E} \tag{10.26}$$

Upon integrating this expression over a volume designating the problem of interest,

$$\iiint_V \nabla \bullet (\boldsymbol{E} \times \boldsymbol{H}) dV = -\iiint_V \boldsymbol{H} \bullet \frac{\partial \boldsymbol{B}}{\partial t} dV - \iiint_V \boldsymbol{J} \bullet \boldsymbol{E} dV \tag{10.27}$$

Finally, from Gauss' law, equation (1.35) after some rearranging becomes,

$$\oint_S \boldsymbol{E} \times \boldsymbol{H} \bullet d\boldsymbol{S} + \iiint_V \boldsymbol{E} \bullet \boldsymbol{J} dV + \iiint_V \boldsymbol{H} \bullet \frac{\partial \boldsymbol{B}}{\partial t} dV = 0 \tag{10.28}$$

where the surface S encloses the volume V. The first term in this expression represents the power which is entering or escaping from the volume V. The surface S can be considered as being composed of three surfaces: first the x,y plane under investigation; second, an x,y plane parallel to the first plane separated by a distance l; and third, a cylindrical shaped body connecting these two planes. Because of the z-directed current density vector $\boldsymbol{J}$, both $\boldsymbol{E}$ and $\boldsymbol{H}$ vectors lie in the two parallel planes and do not have z-directed components. The cross product of $\boldsymbol{E} \times \boldsymbol{H}$ thus has only z-directed components. The vector corresponding to $\boldsymbol{E} \times \boldsymbol{H}$ has the same direction in both planes whereas the unit normal to the two planes is oppositely directed. Finally, since $\boldsymbol{E} \times \boldsymbol{H}$ does not have a component normal to the cylindrical surface this term is identically zero. Thus one can state that,

$$\begin{aligned}\oint_S (\boldsymbol{E} \times \boldsymbol{H}) \bullet d\boldsymbol{S} = &\int_{(x,y,z=0\text{ plane})} (\boldsymbol{E} \times \boldsymbol{H}) \bullet \boldsymbol{u}_z dS + \int_{(x,y,z=l\text{ plane})} (\boldsymbol{E} \times \boldsymbol{H}) \bullet (-\boldsymbol{u}_z)\, dS \\ &+ \int_{\text{cylinder}} (\boldsymbol{E} \times \boldsymbol{H}) \bullet (-\boldsymbol{u}_r)\, dS\end{aligned} \tag{10.29}$$

$$= P_{\text{surf}}(x, y) - P_{\text{surf}}(x, y) + 0 = 0 \tag{10.30}$$

Equation (10.28) reduces to

$$\iiint_V \boldsymbol{E} \bullet \boldsymbol{J} dV + \iiint_V \boldsymbol{H} \bullet \frac{\partial \boldsymbol{B}}{\partial t} dV = 0 \tag{10.31}$$

The second term of this expression is clearly related to the time rate of change of energy stored in the magnetic field. It is shown in the appendix at the end of this chapter that this term is precisely the time derivative of the volume integral of equation (10.22). That is,

$$P_{\text{mag}} = \iiint_V \boldsymbol{H} \bullet \frac{\partial \boldsymbol{B}}{\partial t} dV = \frac{\partial}{\partial t} \iiint_V \left[\int_0^B \boldsymbol{H} \bullet d\boldsymbol{B} \right] dV \tag{10.32}$$

The first term in equation (10.31) represents the electrical power entering the system, P_e. The input power may be expressed in terms of $\boldsymbol{A}$ rather than $\boldsymbol{B}$ by using the definition of $\boldsymbol{A}$

$$\boldsymbol{B} = \nabla \times \boldsymbol{A} \tag{10.33}$$

together with Faraday's law,

$$\nabla \times \boldsymbol{E} = -\frac{\partial}{\partial t} \boldsymbol{B} \tag{10.34}$$

Combining these two equations

$$\nabla \times \boldsymbol{E} = -\frac{\partial}{\partial t} (\nabla \times \boldsymbol{A}) \tag{10.35}$$

Thus

$$\boldsymbol{E} = -\frac{\partial}{\partial t} (\boldsymbol{A} - \nabla \phi) \tag{10.36}$$

The quantity $\nabla \phi$ is the gradient of a potential function which must be added to the expression to be completely general. This is due to the fact that for any scalar potential ϕ

$$\nabla \times (\nabla \phi) = 0 \tag{10.37}$$

The quantity acts somewhat as a constant of integration in integral calculus. If it is assumed that there is negligible electrostatic potential, equation (10.36) becomes

$$\boldsymbol{E} = -\frac{\partial}{\partial t} \boldsymbol{A} \tag{10.38}$$

Substituting this result into the expression for electric power input P_e gives

$$P_e = -\iiint_V \frac{\partial \boldsymbol{A}}{\partial t} \cdot \boldsymbol{J} dV \tag{10.39}$$

Assuming co-directional $\boldsymbol{A}$ and $\boldsymbol{J}$ and applying the same relationship as used for $\boldsymbol{B}$ and $\boldsymbol{H}$, equation (10.132), results in

$$P_e = -\frac{\partial}{\partial t}\iiint_V \left(\int_0^A \boldsymbol{J} \bullet d\boldsymbol{A}\right) dV \tag{10.40}$$

The time integral of equations (10.32) and (10.40) is the net stored magnetic energy and input electrical energy respectively,

$$W_{mag} = \iiint_V \left[\int_0^B \boldsymbol{H} \bullet d\boldsymbol{B}\right] dV \tag{10.41}$$

$$W_e = -\iiint_V \left(\int_0^A \boldsymbol{J} \bullet d\boldsymbol{A}\right) dV \tag{10.42}$$

In order that the power balance equation, equation (10.31), be valid it is apparent that the magnetic plus electrical energy must be a maximum, that is,

$$W_{mag} + W_e = \iiint_V \left[\int_0^B \boldsymbol{H} \bullet d\boldsymbol{B}\right] dV - \iiint_V \left(\int_0^A \boldsymbol{J} \bullet d\boldsymbol{A}\right) dV = W_{max} \tag{10.43}$$

The sum of the electrical and magnetic energy in the system is said to form a *functional* (function of a function) which, when maximized will satisfy the power balance equation, equation (10.28).

In practice, the functional can be simplified somewhat since only a two-dimensional solution is of concern here and the distribution of $\boldsymbol{J}$ is assumed as known. The input energy density is the integrand of the volume integral expressing electrical energy, that is,

$$w_e = -\int_0^A \boldsymbol{J} \bullet d\boldsymbol{A} \tag{10.44}$$

In general, the exact distribution of the current density $\boldsymbol{J}$ is unknown. However, for the quasi-static fields of interest, one can, with reasonable accuracy, assume that the current density is uniform over the conductors carrying the exciting current and zero elsewhere. Hence, if $\boldsymbol{J}$ is a defined input,

$$\begin{aligned} w_s &= -\boldsymbol{J} \bullet \int_0^A d\boldsymbol{A} \\ &= -\boldsymbol{J} \bullet \boldsymbol{A} \end{aligned} \tag{10.45}$$

The fields described by $\boldsymbol{J}$ and $\boldsymbol{A}$ vary only in the x,y plane (and, in this case, are z-directed). Therefore, the required energy functional expression need only involve a surface rather than a volume integral. Including both energy sources and energy storage, the required functional can be written,

$$F = \iint_S \left[\int_0^B \boldsymbol{H} \bullet d\boldsymbol{B} - \boldsymbol{J} \bullet \boldsymbol{A} \right] dS \tag{10.46}$$

or, in terms of components,

$$F = \iint_S \left[\int_0^{B_x} H_x dB_x + \int_0^{B_y} H_y dB_y - J_z A_z \right] dS \tag{10.47}$$

$$= \iint_S \left[\int_0^{B_x} \upsilon_x B_x dB_x + \int_0^{B_y} \upsilon_y B_y dB_y - J_z A_z \right] dS \tag{10.48}$$

where

$$\upsilon_x = \frac{1}{\mu_{rx}\mu_0} \tag{10.49}$$

$$\upsilon_y = \frac{1}{\mu_{ry}\mu_0} \tag{10.50}$$

If the material is linear, then

$$\int_0^{B_x} H_x dB_x = \frac{\upsilon}{2} B_x^2 \quad \int_0^{B_y} H_y dB_y = \frac{\nu}{2} B_y^2 \tag{10.51}$$

so that

$$F = \iint_S \left[\frac{\upsilon}{2} B_x^2 + \frac{\upsilon}{2} B_y^2 - J_z A_z \right] dS \tag{10.52}$$

However,

$$B_x = \frac{\partial A_z}{\partial Y}$$

$$B_y = -\frac{\partial A_z}{\partial x}$$

so that an alternative form of the energy functional for the linear case is

$$F = \iint_S \left[\frac{\upsilon}{2} \left(\frac{\partial A}{\partial x} \right)^2 + \frac{\upsilon}{2} \left(\frac{\partial A}{\partial y} \right)^2 - JA \right] dS \tag{10.53}$$

where the subscript "z" is again now implicit. Note that the units for the energy functional are J/m. It is important to remember that when the material is nonlinear, equation (10.47) does not reduce to (10.52) since

$$\int \boldsymbol{H} \cdot d\boldsymbol{B} \neq \frac{\upsilon}{2}|\boldsymbol{B}|^2$$

10.5 DESCRIPTION OF THE FINITE-ELEMENT METHOD

Since the field region is non-homogeneous and perhaps even anisotropic and/or nonlinear, the minimization of the functional described by equation (10.53) cannot be carried out exactly and it is necessary to construct an approximate set of functions which will satisfy the boundary conditions and yield a sufficiently accurate solution. It is possible to subdivide the field region into a set of triangles in any arbitrary manner as shown in Figure 10.2. The only requirements for the construction of the triangles are that all nodes are connected. A typical triangle *i, j, k* can be considered to prescribe a first-order function to exist within the triangle which is a linear interpolation of the function values at the nodes *i, j,* and *k*. It is useful to define this function as

$$A = \alpha_1 + \alpha_2 x + \alpha_3 y \tag{10.54}$$

over any point within the triangle defined by the nodes *i, j,* and *k*.

It is apparent that, in particular, the vector potential at the three nodes *i, j, k* must, therefore, equal

$$A_i = \alpha_1 + \alpha_2 x_i + \alpha_3 y_i \tag{10.55}$$

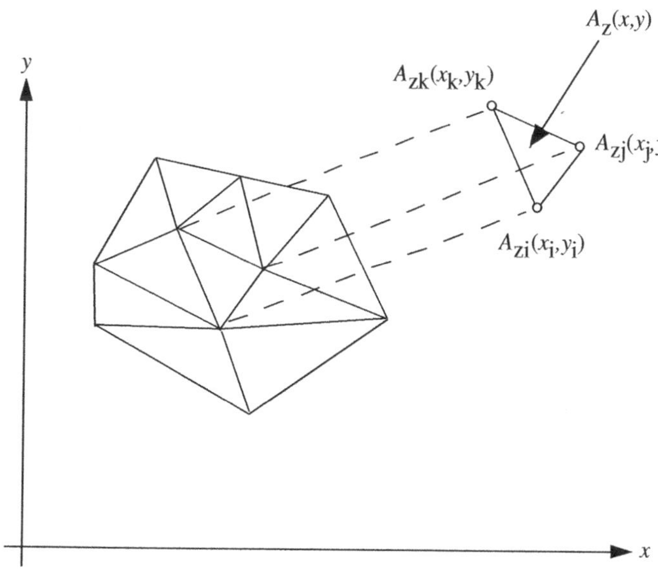

Figure 10.2 Typical triangular grid showing element having nodes *i, j,* and *k*.

$$A_j = \alpha_1 + \alpha_2 x_j + \alpha_3 y_j \tag{10.56}$$

$$A_k = \alpha_1 + \alpha_2 x_k + \alpha_3 y_k \tag{10.57}$$

Equations (10.55), (10.56), and (10.57) can be written in matrix form as

$$\begin{bmatrix} A_i \\ A_j \\ A_k \end{bmatrix} = \begin{bmatrix} 1 & x_i & y_i \\ 1 & x_j & y_j \\ 1 & x_k & y_k \end{bmatrix} \cdot \begin{bmatrix} \alpha_1 \\ \alpha_2 \\ \alpha_3 \end{bmatrix} \tag{10.58}$$

from which one can obtain

$$\begin{bmatrix} \alpha_1 \\ \alpha_2 \\ \alpha_3 \end{bmatrix} = \begin{bmatrix} 1 & x_i & y_i \\ 1 & x_j & y_j \\ 1 & x_k & y_k \end{bmatrix} \cdot \begin{bmatrix} A_i \\ A_j \\ A_k \end{bmatrix} \tag{10.59}$$

The inverse matrix can be written explicitly as

$$\begin{bmatrix} \alpha_1 \\ \alpha_2 \\ \alpha_3 \end{bmatrix} = \frac{1}{2\Delta} \begin{bmatrix} a_i & a_j & a_k \\ b_i & b_j & b_j \\ c_i & c_j & c_k \end{bmatrix} \cdot \begin{bmatrix} A_i \\ A_j \\ A_k \end{bmatrix} \tag{10.60}$$

where

$$a_i = x_j y_k - x_k y_j \tag{10.61}$$

$$b_i = y_j - y_k \tag{10.62}$$

$$c_i = x_k - x_j \tag{10.63}$$

$$a_j = x_k y_i - x_i y_k \tag{10.64}$$

$$b_j = y_k - y_i \tag{10.65}$$

$$c_j = x_i - x_k \tag{10.66}$$

$$a_k = x_i y_j - x_j y_i \tag{10.67}$$

$$b_k = y_i - y_j \tag{10.68}$$

$$c_k = x_j - x_i \tag{10.69}$$

and

$$\Delta = \frac{1}{2}[(x_j y_k - x_k y_j) + x_i(y_j - y_k) + y_i(x_k - x_j)] \tag{10.70}$$

It is interesting to note that Δ is identically equal to the area of the triangle defined by nodes i, j, and k.

Substituting for $\alpha_1, \alpha_2, \alpha_3$ from equation (10.60) into (10.54) yields

$$A = \frac{1}{2\Delta}[(a_iA_i + a_jA_j + a_kA_k) + (b_iA_i + b_jA_j + b_kA_k)x + (c_iA_i + c_jA_j + c_kA_k)y] \tag{10.71}$$

or

$$A = \frac{1}{2\Delta}[(a_i + b_ix + c_iy)A_i + (a_j + b_jx + c_jy)A_j + (a_k + b_kx + c_ky)A_k] \tag{10.72}$$

Equation (10.72) can be written conveniently as

$$A = [N_i, N_j, N_k] \cdot \begin{bmatrix} A_i \\ A_j \\ A_k \end{bmatrix} \tag{10.73}$$

where

$$N_i = \frac{1}{2\Delta}(a_i + b_ix + c_iy)$$
$$N_j = \frac{1}{2\Delta}(a_j + b_jx + c_jy)$$
$$N_k = \frac{1}{2\Delta}(a_k + b_kx + c_ky)$$

The quantities N_i, N_j, N_k are called the *shape functions* of the finite-element discretization process.

Note that the shape functions are linear functions of x and y. The shape functions must be chosen to give appropriate nodal potentials when the coordinates of nodes are inserted. Hence, at the point $(x,y) = (x_i, y_i)$ one has $N_i = 1, N_j = 0, N_k = 0$ while at $(x,y) = (x_j, y_j)$ one has $N_i = 0, N_j = 1, N_k = 0$, and so forth.

10.6 MAGNETIC INDUCTION AND RELUCTIVITY IN THE TRIANGLE ELEMENT

It has been shown earlier that from,

$$\boldsymbol{B} = \nabla \times \boldsymbol{A}$$

one arrives at, for two-dimensional fields,

$$B_x = \frac{\partial A}{\partial y}$$
$$B_y = -\frac{\partial A}{\partial x}$$

Hence, from equation (10.71),

$$B_x = \frac{1}{2\Delta}(c_iA_i + c_jA_j + c_kA_k) \tag{10.74}$$

Similarly

$$B_y = -\frac{1}{2\Delta}(b_i A_i + b_j A_j + b_k A_k) \tag{10.75}$$

Hence,

$$|B| = \sqrt{B_x^2 + B_y^2} \tag{10.76}$$

$$= \frac{1}{2\Delta}\sqrt{(b_i A_i + b_j A_j + b_k A_k)^2 + (c_i A_i + c_j A_j + c_k A_k)^2} \tag{10.77}$$

Note that the magnetic flux density B is constant over the triangle. Since the reluctivity ν is a function of the amplitude B, it is evident that $\nu$is also constant over the region as well.

10.7 FUNCTIONAL MINIMIZATION

Thus far an approximate expression for the vector potential A inside of each triangle has been obtained and expressed in terms of the values of A at the triangle nodes. It is clear from presentation in Section 10.4 that the functional F defined by equation (10.47) must be minimized since, ideally, it should equal zero. The energy within each element is therefore realized by minimizing the energy functional F with respect to each nodal potential, that is,

$$\frac{\partial F}{\partial A_i} = 0; \quad \frac{\partial F}{\partial A_j} = 0; \quad \frac{\partial F}{\partial A_k} = 0 \tag{10.78}$$

With respect to the node i one has

$$\frac{\partial F}{\partial A_i} = 0 = \iint_s \left\{ \frac{\partial}{\partial A_i}\left[\int_0^B \upsilon B dB\right] - \frac{\partial}{\partial A_i} J A_i \right\} dS \tag{10.79}$$

$$= \iint_S \left\{ \frac{\partial}{\partial B}\left[\int_0^B \upsilon B dB\right] \frac{\partial}{\partial A_i}(B) - \frac{\partial}{\partial A_i}(J A_i) \right\} dS \tag{10.80}$$

$$= \iint_S \left[\upsilon B \frac{\partial}{\partial A_i}(B) - \frac{\partial}{\partial A_i}(J A_i) \right] dS \tag{10.81}$$

From equation (10.21), this can be written as

$$\frac{\partial F}{\partial A_i} = \iint_S \left[\upsilon B \frac{\partial}{\partial A_i}\sqrt{\left(\frac{\partial A}{\partial x}\right)^2 + \left(\frac{\partial A}{\partial y}\right)^2} - \frac{\partial}{\partial A_i}(J A_i) \right] dS \tag{10.82}$$

It is convenient to let

$$A_x = \frac{\partial A}{\partial x}$$
$$A_y = \frac{\partial A}{\partial y}$$

Then

$$\frac{\partial F}{\partial A_i} = \iint_S \upsilon B\left\{\frac{\partial}{\partial A_x}\sqrt{A_x^2 + A_y^2}\cdot\frac{\partial A_x}{\partial A_i} + \frac{\partial}{\partial A_y}\sqrt{A_x^2 + A_y^2}\cdot\frac{\partial A_y}{\partial A_i} - \frac{\partial}{\partial A_i}(JA_i)\right\} dS \tag{10.83}$$

$$\iint_S \left[\left(\upsilon B\left(\frac{1}{2}\right)2\frac{A_x}{B}\frac{\partial A_x}{\partial A_i} + \upsilon B\left(\frac{1}{2}\right)2\frac{A_y}{B}\frac{\partial A_y}{\partial A_i}\right) - \frac{\partial}{\partial A_i}(JA)\right] dS \tag{10.84}$$

$$= \iint_S \left\{\upsilon\left[\frac{\partial A}{\partial x}\frac{\partial}{\partial A_i}\left(\frac{\partial A}{\partial x}\right) + \frac{\partial A}{\partial y}\frac{\partial}{\partial A_i}\left(\frac{\partial A}{\partial y}\right)\right] - \frac{\partial}{\partial A_i}(JA)\right\} dS \tag{10.85}$$

However, since

$$A = \frac{1}{2\Delta}[(a_i + b_i x + c_i y)A_i + (a_j + b_j x + c_j y)A_j + (a_k + b_k x + c_k y)A_k]$$

then

$$\frac{\partial A}{\partial x} = \frac{1}{2\Delta}(b_i A_i + b_j A_j + b_k A_k)$$
$$\frac{\partial A}{\partial y} = \frac{1}{2\Delta}(c_i A_i + c_j A_j + c_k A_k)$$

and also

$$\frac{\partial}{\partial A_i}\frac{\partial A}{\partial x} = \frac{b_i}{2\Delta}$$
$$\frac{\partial}{\partial A_i}\frac{\partial A}{\partial y} = \frac{c_i}{2\Delta}$$

Equation (10.85) reduces to

$$\frac{\partial F}{\partial A_i} = \iint_S \left\{\upsilon\left[\frac{b_i A_i + b_j A_j + b_k A_k}{(2\Delta)(2\Delta)}b_i + \frac{c_i A_i + c_j A_j + c_k A_k}{(2\Delta)(2\Delta)}c_i\right]\right.$$
$$\left. - \frac{\partial}{\partial A_i}\frac{J}{2\Delta}((a_i + b_i x + c_i y)A_i + (a_j + b_j x + c_j y)A_j + (a_k + b_k x + c_k y)A_k\right\} dS \tag{10.86}$$

$$= \iint_S \frac{\upsilon}{4\Delta^2}\left[(b_i^2 + c_i^2), (b_i b_j + c_i c_j), (b_i b_k + c_i c_k)\right]\cdot\begin{bmatrix} A_i \\ A_j \\ A_k \end{bmatrix}\cdot dS$$
$$-J\iint_S \frac{(a_i + b_i x + c_i y)}{2\Delta}\cdot dS = 0 \tag{10.87}$$

Similarly, if one derives similar expressions for $\frac{\partial F}{\partial A_j}$ and $\frac{\partial F}{\partial A_k}$, and writes the result in matrix form, the following is obtained.

$$\begin{bmatrix} \frac{\partial F}{\partial A_i} \\ \frac{\partial F}{\partial A_j} \\ \frac{\partial F}{\partial A_k} \end{bmatrix} = \frac{1}{4\Delta^2} \begin{bmatrix} (b_i^2 + c_i^2) & (b_i b_j + c_i c_j) & (b_i b_k + c_i c_k) \\ (b_j b_i + c_j c_i) & (b_j^2 + c_j^2) & (b_j b_k + c_j c_k) \\ (b_k b_i + c_k c_i) & (b_k b_j + c_k c_j) & (b_k^2 + c_k^2) \end{bmatrix} \cdot \begin{bmatrix} A_i \\ A_j \\ A_k \end{bmatrix} \cdot \iint_S \upsilon dS$$

$$- \frac{J}{2\Delta} \begin{vmatrix} \iint_S (a_i + b_i x + c_i y)\, dS \\ \iint_S (a_j + b_j x + c_j y)\, dS \\ \iint_S (a_k + b_k x + c_k y)\, dS \end{vmatrix} = \begin{bmatrix} 0 \\ 0 \\ 0 \end{bmatrix} \tag{10.88}$$

Since, from equation (10.77), B is constant over the triangle, the reluctivity ν can also be assumed constant over the area of the triangle, then

$$\iint_S \upsilon dS = \Delta \cdot \upsilon \tag{10.89}$$

Equation (10.88) can be written compactly as

$$\begin{bmatrix} \frac{\partial F}{\partial A_i} \\ \frac{\partial F}{\partial A_j} \\ \frac{\partial F}{\partial A_k} \end{bmatrix} = \begin{bmatrix} S_{ii} & S_{ij} & S_{ik} \\ S_{ji} & S_{jj} & S_{jk} \\ S_{ki} & S_{kj} & S_{kk} \end{bmatrix} \cdot \begin{bmatrix} A_i \\ A_j \\ A_k \end{bmatrix} - \frac{J}{2\Delta} \begin{bmatrix} \iint_S (a_i + b_i x + c_i y)\, dS \\ \iint_S (a_j + b_j x + c_j y)\, dS \\ \iint_S (a_k + b_k x + c_k y)\, dS \end{bmatrix} \tag{10.90}$$

where

$$S_{ii} = \frac{\upsilon (b_i^2 + c_i^2)}{4\Delta}$$

$$S_{ij} = \frac{\upsilon (b_i b_j + c_i c_j)}{4\Delta} = S_{ji}$$

$$S_{ik} = \frac{\upsilon (b_i b_k + c_i c_k)}{4\Delta} = S_{ki}$$

$$S_{jj} = \frac{\upsilon\left(b_j^2 + c_j^2\right)}{4\Delta}$$

$$S_{jk} = \frac{\upsilon(b_j b_k + c_j c_k)}{4\Delta} = S_{kj}$$

$$S_{kk} = \frac{\upsilon\left(b_k^2 + c_k^2\right)}{4\Delta}$$

It is now necessary to evaluate the area integrals of equation (10.90). The easiest method of evaluating such integrals is by numerical quadrature. The reader is referred to textbooks for such formulas as Gauss, Gauss–Legendre, NewtonCoates and so forth. For the present case, a simple approximation can be made based on the centroid of the triangle.

Since the integration over the area of a triangle is equivalent to integrating a weighted function at its centroid, one can write

$$\iint_S (a_i + b_i x + c_i y)\, dS = \iint_S (a_i + b_i \bar{x} + c_i \bar{y})\, dS \tag{10.91}$$

where

$$\bar{x} = \frac{(x_i + x_j + x_k)}{3} \tag{10.92}$$

$$\bar{y} = \frac{(y_i + y_j + y_k)}{3} \tag{10.93}$$

The quantities $\bar{x}$ and $\bar{y}$ are the x and y coordinates of the "center of gravity" of the triangle. Substituting equations (10.61), (10.62), and (10.63) and equations (10.92) and (10.93) into equation (10.91) it is possible to obtain

$$\iint_S (a_i + b_i x + c_i y)\, dS = \iint_S \Bigg[(x_j y_k - x_k y_j) + (y_j - y_k)\frac{(x_i + x_j + x_k)}{3} + (x_k - x_j)\frac{(y_i + y_j + y_k)}{3}\Bigg]\, dS \tag{10.94}$$

$$= \frac{1}{3}[3x_j y_k - 3x_k y_j + x_i y_j - x_i y_k + x_j y_j - x_j y_k + x_k y_j - x_k y_k + x_k y_i - x_j y_i + x_k y_j - x_j y_j + x_k y_k - x_j y_k] \iint_S dS \tag{10.95}$$

$$= \frac{1}{3}[(x_j y_k - x_k y_j) + x_i(y_j - y_k) + y_i(x_k - x_j)] \iint_S dS \tag{10.96}$$

$$= \frac{2\Delta}{3}\int_S dS \tag{10.97}$$

$$= \frac{2\Delta^2}{3} \tag{10.98}$$

Therefore,

$$\frac{J}{2\Delta}\iint_S (a_i + b_i x + c_i y)\, dS = \frac{J}{2\Delta}\cdot\frac{2\Delta^2}{3} = \frac{J\Delta}{3} \tag{10.99}$$

Similarly

$$\frac{J}{2\Delta}\iint_S (a_j + b_j x + c_j y)\, dS = \frac{J\Delta}{3} \tag{10.100}$$

$$\frac{J}{2\Delta}\iint_S (a_k + b_k x + c_k y)\, dS = \frac{J\Delta}{3} \tag{10.101}$$

Hence, finally for equation (10.88), the matrix equation of functional minimization for the triangle under consideration,

$$\begin{bmatrix} \dfrac{\partial F}{\partial A_i} \\ \dfrac{\partial F}{\partial A_j} \\ \dfrac{\partial F}{\partial A_k} \end{bmatrix} = \begin{bmatrix} S_{ii} & S_{ij} & S_{ik} \\ S_{ji} & S_{jj} & S_{jk} \\ S_{ki} & S_{kj} & S_{kk} \end{bmatrix} \cdot \begin{bmatrix} A_i \\ A_j \\ A_k \end{bmatrix} - \frac{J\Delta}{3}\begin{bmatrix} 1 \\ 1 \\ 1 \end{bmatrix} = \begin{bmatrix} 0 \\ 0 \\ 0 \end{bmatrix} \tag{10.102}$$

Consequently, for each of the triangles, the method of finite elements involves solving a three-dimensional set of equations of the form

$$\begin{bmatrix} S_{ii} & S_{ij} & S_{ik} \\ S_{ji} & S_{jj} & S_{jk} \\ S_{ki} & S_{kj} & S_{kk} \end{bmatrix} \cdot \begin{bmatrix} A_i \\ A_j \\ A_k \end{bmatrix} = \frac{J\Delta}{3}\begin{bmatrix} 1 \\ 1 \\ 1 \end{bmatrix} \tag{10.103}$$

The quantity $J\Delta$ is clearly equal to the current I passing through the triangle under consideration.

10.8 FORMULATION OF THE STIFFNESS MATRIX EQUATION

In general, a particular node i can be a vertex of an arbitrary number of triangles. Consider, for example, the three triangle mesh shown in Figure 10.3, where, for convenience, designation of the nodes by numbers 1, 2, 3, etc. has been used rather than letters i, j, k, etc. For triangle element "1"

$$\begin{bmatrix} S_{11,1} & S_{12,1} & S_{13,1} \\ S_{21,1} & S_{22,1} & S_{23,1} \\ S_{31,1} & S_{32,1} & S_{33,1} \end{bmatrix} \cdot \begin{bmatrix} A_1 \\ A_2 \\ A_3 \end{bmatrix} = \frac{J_1\Delta_1}{3}\begin{bmatrix} 1 \\ 1 \\ 1 \end{bmatrix} \tag{10.104}$$

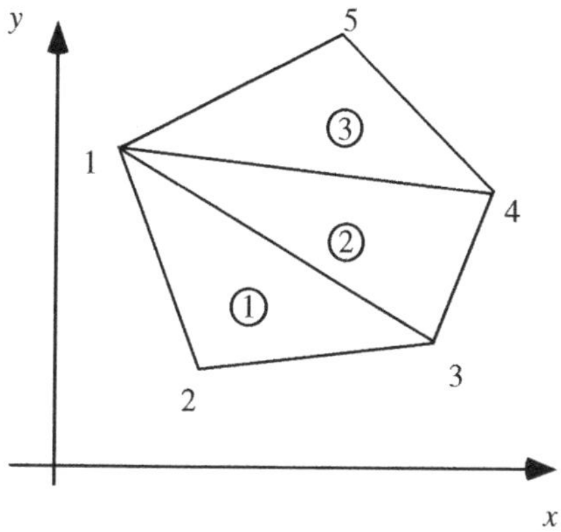

element	node i	j	k
①	1	2	3
②	1	3	4
③	1	4	5

Figure 10.3 Simple triangular grid having five nodes and three triangles.

equation (10.104) can be expressed as

$$\begin{bmatrix} S_{11,1} & S_{12,1} & S_{13,1} & 0 & 0 \\ S_{21,1} & S_{22,1} & S_{23,1} & 0 & 0 \\ S_{31,1} & S_{32,1} & S_{33,1} & 0 & 0 \\ 0 & 0 & 0 & 0 & 0 \\ 0 & 0 & 0 & 0 & 0 \end{bmatrix} \cdot \begin{bmatrix} A_1 \\ A_2 \\ A_3 \\ A_4 \\ A_5 \end{bmatrix} = \frac{J_1\Delta_1}{3} \begin{bmatrix} 1 \\ 1 \\ 1 \\ 0 \\ 0 \end{bmatrix} \tag{10.105}$$

In a similar manner, for triangle element “2”

$$\begin{bmatrix} S_{11,2} & 0 & S_{13,2} & S_{14,2} & 0 \\ 0 & 0 & 0 & 0 & 0 \\ S_{31,2} & 0 & S_{33,2} & S_{34,2} & 0 \\ S_{41,2} & 0 & S_{43,2} & S_{44,2} & 0 \\ 0 & 0 & 0 & 0 & 0 \end{bmatrix} \cdot \begin{bmatrix} A_1 \\ A_2 \\ A_3 \\ A_4 \\ A_5 \end{bmatrix} = \frac{J_2\Delta_2}{3} \begin{bmatrix} 1 \\ 0 \\ 1 \\ 1 \\ 0 \end{bmatrix}$$

For triangle “3”

$$\begin{bmatrix} S_{11,3} & 0 & 0 & S_{14,3} & S_{15,3} \\ 0 & 0 & 0 & 0 & 0 \\ 0 & 0 & 0 & 0 & 0 \\ S_{41,3} & 0 & 0 & S_{44,3} & S_{45,3} \\ S_{51,3} & 0 & 0 & S_{54,3} & S_{55,3} \end{bmatrix} \cdot \begin{bmatrix} A_1 \\ A_2 \\ A_3 \\ A_4 \\ A_5 \end{bmatrix} = \frac{J_3\Delta_3}{3} \begin{bmatrix} 1 \\ 0 \\ 0 \\ 1 \\ 1 \end{bmatrix}$$

Upon adding these three matrix equations,

$$\begin{bmatrix} (S_{11,1}+S_{11,2}+S_{11,3}) & S_{12,1} & (S_{13,1}+S_{13,2}) & (S_{14,2}+S_{14,3}) & S_{15,3} \\ S_{21,1} & S_{22,1} & S_{23,1} & 0 & 0 \\ (S_{31,1}+S_{31,2}) & S_{32,1} & (S_{33,1}+S_{33,2}) & S_{34,2} & 0 \\ (S_{41,2}+S_{41,3}) & 0 & S_{43,2} & (S_{44,2}+S_{44,3}) & S_{45,3} \\ S_{51,3} & 0 & 0 & S_{54,3} & S_{55,3} \end{bmatrix} \cdot \begin{bmatrix} A_1 \\ A_2 \\ A_3 \\ A_4 \\ A_5 \end{bmatrix}$$

$$= \frac{1}{3}\begin{bmatrix} I_1+I_2+I_3 \\ I_1 \\ I_1+I_2 \\ I_2+I_3 \\ I_3 \end{bmatrix} \tag{10.106}$$

where the current $I_1 = J_1\Delta_1$ and so forth.

When this process is extended to all triangles in the region of the field, one obtains a single matrix equation

$$[S]\cdot[A] = [T] \tag{10.107}$$

where $[S]$ represents the matrix formed out of adjoining the appropriate elements S_{ii}, S_{ij}, S_{ik}, etc. of each of the triangular matrices of the network. Having evolved from structural mechanics, the matrix $[S]$ is commonly called the *stiffness matrix*. Since it is, in the case of magnetic fields, proportional to reluctance (times unit depth), it is also called the *reluctivity matrix*. It is important to note that $[S]$ is sparse (many zero entries) and is also symmetric, bounded, and positive definite. The quantity $[T]$ is a similar matrix formed out of right-hand side triangle matrices and has nonzero elements only for nodes associated with triangles enclosing a current. Examination of equation (10.106) indicates that the matrix $[T]$ is obtained by simply taking 1/3 the sum of the currents passing through the triangles having a vertex touching the node corresponding to a particular row (particular nodal vector potential). The matrix $[A]$ is the matrix of unknown vector potentials.

10.9 CONSIDERATION OF BOUNDARY CONDITIONS

Since the magnetic field typically extends indefinitely out into the air surrounding a magnetic device, the solution of the finite-element problem would normally be impossible. In order to make the problem tractable, it is often necessary to limit the boundary around the magnetic device to a "reasonable" region. The boundary is established by simply defining the boundary to be a line of constant vector potential (i.e., a flux line). While any number can be assigned to a vector potential boundary line, the value of zero is most often used. Since the nodes associated with the boundary are known, they do not need to be calculated and the $[S]$ and $[T]$ matrices should be modified accordingly. For example, if node 2 in Figure 10.3 is A_0, then equation (10.106) becomes

$$\begin{bmatrix} (S_{11,1}+S_{11,2}+S_{11,3}) & S_{12,1} & (S_{13,1}+S_{13,2}) & (S_{14,2}+S_{14,3}) & S_{15,3} \\ 0 & 1 & 0 & 0 & 0 \\ (S_{31,1}+S_{31,2}) & S_{32,1} & (S_{33,1}+S_{33,2}) & S_{34,2} & 0 \\ (S_{41,2}+S_{41,3}) & 0 & S_{43,2} & (S_{44,2}+S_{44,3}) & S_{45,3} \\ S_{51,3} & 0 & 0 & S_{54,3} & S_{55,3} \end{bmatrix} \cdot \begin{bmatrix} A_1 \\ A_2 \\ A_3 \\ A_4 \\ A_5 \end{bmatrix} = \frac{1}{3} \begin{bmatrix} I_1+I_2+I_3 \\ 3A_0 \\ I_1+I_2 \\ I_2+I_3 \\ I_3 \end{bmatrix} \tag{10.108}$$

A solution of equation (10.108) will now automatically assign the value A_0 to A_2 while preserving the proper coupling with the remaining nodes.

Another type of boundary condition involves the so-called Neumann condition in which the vector potential or flux lines intersect the boundary at right angles. This condition can usually be observed over lines of symmetry. This condition is automatically satisfied if the potentials at nodes on the boundary are simply left as variables. Only adjacent nodes inside the problem will form the row associated with a boundary node. In effect, the reluctivity of the nodes on the other side of the boundary is allowed to go to zero (infinite permeability) thereby forcing the equipotential lines to intersect this portion of the boundary at right angles.

A third type of boundary condition concerns Neumann conditions on two portions of the boundary, which, because of symmetry, have equal or equal and opposite values. In this case, an additional constraint must be imposed. If, for example, one returns to Figure 10.3 and assumes that an odd Neumann condition applies at the boundary, then the potential of node 4 is equal to node 2 in which case

$$A_4 = -A_2 \tag{10.109}$$

Hence, the fourth row of equation (10.106) need not be solved. It is now possible to write the matrix as

$$\begin{bmatrix} (S_{11,1}+S_{11,2}+S_{11,3}) & S_{12,1} & (S_{13,1}+S_{13,2}) & (S_{14,2}+S_{14,3}) & S_{15,3} \\ S_{21,1} & S_{22,1} & S_{23,1} & S_{24,2} & 0 \\ (S_{31,1}+S_{31,2}) & S_{32,1} & (S_{33,1}+S_{33,2}) & S_{34,2} & 0 \\ 0 & 1 & 0 & 1 & 0 \\ S_{51,3} & 0 & 0 & S_{54,3} & S_{55,3} \end{bmatrix} \cdot \begin{bmatrix} A_1 \\ A_2 \\ A_3 \\ A_4 \\ A_5 \end{bmatrix} = \frac{1}{3} \begin{bmatrix} I_1+I_2+I_3 \\ I_1 \\ I_1+I_2 \\ 0 \\ I_3 \end{bmatrix} \tag{10.110}$$

Note that introducing the constraint in this manner disturbs the symmetry of the matrix [S] so that some of the powerful solution methods for inversion of symmetric matrices cannot be used. It is better to eliminate this node potential entirely from the solution so that the matrix to be solved becomes

$$\begin{bmatrix} (S_{11,1}+S_{11,2}+S_{11,3}) & S_{12,1} & (S_{13,1}+S_{13,2}) & S_{15,3} \\ S_{21,1} & S_{22,1} & S_{23,1} & 0 \\ (S_{31,1}+S_{31,2}) & S_{32,1} & (S_{33,1}+S_{33,2}) & 0 \\ S_{51,3} & 0 & 0 & S_{55,3} \end{bmatrix} \cdot \begin{bmatrix} A_1 \\ A_2 \\ A_3 \\ A_5 \end{bmatrix}$$

$$= \frac{1}{3} \begin{bmatrix} I_1 + I_2 + I_3 \\ I_1 \\ I_1 + I_2 \\ I_3 \end{bmatrix} \tag{10.111}$$

Note that the matrix [S] now remains bounded in the same manner as before. A separate matrix (or vector) can clearly be formed to keep track of nodes eliminated because of symmetry. Constraints with odd rather than even symmetry can be handled similarly.

10.10 STEP-BY-STEP PROCEDURE FOR SOLVING THE FINITE-ELEMENT PROBLEM

Consider now, as an example, the simple iron-cored inductor with an air gap shown in Figure 10.4. Assume that the dimensions of the inductor, with zero gap, are $6 \times 6 \times 1$. The core is wound with copper such that the current density over core window is 1000 A/in.^{2} The return portion of the coil is located opposite the movable iron keeper. The magnetic field is to be determined for a $1/2''$ air gap.

Step 1. Generate Meshes. The total number of triangular elements and their location must first be chosen. In general, the elements should be chosen such that they are denser where the field is denser. This is particularly important in the gap region if nonlinear iron is used. A typical triangular grid is shown in Figure 10.5.

Step 2. Number the Triangular Elements and Nodes. Triangular elements are numbered in any consistent manner. A numbering of the vertices now takes place. For each element, three nodal numbers are assigned. The numbering of the vertices should follow a "zigzag" pattern to keep the "coupling" between elements banded around the major diagonal. The numbers corresponding to each element must be stored in the computer in a consistent clockwise, or counterclockwise manner.

Step 3. Calculate the Coordinates and Area of Each Element. Each triangular element is assigned the following attributes in a computer array: (1) vertex numbers ordered counterclockwise, (2) x and y coordinates of each of the

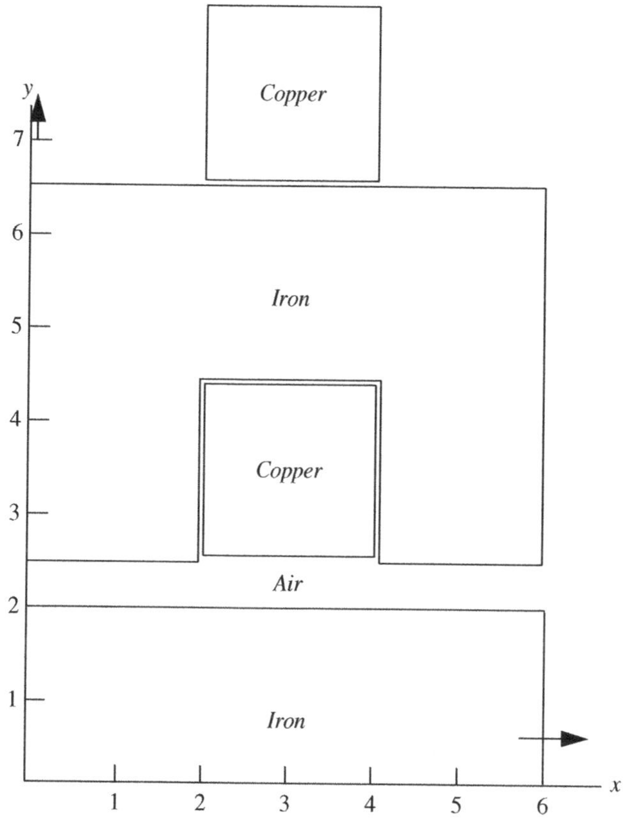

Figure 10.4 Example of an iron-cored inductor with air gap.

three vertices, (3) area of the element (calculated immediately before solution since it remains constant), (4) a number identifying the material within the element (magnetic, or non-magnetic, and what type if magnetic), (5) current density within the element (assumed constant).

Step 4. Define the Boundary Conditions. The boundary conditions are now defined serving to establish the limits over which the problem will be solved. In the example of Figure 10.4, the outer surface of the inductor is defined to be zero potential. It should be noted, however, that this constraint imposes a serious limitation on the solution, namely that flux lines are prohibited from extending from the outer surface of the inductor so that fringing near the air gap and also leakage flux emanating from the copper region outside the inductor iron will be neglected.

Step 5. Set Up the [*S*] and [*T*] Matrices. The [*S*] and [*T*] matrices are now constructed using the information in the array of Step 3. The matrices are built up by incorporating the effect of one triangular element at a time using the example of Section 10.8.

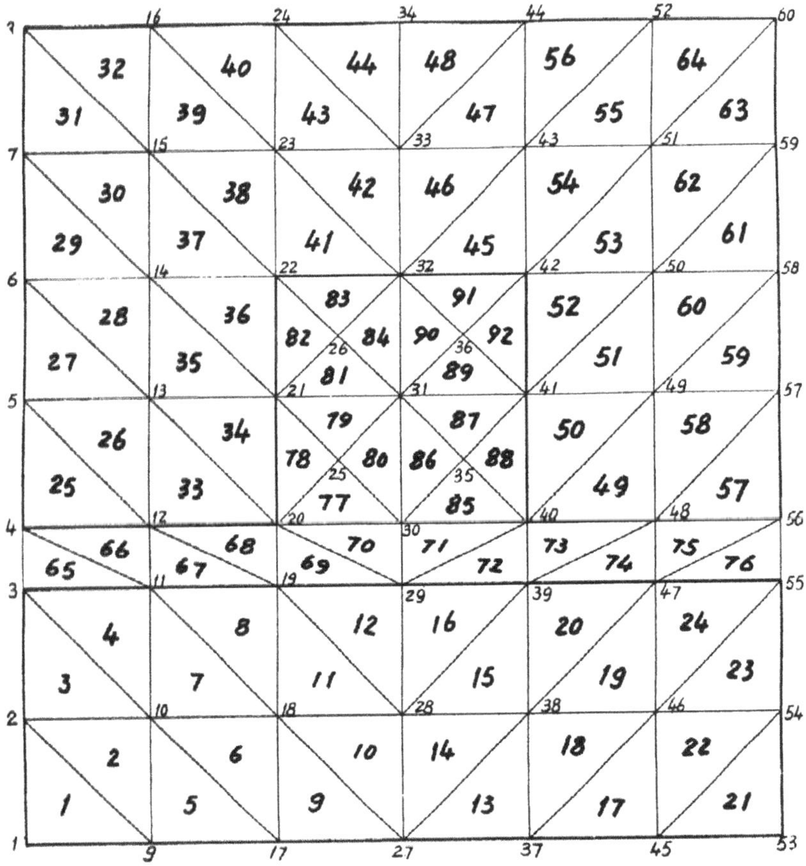

Figure 10.5 Triangular grid corresponding to the example of iron-cored inductor of Figure 10.4.

Step 6. Solution of the Equations. The solution of the equations follows the simple flow chart of Figure 10.6.

- First, equation (10.107) is solved for the nodal potentials $[A]$.
- Second, the flux density of each element is calculated using equations (10.74) and (10.75).

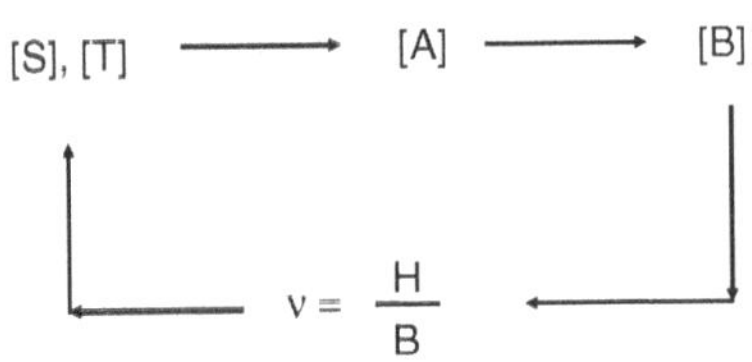

Figure 10.6 Simple flow chart for the solution of the nonlinear finite-element matrix equations.

- Third, the reluctivity of each element is calculated from the slope of the appropriate material *B–H* characteristic, knowing the *B* over each element.
- Fourth, the reluctivities are updated. In general, the updating of the reluctances of each element is done slowly to avoid oscillations in the solution. In practice, the so-called chord method is applied where for each iteration i, the reluctivity υ_i^k assigned to element k is determined according to the formula

$$\upsilon_i^k = \upsilon_{i-1}^k + \kappa_a \left(\upsilon_{i-1(\text{calc})}^k - \upsilon_{i-1}^k \right) \tag{10.112}$$

  where κ_a is an acceleration constant usually taken as 0.10 and $\upsilon_{i-1(\text{calc})}$ the reluctivity calculated from the element flux density found from the (i–1)th iteration vector potential solution. For the first iteration ($i = 1$) the same reluctivity is assigned to all elements of each material type or to speed convergence a reluctivity distribution from a previous nonlinear solution with the same grid (but perhaps different excitation) is assigned to the grid elements.
- Fifth, the [*S*] matrix is recomputed and the algorithm returns to the solution of equation (10.107) for [*A*], that is, Step 1 Generation of Meshes.

The nonlinear iterative solution is said to converge to the correct solution if *both* of the following convergence criteria are met.

$$\frac{\sqrt{\sum_{k=1}^{n_e} \left| B_i^k - B_{i-1}^k \right|^2}}{\sqrt{\sum_{k=1}^{n_e} \left(B_i^k \right)^2}} \leq B_{\text{error}} \tag{10.113}$$

where n_e is the number of elements in the solution region and B_i^k is the calculated flux density of element k for the field solution of iteration i, and

$$\frac{\sqrt{\sum_{k=1}^{n} \left| A_{zi}^k - A_{zi-1}^k \right|^2}}{\sqrt{\sum_{k=1}^{n} \left(A_{zi}^k \right)^2}} \leq A_{\text{error}} \tag{10.114}$$

where n is the number of nodes and A_{zi} is the calculated vector magnetic potential at node k for the solution at iteration i. The quantities B_{error} and A_{error} are typically taken as 0.01. A flow chart for a nonlinear, iterative finite-element solution is shown in Figure 10.7.

Step 7. Plot of the Field. Since lines of flux correspond to lines of constant vector potential, the maximum and minimum values of vector potential are recorded. Then, if N_{ep} is the number of equipotential lines desired,

$$\Delta A = \frac{A_{\max} - A_{\min}}{N_{ep}} \tag{10.115}$$

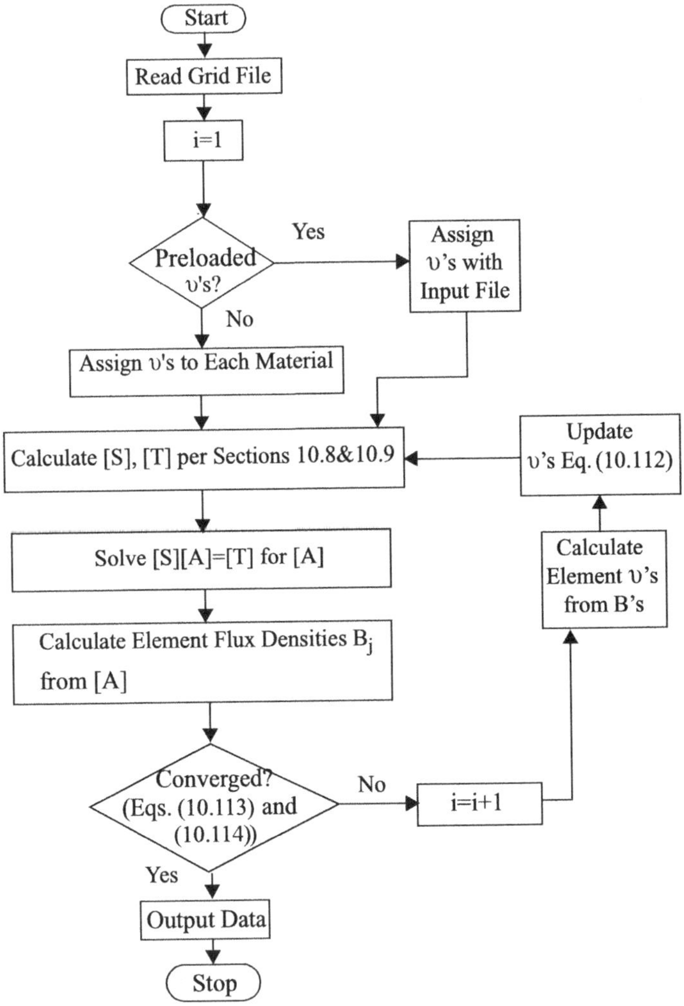

Figure 10.7 Flow chart for solution of the finite-element equations.

For each element, points along each side are located which are an integral multiple of ΔA. Points of equal potential on each of two sides are connected by a straight line as shown in Figure 10.8.

When this procedure is repeated for all elements, a flux plot having the desired number of equipotential lines results. A finished flux plot for the example of Figure 10.4 is shown in Figure 10.9.

Step 8. Determination of Lumped Parameter Values. Using the vector potential, the number of flux lines per unit depth passing through any surface can be determined. For example, in Figure 10.9, the maximum equipotential line

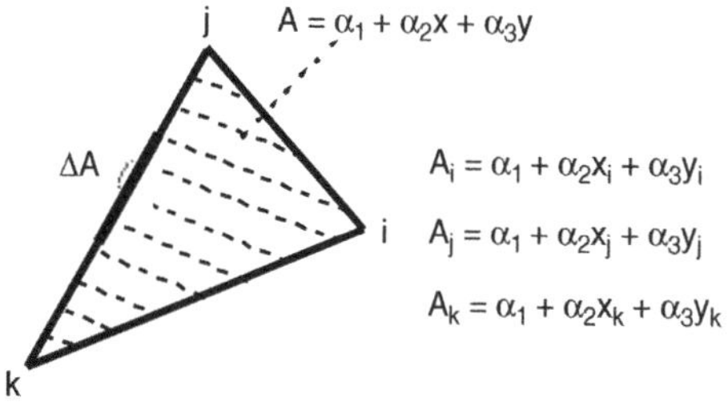

Figure 10.8 Showing connected equipotential lines on the boundary of one triangular element.

A_{max} is 1.51×10^{-2} W/m. If the inductor is one-inch deep then the total flux linking the core body is

$$\Phi = 1.51 \cdot 10^{-2} \cdot \frac{1}{39.37} = 3.84 \cdot 10^{-4} \text{ webers}$$

If the coil has, for example, 100 turns carrying 40 amperes then the inductance of the coil is

$$L = N\frac{\phi}{i} = 100 \cdot \frac{3.84 \cdot 10^{-4}}{40} = 0.96\ mH$$

An example solution of this problem using MATLAB® code is given in Appendix B.

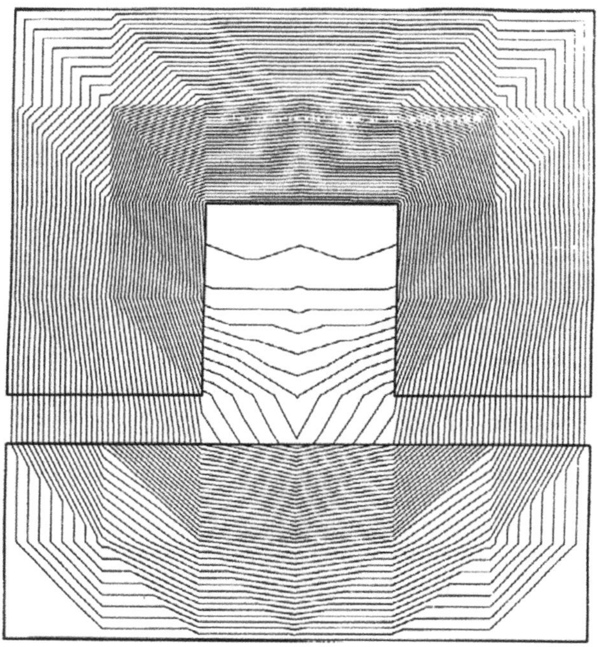

Figure 10.9 Finished flux plot for the iron-cored inductor example of Figure 10.4.

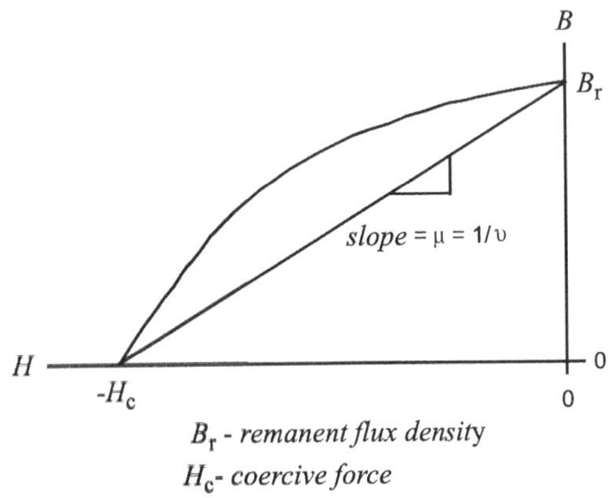

Figure 10.10 B versus H curve for permanent magnet showing second quadrant region of operation.

10.11 FINITE-ELEMENT MODELING OF PERMANENT MAGNETS

The first-order finite-element solution technique described previously assumes that the field excitation is a consequence of source currents in the region. Another possible source of excitation is that of permanent magnets. Figure 10.10 shows the second quadrant of the B–H characteristic of a permanent magnet. This characteristic applies to the field parallel to the direction of magnetization. The magnetic field along this axis can be described by the constitutive equation

$$H = \upsilon B - H_c \tag{10.116}$$

with the reluctivity υ being a function of flux density B and the coercive force H_c defined as a positive constant. Along an axis perpendicular to the magnetization, the coercive force is zero so that the constitutive equation becomes the familiar

$$H = \upsilon_q B \tag{10.117}$$

with υ_q being the reluctivity of the magnet along an axis in quadrature to its direction of magnetization. Binns, Low, and Jabbar [4] have experimentally shown that in a rare-earth permanent magnet sample, υ_q is within 12% of υ. This difference can be neglected without much effect on the solution so that the constitutive equation can be expressed as the single vector relationship given by

$$\boldsymbol{H} = \upsilon \boldsymbol{B} - \boldsymbol{H}_c \tag{10.118}$$

The vector H_c points in the direction of magnetization (from south to north pole) with a magnitude equal to the coercive force. The magnet reluctivity υ can then be calculated from the B–H curve as shown in Figure 10.10. Maxwell's equation for Ampere's law is

$$\nabla \times \boldsymbol{H} = \boldsymbol{J} \tag{10.119}$$

Substituting equation (10.118) into Maxwell's equation

$$\nabla \times [\upsilon \boldsymbol{B} - \boldsymbol{H}_c] = \boldsymbol{J} \quad (10.120)$$

and, from equation (10.4), this becomes

$$\nabla \times (\upsilon \nabla \times \boldsymbol{A}) = \boldsymbol{J} + \nabla \times \boldsymbol{H}_c \quad (10.121)$$

Since $\boldsymbol{H}_c$ is constant within the magnet and is discontinuous at the permanent magnet boundary, its curl is nonzero only along these boundaries.

In the finite-element formulation, the equivalent of equation (10.13) becomes

$$\frac{\partial}{\partial x}\left(\upsilon \frac{\partial A}{\partial x}\right) + \frac{\partial}{\partial y}\left(\upsilon \frac{\partial A}{\partial y}\right) = -J - J_{pm} \quad (10.122)$$

where J_{pm} represents the equivalent currents at the magnet boundaries. Discretizing the region into first-order finite elements and satisfying the functional minimization of the partial differential equations, equation (10.122), results in the following finite-element matrix equation

$$[S] \cdot [A] = [T] + [T_{pm}] \quad (10.123)$$

The quantities $[S]$, $[A]$, and $[T]$ are the same as previously obtained, while $[T_{pm}]$ is a vector which contains equivalent nodal current which provides for the magnet excitation.

The equivalent nodal currents can be derived by considering two adjacent elements in the finite-element solution region with differing values of magnetization $\boldsymbol{H}_{c1}$ and $\boldsymbol{H}_{c2}$ as shown in Figure 10.11. The necessary discontinuity between the tangential $\boldsymbol{H}$ values at the material interface can be produced by a surface current along the

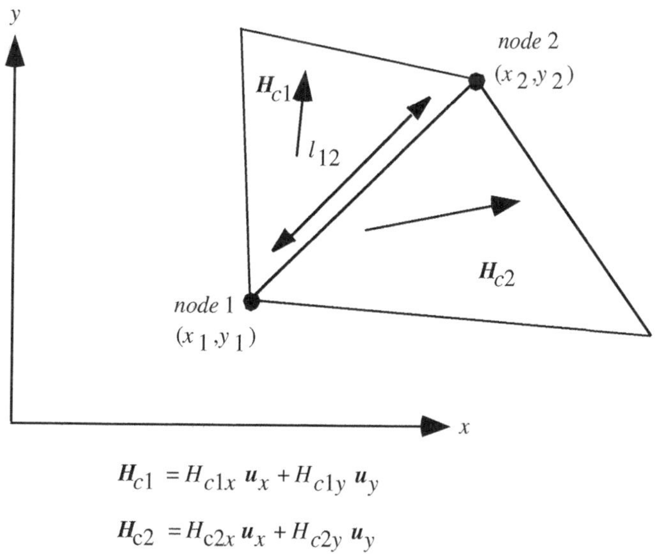

Figure 10.11 Permanent magnet element interface.

common element side. The surface current density required, J_{pm12}, can be found from the tangential components of $\boldsymbol{H}_{c1}$ and $\boldsymbol{H}_{c2}$ along the common element side yielding,

$$J_{pm12} = \frac{(H_{c1x} - H_{c2x})(x_1 - x_2) + (H_{c1y} - H_{c2y})(y_1 - y_2)}{l_{12}} \tag{10.124}$$

The total current along the element boundary I_{12} is simply

$$I_{12} = l_{12} J_{pm12} \tag{10.125}$$

This surface current is evenly split between the two nodes (1 and 2 in Figure 10.11) to obtain nodal currents which contribute to the vector components in $[T_{pm}]$ in equation (10.123). Taking into account of all permanent magnet elements in the solution region, $[T_{pm}]$ can be constructed if a vector is defined such that

$$[T_{pm}] = [T_{pm1}, T_{pm2}, \dots, T_{pmn}]^t \tag{10.126}$$

where

$$T_{pmi} = \sum_k \frac{1}{2} \boldsymbol{H}_{ck} \cdot \boldsymbol{l}_k \tag{10.127}$$

The summation is over all elements with vertices at node i and H_{ck} is the coercive force vector of element k. The length vector is obtained from the sum of the length vectors associated with the two element sides which radiate from node i so that the

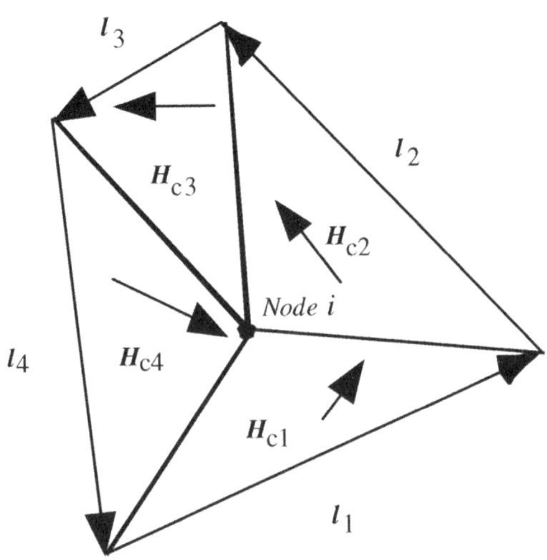

T_{pmi} = *Equivalent Current of Node i Due to PM Excitation*

$$= \sum_{k=1}^{4} \frac{1}{2} H_{ck} \cdot l_k$$

Figure 10.12 Nodal current calculation for finite-element model of permanent magnet.

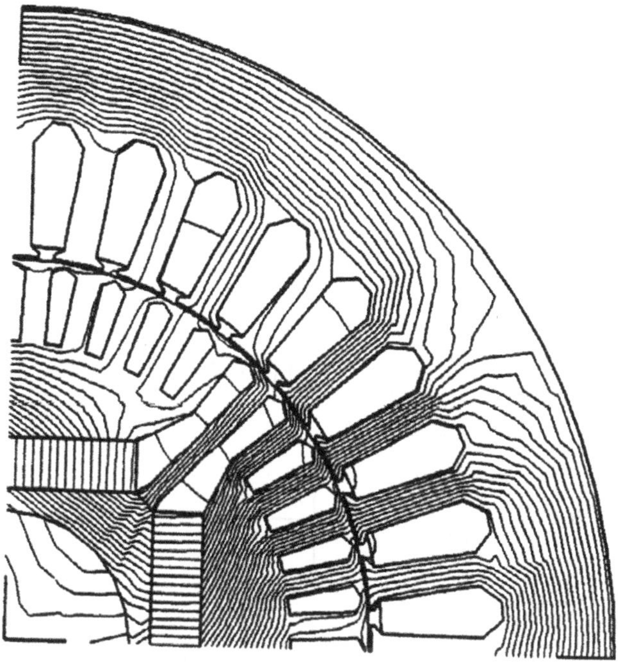

Figure 10.13 Equal vector potential line plot for permanent magnet machine operation under full load and rated voltage.

magnitude of $\bar{l}_k$ is equal to the length of the element side opposite node i and its direction is parallel to this side in a counter-clockwise direction around node i as shown in Figure 10.12. Hence, both types of practical sources, current sources, and permanent magnets, can be incorporated into $[T]$ and $[T_{\text{pm}}]$

A typical example of a finite-element plot of a permanent magnet motor is shown in Figure 10.13 [5]. Note particularly, that the solution is obtained over only one of the four poles of the machine. Hence, the odd symmetry boundary condition constraint over the core at the two sides of the boundary as discussed in Section 10.10.

10.12 CONCLUSION

This chapter has served only as an introduction to the extensive and ongoing work in the field of finite-element models. Undoubtedly, the reader will not program his own code to solve such problems but utilize one of the many source codes, several of which are available without cost. However, the principle of GIGO (garbage in produces garbage out) will always prevail and the use of a "soft black box" (i.e., someone else's computer code) could easily produce such results unless the user has at least a basic understanding of what is inside. It is in this spirit that this chapter has been written.

10.A APPENDIX

In order to relate the third term of equation (10.28) to (10.23), it is necessary to show that

$$\boldsymbol{H} \bullet \frac{\partial \boldsymbol{B}}{\partial t} = \frac{\partial}{\partial t} \int_0^B (\boldsymbol{H} \bullet d\boldsymbol{B}) \tag{10.A.1}$$

First, the chain rule for partial derivatives is used to yield

$$\frac{\partial}{\partial t} \int_0^B (\boldsymbol{H} \bullet d\boldsymbol{B}) = \left[\frac{\partial}{\partial B} \int_0^B (\boldsymbol{H} \bullet d\boldsymbol{B}) \right] \frac{\partial B}{\partial t} + \left[\frac{\partial}{\partial H} \int_0^B (\boldsymbol{H} \bullet d\boldsymbol{B}) \right] \frac{\partial H}{\partial t} \tag{10.A.2}$$

$$= \boldsymbol{H} \bullet \frac{\partial \boldsymbol{B}}{\partial t} + \left[\frac{\partial}{\partial H} \int_0^B (\boldsymbol{H} \bullet d\boldsymbol{B}) \right] \frac{\partial H}{\partial t} \tag{10.A.3}$$

Integrating the remaining integral by parts,

$$\frac{\partial}{\partial t} \int_0^B (\boldsymbol{H} \bullet d\boldsymbol{B}) = \boldsymbol{H} \bullet \frac{\partial \boldsymbol{B}}{\partial t} + \left[\frac{\partial}{\partial H} \left(\boldsymbol{H} \bullet \boldsymbol{B} - \int_0^H (\boldsymbol{B} \bullet d\boldsymbol{H}) \right) \right] \tag{10.A.4}$$

$$= \left(\boldsymbol{H} \bullet \frac{\partial \boldsymbol{B}}{\partial t} \right) + (\boldsymbol{B} - \boldsymbol{B}) \tag{10.A.5}$$

so that finally

$$= \boldsymbol{H} \bullet \frac{\partial \boldsymbol{B}}{\partial t} \tag{10.A.6}$$

Utilizing equation (10.A.6), equation (10.28) becomes

$$\boldsymbol{E} \times \boldsymbol{H} \bullet d\boldsymbol{S} + \iiint_V \boldsymbol{E} \bullet \boldsymbol{J} dV + \iiint_V \left[\frac{\partial}{\partial t} \int_0^B (\boldsymbol{H} \bullet d\boldsymbol{B}) \right] dV = 0 \tag{10.A.7}$$

Interchanging the order of the integration and differentiation in the last term

$$\int_S \boldsymbol{E} \times \boldsymbol{H} \bullet d\boldsymbol{S} + \iiint_V \boldsymbol{E} \bullet \boldsymbol{J} dV + \frac{\partial}{\partial t} \iiint_V \left[\int_0^B (\boldsymbol{H} \bullet d\boldsymbol{B}) \right] dV = 0 \tag{10.A.8}$$

The third integral on the left-hand side is the time rate of change of increase in the stored magnetic energy. Upon integrating this term the stored magnetic energy is

$$W_{\mathrm{m}} = \iiint \left(\int_0^B (\boldsymbol{H} \bullet d\boldsymbol{B}) \right) dV \tag{10.A.9}$$

Note that the integrand of the volume integral is clearly the magnetic energy density or,

$$w_{\mathrm{m}} = \int_0^B (\boldsymbol{H} \bullet d\boldsymbol{B}) \tag{10.A.10}$$

which is the same as equation (10.23). Hence, the validity of this expression is also proven.

REFERENCES

[1] A. Stratton, *Electromagnetic Theory*, McGraw-Hill, New York, 1941, pp. 131–132.

[2] P. Sylvester and M. V. K. Chari, "Finite element solution of saturable magnetic field problems," *IEEE Transactions on Power Apparatus and Systems*, vol. PAS-89, no. 7, September/October 1970, pp. 1642–1652.

[3] J. R. Brauer, "Saturated Magnetic Energy Functional for Finite Element Analysis of Electric Machines," IEEE PES Winter Meeting, Paper C75- 151-6, January 26–31, 1975, p.

[4] K. J. Binns, T. S. Low, and M. A. Jabbar, "Behaviour of a polymer-bonded rare-earth magnet under excitation in two directions at right angles," *IEE Proceedings*, vol. 130, Pt. B, no. 1, January 1983, pp. 25–32.

[5] R. Schiferl, "Design Considerations for Salient Pole, Permanent Magnet Synchronous Motors in Variable Speed Drive Applications," Ph.D. Thesis, University of Wisconsin, September 1987.

APPENDIX A

COMPUTATION OF BAR CURRENT

```
% Program for plotting speed/torque curves for an induction machine
% utilizing the per-phase equivalent circuit model. The skin effect is
% considered for calculation of referred rotor resistance and leakage
% inductance parameters from the rotor bar geometry.

for index=1:3      % the index loop iterates the entire process for the
                        % three cases of rotor dimensions.
% Induction machine parameters
r1 = 0.435;                 % equivalent circuit stator resistance
l1 = 3.32/377;              % equivalent circuit stator leakage
                            % reactance
lm = 81.4/377;              % equivalent circuit magnetizing reactance
re = 0.78e-06;              % resistance of end ring portion of one
                            % rotor slot pitch
leff = 9.06*2.54;           % effective length of rotor bars under
                            % stator iron (cm)
lgross = 13.95*2.54;        % gross bar length including extensions and
                            % ducts (cm)
rslots = 97.;               % number of rotor slots
ns = 240;                   % number of stator turns
lend = 1.986e-08;           % leakage reactance of rotor ring
lbhar = 0.259e-6;           % leakage inductance of rotor harmonics for
                            % one bar
lbtop = 0.947e-6;           % leakage inductance of top air portion of
                            % bar
rho = 2.1e-06;              % resistivity of rotor bar material (ohm-
                            % cm)
v = 2400/sqrt(3);           % applied voltage
u = 4*pi*1.e-09;            % Henries per cm
bardepthbase = .5625*2.54;  % rotor bar depth in cm
barwidthbase = .375*2.54;   % rotor bar width in cm
k =.91;                     % winding factor (pitch,distribution,skew)
p =8.;                      % number of machine poles
fop=60;
we = 2*pi*fop;
slices = 100;
```

Introduction to AC Machine Design, First Edition. Thomas A. Lipo.

```
bardepth = bardepthbase*index/2;
barwidth = 2*barwidthbase/index;
% Induction Machine Rotor Matrix, elements arranged (height, width)
                              % all in cm
for i=1:slices
rm(i,1)=bardepth/slices;
rm(i,2)=barwidth;
end

% determine number of "slices" of the rotor bar
number=slices;

for count=1:100    % count iterates the calculation of rotor parameters
                   % as well as equivalent circuit currents and
                   % machine output torque for slip ranging from
                   % 0.01 to 1.0

slip = 0.01*count;
s(count)=0.01*count;

% determine resistance for each "slice" of the rotor bar
for n= 1:number
resist(n)=(lgross*rho)/(rm(n,1)* rm(n,2));
end

% determine the leakages for each slice of the rotor bar
for n=1:number
leak(n)=2*pi*slip*fop*u*leff*rm(n,1)/ rm(n,2);
end

% intial current is set to 0.001 amp, everything will eventually be
% scaled to the applied voltage.
current(1)=0.001;
currentsum=0.001;
for n=2:number
current(n)=resist(n-1)*current(n-1)/resist(n) + j*leak(n-1)*currentsum/
  resist(n);
currentsum=currentsum+current(n);
end
% determine voltage seen at air gap
magvolts= resist(number)*current(number) +
j*(leak(number)+slip*we*lbtop)*currentsum;

% determine equivalent bar resistance and reactance
z = magvolts/currentsum;
resistbar = real(z);
reactbar = imag(z);

%final determination of rotor resistance reflected onto the stator
rbe = (resistbar + re/(2*(sin((pi*p)/(2*rslots)))^2));
rr(count) = 12*k*k*ns*ns*(2*rbe)/rslots;
rrref=rr(count);
```

```
% final determination of rotor reactance reflected onto the stator
llr= (reactbar/(2*pi*slip*fop)) + lend/(2*(sin((pi*p)/(2*rslots)))^2 + lbhar);
lr(count) = 12*k*k*ns*ns*(2*llr)/rslots;
lrref=lr(count);

% induction motor equivalent circuit model

% combine rotor and magnetizing reactances
zeq(count)=((j*we*lm*rr(count)/s(count))-(we*we*lr(count)*lm))
  /(rr(count)/s(count) +
j*we*(lm+lr(count)));

% determine stator current
is(count) = v / (r1 + j*we*l1 + zeq(count));

% determine voltage across the magnetizing reactance
vm(count) = v - is(count)*(r1 + j*we*l1);

% determine rotor current
ir(count) = vm(count)/(j*we*lr(count) + rr(count)/s(count));

% determine mechanical shaft speed
speed(count) = 3600*(2/p) *(1-s(count));
power(count) = 3*abs(ir(count))^2*rr(count)*(1-s(count))/s(count);
powert(count) = p*power(count)/(2*we);

% determine output torque
torque(count) = 3*(abs(ir(count))^2)* rr(count) / (s(count)*2*we/p);
if index==1
              rout1 (count)=rr(count);
              lout1 (count)=lr(count);
              tout1 (count)=torque(count);
elseif index==2
              rout2(count)=rr(count);
              lout2(count)=lr(count);
              tout2(count)=torque(count);
else
              rout3(count)=rr(count);
end
end
end
```

APPENDIX **B**

FEM EXAMPLE

```
% FEM code (MATLAB) for the example in Chapter 10
% Generated by Wen Ouyang
% The mesh is generated in a straightforward manner for clarity
% Physical model and X,Y notation:
```

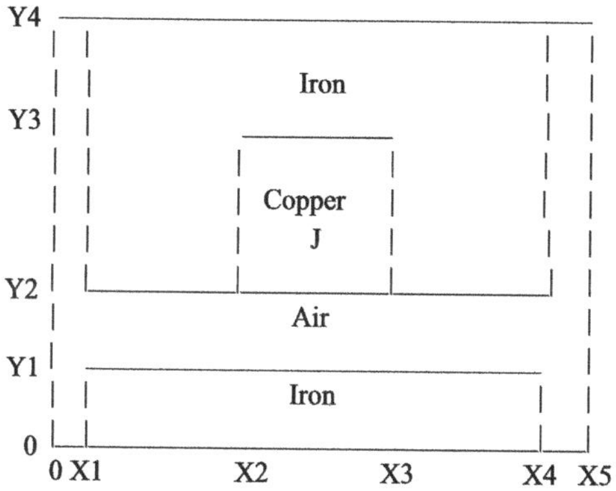


```
clear all; close all; clc;

% ------------------------------------------------------------------(1)
% Parameters

% define BH curve of 1010 steel in Amperes/m.
BH=[    0           0;      0.2003    238.7;     0.3204    318.3;
        0.40045   358.1;    0.50055   437.7;     0.5606    477.5;
        0.7908    636.6;    0.931     795.8;     1.1014   1114.1;
        1.2016   1273.2;    1.302    1591.5;     1.4028   2228.2;
        1.524    3183.1;    1.626    4774.6;     1.698    6366.2;
        1.73     7957.7;    1.87    15915.5;     1.99      31831;
        2.04    47746.5;    2.07      63662;     2.095   79577.5;
        2.2      159155;    2.4      318310];

% Constants

    u0=4*pi*1e-7;            % free space permeability
    ur_air=1.0000004;        % Air relative permeability
    ur_cop=0.999991;         % Copper relative permeability
    acc=0.1;                 % acceleration factor
```

```
   Berror=0.01;          % error control factor for B
   Aerror=0.01;          % error control factor for A
   Bmax=2;               % saturation level limit
   inTome=2.54/100;      % inch to meter ratio

% Dimension in RELATIVE value to the previous point

  X1 = 0.5;              % inches, left air region width
  X2 = 2;                % inches, left tooth width
  X3 = 2;                % inches, slot width
  X4 = 2;                % inches, right tooth width
  X5 = 0.5;              % inches, right air region width

  Y1 = 2;                % inches, bottom iron height
  Y2 = 0.5;              % inches, air gap height
  Y3 = 2;                % inches, coil height
  Y4 = 2;                % inches, yoke height

% initial values

 Binit=0.5;              % Tesla, initial value
 vinit=interp1(BH(:,1),BH(:,2),Binit,'linear')/Binit;

 Aboundary=0;            % boundary condition set to zero
 J=1000;                 % Amp/in^2, current density
 J = J*2^2/(2*2.54/100)^2 %Amp/m^2, current density

 status=0;               % check if status=1, solution converged
 Loop=0;                 % loop counts

%----------------------------------------------------------(2)
% Generate meshes and number the nodes of meshes.

dx = 0.5/3;    % X direction mesh step, it is a global setting.
% check dx to match it with X1=0.5,X2=X3=X4=2,X5=0.5.

dy1 = 0.4;    % Bottom iron Y direction mesh step, note Y1=2.
dy2 = 0.1;    % Air gap Y direction mesh step, notice airgap=0.5.
dy3 = 0.4;  % Top iron tooth Y direction mesh step, note Y3=2.
dy4 = 1.0;  % Top iron yoke Y direction mesh step, note Y4=2.

% Node number information

x1=round(X1/dx+1);  x2=round(x1+X2/dx);     x3=round(x2+X3/dx);
x4=round(x3+X4/dx); x5=round(x4+X5/dx);

NodeX=x5;      % Nodes in X direction

y1=round(Y1/dy1+1); y2=round(y1+Y2/dy2);    y3=round(y2+Y3/dy3);
y4=round(y3+Y4/dy4);

NodeY=y4;      % Nodes in Y direction

TotalNodes=round(NodeX*NodeY);
TotalMeshes=round((NodeX-1)*(NodeY-1)*2);
```

```
% Develop nodal matrix for nodes with actual node coordinates in meters.
% Each node corresponding to each row, node is numbered from bottom to top,
% from left to right.

Nodes=zeros(TotalNodes,2);

% from the bottom to top, from left to right
for i=1:NodeY;              % Y-coordinate
    for j=1:NodeX;          % X-coordinate
        if i<=y1;           % Bottom iron region
            Nodes((i-1)*NodeX+j,:)=[(j-1)*dx,(i-1)*dy1]*inTome;
        end
        if (i>y1)&(i<=y2);  % air region
            Nodes((i-1)*NodeX+j,:)=[(j-1)*dx,Y1+(i-y1)*dy2,]*inTome;
        end
        if (i>y2)&(i<=y3);  % Top iron tooth region
            Nodes((i-1)*NodeX+j,:)=[(j-1)*dx,Y1+Y2+(i-y2)*dy3]*inTome;
        end
        if (i>y3);             % Top yoke region
                         Nodes((i-1)*NodeX+j,:)=[(j-1)*dx,Y1+Y2+Y3+(i-y3)
                         *dy4]*inTome;
        end
    end
end

%-------------------------------------------------------------------(3)
% Number the meshes and vertices.
% Assign material number and the initial reluctance.
% Calculate the triangle area.

A=zeros(TotalNodes,1);
B=zeros(TotalMeshes,1);
delta=zeros(TotalMeshes,1);

% Develop mesh element matrix(MEM), each row is for each mesh with
% 3 triangle nodal numbers, material number, and reluctance for
% each element.
% Material 0: Air
%          1: Iron
%          2: Copper

MEM=zeros(TotalMeshes,5);

for i=1:(NodeY-1);
    for j=1:(NodeX-1);
        if i<y1                 % air & iron
            if (j<x1)|(j>=x4)   % air
                Material=0;
                v=1/(u0*ur_air);
            else                % iron
                Material=1;
                v=vinit;
            end
        end
```

```
        if (i>=y1)&(i<y2)            % air
            Material=0;
            v=1/(u0*ur_air);
        end

        if (i>=y2)&(i<y3)            % air, iron & copper
            if (j<x1)|(j>=x4)        % air
                Material=0;
                v=1/(u0*ur_air);
            elseif (j>=x2)&(j<x3)    % copper
                Material=2;
                v=1/(u0*ur_cop);
            else                     % iron
                Material=1;
                v=vinit;
            end
        end

        if (i>=y3)&(i<y4)            % air & iron
            if (j<x1)|(j>=x4)        % air
                Material=0;
                v=1/(u0*ur_air);
            else                     % iron
                Material=1;
                v=vinit;
            end
        end

        elemBot=((j-1)*2+1)+((i-1)*2*(NodeX-1));
                                % serial number of bottom triangle
        elemTop=elemBot+1;      % serial number of top triangle

        vert1=j+(i-1)*NodeX;    % Nodal number of vertex at bottom left
        vert2=vert1+1;          % Nodal number of vertex at bottom right
        vert3=vert1+NodeX;      % Nodal number of vertex at top left
        vert4=vert3+1;          % Nodal number of vertex at top right

        MEM(elemBot,:)=[vert1,vert2,vert3,Material,v];
        MEM(elemTop,:)=[vert3,vert2,vert4,Material,v];

        B(elemBot)=Binit;       % initial value for element B
        B(elemTop)=Binit;
    end
end

% Calculate the area of triangle, store in delta vector indexed
% by element number

  for index=1:TotalMeshes


   i=MEM(index,1);     j=MEM(index,2);     k=MEM(index,3);

    xi=Nodes(i,1);     yi=Nodes(i,2);      xj=Nodes(j,1);
    yj=Nodes(j,2);     xk=Nodes(k,1);      yk=Nodes(k,2);
```

```
        area=((xj*yk-xk*yj)+xi*(yj-yk)+yi*(xk-xj))/2;
        delta(index)=area;
    end

% --------------------------------------------------------------(4)
% [S] & [T] Matrices and boundary conditions

S=zeros(TotalNodes,TotalNodes);
T=zeros(TotalNodes,1);

for index=1:TotalMeshes

    i=MEM(index,1);     j=MEM(index,2);     k=MEM(index,3);

    xi=Nodes(i,1);      yi=Nodes(i,2);      xj=Nodes(j,1);
    yj=Nodes(j,2);      xk=Nodes(k,1);      yk=Nodes(k,2);

    ai=xj*yk-xk*yj;         bi=yj-yk;           ci=xk-xj;
    aj=yi*xk-xi*yk;         bj=yk-yi;           cj=xi-xk;
    ak=xi*yj-xj*yi;         bk=yi-yj;           ck=xj-xi;

    S(i,i)=S(i,i)+MEM(index,5)*(bi^2+ci^2)/(4*delta(index));
    S(i,j)=S(i,j)+MEM(index,5)*(bi*bj+ci*cj)/(4*delta(index));
    S(i,k)=S(i,k)+MEM(index,5)*(bi*bk+ci*ck)/(4*delta(index));
    S(j,i)=S(i,j);
    S(j,j)=S(j,j)+MEM(index,5)*(bj^2+cj^2)/(4*delta(index));
    S(j,k)=S(j,k)+MEM(index,5)*(bj*bk+cj*ck)/(4*delta(index));
    S(k,i)=S(i,k);
    S(k,j)=S(j,k);
    S(k,k)=S(k,k)+MEM(index,5)*(bk^2+ck^2)/(4*delta(index));
end

% boundary condition, boundaries all set to zero

% bottom and top lines:
    for j=1:NodeX

        ZeroNode1=j;
        ZeroNode2=NodeX*(NodeY-1)+j;

        S(ZeroNode1,:)=zeros(1,size(Nodes,1));
                                                % entire row set to zeros
        S(ZeroNode1,ZeroNode1)=1;               % set the diagonal to one
        T(ZeroNode1)=Aboundary;                 % A=0

        S(ZeroNode2,:)=zeros(1,size(Nodes,1));
                                                % entire row set to zeros
        S(ZeroNode2,ZeroNode2)=1;               % set the diagonal to one
        T(ZeroNode2)=Aboundary;                 % A=0
    end

% right and left lines:
    for i=1:NodeY

        ZeroNode1=i*NodeX;
```

```
        ZeroNode2=(i-1)*NodeX+1;

        S(ZeroNode1,:)=zeros(1,size(Nodes,1));
                                                % entire row set to zeros
        S(ZeroNode1,ZeroNode1)=1;               % set the diagonal to one
        T(ZeroNode1)=Aboundary;                 % A=0

        S(ZeroNode2,:)=zeros(1,size(Nodes,1)); % entire row set to zeros
        S(ZeroNode2,ZeroNode2)=1;               % set the diagonal to one
        T(ZeroNode2)=Aboundary;                 % A=0
    end

% current density on copper area
    for N=1:TotalNodes
        I=0;
        for index=1:TotalMeshes
            i=MEM(index,1);
            j=MEM(index,2);
            k=MEM(index,3);
            if ((N==i)|(N==j)|(N==k))&(MEM(index,4)==2)
                I=I+J*delta(index);
            end
        end
        if (I~=0)
            T(N)=I/3;
        end
    end

%-------------------------------------------------------------------(5)
% Calculating until B & A converge
 while ~status                          % continue until convergence

    Loop=Loop+1;      Aold=A;     Bold=B;

    A=S\T;                              % calculate new A by inv(S)*T

%update the reluctivity of mesh elements and calculate the field energy

    for index=1:TotalMeshes

        i=MEM(index,1);        j=MEM(index,2);         k=MEM(index,3);

        xi=Nodes(i,1);         yi=Nodes(i,2);          xj=Nodes(j,1);
        yj=Nodes(j,2);         xk=Nodes(k,1);          yk=Nodes(k,2);

        ai=xj*yk-xk*yj;            bi=yj-yk;               ci=xk-xj;
        aj=yi*xk-xi*yk;            bj=yk-yi;               cj=xi-xk;
        ak=xi*yj-xj*yi;            bk=yi-yj;               ck=xj-xi;

 % update flux density in each element

 Bx=(1/(2*delta(index)))*(ci*A(MEM(index,1))+cj*A(MEM(index,2))+ck*A
 (MEM(index,3)));
 By=(-1/(2*delta(index)))*(bi*A(MEM(index,1))+bj*A(MEM(index,2))+bk*A
 (MEM(index,3)));
```

```
        B(index)=sqrt(Bx^2+By^2);
        if B(index)>=Bmax              % saturation control
            B(index)=Bmax;
        end

        % update the reluctivity,v

        if (MEM(index,4)==1)           % iron
           vcalc=interp1(BH(:,1),BH(:,2),B(index),'linear')/B(index);

        elseif (MEM(index,4)==2)    % copper
            vcalc=1/(ur_cop*u0);
        else
            vcalc=1/(ur_air*u0);    % air
        end

        v=MEM(index,5)+acc*(vcalc-MEM(index,5));
        MEM(index,5)=v;

        % Update [S]
        S=zeros(TotalNodes,TotalNodes);
        for index_1=1:TotalMeshes

           i=MEM(index_1,1);    j=MEM(index_1,2);    k=MEM(index_1,3);

           xi=Nodes(i,1);     yi=Nodes(i,2);      xj=Nodes(j,1);
           yj=Nodes(j,2);     xk=Nodes(k,1);      yk=Nodes(k,2);

           ai=xj*yk-xk*yj;          bi=yj-yk;           ci=xk-xj;
           aj=yi*xk-xi*yk;          bj=yk-yi;           cj=xi-xk;
           ak=xi*yj-xj*yi;          bk=yi-yj;           ck=xj-xi;

S(i,i)=S(i,i)+MEM(index_1,5)*(bi^2+ci^2)/(4*delta(index_1));
S(i,j)=S(i,j)+MEM(index_1,5)*(bi*bj+ci*cj)/(4*delta(index_1));
S(i,k)=S(i,k)+MEM(index_1,5)*(bi*bk+ci*ck)/(4*delta(index_1));
S(j,i)=S(i,j);
S(j,j)=S(j,j)+MEM(index_1,5)*(bj^2+cj^2)/(4*delta(index_1));
S(j,k)=S(j,k)+MEM(index_1,5)*(bj*bk+cj*ck)/(4*delta(index_1));
S(k,i)=S(i,k);
S(k,j)=S(j,k);
S(k,k)=S(k,k)+MEM(index_1,5)*(bk^2+ck^2)/(4*delta(index_1));
      end

      % Energy calculation for each element

      Energy(index)=B(index)^2*delta(index)*MEM(index,5)/2;

  end % end of index loop

% Reapply boundary condition.

  % left and right lines:
  for i=1:NodeY

      ZeroNode1=i*NodeX;
```

```
        ZeroNode2=(i-1)*NodeX+1;

        S(ZeroNode1,:)=zeros(1,size(Nodes,1));
                                                    % entire row set to zeros
        S(ZeroNode1,ZeroNode1)=1;                   % set the diagonal to one
        T(ZeroNode1)=Aboundary;                     % A=0

        S(ZeroNode2,:)=zeros(1,size(Nodes,1));
                                                    % entire row set to zeros
        S(ZeroNode2,ZeroNode2)=1;                   % set the diagonal to one
        T(ZeroNode2)=Aboundary;                     % A=0
    end

    % bottom and top lines:
    for j=1:NodeX

        ZeroNode1=j;
        ZeroNode2=NodeX*(NodeY-1)+j;

        S(ZeroNode1,:)=zeros(1,size(Nodes,1));
                                                    % entire row set to zeros
        S(ZeroNode1,ZeroNode1)=1;                   % set the diagonal to one
        T(ZeroNode1)=Aboundary;                     % A=0

        S(ZeroNode2,:)=zeros(1,size(Nodes,1));
                                                    % entire row set to zeros
        S(ZeroNode2,ZeroNode2)=1;                   % set the diagonal to one
        T(ZeroNode2)=Aboundary;                     % A=0
    end
    deltaB=abs(B-Bold);                             % Check convergence of B
    deltaA=abs(A-Aold);                             % Check convergence of A

                    status=(sqrt(sum(deltaB.^2))/sqrt(sum(B.^2))<=Berror)&
                    (sqrt(sum(deltaA.^2))/sqrt(sum(A.^2))<=Aerror);

    disp('Loop='); disp(Loop);
    if Loop >=50          % Loop control
        break;
    end
end                           % end of while Loop

%----------------------------------------------------------------------(6)
% Draw flux lines, model, meshes

    TotalEnergy=sum(Energy);
    TotalFlux=max(A);

    Az=zeros(NodeY,NodeX);      % rearrange the Az distribution in the model
                                format
    Xaxis=zeros(NodeX,1);
    Yaxis=zeros(NodeY,1);

    for i=1:NodeY
```

```
        for j=1:NodeX
            Az(i,j)=A(j+(i-1)*NodeX);
        end
    end

% Find x coordinates values for material critical nodes in INCH.

    for j=1:NodeX
        if(j<=x1)               Xaxis(j)=(j-1)*dx;end
        if(j>x1)&(j<=x2)        Xaxis(j)=(j-x1)*dx+X1;end
        if(j>x2)&(j<=x3)        Xaxis(j)=(j-x2)*dx+X1+X2; end
        if(j>x3)&(j<=x4)        Xaxis(j)=(j-x3)*dx+X1+X2+X3; end
        if(j>x4)&(j<=x5)        Xaxis(j)=(j-x4)*dx+X1+X2+X3+X4; end
    end

    for i=1:NodeY
        if(i<=y1)               Yaxis(i)=(i-1)*dy1; end
        if(i>y1)&(i<=y2)        Yaxis(i)=(i-y1)*dy2+Y1; end
        if(i>y2)&(i<=y3)        Yaxis(i)=(i-y2)*dy3+Y1+Y2; end
        if(i>y3)&(i<=y4)        Yaxis(i)=(i-y3)*dy4+Y1+Y2+Y3; end
    end

% Plot model
    PP(1,:)=[0,                 0];
    PP(2,:)=[X1,                0];
    PP(3,:)=[X1+X2+X3+X4,       0];
    PP(4,:)=[X1+X2+X3+X4+X5,0];
    PP(5,:)=[X1,                Y1];
    PP(6,:)=[X1+X2+X3+X4,       Y1];
    PP(7,:)=[X1,                Y1+Y2];
    PP(8,:)=[X1+X2,             Y1+Y2];
    PP(9,:)=[X1+X2+X3,          Y1+Y2];
    PP(10,:)=[X1+X2+X3+X4,      Y1+Y2];
    PP(11,:)=[X1+X2,            Y1+Y2+Y3];
    PP(12,:)=[X1+X2+X3,         Y1+Y2+Y3];
    PP(13,:)=[0,                Y1+Y2+Y3+Y4];
    PP(14,:)=[X1,               Y1+Y2+Y3+Y4];
    PP(15,:)=[X1+X2+X3+X4,      Y1+Y2+Y3+Y4];
    PP(16,:)=[X1+X2+X3+X4+X5,Y1+Y2+Y3+Y4];

% Plot iron
    IronX1=[PP(2,1) PP(5,1) PP(6,1) PP(3,1)];
    IronY1=[PP(2,2) PP(5,2) PP(6,2) PP(3,2)];
    IronX2=[PP(7,1)PP(8,1)PP(11,1) PP(12,1),PP(9,1),PP(10,1), PP(15,1),
PP(14,1)];
    IronY2=[PP(7,2)PP(8,2)PP(11,2),PP(12,2),PP(9,2),PP(10,2),
PP(15,2),PP(14,2)];

% Plot copper
    CopX1=[PP(8,1) PP(9,1) PP(12,1) PP(11,1)];
    CopY1=[PP(8,2) PP(9,2) PP(12,2) PP(11,2)];

    figure(1);
    fill(IronX1,IronY1,'y'); hold on;
    fill(IronX2,IronY2,'y'); hold on;
```

```
    fill(CopX1,CopY1,'r');hold on;
    axis equal; axis([0 7 0 6.5]);

    contour(Xaxis,Yaxis,Az,30,'b');
    disp('Amax='); disp(TotalFlux);

% Plot Meshes
    figure(2);

    fill(IronX1,IronY1,'y'); hold on;
    fill(IronX2,IronY2,'y'); hold on;
    fill(CopX1,CopY1,'r');   hold on;
    axis equal; axis([0 7 0 6.5]);

    for index=1:TotalMeshes

        i=MEM(index,1);          j=MEM(index,2);          k=MEM(index,3);

  xi=Nodes(i,1)/inTome; yi=Nodes(i,2)/inTome;   xj=Nodes(j,1)/inTome;
  yj=Nodes(j,2)/inTome; xk=Nodes(k,1)/inTome; yk=Nodes(k,2)/inTome;

line([xi,xj],[yi,yj]); line([xj,xk],[yj,yk]);  line([xk,xi], [yk,yi]);

    end
```

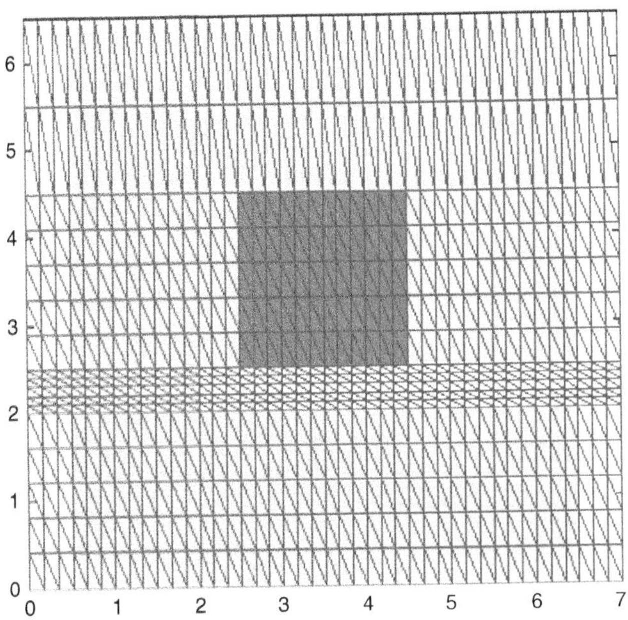

Figure B.1 Mesh generated.

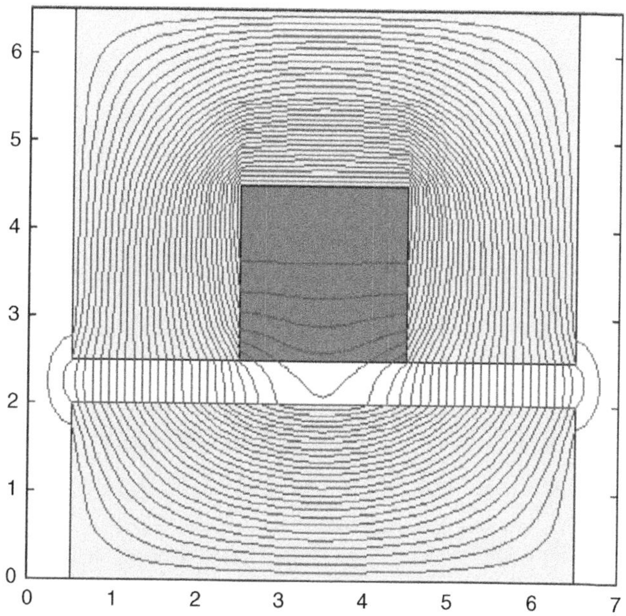

Figure B.2 FEM solution.

INDEX

Introduction to AC Machine Design, First Edition. Thomas A. Lipo.

IEEE Press Series on Power Engineering

Series Editor: **M. E. El-Hawary,** Dalhousie University, Halifax, Nova Scotia, Canada

The mission of IEEE Press Series on Power Engineering is to publish leading-edge books that cover the broad spectrum of current and forward-looking technologies in this fast-moving area. The series attracts highly acclaimed authors from industry/academia to provide accessible coverage of current and emerging topics in power engineering and allied fields. Our target audience includes the power engineering professional who is interested in enhancing their knowledge and perspective in their areas of interest.

1. *Principles of Electric Machines with Power Electronic Applications, Second Edition*
M. E. El-Hawary

2. *Pulse Width Modulation for Power Converters: Principles and Practice*
D. Grahame Holmes and Thomas Lipo

3. *Analysis of Electric Machinery and Drive Systems, Second Edition*
Paul C. Krause, Oleg Wasynczuk, and Scott D. Sudhoff

4. *Risk Assessment for Power Systems: Models, Methods, and Applications*
Wenyuan Li

5. *Optimization Principles: Practical Applications to the Operations of Markets of the Electric Power Industry*
Narayan S. Rau

6. *Electric Economics: Regulation and Deregulation*
Geoffrey Rothwell and Tomas Gomez

7. *Electric Power Systems: Analysis and Control*
Fabio Saccomanno

8. *Electrical Insulation for Rotating Machines: Design, Evaluation, Aging, Testing, and Repair, Second Edition*
Greg Stone, Edward A. Boulter, Ian Culbert, and Hussein Dhirani

9. *Signal Processing of Power Quality Disturbances*
Math H. J. Bollen and Irene Y. H. Gu

10. *Instantaneous Power Theory and Applications to Power Conditioning*
Hirofumi Akagi, Edson H. Watanabe, and Mauricio Aredes

11. *Maintaining Mission Critical Systems in a 24/7 Environment*
Peter M. Curtis

12. *Elements of Tidal-Electric Engineering*
Robert H. Clark

13. *Handbook of Large Turbo-Generator Operation and Maintenance, Second Edition*
Geoff Klempner and Isidor Kerszenbaum

14. *Introduction to Electrical Power Systems*
Mohamed E. El-Hawary

15. *Modeling and Control of Fuel Cells: Distributed Generation Applications*
M. Hashem Nehrir and Caisheng Wang

16. *Power Distribution System Reliability: Practical Methods and Applications*
Ali A. Chowdhury and Don O. Koval

17. *Introduction to FACTS Controllers: Theory, Modeling, and Applications*
Kalyan K. Sen and Mey Ling Sen

18. *Economic Market Design and Planning for Electric Power Systems*
James Momoh and Lamine Mili

19. *Operation and Control of Electric Energy Processing Systems*
James Momoh and Lamine Mili

20. *Restructured Electric Power Systems: Analysis of Electricity Markets with Equilibrium Models*
Xiao-Ping Zhang

21. *An Introduction to Wavelet Modulated Inverters*
S.A. Saleh and M.A. Rahman

22. *Control of Electric Machine Drive Systems*
Seung-Ki Sul

23. *Probabilistic Transmission System Planning*
Wenyuan Li

24. *Electricity Power Generation: The Changing Dimensions*
Digambar M. Tigare

25. *Electric Distribution Systems*
Abdelhay A. Sallam and Om P. Malik

26. *Practical Lighting Design with LEDs*
Ron Lenk and Carol Lenk

27. *High Voltage and Electrical Insulation Engineering*
Ravindra Arora and Wolfgang Mosch

28. *Maintaining Mission Critical Systems in a 24/7 Environment, Second Edition*
Peter Curtis

29. *Power Conversion and Control of Wind Energy Systems*
Bin Wu, Yongqiang Lang, Navid Zargari, and Samir Kouro

30. *Integration of Distributed Generation in the Power System*
Math H. Bollen and Fainan Hassan

31. *Doubly Fed Induction Machine: Modeling and Control for Wind Energy Generation Applications*
Gonzalo Abad, Jesus Lopez, Miguel Rodrigues, Luis Marroyo, and Grzegorz Iwanski

32. *High Voltage Protection for Telecommunications*
Steven W. Blume

33. *Smart Grid: Fundamentals of Design and Analysis*
James Momoh

34. *Electromechanical Motion Devices, Second Edition*
Paul C. Krause, Oleg Wasynczuk, and Steven D. Pekarek

35. *Electrical Energy Conversion and Transport: An Interactive Computer-Based Approach, Second Edition*
George G. Karady and Keith E. Holbert

36. *ARC Flash Hazard and Analysis and Mitigation*
J. C. Das

37. *Handbook of Electrical Power System Dynamics: Modeling, Stability, and Control*
Mircea Eremia and Mohammad Shahidehpour

38. *Analysis of Electric Machinery and Drive Systems, Third Edition*
Paul Krause, Oleg Wasynczuk, S. D. Sudhoff, and Steven D. Pekarek

39. *Extruded Cables for High-Voltage Direct-Current Transmission: Advances in Research and Development*
Giovanni Mazzanti and Massimo Marzinotto

40. *Power Magnetic Devices: A Multi-Objective Design Approach*
S. D. Sudhoff

41. *Risk Assessment of Power Systems: Models, Methods, and Applications, Second Edition*
Wenyuan Li

42. *Practical Power System Operation*
Ebrahim Vaahedi

43. *The Selection Process of Biomass Materials for the Production of Bio-Fuels and Co-Firing*
Najib Altawell

44. *Electrical Insulation for Rotating Machines: Design, Evaluation, Aging, Testing, and Repair, Second Edition*
Greg C. Stone, Ian Culbert, Edward A. Boulter, and Hussein Dhirani

45. *Principles of Electrical Safety*
Peter E. Sutherland

46. *Advanced Power Electronics Converters: PWM Converters Processing AC Voltages*
Euzeli Cipriano dos Santos Jr. and Edison Roberto Cabral da Silva

47. *Optimization of Power System Operation, Second Edition*
Jizhong Zhu

48. *Power System Harmonics and Passive Filter Designs*
J. C. Das

49. *Digital Control of High-Frequency Switched-Mode Power Converters*
Luca Corradini, Dragan Maksimovic, Paolo Mattavelli, and Regan Zane

50. *Industrial Power Distribution, Second Edition*
Ralph E. Fehr, III

51. *HVDC Grids: For Offshore and Supergrid of the Future*
Dirk Van Hertem, Oriol Gomis-Bellmunt, and Jun Liang

52. *Advanced Solutions in Power Systems: HVDC, FACTS, and Artificial Intelligence*
Mircea Eremia, Chen-Ching Liu, and Abdel-Aty Edris

53. *Power Grid Operation in a Market Environment: Economic Efficiency and Risk Mitigation*
Hong Chen

54. *Current Signature Analysis for Condition Monitoring of Cage Induction Motors: Industrial Application and Case Histories*
William T. Thomson and Ian Culburt

55. *Model Predictive Control of Wind Energy Conversion Systems*
Venkata Yaramasu and Bin Wu

56. *Electric Power System Basics for the Nonelectrical Professional, Second Edition*
Steven W. Blume

57. *Energy Production Systems Engineering*
Thomas H. Blair

58. *Understanding Symmetrical Components for Power System Modeling*
J. C. Das

59. *High-Power Converters and AC Drives, Second Edition*
Bin Wu and Mehdi Narimani

60. *Practical Lighting Design with LEDs, Second Edition*
Ron Lenk and Carol Lenk

61. *Introduction to Electric Power and Drive Systems*
Paul Krause, Oleg Wasynczuk, Timothy O'Connell, and Maher Hasan

62. *Instantaneous Power Theory and Applications to Power Conditioning, Second Edition*
Hirofumi Akagi, Edson Hirokazu Watanabe, and Mauricio Aredes

63. *Introduction to AC Machine Design*
Thomas A. Lipo